The Mycotoxin Blue Book

The Mycotoxin Blue Book

Edited by

D.E. Diaz

NOTTINGHAM
University Press

Nottingham University Press
Manor Farm, Main Street, Thrumpton
Nottingham, NG11 0AX, United Kingdom

NOTTINGHAM

First published 2005
Reprinted 2005

British Library Cataloguing in Publication Data
The Mycotoxin Blue Book
I. Diaz, D.E.

ISBN 1-904761-19-4

Disclaimer

Every reasonable effort has been made to ensure that the material in this book is true, correct, complete and appropriate at the time of writing. Nevertheless, the publishers and authors do not accept responsibility for any omission or error, or for any injury, damage, loss or financial consequences arising from the use of the book.

Typeset by Nottingham University Press, Nottingham
Printed and bound by The Bath Press, Bath, England

PREFACE

The Food and Agriculture Organization of the United Nations stated, in a recent report, that at least 99 countries around the world, representing 87% of the world's inhabitants, had some type of mycotoxin guidelines for foodstuffs and/or feedstuffs. Such worldwide interest is also reflected in the great volume of research published since the discovery of aflatoxin B_1 as the causative agent in Turkey X disease in the early 1960s. Due to the ubiquitous nature of mycotoxin-producing moulds and our general inability to prevent many of the conditions favorable for their growth and mycotoxin production, mycotoxin contamination of foods and feeds is an inevitable part of modern agriculture. The increasing irregularity of meteorological events related to the ever-changing global climate will further challenge us in the years to come. Although the true impact of moulds and mycotoxins is impossible to measure, the risks associated with their presence is well established.

The main aim of this book is to highlight the practical aspects of mycotoxin contamination. Careful attention is given to the understanding of toxicity and its impact on performance of animals exposed to mycotoxin contaminated feedstuffs. Chapters within this volume will also discuss the basis of proper feedstuff sampling and the principles of mycotoxin analysis. Additionally, this book will attempt to review concepts of mould growth and mycotoxin production, mycotoxin contamination of forages, and the contamination of human foods. Other chapters will focus on the mechanisms of toxicity, and on the specific effects of mycotoxins on antioxidant status. Therefore information provided in this book will be useful to professionals associated with the fields of veterinary and medical sciences, biotechnology, physiology, immunology, food technology, toxicology and nutrition, as well as students interested in this rapidly-developing field.

The Mycotoxin Blue Book owes its existence to the labour and commitment of many contributors to whom I am indeed grateful. I want to particularly acknowledge Professor Lon Whitlow for his support and encouragement.

Duarte E Diaz

CONTENTS

1

SAMPLING FEEDS FOR MYCOTOXIN ANALYSIS

Thomas B. Whitaker, Andrew B. Slate and Anders Sture Johansson
US Department of Agriculture, Agricultural Research Service, North Carolina State University, Raleigh, North Carolina, USA

Introduction

Mycotoxins are toxic and/or carcinogenic compounds produced by various fungal species that grow on various agricultural commodities (Cullen and Newberne, 1994). There are many different mycotoxins (CAST, 2003) and a partial list is shown in Table 1.

Table 1. Partial listing of several mycotoxins produced by various fungi.

- Aflatoxins
- Fumonisins
- Ochratoxin A
- Deoxynivalenol
- Patulin
- Zearalenone

Commodities can be contaminated either in the field or in storage. Pre- and post-harvest strategies to prevent crop contamination include yearly crop rotations, irrigation in hot and dry weather, use of pesticides to reduce the insect population, drying crops to a safe moisture level, and providing protective storage (Phillips *et al.*, 1994). Because mycotoxins are toxic and carcinogenic in animals, many countries regulate the maximum level that can occur in foods and feeds. Most regulations are concerned with controlling aflatoxin because it is considered the most toxic and carcinogenic of the naturally occurring mycotoxins. A recent FAO/WHO survey indicated that almost 100 countries regulate aflatoxin in foods and feeds (FAO, 1995). However, maximum levels differ widely from country to country because of a lack of agreement on what constitutes a safe maximum level for humans. Some of the maximum levels found in the FAO/WHO survey for aflatoxin are shown below in Table 2 (FAO, 1995).

Table 2. Examples of aflatoxin legal limits found in various countries.

Country	*Aflatoxin B_1 (ppb)*	*Total aflatoxin (ppb)*
US		20
EU	8	15
Australia		15
Canada		15
Egypt		
Maize		20
Peanuts		10
Nigeria	20	
Philippines		20
South Africa		10

It is important to be able to detect and quantify the mycotoxin concentration in foods and feeds destined for human and animal consumption. In research, quality assurance, and regulatory activities, correct decisions concerning the fate of commercial lots can only be made if the mycotoxin concentration in the lot can be determined with a high degree of accuracy and precision. The mycotoxin concentration of a bulk lot is usually estimated by measuring the mycotoxin concentration in a small portion of the lot or a sample taken from the lot (Figure 1).

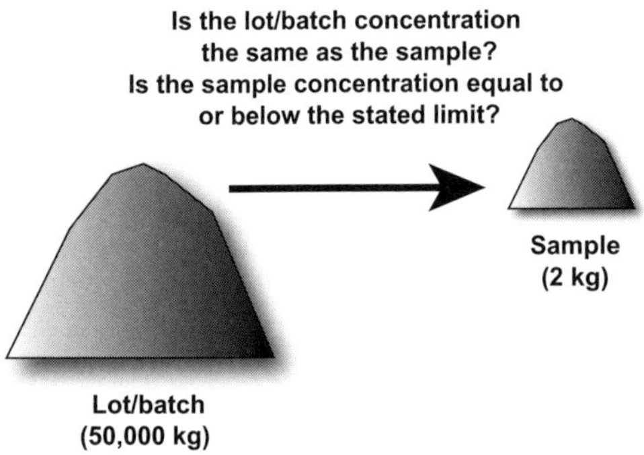

Figure 1. Lot mycotoxin concentration is assumed to equal the mycotoxin concentration measured in a small sample.

The mycotoxin concentration in the bulk lot is assumed to be the same as the measured mycotoxin concentration in the sample. Based on the measured sample concentration, some decision is made about the edible quality of the bulk lot or the effect of a treatment or a process on reducing aflatoxin in the lot. For example, in a regulatory environment, decisions will be made to classify the lot as acceptable or unacceptable based upon a comparison of the measured sample concentration to a legal limit. If the sample concentration does not accurately reflect the lot concentration, then the lot may be misclassified and there may be undesirable economic and/or health consequences. Fortunately, sampling plans can be designed to minimise the misclassification of lots and reduce the undesirable consequences associated with regulatory decisions about the fate of bulk lots. In this chapter, sampling plans will be defined, sources of uncertainty associated with a mycotoxin sampling plan will be identified, risks associated with misclassifying lots will be discussed, and methods that reduce misclassification of lots will be described.

Definition of sampling plan

A mycotoxin sampling plan is defined by a mycotoxin test procedure and a defined accept/reject limit. A mycotoxin test procedure is a multi-stage process (Figure 2) and generally consists of three steps: sampling, sample preparation, and analysis (quantification).

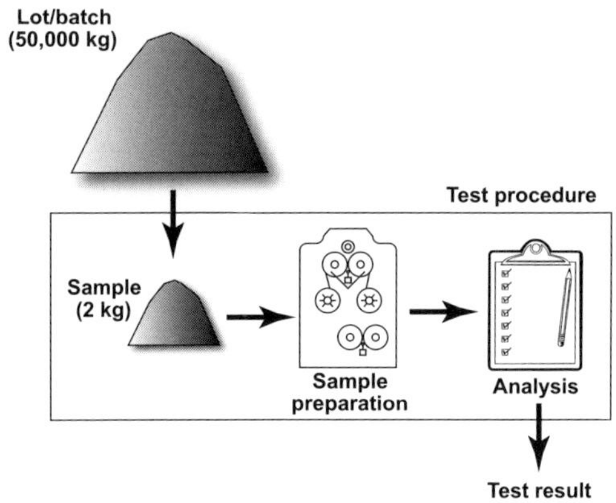

Figure 2. A mycotoxin test procedure usually consists of a sampling, sample preparation and analytical step.

The sampling step specifies how the sample will be selected or taken from the bulk lot and the size of the sample. For granular products, the sample preparation step is also a two-part process where the sample is ground in a mill to reduce particle size and a subsample is removed from the comminuted sample. Finally in the analytical step, the mycotoxin is solvent-extracted from the comminuted subsample and quantified using approved procedures.

The measured mycotoxin concentration in the sample is used to estimate the true mycotoxin concentration in the bulk lot or compared to a defined accept/reject limit that is usually equal to a maximum level or regulatory legal limit. Comparing the measured concentration to an accept/reject limit is often called **acceptance sampling** because the measured concentration value is not as important as whether the measured concentration (and thus the lot concentration) is above or below a legal limit. In quality assurance and research activities, a precise and accurate estimate of the true lot mycotoxin concentration becomes important.

UNCERTAINTY

There is always some level of uncertainty associated with a sampling plan. Because of the uncertainty associated with a mycotoxin sampling plan, the true mycotoxin concentration of a bulk lot cannot be determined with 100% certainty; nor can all lots be correctly classified into good and bad categories (based upon some legal limit) with 100% accuracy. Accuracy and precision are two types of uncertainties associated with a sampling plan (Cochran and Cox, 1957).

Accuracy

Accuracy is defined as the closeness of measured values to the true value. Another term associated with accuracy is **bias**. A bias is some force or influence that makes the measured values deviate from the true value in a consistent manner on the average. Using target practice as an example, the center of the target is analogous to the true value and holes in the target represent the measured values (Figure 3). Figure 3 shows that the rifle used on the left is not as accurate as the rifle used on the right where the average of the cluster of shots is around the center of the target.

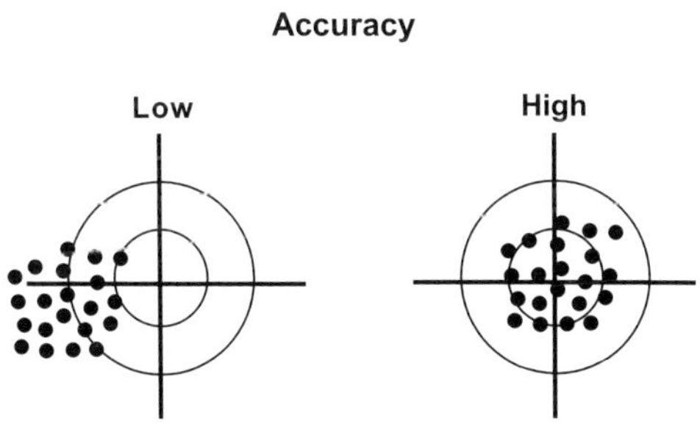

Figure 3. Low and high accuracy using target practice as an example.

Mathematically, accuracy (A) is the difference between the true value (U) and the average of the n measured values (X_i) (Equation 1).

$$A = U - [SUM(X_i)/n] \qquad (1)$$

Accuracy is associated with a bias. Biases have the potential to occur in the sample selection process, sample preparation, and in the quantifiation steps of the test procedure. Biases should be the easiest to control and reduce to acceptable levels, but methods to reduce bias are difficult to evaluate because of the difficulty in knowing the true mycotoxin concentration of the lot. Sample selection, sample preparation equipment, and analytical methods are continuously performance tested to minimise any biases.

Precision

Precision is defined as the closeness of measured values to each other. Another term for precision is **variability**. The definition of precision makes no mention about how close the measured values are to the true value. Using target practice to illustrate precision, the closeness of the holes to each other is a measure of precision (Figure 4).

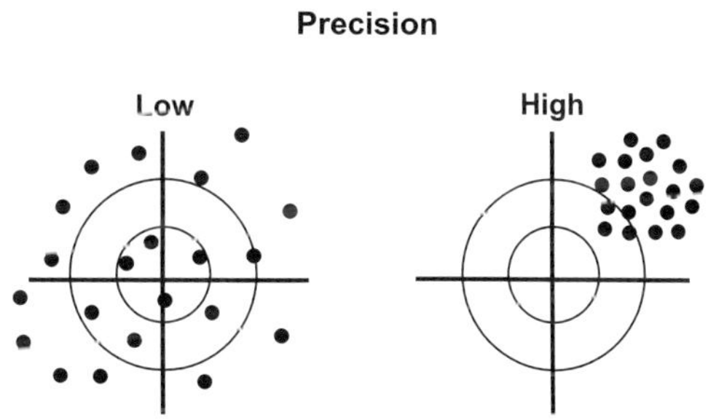

Figure 4. Low and high precision using target practice as an example.

Three statistical measures of variability, **variance** (V), **standard deviation** (S), or **coefficient of variation** (CV) can be used as a measure of precision (P) (Equations 2-4).

$$V = [\Sigma_i (x_i - m)^2/(n-1)]$$
$$\text{for I} = 1,2,\ldots,n \qquad (2)$$

$$S = \sqrt{V} \qquad (3)$$

The CV, expressed as a percent:

$$CV = 100 \times (S/m) \qquad (4)$$

Where x_i is the measured value and m is the mean of the n x_i values. Precision is associated with variability, which can occur with each step of the mycotoxin test procedure.

When describing the uncertainty of a process, one must consider the various combinations of accuracy and precision that may occur. As shown in Figure 5, there are four extreme combinations of accuracy and precision: low precision and low accuracy, low precision and high accuracy, high precision and low accuracy, and high precision and high accuracy.

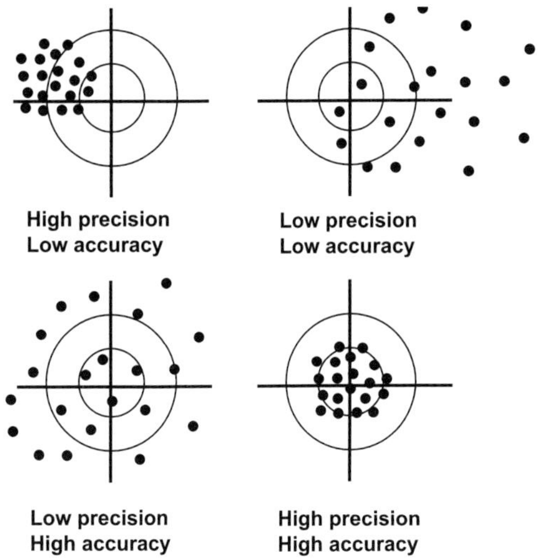

Figure 5. The four extreme combinations of uncertainty that can occur with a sampling plan.

The worst possible situation is to have a process with low precision and low accuracy. The best possible situation is to have a process that has both high precision and high accuracy. The goal associated with detecting a mycotoxin in a bulk shipment is to design a mycotoxin test procedure or sampling plan that has both high precision and high accuracy.

Sample selection

Procedures used to take a sample from a bulk lot are extremely important. Every individual item in the lot should have an equal chance of being chosen (called random sampling). Biases are introduced by sample selection methods if equipment and procedures used to select the sample prohibit or reduce the chances of any item in the lot from being chosen. Examples of bias in the sample selection process, shown in Figure 6, are the use of a sampling probe that does not allow larger particles into the probe, a probe that does not reach every location in the shipment, and use of a single probing point in a poorly mixed lot.

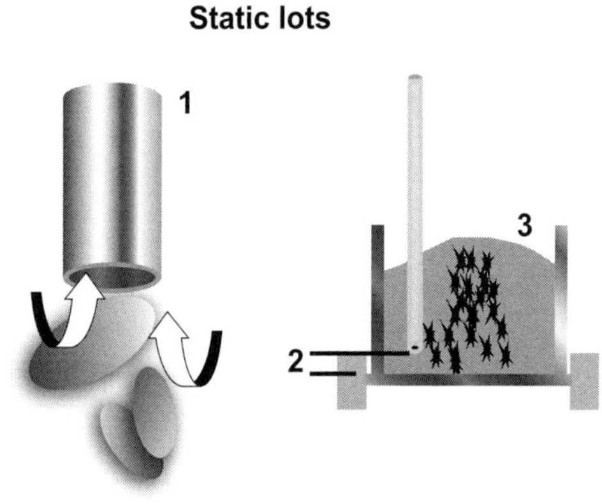

Figure 6. Different types of biases associated with selecting samples from bulk lots. (1) Particles larger than probe opening; (2) Some particles cannot be reached; (3) Using a single probing point with an unmixed lot.

If the lot has been blended thoroughly from the various material handling operations, then the contaminated particles are assumed to be distributed uniformly throughout the lot (Williams, 1991). In this situation, it is probably not too important from what location in the lot the sample is drawn. However, if the lot is contaminated because of moisture leaks that cause high moisture clumps or for other localized reasons, then the mycotoxin-contaminated particles may be located in isolated pockets in the lot (Shotwell *et al.*, 1975). If the sample is drawn from a single location, the contaminated particles may be missed or too many contaminated particles may be collected (Figure 7).

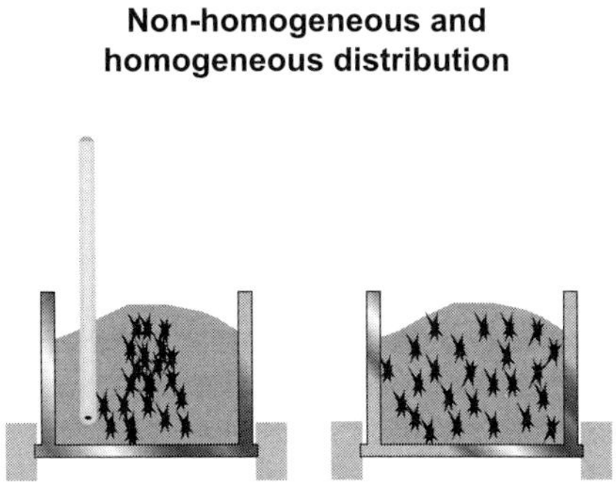

Figure 7. Multiple probing points should be used with a non-homogeneous lot.

Because contaminated particles may not be distributed uniformly throughout the lot, the sample should be an accumulation of many small portions taken from many different locations throughout the lot (Bauwin and Ryan, 1982; Hurburgh and Bern, 1983). FAO/WHO recommends that each incremental portion be about 200 g and one incremental portion be taken for every 200 kg of product (FAO, 2001). The accumulation of many small incremental portions is called a **bulk sample**. If the bulk sample is larger than desired, the bulk sample should be blended and subdivided until the desired sample size is achieved (Figure 8).

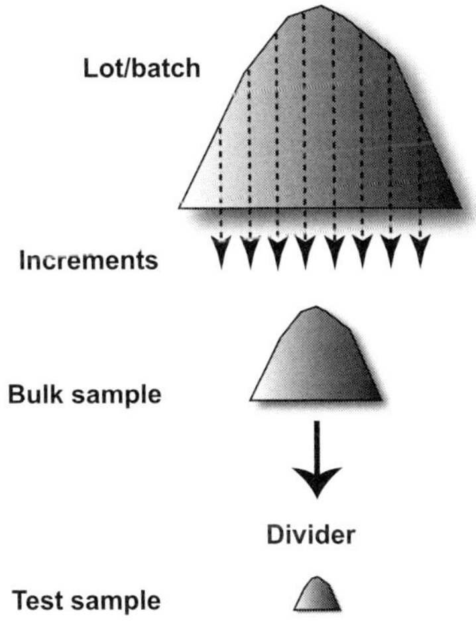

Figure 8. A test sample is removed from a bulk sample. A bulk sample is the accumulation of many small incremental portions taken from many different locations in the lot.

The smallest sample size that is subdivided from the bulk sample and comminuted in a grinder in the sample preparation step is called the **test sample**. It is generally more difficult to obtain a representative (lack of bias) test sample from a lot at rest (static lot) than from a moving stream of the product (dynamic lot) as the lot is moved from one location to another. Sample selection methods differ depending on whether the lot is static or dynamic.

STATIC LOTS

Examples of static lots are commodities contained in storage bins, rail cars, or many small containers such as sacks. When drawing a sample from a bulk container, a probing pattern should be developed so that product can be collected from different locations in the lot. An example of several probing patterns used by the USDA to collect samples from peanut lots is shown in Figure 9 (USDA, 1975; Parker *et al.*, 1982; Whitaker and Dowell, 1995).

The sampling probe should be long enough to reach the bottom of the container when possible. Attempts should be made to use a sampling rate similar to the 200 g per 200 kg mentioned above. However, it may not be possible to achieve the suggested sampling rate because of the design of the sampling equipment, size of the individual containers, and the size of the lot. As an example, a test sample (TSS) of 5,000 g is to be taken from a lot (LS) of 25,000 kg. The preferred increment size (ISS) is 200 g. The minimum number of increments needed to provide a test sample of 5,000 g is TSS/ISS or 25 incremental portions of 200 g each. If a total of 25 incremental portions are to be taken from the lot of 25,000 kg, then an increment is taken for every 1,000 kg (25,000/25) of lot (taking an increment every 1,000 kg of lot is larger than the recommended 200 kg of lot). If a 200 g incremental portion is taken every 200 kg of lot, then a total of 125 (25,000/200) increments of 200 g each will be taken and the bulk sample size is 25,000 g or five times bigger than the needed test sample size of 5,000 g. The 25,000

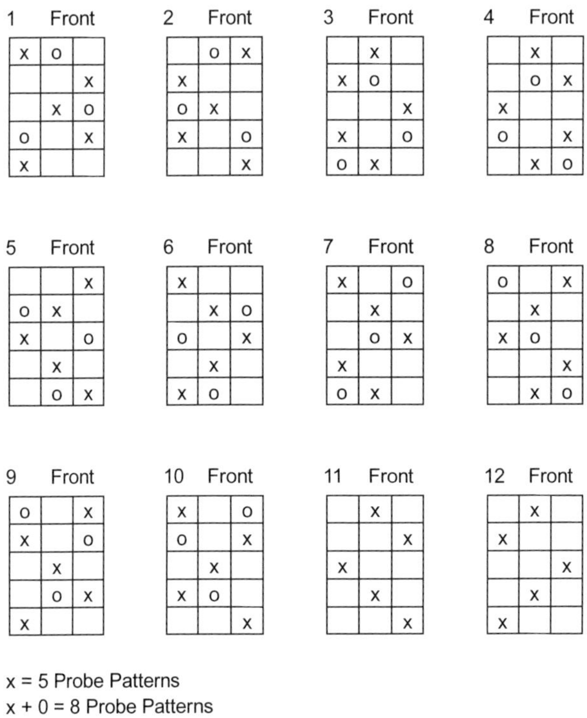

Figure 9. Example of several 5- and 8-probe patterns used by the US Department of Agriculture to sample large peanut containers (trucks and wagons) for grade.

g bulk sample must be subdivided to obtain the 5,000 g test sample. A flow diagram showing the interactions among all the variables is shown in Figure 10.

When sampling a static lot in separate containers such as sacks or retail containers, the sample should be taken from many containers dispersed throughout the lot. When storing sacks in a storage facility, access lanes should be constructed in order to allow access to sacks at interior locations. The recommended number of containers sampled can vary from one in four in small lots (< 20 metric tonnes) to the square root of the total number of containers for large (> 20 metric tonnes) lots (FAO, 2001).

If the lot is in a container where access is limited, the sample should be drawn when the product is either being removed from or being placed into the container. If the accumulated bulk sample is larger than required, the bulk sample should be thoroughly blended and reduced to the required test sample size using a suitable divider that randomly removes a test sample from the bulk sample.

DYNAMIC LOTS

True random sampling can be more nearly achieved when selecting a bulk sample from a moving stream as the product is transferred (i.e. conveyor belt) from one location to another. When sampling from a moving stream, small increments of product should be taken along the entire length of the moving stream (Figure 11). Small increments of product should be taken across the entire cross section of the moving stream; composite all the increments of product to obtain a bulk sample. If the bulk sample is larger than required, then blend and subdivide the bulk sample to obtain the desired size test sample.

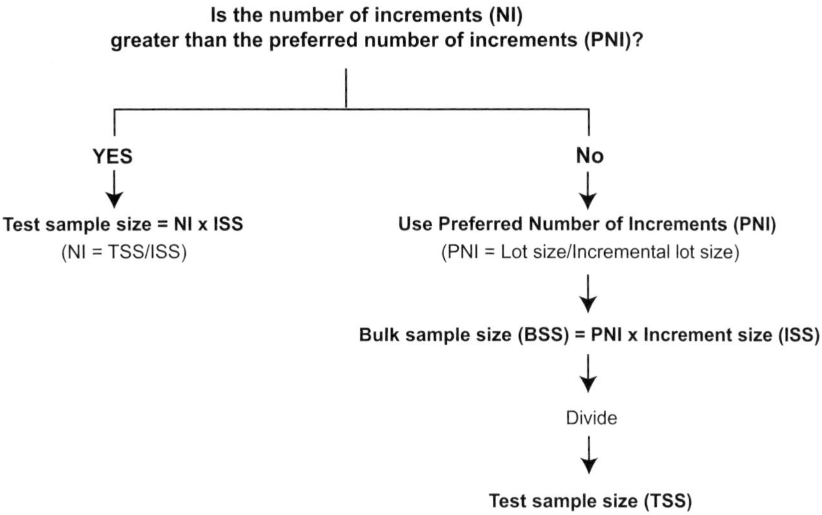

Figure 10. Interaction between lot size, increment size, and test sample size.

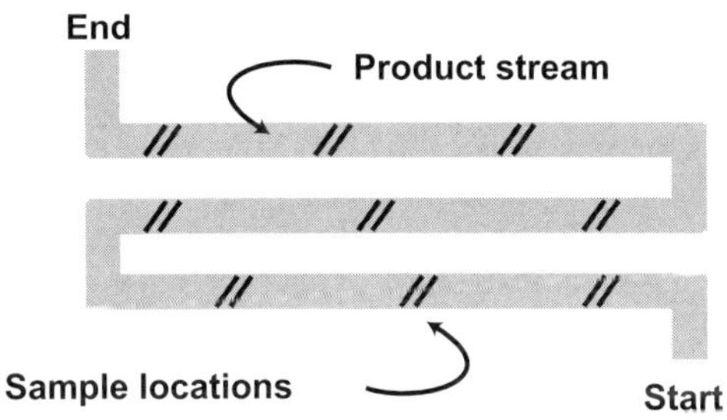

Figure 11. Sample selection from a moving stream of product should be the accumulation of many small incremental portions taken from the beginning to the end of the product stream.

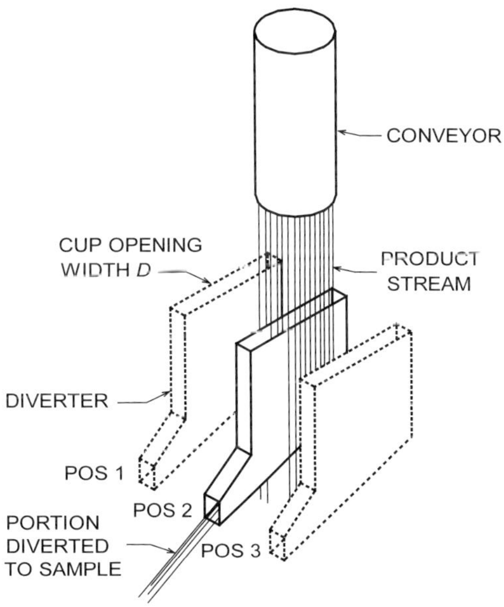

Figure 12. The automatic sampler cup should move at a constant velocity and cut through the entire stream of product.

Automatic sampling equipment such as cross-cut samplers (Figure 12) are commercially available with timers that automatically pass a diverter cup through the moving stream at predetermined and uniform intervals.

When automatic equipment is not available, a person can be assigned to manually pass a cup though the stream at periodic intervals to collect the bulk sample. Whether using automatic or manual methods, small increments of product should be collected and composited at frequent and uniform intervals throughout the entire time product flows past the sampling point.

Cross-cut samplers should be installed in the following manner: (a) the plane of the opening of the sampling cup should be perpendicular to the direction of flow; (b) the sampling cup should pass through the entire cross sectional area of the stream; and (c) the opening of the sampling cup should be wide enough to accept all items of interest in the lot. As a general rule, the width of the sampling cup opening should be two to three times the largest dimensions of the items in the lot.

The size of the bulk sample, S in kg, taken from a lot by a cross-cut sampler is

$$S = (D)(L) / (T)(V), \quad (5)$$

where D is the width of the sampling cup opening in cm, L is the lot size in kg, T is interval or time between cup movement through the stream in seconds, and V is cup velocity in cm/sec.

Equation 5 can also be used to compute other terms of interest such as the time between cuts, T. For example, the required time, T, between cuts of the sampling cup to obtain a 10 kg sample from a 30,000 kg lot where the sampling cup width is 5.08 cm (2 inches), and the cup velocity through the stream 30 cm/sec. Solving for T in Equation 5,

$$T = (5.08 \text{ cm} \times 30{,}000 \text{ kg})/(10 \text{ kg} \times 30 \text{ cm/sec}) = 508 \text{ sec}$$

If the lot is moving at 1000 kg per minute, the entire lot will pass through the sampler in 30 minutes and only three or four cuts will be made by the cup through the lot. This may be considered too infrequent, because too much product passes the sampling point between the times the cup cuts through the stream. The interaction among the variables in Equation 5 needs to be fully understood in terms of the amount of sample accumulated and the frequency of cuts through the product.

BULK VERSUS TEST SAMPLE

Because contaminated particles may not be uniformly dispersed, many incremental portions are taken from many different locations throughout the lot and accumulated to form a bulk sample. As a result, the bulk sample is usually larger than the desired test sample size used to estimate the lot mycotoxin concentration. For granular material, the test sample is the smallest sample of granular product ground in a mill in the sample preparation step. For finely ground materials (corn flour) or liquids (milk), the test sample is the smallest sample used in the analytical step to quantify the mycotoxin. When the bulk sample is larger than the test sample, mechanical dividers such as a Boerner or riffle divider should be used to remove the desired test sample from the bulk sample. Mechanical dividers are considered to produce random divisions (Parker *et al.*, 1982); therefore, the bulk sample does not have to be blended before the test sample is removed. However, if the test sample is to be removed from the bulk sample using quartering or a manual device such as a cup or scoop, then the bulk sample should be blended before the test sample is removed.

If the test sample is a granular product such as shelled corn or nuts, then the test sample should not be further reduced in size before grinding the sample in the sample preparation step. As the test sample becomes smaller, the uncertainty associated with estimating the true lot mycotoxin concentration becomes greater. As will be shown later, the size of the test sample put through the grinder should be as large as possible. Recommended sample sizes for various commodities are shown in Table 3.

Sample preparation

Once a sample has been taken from the lot, the test sample must be prepared for mycotoxin quantification. Since it is not practical to extract the mycotoxin from a large test sample, the mycotoxin is usually extracted from a much smaller portion of product **(subsample)** taken from the test sample. If the commodity is a granular product such as shelled corn, it is

Table 3. Product sample sizes used by the United States Food and Drug Administration.

Product	*Description*	*Package type*	*Lot size*	*Number of sample units*	*Unit size*	*Sample (lbs)*
Peanut butter	Smotth	Consumer &		24	0.5	12
		bulk		12	1	12
Peanuts	Crunchy, butter, raw roasted, ground topping	Consumer & bulk		48	1	48
Tree nuts	In-shell, shelled	Consumer &		10	1	10
	slices or flour	bulk		50	1	50
	paste			12	1	12
Brazil nuts	In-shell in import	Bulk	<200 bags	20	1	20
	status		201-800 bags	40	1	40
			801-2000 bags	60	1	60
Pistachio nuts	In-shell in import	Bulk	75,000 lbs	20% of units		50
	status		<75,000 lbs	20% of units		25
Corn	Shelled, meal, flour grits	Consumer & bulk		10	1	10
Cottonseed		Bulk		15	4	60
Oilseed meals	Peanut, cottonseed	Bulk		20	1	20
Edible seeds	Pumpkin, melon, sesame etc	Bulk		50	1	50
Ginger root	Dried, whole ground	Bulk &	'n' units	$\sqrt{n}$		15
		consumer		10	10 × 0.06	10
Milk	Whole, low fat, skim	Bulk &				10
		consumer		10	1	10
Small grains	Sorghum, wheat, barley etc	Bulk		10	1	10
Dried fruit	i.e. figs	Consumer & bulk		50	1	50
Mixtures	Commodity particles large	Consumer & bulk		50	1	50
	Commodity particles finely ground			10	1	10

essential that the entire test sample be comminuted in a suitable mill before a subsample is removed from the test sample (Dickens and Whitaker, 1982; Campbell *et al.*, 1986). Removing a subsample of whole seed from the test sample before the comminuting process would eliminate the benefits associated with the larger size test sample of granular product. After the test sample has been comminuted, a subsample is removed from the comminuted test sample for mycotoxin extraction (Figure 13).

Grinders should be used that reduce the particle size of the seed in the test sample to the smallest size possible. Grinders that produce small particles provide a more homogeneous test sample (Figure 13). As a result the mycotoxin concentration of the subsample will more closely reflect the true mycotoxin concentration of the test sample. Some grinders such as the Romer mill (Malone, 2000) and the USDA peanut mill (Dickens and Satterwhite, 1969) are designed to automatically provide a subsample during the grinding process. If the mill does not provide a subsample, the subsample can be obtained using a riffle divider. If the subsample is obtained using a manual device such as a scoop, blend the comminuted test sample before scooping out a subsample.

Normally, there will be no sample preparation step associated with samples of non-granular products such as liquids (milk) or paste

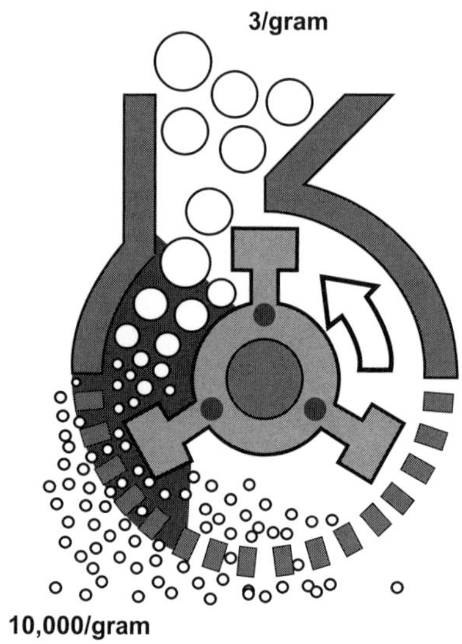

Figure 13. A test sample of granular product should be ground in a mill to reduce particle size.

(peanut butter). A small portion of the sample may have to be removed for mycotoxin analysis because the entire sample cannot be analysed. However, it is important to blend or mix liquid samples and paste samples before removing a small portion for mycotoxin analysis.

Subsample sizes vary, but usually are on the order of 25 to 1000 g depending on particle size. The smaller the particle size, the smaller the subsample size can be without increasing error or uncertainty.

Analytical quantification

Once the subsample is removed from the ground test sample, the mycotoxin is extracted by blending a solvent with the comminuted subsample. Before the mycotoxin can be quantified in the solvent extract, analytical methods usually consist of several steps related to removing interfering compounds (i.e. oils) and concentrating the mycotoxins for quantification. These steps include centrifugation, filtration, drying, and dilution (Steyn *et al.*, 1991). Three types of methods can be used to quantify the mycotoxin extracted from the subsample: thin layer chromatography (TLC), ELISA methods that use antibody technology, and high performance liquid chromatography (HPLC). Organizations such as the Association of Official Analytical Chemists (AOAC) evaluate the performance of analytical methods using collaborative studies.

There are several sources of bias associated with analytical methods. Less than 100% of the mycotoxin may be extracted from the subsample by the solvents; compounds other than mycotoxins may be extracted into the solvent and mistakenly quantified as a mycotoxin; mycotoxin standards used in quantification may not be exact; and instruments to measure the mycotoxin may not be correctly calibrated.

Accept/reject limit

Once the mycotoxin concentration is quantified, the sample value is used to estimate the true shipment concentration or is compared to an accept/reject limit (ARL). The ARL is a predefined threshold value, usually equal to a legal limit used in regulatory applications. If the sample mycotoxin value is less than or equal to the ARL, the lot is accepted. If otherwise, the lot is rejected. When lots are inspected by regulatory agencies, the ARL is usually set equal to the legal limit. However, manufacturers of consumer-ready products will often use an ARL less than the legal limit to reduce the chances that consumer-ready products will be found by regulatory agencies with mycotoxin concentrations above the legal limit. Often private industry will use an ARL that is about half the legal limit.

Many countries agree on the need to establish legal limits, but often disagree on the value of the limit. A survey by FAO in 1995 (FAO, 1995) showed that some countries have aflatoxin legal limits based upon B_1 only and some countries use total ($B_1+B_2+G_1+G_2$) aflatoxin and these legal limits vary widely. The CODEX Committee on Food Additives and Contaminates has established a standard aflatoxin limit for peanuts at 15 ppb total aflatoxin for peanuts traded on the international market (FAO, 2001). This limit does not infringe on any nation's internal limits.

Random variation

Even when using accepted sampling, sample preparation, and analytical procedures (Campbell *et al.*, 1986; AOAC, 1990; Nesheim, 1979), there are errors (the term **error** will be used to denote variability) associated with each of the above steps of the mycotoxin test procedure (Whitaker *et al.*, 1974). Because of these errors, the true mycotoxin concentration in the lot cannot be determined with 100% certainty by measuring the mycotoxin concentration in a test sample taken from the lot. For example, 10 replicated aflatoxin test results from each of six contaminated shelled peanut lots are shown in Table 4 (Whitaker *et al.*, 1972). For each test result in the table, the mycotoxin test procedure consisted of (a) comminuting a 5.45 kg test sample of peanut kernels in a USDA subsampling mill developed by the US Department of Agriculture (Dickens and Satterwhite, 1969), (b) removing a 280 g subsample from the comminuted test sample, (c) solvent extracting aflatoxins from a 280 g subsample as described by AOAC Method II (AOAC, 1990), and (d) quantifying the aflatoxins densitometrically using thin layer chromatography (TLC). The 10 aflatoxin test results from each lot are ranked from low to high to demonstrate several important characteristics about replicated aflatoxin test results taken from the same contaminated lot.

First, the wide range among replicated sample results from the same lot reflects the large variability associated with estimating the true mycotoxin content of a bulk lot. In Table 4, the variability is described by the standard deviation (SD) and the coefficient of variation (CV). The maximum sample result can be four to five times the lot concentration (the average of the 10 sample results is the best estimate of the lot concentration). Secondly, the amount of variation among the 10 sample results appears to be a function of the lot concentration. As the lot concentration increases, the standard deviation among sample results increases, but the standard deviation relative to the lot mean, as measured by the CV, decreases. Thirdly, the distribution of the 10 sample results for each lot in Table 4 is not always symmetrical about the lot concentration. The distributions are **positively skewed**, meaning that more than half of the sample results are below the true lot concentration. However, the distribution of sample test results becomes more symmetrical as the lot concentration increases. This skewness can be observed by counting the number of aflatoxin test results above and below the lot concentration in Table 4 (average of the 10 sample test results). If a single sample is tested

Table 4. Distribution of aflatoxin test results for ten 5.4 kg samples from each of six lots of shelled peanuts[a,b]

Lot number	*Sample test result (ng/g)*										*Mean (ppb)*	*SD[c] (ppb)*	*CV[d] (%)*
1	0	0	0	0	2	4	8	14	28	43	10	15	150
2	0	0	0	0	3	13	19	41	43	69	19	24	126
3	0	6	6	8	10	50	60	62	66	130	40	42	105
4	5	12	56	66	70	92	98	132	141	164	84	53	63
5	18	50	53	72	82	108	112	127	182	191	100	56	56
6	29	37	41	71	95	117	168	174	183	197	111	66	59

[a] Whitaker *et al.*, 1972.
[b] Aflatoxin test results are ordered by aflatoxin concentration, (ng/g).
[c] SD = Standard Deviation
[d] CV = Coefficient of Variation = SD × 100/mean

from a contaminated lot, there is more than a 50% chance that the sample test result will be lower than the true lot concentration. While it cannot be shown in Table 4, the skewness is greater for small sample sizes and the distribution becomes more symmetrical as sample size increases (Whitaker *et al.*, 1972). The above characteristics described by Table 4 for aflatoxin in peanuts are also generally found for other mycotoxins and other commodities (Dickens *et al.*, 1979; Whitaker *et al.*, 1993; Whitaker *et al.*, 1998).

The sources of the variability among mycotoxin test results in Table 4 are associated with each step of the mycotoxin test procedure. The sampling, sample preparation, and analytical steps of the mycotoxin test procedure each contribute to the total variability observed among mycotoxin test results. As shown in Figure 14, the total error or variability is the sum of the sampling, sample preparation, and analytical variability.

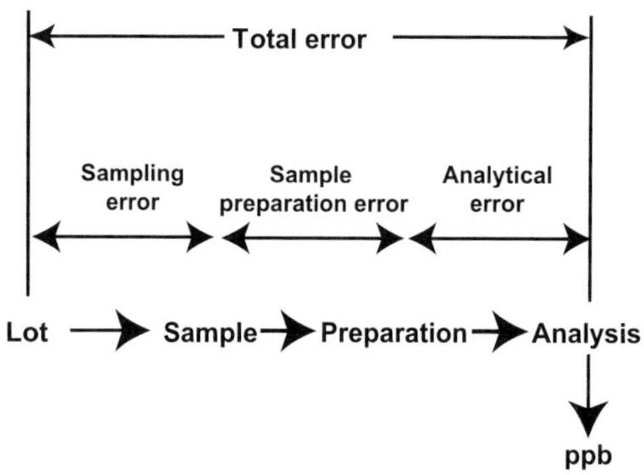

Figure 14. Total error of the mycotoxin test procedure is the sum of sampling, sample preparation, and analytical errors.

Among the statistical measures of variability shown in Equations 2 to 4, only the variance is additive. Therefore, it is assumed that the total variance associated with a mycotoxin test procedure is the sum of the sampling, sample preparation, and analytical variances (Equation 6).

$$VT = VS + VSS + VA \qquad (6)$$

Reasons why each step of the mycotoxin test procedure contributes to the overall variability are discussed below. An example of the magnitude of the contribution each step contributes to the total variability is also shown when testing shelled corn (maize) for aflatoxin.

SAMPLING VARIABILITY

Studies by researchers on a wide variety of agricultural products (peanuts, cottonseed, shelled corn and pistachio nuts) indicate that, especially for small sample sizes, the sampling step is usually the largest source of variability associated with the mycotoxin test procedure (Dickens *et al.*, 1979). Even when using accepted sample selection equipment and random sample selection procedures, sampling error is large because of the extreme distribution among contaminated particles within a lot. Studies by researchers on a wide variety of agricultural products such as peanuts and shelled corn (Johansson *et al.*, 2000a; Cucullu *et al.*, 1986; Cucullu *et al.*, 1977) indicate that a very small percentage (0.1%) of the kernels in the lot are contaminated and the concentration on a single kernel may be extremely high. Cucullu *et al.*, (1986) reported aflatoxin concentrations in excess of 1,000,000 ng/g (ppb) for individual peanut kernels and 5,000,000 ng/g for cottonseed. Shotwell *et al.* (1974) reported finding over 400,000 ng/g aflatoxin in a corn kernel.

Because of this extreme range in aflatoxin concentrations among a few contaminated kernels in a lot, variation among replicated sample test results tends to be large. As an example, the sampling variance, VS, associated with testing shelled corn was estimated empirically (Johansson *et al.*, 2000a) and is shown in Equation 7 for any sample size, ns.

$$VS = (12.95/ns)\, M^{0.98} \qquad (7)$$

where M is the aflatoxin concentration in the lot in nanograms of total aflatoxin per g of corn, ns is the mass of shelled corn in the sample in kg (kernel count per gram was 3.0). From Equation 7 one can see that the sampling variance is a function of the lot aflatoxin concentration M and sample size ns. The sampling variance among replicated 0.91 kg (2 lb) samples taken from a lot of shelled corn at 20 ppb is 268.1. The coefficient of variation is 81.8%.

Researchers have developed equations to describe the sampling variance for several commodities and mycotoxins (Whitaker *et al.*, 1974; 1993; 1998; Johansson *et al.*, 2000a). The equations are specific for the type of mycotoxin and the type of product studied, but generally show that sampling variance increases with an increase in concentration, and decreases with an increase in sample size.

SAMPLE PREPARATION VARIABILITY

Once the test sample has been taken from the lot, the sample must be prepared for mycotoxin quantification. Since it is not practical to extract the mycotoxin from a large test sample, the test sample is comminuted in a mill and the mycotoxin is extracted from a small subsample taken from the comminuted test sample. If the commodity is a granular product such as shelled corn, it is essential that the entire test sample be comminuted in a suitable mill before a subsample is removed from the test sample (Campbell *et al.*, 1986). Removing a subsample of whole seed from the test sample before the comminuting process is simply a sample size reduction process and eliminates the benefits associated with the larger size sample of granular product. After the sample has been comminuted in a mill to reduce particle size, a subsample is removed for mycotoxin extraction. It is assumed that the mycotoxin distribution of contaminated particles in the comminuted sample is similar to the distribution among contaminated kernels found in the lot. As a result, there is also variability among replicated subsamples taken from the same test sample. However, the sample preparation variance is not as large as the sampling variance due to the large number of comminuted particles in the subsample. An example of sample preparation variance for aflatoxin and shelled corn, VSS, is shown below in Equation 8 for any subsample size nss (Johansson *et al.*, 2000a).

$$VSS = (62.70/nss)\, M^{1.27} \qquad (8)$$

where M is the aflatoxin concentration in the test sample in ppb, nss is the mass of shelled corn in the subsample in grams. The variance in Equation 8 also reflects the use of a Romer mill that produces a particle size where most of the particles will pass through a No. 20 screen. From Equation 8, it can be seen that the sample preparation variance is also a function of the aflatoxin concentration in the sample and the subsample size. The sample preparation variance associated with a 50 g subsample taken from a sample at 20 ppb is 56.3 and the CV is 37.5%.

Researchers have developed equations to describe the sample preparation variance for several commodities, mills, and mycotoxins (Whitaker *et al.*, 1993; 1998; Johansson *et al.*, 2000a). The equations are specific for the type of mycotoxin, type of mill (particle size), and the type of product used in the study. The type of mill affects the particle size distribution. If the average particle size decreases (number of particles per unit mass increases), then the subsampling variances for a given size subsample decreases.

ANALYTICAL VARIABILITY

Once the subsample is removed from the comminuted test sample, the mycotoxin is solvent-extracted. Analytical methods usually involve several steps such as solvent extraction, centrifugation, drying, dilution, and quantification (Steyn *et al.*, 1991). As a result, there can be considerable variation among replicated analyses on the same subsample extract. The analytical variance, VAh, associated with HPLC techniques used to measure aflatoxin in shelled corn is given by Equation 9 (Whitaker *et al.*, 1996) for any number of aliquots, na.

$$VAh = (0.143/na)\ M^{1.16} \quad (9)$$

where M is the aflatoxin concentration in the subsample in ppb, na is the number of aliquots quantified by HPLC methods. The analytical variance and CV associated with using HPLC to measure aflatoxin in a comminuted subsample of corn at 20 ng/g, is 4.6 and 10.7%, respectively.

High performance liquid chromatography tends to have less variability than other analytical technologies such as TLC and immunoassay (ELISA) methods (Whitaker *et al.*, 1996). Using precision estimates from collaborative studies, the analytical variances associated with TLC (VAt) and ELISA (VAe) methods to measure aflatoxin in corn are shown in Equations 10 and 11, respectively.

$$VAt = (0.316/na)\ M^{1.744} \quad (10)$$

$$VAe = (0.631/na)\ M^{1.293} \quad (11)$$

The coefficients of variation associated with measuring aflatoxin in a corn subsample at 20 ppb with the TLC and ELISA methods are 38.3 and 27.5%, respectively. The variability associated with HPLC, 10.7%, (Equation 9) is lower than either TLC or ELISA.

All of the analytical variance information described above reflects results from single laboratories and does not reflect among-laboratory variances. As a result, some laboratories may have higher or lower variances than those reported in Equations 9, 10, and 11. Among-laboratory variance is about double the within-laboratory variance (Whitaker *et al.*, 1996).

TOTAL VARIABILITY

As shown in Table 5 and Equation 6, the total variability, VT, (using variance as the statistical measure of variability) associated with a mycotoxin test procedure is equal to the sum of the sampling (VS), sample preparation (VSS), and analytical (VA) variances associated with each step of the mycotoxin test procedure. The total variability associated with testing shelled corn for aflatoxin, grinding the test sample in a Romer mill, and quantifying aflatoxin by immunoassay is the sum of Equations 7, 8, and 11 (Equation 12).

$$VT = (12.95/ns)\ M^{0.98} + (62.70/nss)\ M^{1.27} + (0.631/na)\ M^{1.293} \quad (12)$$

Table 5. The variability measured by the variance associated with a 0.91 kg sample, 50 g subsample, measuring aflatoxin in 1 aliquot by immunoassay in a lot of shelled corn at 20 ppb aflatoxin[1].

	Variance	*Ratio (%)*
Sample = 0.91 kg	268.1	75.5
Subsample[2], 50 g	56.3	15.9
Immunoassay, 1 aliquot	30.4	8.6
Total	354.8	100.0

[1]Sampling, sample preparation, and analysis errors account for about 75.5, 15.9, and 8.6% of the total error, respectively.

[2]Romer mill used to grind

Using Equation 12, the total, sampling, sample preparation, and analytical variances associated with testing shelled corn over a range of lot concentrations (M) when using a 0.91 kg sample (ns), grinding the test sample in a Romer mill, taking a 50 g subsample (nss) from a comminuted sample, and quantifying aflatoxin in one aliquot (na) by immunoassay methods are shown in Figure 15.

When sampling a shipment of shelled corn at 20 ppb, the magnitude of the variance associated with each step of the above aflatoxin test procedure (Equation 12) is shown below in Equation 13.

$$VT = 268.1 + 56.3 + 30.4 = 354.8 \quad (13)$$

As shown in Table 5, the sampling, subsampling, and analytical variances account for 75.5, 15.9, and 8.6% of the total mycotoxin testing variance, respectively.

As the above example demonstrates, the sampling step accounts for most of the variability (uncertainty) associated with the total variability of a mycotoxin test procedure because of the

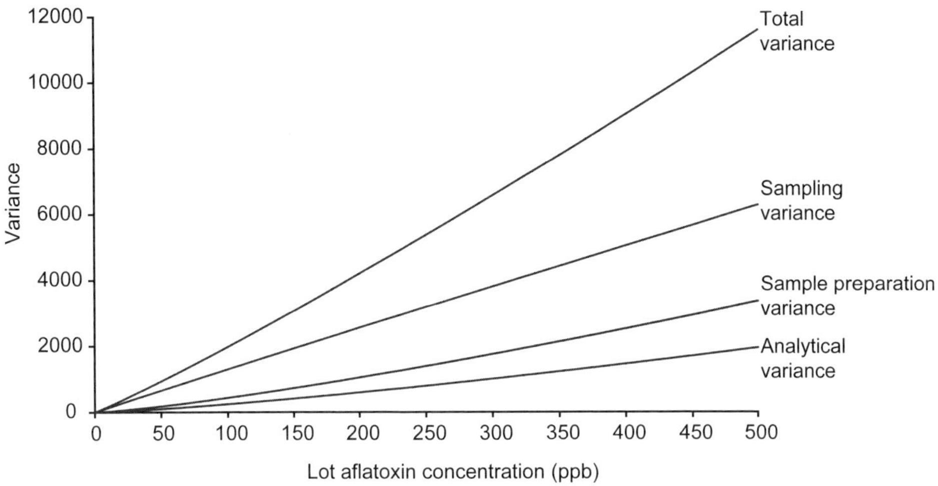

Figure 15. Variability of each step of the aflatoxin test procedure, as measured by the variance (V), increases with aflatoxin concentration. The total variance, VT, is the sum of sampling variance,VS, sample preparation variance, VSS, and analytical variance, VA.

extreme distribution among contaminated seed in a lot. For shelled corn, it is estimated that only six kernels in 10,000 are contaminated in a lot at 20 ppb (Johansson *et al.*, 2000b). Because of this extreme mycotoxin distribution among seed in a contaminated lot, it is easy to miss the contaminated seed with a small sample and underestimate the true lot concentration. On the other hand, if the test sample contains one or more highly contaminated seeds, then the test sample will overestimate the true mycotoxin contamination in the lot. Even using proper sample selection techniques, the variation among test sample concentrations is large due to the mycotoxin distribution described above.

Reducing variability of a mycotoxin test procedure

The only way to achieve a more precise estimate of the true lot concentration is to reduce the total variability of the test procedure. The total variability of the test procedure can be reduced by reducing the variability associated with each step of the mycotoxin test procedure. Increasing the size of the sample can reduce the sampling variability. The sample preparation variability can be reduced either by increasing the size of the subsample and/or by increasing the degree of comminuting (increasing the number of particles per unit mass in the subsample). The analytical variance can be reduced by either increasing the number of aliquots quantified by the analytical method and/or using a more precise quantification method (i.e. using HPLC instead of TLC). If the variability associated with one or more of these steps can be reduced, then the total variability associated with a mycotoxin test result can be reduced (Equation 6).

Decreasing the total variability (improving precision) associated with a mycotoxin test procedure will decrease the range of possible aflatoxin test results when replicated tests are made on the same lot. The range of mycotoxin test results associated with any size sample and subsample, and number of analyses about the lot concentration M can be estimated from the total variance, VT, or standard deviation, S, (square root of the total variance) associated with the mycotoxin test procedure. Approximately 95% of all test results will fall between a low of $(M - 1.96 \times S)$ and a high of $(M + 1.96 \times S)$.

As an example, when sampling a lot of shelled corn at 20 ppb using a 0.91 kg sample (ns), grinding the test sample in a Romer mill, taking a 50 g subsample (nss) from a comminuted sample, and quantifying aflatoxin in one aliquot (na) using an immunoassay method, Equation 13 shows that the total variance and standard deviation are 354.8 and 18.8, respectively. The range of aflatoxin test results should fall between $20 \pm (1.96 \times 18.8)$ or 20 ± 37 or 0 and 57 ppb (Table 5).

The calculated range of aflatoxin test results is only valid for a normal distribution where test results are symmetrical about the mean. The distribution among aflatoxin test results is usually skewed, but will approach a symmetrical distribution as sample size becomes large.

SAMPLE SIZE

The effect of increasing sample size on reducing the total variability and the range of mycotoxin test results when testing a contaminated lot of shelled corn at 20 ppb aflatoxin is shown in Table 6 when increasing sample size from 0.91 to 4.54 kg.

Table 6. Effect of increasing sample size on reducing the sampling variability[1].

	Sample size (kg)	
	0.91 kg	*4.54 kg*
Variance	266.5	53.3
Subsample[2], 50 g	56.3	56.3
TLC, 1 aliquot	27.9	27.9
Total	**350.7**	**137.5**
Range	20 ± 37	20 ± 23

[1]Lot shelled corn at 20 ppb
[2]Romer mill used to grind

Increasing sample size by a factor of five from 0.91 to 4.54 kg will cut the sampling variance in Equation 13 by a factor of five from 266.3 to 53.3. The total variance is reduced from 350.7 to 137.5 (Equation 14).

$$VT = 53.3 + 56.3 + 27.9 = 137.5 \qquad (14)$$

The range of aflatoxin test results is reduced from 20 ± 37 to 20 ± 23 ppb as sample size is increased from 0.91 to 4.54 kg.

SUBSAMPLE SIZE

The effect of increasing subsample size from 50 to 100 g on reducing the sample preparation variance is shown in Table 7.

Table 7. Effect of increasing subsample size on reducing sample preparation variability[1].

	Subsample[2] size (g)	
	50 g	*100 g*
Variance[2]	266.5	266.5
Subsample[3]	56.3	28.2
TLC, 1 aliquot	27.9	27.9
Total	**350.7**	**322.6**
Range	20 ± 37	20 ± 36

[1]Lot shelled corn at 20 ppb
[2]Sample size = 0.91 kg
[3]Romer mill used to grind

The sample preparation variance is cut in half and is reduced from 56.3 to 28.2. The total variance is reduced from 350.7 to 322.6. The range of aflatoxin test results is reduced from 20 ± 37 to 20 ± 36.

NUMBER OF ALIQUOTS QUANTIFIED

The effect of increasing number of aliquots quantified in the analytical step from 1 to 2 on reducing the analytical variance for immunoassay type method is shown in Table 8.

The analytical variance is cut in half and is reduced from 27.9 to 14.0. The total variance is reduced from 350.7 to 336.8. The reduction is so small that the range of aflatoxin test results is not significantly affected.

There are different costs associated with reducing the variability of each step of a mycotoxin test procedure. One needs to try and maximize the variance reduction for a given cost. Increasing sample size is usually the best use of resources when reducing the total variability of mycotoxin test results.

Table 8. Effect of increasing number of aliquots quantified for aflatoxin on reducing analytical variability[1].

	Number of aliquots 1	2
Variance[2]	266.5	266.5
Subsample = 50 g	56.3	56.3
TLC	27.9	14.0
Total	**350.7**	**336.8**
Range	20 ± 37	20 ± 37

[1]Lot shelled corn at 20 ppb
[2]Sample size = 0.91 kg

Designing mycotoxin sampling plans

Because of the variability among mycotoxin test results, two types of mistakes are associated with any mycotoxin sampling plan. First, good lots (lots with a concentration less than or equal to the legal limit) will test bad and be rejected by the sampling plan. This type of mistake is often called the **seller's risk** (false positives) since these lots will be rejected at an unnecessary cost to the seller of the product. Secondly, bad lots (lots with a concentration greater than the legal limit) will test good and be accepted by the sampling program. This type of mistake is called the **buyer's risk** (false negatives) since contaminated lots will be processed into feed or food causing possible health problems and/or economic loss to the buyer of the product. In order to maintain an effective regulatory and/or quality control program, the above two risks associated with a sample design must be evaluated (Figure 16). Based upon these evaluations, the costs and benefits (benefits refers to removal of mycotoxin contaminated lots) associated with a sampling program need to be evaluated.

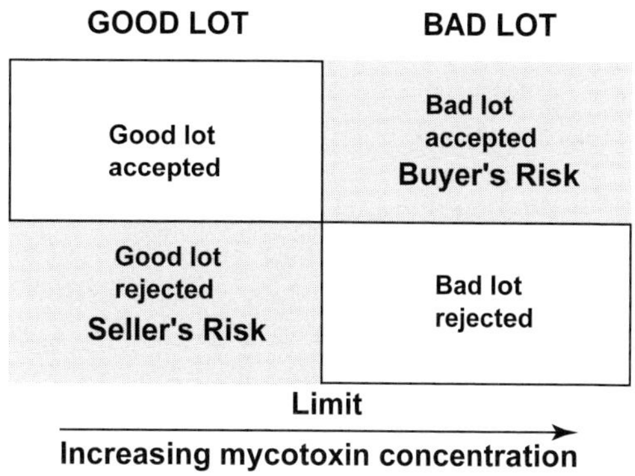

Figure 16. Four possible outcomes when classifying lots as good or bad. Good lots rejected and bad lots accepted are incorrect decisions. Good lots accepted and bad lots rejected are correct decisions.

A lot is termed bad when the sample test result X is above some predefined ARL X_c and the lot is termed good when X is less than or equal to X_c. While X_c is usually equal to the legal limit Mc, X_c can be greater than or less than Mc. For a given sample design, lots with a mycotoxin concentration M will be accepted with a certain probability $P(M) = prob(X < X_c | M)$ by the sampling plan. A plot of P(M) versus the lot concentration M is called an **operating characteristic (OC) curve**. Figure 17 depicts the general shape of an OC curve.

As M approaches 0, P(M) approaches 1 or 100%, and as M becomes large, P(M) approaches zero. Lots with little to no contamination (M=0) are accepted by the sampling plan 100% of the time; lots with very high levels of contamination (M= large) are never accepted (rejected 100% of the time) by the sampling plan; lots with contamination levels near the accept/reject limit are accepted by the sampling plan less than 100% of the time. The shape of the OC curve is uniquely defined for a particular sampling plan design with designated values of sample size, degree of comminution, subsample size, type of analytical method, and number of analyses, and the accept/reject limit X_c.

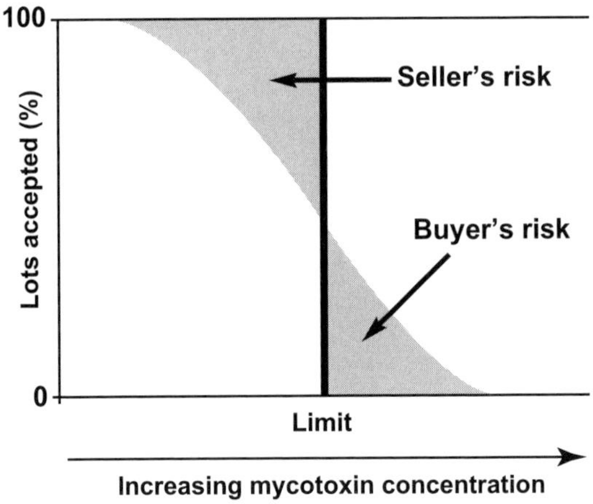

Figure 17. General shape of an operating characteristic (OC) curve. The shape of the OC curve is unique for an aflatoxin test procedure and indicates the magnitude of the buyer's and seller's risks.

CALCULATION OF ACCEPTANCE PROBABILITY

The acceptance probability P(M) associated with sampling a commodity for a mycotoxin can be computed if the distribution among replicated sample test results can be described and if the appropriate variance relationships are known for a mycotoxin test procedure. Several skewed distributions such as the negative binomial and compound gamma have been shown to adequately describe the observed mycotoxin distribution of sample test results for several commodities and several mycotoxins. An example of the probability of accepting and rejecting shelled corn lots over a range of lot concentrations for the sampling plan where ns = 4.54 kg, Romer mill, nss = 50 g, immunoassay analytical method, na = 1 aliquot, and accept/reject limit X_c = 20 ppb is shown in Table 9.

Table 9 shows that most of the lots below 5 ppb are accepted by the sampling plan and most of the lots above 60 ppb are rejected by the sampling plan. For example, 95% and 2% of the lots at 5 and 60 ppb are accepted by the sampling plan, respectively. As lot concentration M increases, the percentage lots accepted by the sampling plan decreases. The acceptance probabilities in Table 9 are plotted in Figure 18 and a smooth curve forced through the points.

Table 9. Probability of accepting and rejecting lots of shelled corn over a range of lot aflatoxin concentrations for a sampling plan that uses 4.54 kg sample size, Romer mill, 50 g subsample, ELISA method, 1 aliquot, and accept/reject limit of 20 ppb.

Lot concentration M (ppb)	*Probability of accepting lot at M* $P(X \le X_c \mid M)$	*Probability of rejecting lot at M* $1 - P(X \le X_c \mid M)$
0	1.000	0.000
5	0.983	0.017
10	0.887	0.113
15	0.726	0.274
20	0.551	0.449
25	0.396	0.604
30	0.274	0.726
35	0.185	0.815
40	0.122	0.878
45	0.079	0.921
50	0.051	0.949
55	0.032	0.968
60	0.020	0.980

For a given sampling plan, the OC curve indicates the magnitudes of the buyer's and seller's risk. When Mc is defined as the legal limit or the maximum lot concentration acceptable, lots with M > Mc are bad and lots with M = Mc are good. In Figure 17, the area under the OC curve for M > Mc represents the buyer's risk (bad lots accepted) while the area above the OC curve for M < Mc represents the seller's risk (good lots rejected) for a particular sampling plan. Using the example in Table 9, if lots at 20 ppb or less are considered good lots and those greater than 20 ppb are consider bad lots, then lots rejected below 20 ppb are considered a measure of the seller's risk (good lots rejected) and the lots accepted above 20 ppb are considered the buyer's risk (bad lots rejected).

Because the shape of the OC curve is uniquely defined by the sample size, degree of comminution, subsample size, the number of analyses and the accept/reject limit, these parameters can be used to reduce the buyer's and seller's risks associated with a sampling plan.

SAMPLE SIZE EFFECT ON RISKS

The effect of increasing sample size on the shape of the OC curve when testing shelled corn lots for aflatoxin is shown in Figure 19 where the accept/reject limit is equal to the legal limit of 20 ppb. As sample size increases from 0.91 to 9.07 kg, the slope of the OC curve about the legal limit increases forcing the two areas associated with each risk to decrease. As a result, increasing the sample size decreases both the buyer's and seller's risks. The same effect can be obtained by increasing either the degree of sample comminution, subsample size or number of analyses.

ACCEPT/REJECT LIMIT EFFECTS ON RISKS

The effect of changing the ARL, relative to the legal limit, on the two risks when testing shelled corn lots for aflatoxin is shown in Figure 20. If the legal limit is assumed to be 20 ppb, then changing X_c to a value less than 20 ppb shifts the OC curve to the left. Compared to the sampling plan where $X_c = 20$, the buyer's risk decreases, but the seller's risk increases. If X_c becomes larger than 20, the OC curve shifts

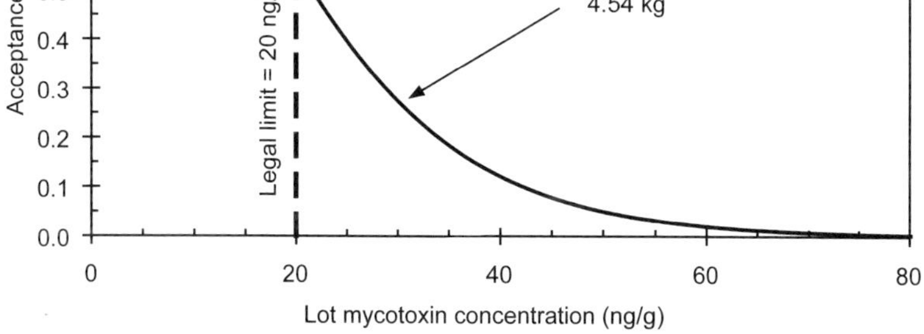

Figure 18. Operating characteristic curve for a sampling plan that uses a 4.54 kg sample, Romer mill, 50 g subsample, ELISA method, 1 aliquot, and a 20 ppb (ng/g) accept/reject limit.

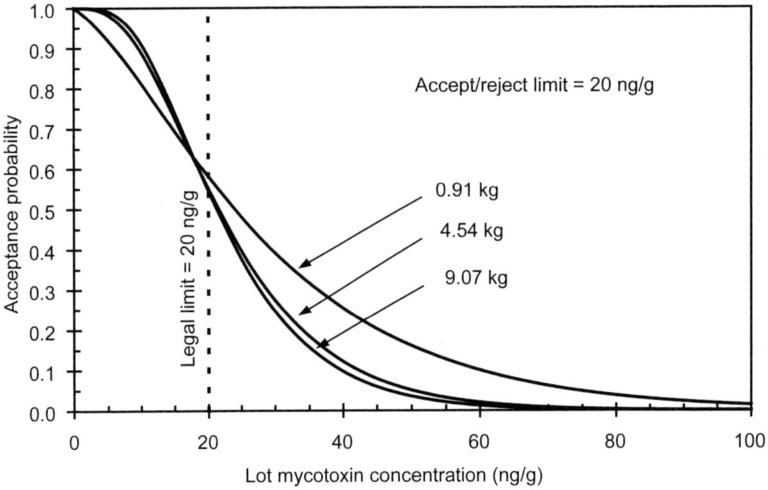

Figure 19. Three operating characteristic curves showing that increasing sample size reduces both buyer's and seller's risks.

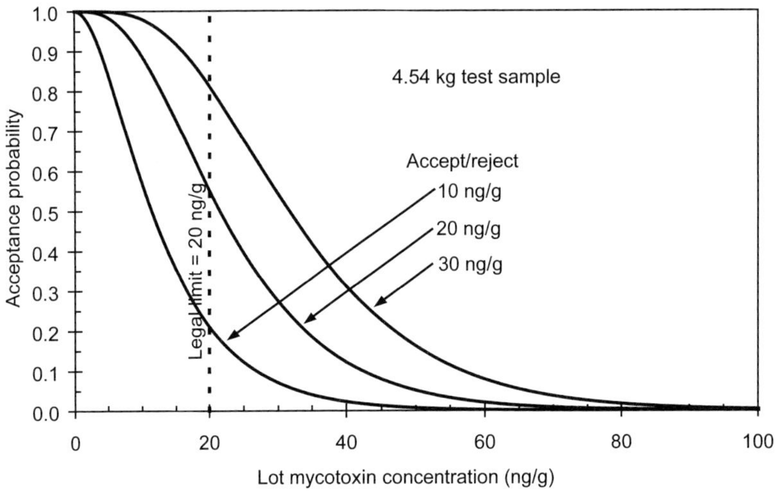

Figure 20. Three operating characteristic curves for three accept/reject limits. If the accept/reject limit (10 ng/g) is less than the legal limit (20 ng/g), the seller's risk increases and the buyer's risk decreases. If the accept/reject (30 ng/g) is greater than the legal limit (20 ng/g), the seller's risk decreases and the buyer's risk increases.

to the right. As a result, the seller's risk decreases but the buyer's risk increases. Changing the ARL relative to the legal limit can reduce only one of the two risks, because reducing one risk will automatically increase the other risk.

Methods have been developed to predict the seller's and buyer's risks, the total number of lots accepted and rejected, the amount of mycotoxin in the accepted and rejected lots, and the costs associated with mycotoxin inspection programmes for several commodities (FAO, 1993; Whitaker and Dickens, 1979; Johansson *et al.*, 2000c; Whitaker *et al.*, 1995). These methods have been used by the UDSA/AMS and the peanut industry to design aflatoxin testing programmes for shelled peanuts

(Whitaker *et al.*, 1995) and by the FAO (FAO, 1993) to design the aflatoxin testing plan for raw shelled peanuts.

MULTIPLE SAMPLES

Increasing the number of samples of a given size taken from a contaminated lot can reduce the risks associated with classifying lots. If the mycotoxin among all samples is averaged, the effect is the same as that described in Figure 19 for the effect of increasing sample size. However, if all multiple sample test results are required to test less than some ARL, the effect is the same as shown in Figure 20 for changing the ARL relative to the legal limit. Three sampling plans showing the effect of requiring either one, two, or three 4.54 kg samples to all test less than or equal to 20 ppb is shown in Figure 21.

As the number of samples required to test less than or equal to the ARL increases, the OC curve shifts to the left reducing the buyer's risk but increasing the seller's risk. The result is similar to reducing the ARL. This type of sampling plan is often used late in the marketing system on finished product destined for animals or humans that have little chance of product containing mycotoxin above the legal limit. The buyer is placing most of the risk on the seller.

Conclusions

Because of the uncertainties (biases and variability) associated with a mycotoxin test procedure, it is difficult to determine with 100% certainty the true concentration of a bulk lot. Even when the sample is correctly selected (no biases), there will be variability associated with the mycotoxin test procedure. The variance associated with a mycotoxin test procedure is the sum of sampling, sample preparation, and analytical variances. For small sample sizes, sampling is usually the largest source of variability. Increasing sample size, the degree of sample comminution, subsample size, and the number of aliquots quantified can reduce the variability associated with a mycotoxin test procedure. Reducing variability of the mycotoxin test procedure will reduce the number of lots misclassified by the sampling plan.

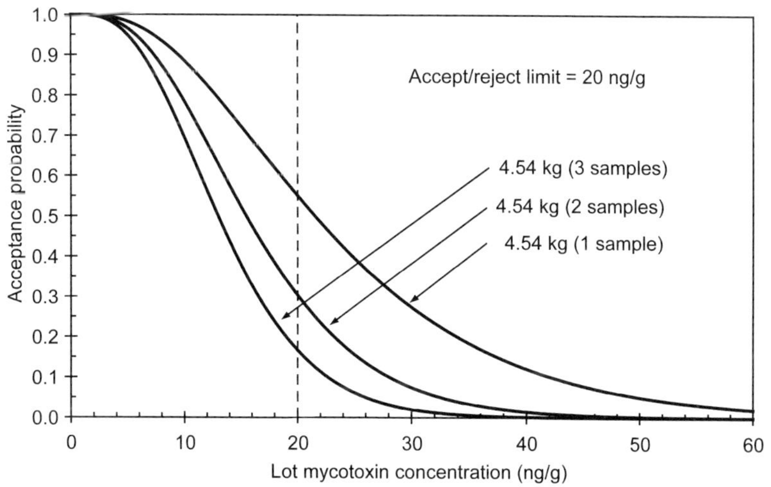

Figure 21. Three operating characteristic curves showing the effect of requiring either one, two, or three 4.54 kg samples of shelled corn to all test less than or equal to the accept/reject limit of 20 ng/g total aflatoxin to accept the lot.

References

Association of Official Analytical Chemists. 1990. In: *Official methods of analysis of the Association of Official Analytical Chemists, 15th edition* (K. Helrich, ed). Association of Official Analytical Chemists, Inc., Arlington, Va.

Bauwin, G.R. and H.L. Ryan. 1982. Sampling inspection and grading of grain. In: *Storage of Cereal Grains and Their Products*, Vol 5 (C.M. Christensen, ed). Am. Assoc. Cereal Chem., St. Paul, Minnesota, p. 115.

Campbell, A.D., T.B. Whitaker, A.E. Pohland, J.W. Dickens and D.L. Park. 1986. Sampling, sample preparation, and sampling plans for foodstuffs for mycotoxin analysis. Pure Appl. Chem. 58:305-314.

Cochran, W.G. and G.M. Cox. 1957. In: *Experimental Design*. John Wiley and Sons, Inc, New York, NY, pp. 15-16.

Council for Agricultural Science and Technology. 2003. Mycotoxins: risks in plant animal, and human systems. Task Force Report Number 139. Ames, IA, pp. 199.

Cucullu, A.F., L.S. Lee, R.Y. Mayne and L.A. Goldblatt. 1986. Determination of aflatoxin in individual peanuts and peanut sections. J. Am. Oil Chem. Soc. 43:89.

Cucullu, A F., L.S. Lee and W.A. Pons. 1977. Relationship of physical appearance of individual mold damaged cottonseed to aflatoxin content. J. Am. Oil chem. Soc. 54:235A.

Cullen, J.M. and P.M. Newberne. 1994. Acute Hepatotoxicity of aflatoxin. In: *The Toxicology of Aflatoxin* (D.L. Eaton and J.D. Groopman, eds). Academic Press, Inc., San Diego, CA, pp. 3-26.

Dickens, J.W. and J.B. Satterwhite. 1969. Subsampling mill for peanut kernels. Food Technol. 23:90-92.

Dickens J.W. and T.B. Whitaker. 1982. Sampling and sampling preparation. In: *Environmental Carcinogens-Selected Methods of Analysis: Some Mycotoxins, Vol. 5* (H. Egan, L. Stoloff, P. Scott, M. Costegnaro, I.K. O'Neill and H. Bartsch, eds). ARC, Lyon, p. 17.

Dickens, J.W., T.B. Whitaker, R.J. Monroe and J.N. Weaver. 1979. Accuracy of subsampling mills for granular material. J. Am. Oil Chem. 56:842.

Food and Agriculture Organization/World Health Organization. 2001. Proposed Draft Revised Sampling Plan for Total Aflatoxin in Peanuts Intended for Further Processing. Joint FAO/WHO Food Standards Program, CODEX Alimentarus Commission, 24th Session, Geneva, Switzerland, July 2-7, 2001. FAO/WHO Joint Office, Viale della Terme di Caracalla, 00100, Rome, Italy. pp. 276-280.

Food and Agriculture Organization. 1995. Worldwide Regulations for Mycotoxins 1995. FAO Food and Nutrition Paper 64. FAO, Viale della Terme di Caracalla, 00100, Rome, Italy. pp. 43.

Food and Agriculture Organization. 1993. Sampling plans for aflatoxin analysis in peanuts and corn. FAO Food and Nutrition Paper 55. FAO, Viale della Terme di Caracalla, 00100, Rome, Italy.

Hurburgh, C.R. and C.J. Bern. 1983. Sampling corn and soybeans. 1. Probing method, Trans. Am. Soc. Agric. Engineers 26:930.

Johansson, A.S., T.B. Whitaker, W.M. Hagler, Jr., F.G. Giesbrecht, J.H. Young and D.T. Bowman. 2000a. Testing shelled corn for aflatoxin. Part I: Estimation of variance components. J. Assoc. Off. Anal. Chem. Int. 83:1264-1269.

Johansson, A.S., T.B. Whitaker, W.M. Hagler, Jr., F.G. Giesbrecht and J.H. Young. 2000b. Testing shelled corn for aflatoxin. Part II: Modeling the distribution of aflatoxin test results. Assoc. Off. Anal. Chem. Int. 83:1270-1278.

Johansson, A.S., T.B. Whitaker, W.M. Hagler, Jr., F.G. Giesbrecht and J.H. Young. 2000c. Testing shelled corn for aflatoxin. Part III: Evaluating the performance of aflatoxin sampling plans. J. Assoc. Off. Anal. Chem.

Int. 83:1279-1284.

Malone, B. 2000. Solution fluorometric method for deoxynivalenol in grains. In: *Mycotoxin Protocols* (M.W. Trucksess and A.E. Pohland, eds). Humana Press, Totowa, NJ, pp. 97-113.

Nesheim, S. 1979. Methods of aflatoxin analysis. NBS Spec. Publ. (US) No. 519, pp. 355.

Parker, P.E., G.R. Bauwin and H.L. Ryan. 1982. Sampling, inspection and grading of grain. In: *Storage of Cereal Grains and Their Products* (C.M. Christensen, ed). American Association of Cereal Chemists, St. Paul, Minnesota, pp. 1-35.

Phillips, T.D., B.A. Clement and D.L. Park. 1994. Approaches to reduction of aflatoxin in foods and feeds. In: *The Toxicology of Aflatoxin* (D.L. Eaton and J.D. Groopman, eds). Academic Press, Inc., San Diego, CA, pp. 383-406.

Shotwell, O.L., M.L. Goulden, R.J. Botast and C.W. Hasseltine. 1975. Mycotoxins in hot spots in grains. 1. Aflatoxin and zearalenone occurrence in stored corn. Cereal Chem. 52:687.

Shotwell, O.L., M.L. Goulden and C.W. Hessletine. 1974. Aflatoxin: Distribution in contaminated corn. Cereal Chem. 51:492.

Steyn, P.S., P.S. Thiel and D.W. Trinder. 1991. Detection and quantification of mycotoxins by chemical analysis. In: *Mycotoxins and Animal Foods* (J.E. Smith and R.S. Henderson, eds). CRC Press, Inc., Boca Raton, FL, pp. 165-221.

United States Department of Agriculture. 1975. In: *Inspectors Instructions*. Agricultural Marketing Service, Washington, D.C., pp. 22.

T.B. Whitaker and J.W. Dickens. 1979. Evaluation of the Peanut Administrative Committee testing program for aflatoxin in shelled peanuts. Peanut Sci. 6:7-9.

T.B. Whitaker, J.W. Dickens and R.J. Monroe. 1974. Variability of aflatoxin test results. J. Amer. Oil Chem. Soc. 51:214-218.

T.B. Whitaker, J.W. Dickens, R.J. Monroe and E.H. Wiser. 1972. Comparison of the observed distribution of aflatoxin in shelled peanuts to the negative binomial distribution. J. Am. Oil Chem. Soc. 49:590-593.

T.B. Whitaker and F.E. Dowell. 1995. Sampling methods to measure aflatoxin and grade factors of peanuts. In: *Advances in Peanut Science* (H.E. Pattee and H.T. Stalker, eds). Am. Peanut Res. Educ. Soc., Stillwater, OK.

T.B. Whitaker, F.E. Dowell, W.M. Hagler, Jr., F.G. Giesbrecht and J. Wu. 1993. Variability associated with sampling, sample preparation, and chemically testing farmers stock peanuts for aflatoxin. J. Assoc. Off. Anal. Chem. Int. 77:107-116.

T.B. Whitaker, W. Horwitz, R. Albert and S. Nesheim. 1996. Variability associated with methods used to measure aflatoxin in agricultural commodities. J. Assoc. Off. Anal. Chem. Int. 79:476-485.

T.B. Whitaker, J. Springer, P.R. Defize, W.J. deKoe and R. Coker. 1995. Evaluation of sampling plans used in the United States, United Kingdom and The Netherlands to test raw shelled peanuts for aflatoxin. J. Assoc. Off. Anal. Chem. Int. 78:1010-1018.

T.B. Whitaker, M.W. Truckess, A.S. Johansson, F.G. Giesbrecht, W.M. Hagler, Jr. and D.T. Bowman. 1998. Variability associated with testing shelled corn for fumonisin. J. Assoc. Off. Anal. Chem. Int. 81:1162-1168.

William, P.C. 1991. Storage of grains and seeds. In: *Mycotoxins and Animal Foods* (J.E. Smith and R.S. Henderson, eds). CRC Press, Inc., Boca Raton, FL, pp. 721-746.

2

MYCOTOXINS: THEIR EFFECTS IN POULTRY AND SOME PRACTICAL SOLUTIONS

G. Devegowda and T.N.K. Murthy
Division of Animal Sciences, College of Veterinary Medicine, Bangalore, India

Introduction

During the past few decades there has been a steady increase in global production of poultry meat and eggs. Although the high nutritive value of eggs and poultry meat has resulted in increasing demand, food quality and safety factors are becoming increasingly significant in determining market value of poultry products. As mycotoxins are one of the major factors suppressing poultry productivity and also product quality, control of their impact is critical.

According to the United Nation's Food and Agriculture Organization (FAO), approximately 25% of the world's grain supply is contaminated with mycotoxins. The greatest economic impact of mycotoxin contamination is felt by crop and poultry producers, as well as food and feed producers (CAST, 2003). Worldwide, the economic losses due to mycotoxins in poultry feeds can be as much as several hundred million USD annually. The adverse effects of mycotoxin-contaminated diets on performance range from undetectable to devastating in terms of reduced egg production in layers and breeders, and growth depression in broilers, turkeys, ducks etc.

The threat of mycotoxins to humans was described during World War II when Russian soldiers suffered severe dermal necrosis, haemorrhage and destruction of bone marrow after eating mouldy grains (*Fusarium*-contaminated). However, it was not until 1960, when the entire turkey population of Britain was decimated by a fatal liver disease called **Turkey X Disease**, that the scientific community recognized the negative effects associated with mycotoxins. British agriculture officials later traced the source of the outbreak to aflatoxin in a shipment of peanut (groundnut) meal that originated in Brazil.

Mycotoxins of importance in poultry are mainly produced by fungi of the genera *Aspergillus, Fusarium* and *Penicillium* either pre-harvest, during harvest, or in storage or during feed processing whenever conditions are favorable. No region of the world escapes these silent killers; and their negative impact on poultry productivity and human health is enormous.

Geographic distribution of mycotoxins

At the First Food and Agricultural Organization (FAO) - World Health Organization (WHO) - United Nations Environment Programme (UNEP) Conference on mycotoxins in 1977, a review was presented on the occurrence of mycotoxins in various commodities throughout the world. Only seven mycotoxins were reported to occur significantly in naturally contaminated foods and feeds: aflatoxins, ochratoxin A, patulin, zearalenone, trichothecenes, citrinin and penicillic acid (Jelinek *et al.*, 1989). The occurrence of mycotoxins in nature is considered a global problem. However, in certain regions of the world, some mycotoxins are produced more commonly than others. In colder, more temperate, regions such as Canada, the northern US and most of the European countries, aflatoxins are not considered to be a major problem except in imported feedstuffs grown in warmer southern climates. Economically, in these regions the most important mycotoxins are deoxynivalenol (vomitoxin, DON), zearalenone (ZEA), ochratoxin A (OTA), diacetoxyscirpenol (DAS), T-2 toxin and HT-2 toxin. DON contamination of corn and corn products in Canada, South Africa, Taiwan, the UK and the US has been reported to be more than 1 mg/kg in some lots of grain tested (Devegowda *et al.*, 1998).

In Europe, the difference in climatic conditions among the northern, middle and southern parts, favours the development of different fungal species. Accordingly, different mycotoxins have gained considerable importance in respective European regions during the last two decades (Leibetseder, 1995). In the maize-growing areas of southern and middle Europe (Sweden, Austria and Hungary) it is mainly fusariotoxins (DON, ZEA, T-2 toxin) that cause illness and poor performance of livestock and poultry whereas OTA is of greater importance in the northern part of Europe (Denmark and Poland).

Investigations in Germany demonstrated that 8.4% of 1,000 feed samples tested positive for trichothecenes. Fifteen grain samples out of 966 contained ZEA (Gedek, 1988). A recent large scale survey, where 1431 samples of wheat, barley, rye, oats and bran were analysed for OTA in Denmark, found 40% of the samples were contaminated with OTA. Among the commodities analysed, bran products showed a higher OTA contamination than cereal kernels. For example OTA was detected in wheat in 62% of the bran samples and only 30% of the samples of kernels (Hohler, 1999). These findings are typical for most of the mycotoxins, where cereal by-products, especially bran, usually contain higher mycotoxin levels than the whole cereal.

Aflatoxins, the most widespread of all the mycotoxins, are common in warm and humid climatic conditions like those existing in Latin American, Asian and African countries, southern regions of US and certain parts of Australia. Aflatoxins occur in about 20% of the corn grain grown in the southern US even in non-crisis years; and in years with extreme weather conditions occurrence may be higher.

Extensive surveys conducted in India, Pakistan, Egypt and South Africa suggested that aflatoxins are often present in substantial levels in mixed feeds and ingredients. The evidence comes from the fact that in southern parts of India, where hot and humid climatic conditions prevail during most of the year, certain feed ingredients and finished feeds were found to be positive for the presence of aflatoxin, OTA and T-2 toxin (Chandrasekaran, 1996). However, during the winter season high moisture conditions in these countries may result in other mycotoxins such as ZEA, DON, T-2 toxin, OTA, etc.

A survey of Chinese compound feeds and ingredients revealed that 88, 84, 77 and 60% of corn samples contained T-2 toxin, aflatoxin, fumonisin and OTA, respectively. All corn samples contained ZEA and DON. The six major toxins were found in more than 90% of compound feeds (Wang *et al.*, 2003).

In Latin American countries including Brazil, Peru, Mexico, Colombia, Venezuela and Argentina, reports exist on the presence of aflatoxins, OTA, T-2 toxin, DON, ZEA and other *Fusarium* toxins. In a recent Canadian survey for ergot alkaloids in various cereal products, individual alkaloids were found at low levels (<200 μg/kg) (Devegowda *et al.*, 1998). A maximum of 0.05% ergot alkaloid has been suggested in Canada as an acceptable level in wheat used for production of flour. Ergot poisoning is also seen in warm and humid regions of India, Ethiopia and the US.

Due to the increase in global trading of

feedstuffs, mycotoxins are no longer restricted by climate and geography and are now more widely distributed (Devegowda *et al.*, 1998a).

Mycotoxins significantly affecting poultry

Some fungal strains are capable of producing more than one mycotoxin and also a single mycotoxin is produced by more than one fungus. The most important mycotoxins for poultry and the fungi that produce them are given in Table 1.

Table 1. Moulds and mycotoxins important in intensive poultry production.

Mycotoxins	*Moulds*
Aspergillus toxins	
Aflatoxins B_1, B_2, G_1 and G_2	*Aspergillus flavus* and *Aspergillus parasiticus*
Cyclopiazonic acid	*Aspergillus flavus*
Ochratoxins	*Aspergillus ochraceus*
Sterigmatocystin	*Aspergillus versicolor*
Pencillium toxins	
Ochratoxins	*Penicillium viridicatum*
Citrinin	*Pencillium citrinum*
Fusarium toxins	
T-2, HT-2, DAS, MAS, DON	*Fusarium tricinctum, Fusarium graminearum Fusarium solani*
Fumonisins B_1, B_2	*Fusarium moniliforme Fusarium proliferatum*
Moniliformin	*Fusarium moniliforme*
Zearalenone	*Fusarium graminearum, Fusarium roseum*

The effects of individual mycotoxins in poultry are discussed in the following sections. However, it should be noted that combinations of toxins, as found naturally in mouldy feeds, cause greater adverse effects than feeds containing a single mycotoxin.

Aflatoxins

The most widespread and the most studied group of all the mycotoxins, aflatoxins are of concern in warm and humid climatic conditions. Although aflatoxins are not considered to be a major problem in cold or temperate regions, caution must be exercised even in colder climates when using feedstuffs imported from warm and humid countries.

Aflatoxins are primarily produced by fungi of the genus *Aspergillus* (*A. flavus* and *A. parasiticus*) and are found in poultry feeds and human food products. Major forms of aflatoxins include B_1, B_2, G_1 and G_2 with aflatoxin B_1 (Figure 1) being the most common and biologically active toxin. Henceforth in this chapter, the term aflatoxin is used to refer to aflatoxin B_1 unless specified otherwise.

IMPACT ON POULTRY MEAT PRODUCTION

Economic losses associated with aflatoxin exposure include poor growth and feed conversion, increased mortality, decreased egg production, leg problems, and carcass condemnations. The frequent aflatoxin contamination of agricultural commodities and the chronic exposure of poultry to these toxins can mean the difference between profit and loss to a poultry operation (Hamilton, 1984; Devegowda *et al.*, 1998b).

Among poultry, ducks are the most susceptible to aflatoxins followed by turkeys, broilers, laying hens and quail. Goslings and pheasants are more sensitive to aflatoxins than chicks and less sensitive than ducklings and turkey poults. Clinically, the signs of aflatoxin toxicity include anorexia, decreased weight gain, decreased egg production, specific visceral haemorrhage, embryo toxicity, and

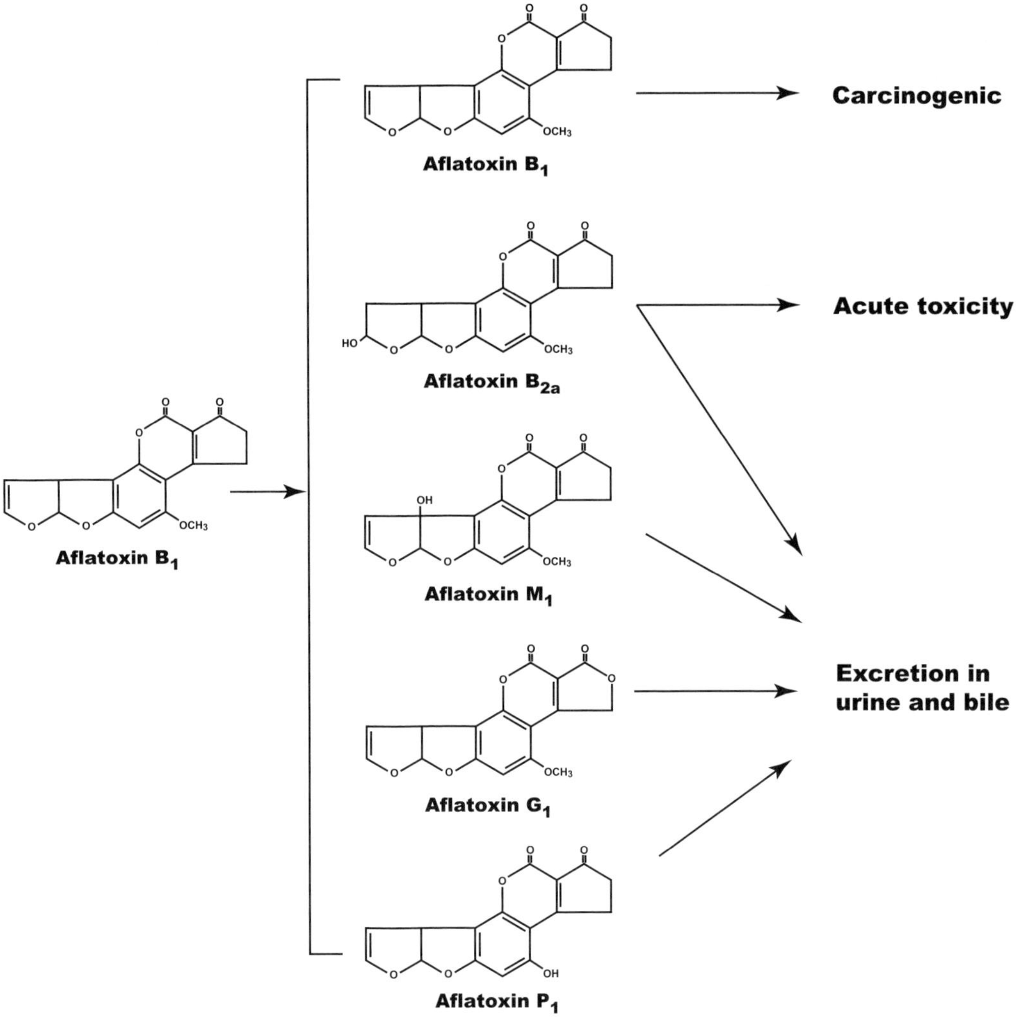

Figure 1. The primary forms of aflatoxin.

increased susceptibility to environmental and microbial stressors. Histopathology reveals fatty liver, liver necrosis and bile duct hyperplasia as the most common findings in birds affected with aflatoxicosis.

Aflatoxin can have a wide range of effects on metabolism in poultry. Aflatoxin decreases the activities of several enzymes important in the digestion of starch, proteins, lipids and nucleic acids. In addition, the activities of serum glutamate pyruvate transaminase (SGPT), serum glutamate oxaloacetate transferase (SGOT) and γ-glutamyl transferase (GGT) are increased, primarily indicating hepatic damage. Reduction in total protein, cholesterol and blood urea nitrogen are also toxic effects of aflatoxin in broilers (Raju and Devegowda, 2000; Aravind *et al.*, 2003) (Table 2).

Aflatoxin is also known to interfere with vitamin D metabolism, contributing to reduced bone strength and leg weakness (Hamilton, 1987). Additionally, metabolism of several minerals including iron (causing hemolytic anemia), phosphorus (causing leg weakness) (Table 3) and copper are also

Table 2. Effect of feeding an aflatoxin contaminated diet on serum biochemical parameters in broilers at five weeks of age.

Aflatoxin 300 ppb	*Total proteins (g %)*	*Total cholesterol (mg %)*	*Blood urea nitrogen (mg %)*
-	2.73[a]	114.8[a]	2.22[a]
+	1.45[b]	57.3[b]	1.01[b]

Raju and Devegowda, 2000
[ab]Means in a column differ ($P<0.01$).

Table 3. Adverse effects of aflatoxin on 0 – 21-day-old ducks.

Aflatoxin (µg/kg)	*Weight gain (g/100g BW)*	*Leg deformity score*	*Liver weight (g/100g BW)*
60	602[a]	1.73[b]	5.72[b]
120	417[b]	2.21[a]	7.81[a]

Khajarern and Khajarern, 1999
[ab]Means in a column differ ($P<0.05$).

affected by aflatoxin. Suppression of hepatic protein synthesis is the main factor resulting in growth suppression and reduced egg production. Owing to reduced bile salt production, absorption of fat and carotenoid pigments is reduced in affected birds. The resultant poor pigmentation of skin and egg yolk is generally referred to as 'pale bird syndrome', which may also be seen in ochratoxicosis.

Bruising

Bruising is a significant cause of downgrading of broiler meat (Hess and Biligili, 1997), accounting for approximately 25% of all downgrades recorded. Mycotoxins, especially aflatoxins, are associated with bruising in chickens (Girish and Devegowda, 2004a). Aflatoxins increase the fragility of capillaries, reducing prothrombin levels thereby drastically increasing the incidence of bruising in carcasses (Tung *et al.*, 1971) (Table 4).

Table 5 shows a study involving several poultry companies where the growers were grouped into good, moderate, or poor categories based on performance (Jones *et al.*, 1982). A review of the records for the incidence of aflatoxin positive feed samples and the levels detected brought out a conclusive finding of the adverse effects of aflatoxin on growth of poultry at all levels of occurrence. It is worthwhile noting that even at low levels (6-14 µg/kg), which may be considered 'practically safe' in some parts of the world, aflatoxin can cause considerable damage to bird health and performance and consequently operation profitability.

Table 4. Body weight and bruising during experimental aflatoxicosis.

Aflatoxin (µg/g)	*Body weight (g)*	*Bruising energy (µjoules)*
0	552	441
0.625	558	365*
1.25	547	314*
2.50	481*	229*
5.00	385*	212*
10.00	327*	206*

Tung, *et al.*, 1971
*Values differ ($P<0.05$) from corresponding controls.

Table 5. Aflatoxin effects on production and processing.

Parameters	*Grower classification*		
	Good	*Moderate*	*Poor*
No. of growers	10	10	10
No. of birds	122,052	110,290	105,267
Average weight, lbs	3.88	3.83	3.79
Average FCR	2.13	2.15	2.16
Livability, %	95.98	95.61	92.78
Aflatoxin, µg/kg	6.13	6.49	13.99
Aflatoxin positive samples, %	18.04	22.08	31.25

Jones *et al.*, 1982

BROILER BREEDERS

Feeding aflatoxin to broiler breeder hens has been reported to decrease egg production and hatchability (Afzali and Devegowda, 1999). Qureshi *et al.* (1998) reported an increase in the concentration of aflatoxin and its metabolites in the serum of hens and in eggs when high levels of aflatoxin were fed. Further, drastic decreases in fertility and hatchability were observed within the first four days of feeding the contaminated feed (Table 6).

Table 6. Effect of aflatoxin on broiler breeder progeny performance.

Aflatoxin[1] *(mg/kg)*	*Fertile eggs (%)*	*Hatchability (%)*	*Phagocytic macrophages*[2] *(%)*
0	98.6[a]	82.8	35.8[a]
10	92.4[b]	35.3	9.7[b]

Qureshi *et al.*, 1998
[ab]Means in a column differ ($P<0.05$)
[1]In the diet of the hen.
[2]To sheep red blood cells.

AFLATOXIN STUDIES IN OTHER SPECIES OF POULTRY

Goslings and pheasants showed reduced weight gains as the level of aflatoxin was increased. All of the goslings fed 4 μg/g died by 2 weeks of age and 90% of the pheasants on the 4 μg/g feed died by 3 weeks of age (Muller *et al.*, 1970). In another study involving poults, all the poults receiving the 1, 2 and 4 μg/g diets were dead by the end of experiment (Muller *et al.*, 1970). As early as 1 week of age, high mortality and reduced weight gains were noted with higher levels of aflatoxin. Ducklings are the most susceptible to aflatoxin injury of the species tested. Mortality was 100% at 2 weeks of age for the birds receiving either 2 or 4 μg/g and at 3 weeks of age for the birds receiving the 1 μg/g diet (Muller *et al.*, 1970).

IMMUNOSUPPRESSION

A major concern of poultry producers is immunosuppression caused by aflatoxin and the consequent increase in susceptibility to a wide variety of infectious diseases along with the accompanying vaccine failure and therapeutic drug failures. Aflatoxin-induced immunosuppression produces a cascading effect starting with depressed protein synthesis, lowered serum albumin and globulin levels, a reduction in circulating antibody levels, impairment of the reticulo-endothelial system, reduced cell-mediated immunity, and abnormal development of the thymus and bursa of Fabricius (Swamy and Devegowda, 1998) (Table 7). There is a large body of evidence to suggest that aflatoxicosis (and also ochratoxicosis and T-2 toxicosis) can increase the susceptibility of birds to salmonella infection, largely owing to a lesser number of circulating leukocytes and their reduced phagocytic ability. Aflatoxicosis can also induce therapeutic failures. It has been found that aflatoxin can reduce the serum concentration of antibiotics (Miller and Wyatt, 1985), which may be attributed to the stepped-up efforts by the liver to maximize systemic detoxification.

Table 7. Effect of aflatoxin B_1 on antibody titres against infectious bursal disease (IBD) and Newcastle disease (ND) in broiler chickens at 6 weeks of age.

Aflatoxin µg/kg	*IBD*	*ND*
0	6180 ± 195[a]	5800 ± 199[a]
100	3800 ± 212[b]	3025 ± 208[b]
200	3046 ± 220[c]	2650 ± 214[c]
400	2200 ± 225[d]	1850 ± 217[c]

Swamy and Devegowda, 1998
[ab]Means in a column differ ($P<0.05$).

In poultry, the bursa of Fabricius, thymus and spleen and to a lesser extent the caecal tonsils and bone marrow constitute the organs contributing to humoral and cellular immunity. Various mycotoxins (Table 8) are known to adversely affect the immune system by depressing both cell-mediated and humoral immune response.

Table 8. Mycotoxins that cause immunosuppression (in decreasing order of severity).

1. Aflatoxin
2. Deoxynivalenol, T-2 toxin, HT-2 toxin
3. Ochratoxin
4. Fumonisins

Aflatoxin has a direct adverse effect on both cellular and humoral immunity. While low levels of dietary aflatoxin may affect cellular immunity, immunoglobulin production and antibody response are suppressed at higher levels (Afzali and Devegowda, 1999). Aflatoxin is considered to be a major contributor to immunosuppression with many disease outbreaks, vaccination failures and poor antibody titres being attributed to it (Raju and Devegowda, 2002) (Table 9).

Recent studies have implicated aflatoxin as a cause of immunosuppression at the embryonic stage by instigating significant DNA damage to B- and T-lymphocytes at a stage when they are in rapid proliferation and differentiation (Qureshi *et al.*, 1998).

RESIDUES

Aflatoxins are well absorbed and rapidly metabolised, producing reactive epoxides that can further produce carcinogenic metabolites. The rate of aflatoxin residue accumulation into poultry meat is much lower than its tranfer into ruminant milk. Residue level in eggs is dose dependent with more residues concentrated in yolk than albumin (Table 10) (Jacobson and Wiseman, 1974). Due to a very low rate of exposure to aflatoxins in Europe in line with the stringent regulations, the risk of undesirable residues in edible tissues is negligible.

Table 9. Impact of aflatoxins on the immune system.

Reduction in size of bursa of Fabricius and thymus
Reduction in T-lymphocyte, B-lymphocyte and white blood cell counts
Reduction in total serum proteins and immunoglobulins
Reduction in antibody titers
Reduction in serum concentration of antibiotics

Table 10. The relationship between the aflatoxin content of layer feed and the aflatoxin concentration in eggs.

Aflatoxin in feed (μg/kg)	*Aflatoxin in eggs (μg/kg)*
100	0.23
200	0.78
400	1.40

Jacobson and Wiseman, 1974

PERMISSIBLE LEVELS IN FEEDS

A majority of countries throughout the world have established regulations for aflatoxins in poultry feeds. Many countries follow a maximum permissible level of 20 μg/kg for aflatoxins in poultry feeds (CAST, 2003). The maximum permissible level in the European Union for total aflatoxins in agricultural commodities is 4 μg/kg and the maximum permissible level of aflatoxin B_1 is 2 μg/kg. In Japan, the regulation for aflatoxin is 10 μg/kg for young chicks and 20 μg/kg for adult chickens.

Cyclopiazonic acid

Cyclopiazonic acid (CPA) is primarily produced by *Aspergillus flavus* and is commonly found in combination with aflatoxin (Figure 2). Urano and co-workers (1992) reported the co-occurrence of CPA and aflatoxins in corn and peanuts in the US. In Indonesia it was reported that CPA occurred in approximately 81% of samples analysed from poultry feed mills at levels up to 9 mg/kg. The toxic effects of CPA have been summarised by Smith *et al.* (1992). Recent toxicological studies have shown that the principal target organs of this mycotoxin are the liver, kidney and digestive system. Feeding trials with chickens have produced mucosal necrosis and inflammation in the crop, proventriculus and gizzard. In these chickens the kidneys were often swollen or distended (CAST, 2003). CPA is associated with haemorrhagic lesions seen on thigh muscle and is thought to be responsible for many haemorrhagic syndromes. CPA has been reported to be markedly distributed in skeletal tissue and has neurotoxic effects in poultry.

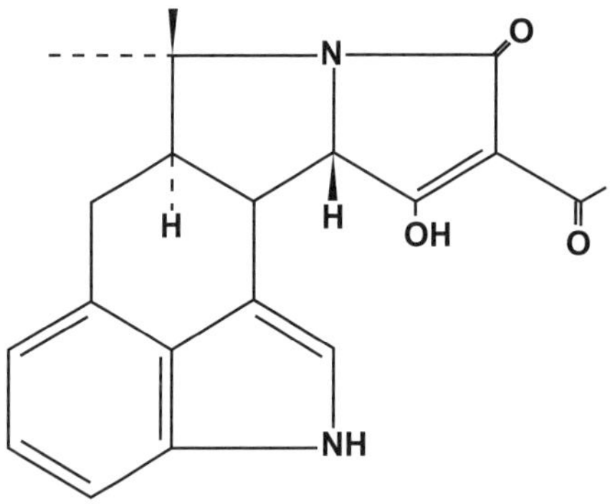

Figure 2. Structure of cyclopiazonic acid (CAST, 2003).

Cole (1986) has reappraised the etiology of the classical Turkey X disease syndrome and has suggested that the arching of the neck, the drawn back head (opisthotonus), and the legs extended fully backwards are effects produced by CPA, but not by aflatoxins. Thus he concluded that this mycotoxin occurred with the aflatoxins in the Brazilian groundnut cake responsible for Turkey X disease.

LAYING HENS

CPA chelates metal cations due to the presence of a tetramic acid moiety. The chelation of cations such as calcium, magnesium and iron is thought to be the main mechanism of CPA toxicity. Hens fed diets contaminated with CPA lay eggs with thin or visibly cracked shells due to impaired egg shell formation (Bryden, 1991).

RESIDUES

Dorner and coworkers (1994) reported that CPA was found in eggs after oral administration of CPA to laying hens. These findings indicate a potential source of human exposure via ingestion of eggs and meat containing residues of CPA.

Sterigmatocystin

This mycotoxin is produced by *Aspergillus versicolor*. It is also produced by toxigenic strains of *A. flavus* and *A. parasiticus*. Sterigmatocystin contains a xanthone nucleus and is considered to be a biosynthetic precursor of aflatoxin (Figure 3).

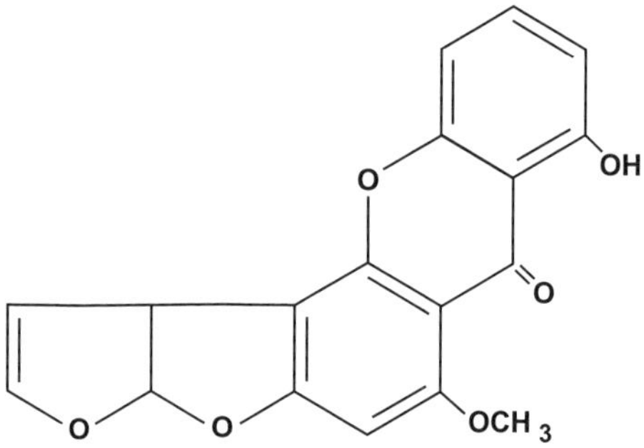

Figure 3. Structure of sterigmatocystin.

Like aflatoxin, it contains the bis-dihydrofuran ring (the vinyl ether); and as would be expected is a carcinogen. The target organ for toxicity is the liver. Sterigmatocystin is found in commodities that are heavily contaminated with aflatoxin. Sterigmatocystin is also found in heavily moulded grains, rice bran, sunflower meal, soyabean meal, peanut meal, etc.

Ochratoxin A

Ochratoxin, first identified by South African researchers in 1965, is one of the most toxic mycotoxins to poultry. Several species of *Aspergillus* and *Pencillium* fungi produce ochratoxin. Three types of ochratoxin occur, with the most common being ochratoxin A (OTA), a common contaminant in a variety of feedstuffs in many countries (Figure 4). In 2001 the World Health Organisation proposed a general maximum limit of OTA of 5 μg/kg in cereals and cereal products. Since 2001 several countries have adopted these levels as the official maximum limit for foods and feeds.

Ochratoxins are important storage toxins in temperate as well as in tropical areas. In tropical and subtropical countries OTA is mainly produced by *Aspergillus* species, whereas in temperate countries, *Penicillium* species (*P. viridicatum)* are the major producers of ochratoxins (Devegowda et *al.*, 1998c).

Ochratoxin A is approximately three times more toxic to young broiler chicks than aflatoxin (Huff *et al.*, 1974), and has been implicated in significant field outbreaks of mycotoxicosis in poultry (Hamilton *et al.*, 1982).

IMPACT ON MEAT AND EGG PRODUCTION

Ochratoxin A is primarily a nephrotoxin that significantly suppresses feed intake, growth, feathering, egg production and feed conversion efficiency (Hamilton et *al.*, 1982). Feeding an ochratoxin-contaminated diet significantly reduced body weight gain (-17.6%), increased GGT activity (+38.5%), and decreased total protein (-16.4%) and hematocrit values (-11.3%) (Aravind and Devegowda, 2003) (Table 11). Although all mycotoxins can suppress body weight gain, ochratoxin produces the greatest body weight gain depression. Birds severely affected with ochratoxicosis show urate deposits in joints and in the abdominal cavity. Acute OTA toxicity results in acute renal failure leading to death. Polyurea leading to large volumes of wet faeces is another common observation with OTA.

Table 11. The effects of dietary ochratoxin on haematological and serum biochemical parameters.

Ochratoxin (500 ppb)	*GGT (IU/L)*	*Proteins (g/100 ml)*	*Hematocrit (%)*
-	8.3[a]	2.62[b]	36.0[b]
+	11.5[b]	2.19[a]	31.9[a]

Aravind and Devegowda, 2003
[ab]Means in a column differ ($P<0.05$).

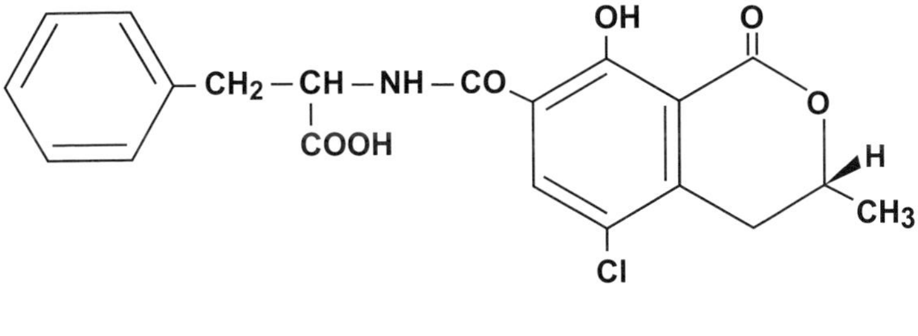

Figure 4. Structure of ochratoxin A.

Poor general resistance to disease and increased mortality are also observed in birds fed OTA-contaminated diets. Young chicks in the initial few weeks of life are highly sensitive. Though older birds are less sensitive, laying hens show reduced egg production in a dose-related fashion with a characteristic yellow staining of egg shells. Higher levels of OTA result in poor egg shell quality and higher incidence of eggs with blood spots (Shirley and Tohala, 1983).

Decreased embryo viability and poor hatchability are observed in eggs from hens fed OTA-contaminated feed. Chicks from such eggs continue to perform below average. Delayed sexual maturity is commonly seen in chicks from hens exposed to OTA.

Since the main consumer of poultry products are humans, it becomes relevant to also view the problem of mycotoxin residues in poultry products from a human health standpoint (Table 12).

Table 12. Occurrence of mycotoxin residues in poultry products.

Mycotoxin	*Poultry production*	*Carry over in meat and eggs*
Aflatoxin B_1	+	Liver
Ochratoxin A	+	Hatching eggs
Cyclopiazonic acid	+	Meat and eggs
Deoxynivalenol	+	Hatching eggs
Zearalenone	+	Eggs
T-2 toxin	-	-
Diacetoxyscripenol	-	-
Fusarochromanone	+	Hatching eggs
Aurofusarin	+	Eggs

TURKEYS

High levels of OTA can have devastating effects on turkeys. Table 13 shows the impact of ochratoxicosis on a commercial turkey farm (Hamilton *et al.*, 1982). The farm lost over 50% of the flock in just nine days. The level of ochratoxin in the feed was very high; and it was reported that even wild birds foraging on spilled feed were dying.

Table 13. Turkey mortality during an outbreak of ochratoxicosis.

Age (Days)	*Mortality (Number of poults)*
6	19
7	12
8	57
9	586
10	246
11	983
12	863
13	370
14	547
15	1033
16	826

Hamilton *et al.*, 1982

IMMUNE STATUS

Ochratoxin is known to cause regression and cellular depletion (lymphocytes) of major lymphoid organs, significantly affecting cellular immunity in poultry. A lesser secondary suppression of humoral immunity with lower circulating immunoglobulins is observed in ochratoxicosis as a result of depletion of immune system effector cells, especially the macrophages.

SIGNIFICANCE OF THE PROBLEM

Ochratoxin has been detected in a number of foods and feeds in various countries (Van Egmond and Spijers, 1994). A recent survey, where 1431 samples of wheat, barley, rye, oats and bran were analysed for OTA in Denmark (Hohler, 1999), showed that 40% of feed samples were contaminated with OTA. Among the commodities analysed, bran products had higher OTA levels than cereal kernels.

RESIDUES

Ochratoxin A, CPA and citrinin bind to plasma proteins once absorbed, resulting in accumulation and a prolonged residence time in the body. Total body clearance varies considerably among species. Even though some EU member states specify regulations for ochratoxin levels in meat, it is generally accepted that the exposure of consumers to OTA, CPA and citrinin originates predominantly from plant-derived foods and only a relatively small percentage originates from animal-derived foods (Veldman, 2003).

Citrinin

Citrinin is produced primarily by *Penicillium citrinum* and some species of *Aspergillus*. Citrinin is nephrotoxic and is often seen in combination with ochratoxin in cereals such as maize, wheat, barely, oats, rice bran, sunflower and soybean meal. The major effects of citrinin toxicity are increased water consumption and diarrhoea (Table 14). Reduced body weight with postmortem lesions including mottled liver and enlarged kidneys are the usual findings in affected birds. Studies to date do not indicate any relation between citrinin and immunity in poultry.

Table 14. Individual and combined effects of ochratoxin and citrinin on water consumption of broilers between 2 and 3 weeks of age.

Ochratoxin (mg/g)	*Citrinin (mg/g)*	*Water consumed (ml/bird)*
0	0	631 ± 26[c]
0	400	2289 ± 53[a]
2.0	0	700 ± 18[c]
2.0	400	1181 ± 66[b]*

Manning et *al*., 1985
[a,b,c]Values with different superscripts differ ($P<0.5$)
*Significant ($P<0.5$) interaction.

Fusarium mycotoxins

Fusarium mycotoxins are economically the most significant mycotoxins in foods and feeds on a global scale. Although more than 100 *Fusarium* mycotoxins are known, those fusariotoxins of most concern based on toxicity and occurrence on a worldwide basis are the trichothecenes (T-2 toxin, diacetoxyscripenol, monoacetoxyscirpenol, HT-2 toxin, deoxynivalenol), zearalenone, fumonisins and moniliformin (Hussein and Brasel, 2001). Furthermore, the trichothecene mycotoxins themselves comprise a vast group of over 100 fungal compounds with the same basic structure, and in many cases occur in contaminated feed with additive toxic effects (Raju and Devegowda, 2000; Swamy *et al.*, 2002).

Trichothecenes (T-2 toxin, diacetoxyscirpenol, monoacetoxyscirpenol, HT-2 toxin and deoxynivalenol)

Trichothecenes (Figure 5) are proven tissue irritants with the major observation associated with their ingestion being oral lesions, dermatitis and intestinal irritation. The major physiological response to trichothecene mycotoxins is loss of appetite, thus earning them the name, **feed refusal toxins**. Pigs are more sensitive to deoxynivalenol (DON) whereas poultry are more sensitive to T-2 toxin and diacetoxyscirpenol (DAS). Most trichothecenes cause necrosis and inflammation of the oral cavity (Smith *et al.*, 2001). 'Helicopter disease' is a manifestation of altered feather structure, growth and appearance and is commonly observed in chickens fed trichothecene-contaminated feeds.

Trichothecenes are the second most immunosuppressive mycotoxins after aflatoxin. They primarily affect cellular immune response by direct effects on bone marrow, spleen, lymphoid tissues, thymus and intestinal mucosa, where actively dividing cells are damaged.

T-2 TOXIN

T-2 toxin is a naturally occurring mycotoxin

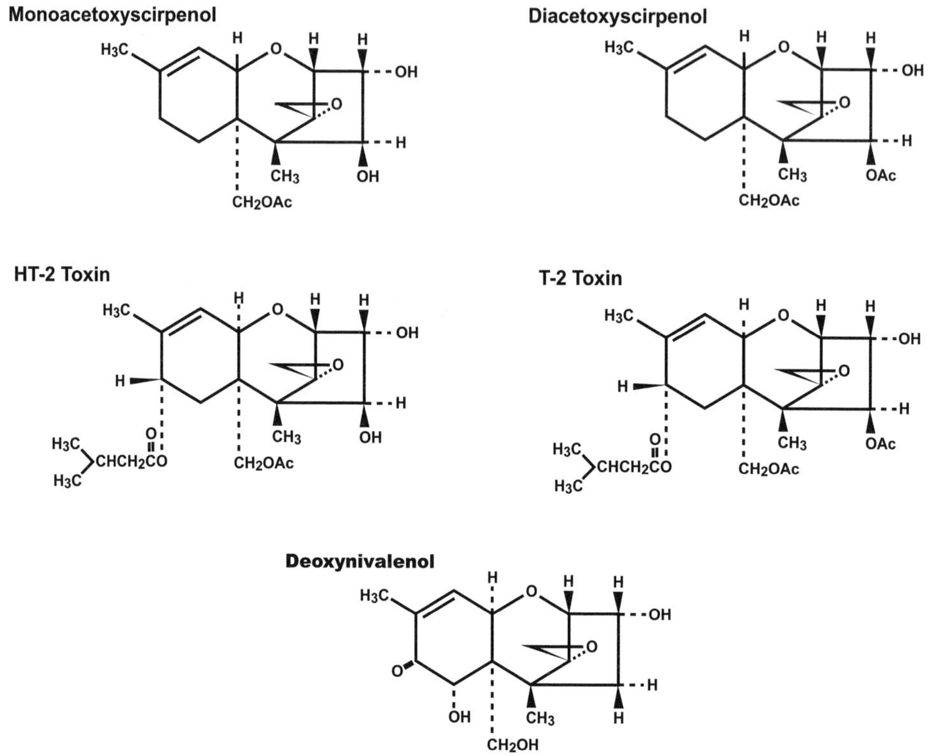

Figure 5. The trichothecene mycotoxins.

produced by several species of fungi in the genus *Fusarium* that are found in many cereals, cereal by-products and feeds.

Indicators of T-2 toxicosis

Mouth lesions and associated losses in productivity caused by T-2 toxin are a major concern to the poultry industry (Wyatt *et al.*, 1975; Raju and Devegowda, 2000). As an indicator of T-2 toxicosis, oral lesions are about four times more sensitive than growth inhibition. The lesions look like circumscribed proliferate yellow caseous plaques occurring at the margin of the beak, mucosa of the hard palate and the angle between the mouth and the tongue (Figure 6).

T–2 toxin at concentrations as low as 0.4 mg/kg causes oral lesions by affecting the epithelial cells of oral mucous membranes. T-2 toxin is highly irritant in nature and is known to cause gizzard erosion and necrosis of proventricular mucosa. Recently it has been observed that T-2 toxin can also cause tibial dyschondroplasia in broiler chickens (Nascimento *et al.*, 2001).

Impact on production

Symptoms of T-2 toxicosis include decreased feed intake, body weight, egg production and poor shell quality, regression of ovaries and increased liver weight in laying hens (Manoj and Devegowda, 2000) (Table 15). Additionally, changes in plasma enzyme profile have also been observed in association with loss of body weight and regression of ovaries in laying birds (CAST, 2003). Further, T-2 toxin is found to damage ovarian function in geese (Vanyi *et al.*, 1994) with resultant decreased egg production, reduced hatchability and high mortality of

Figure 6. Severe oral lesions in a broiler chick caused by T-2 toxin.

goslings and turkey poults (Fazekas *et al.*, 1993). T-2 toxin reduced body weight, altered serum and plasma chemical values and haematological parameters in white Pekin ducks (Rafai *et al.*, 2000).

Table 15. Effect of T-2 toxin on laying hen performance.

T-2 toxin (ppm)	*Egg production (%)*	*Egg weight (g)*	*Body weight (g)*
0.0	96.29^{d}	52.45^{b}	1331.8^{c}
0.5	93.81^{c}	51.77^{ab}	1312.5^{b}
1.0	91.75^{b}	51.35^{a}	1286.0^{a}
2.0	86.65^{a}	51.33^{a}	1284.9^{a}

Manoj and Devegowda, 2001
^{ab}Means in a column differ ($P<0.05$).

A field case of T-2 toxicosis narrated here may be able to properly describe the adverse effects of T-2 toxin on productivity of poultry. In a flock of laying hens affected with T-2 toxin, egg production decreased by about 22% and the percentage of cracked eggs went up by 12%. The incidence of egg breakage during transit was as high as 18% with incidence of blood spots increasing from 0 to 3%. Oral lesions were observable in more than 85% of the birds; and the farm had to be closed (Hamilton, 1987).

Muscovy ducklings are exceptionally sensitive to the dermatonecrotic effects of T-2 toxin. This extreme sensitivity renders them useful as bioassay animals to detect mycotoxins in contaminated grains. The findings of Shlosberg *et al.* (1986) were reinforced by the studies of Salyi and Glavits (1995) on Muscovy ducklings that were fed two dietary levels of T-2 toxin (0.75 and 1.5 mg/kg) between 4 and 26 days of age (cited in Rafai *et al.*, 2000).

Inclusion of T-2 toxin in a quail diet significantly decreased α-tocopherol, carotenoid and ascorbic acid (Table 16) accumulation in the liver (Dvorska and Surai, 2001).

Table 16. Effect of T-2 toxin on antioxidants in the quail liver (mg/g).

Antioxidant	*Control*	*T-2 toxin*
α-Tocopherol	18.4 ± 1.33^{a}	10.2 ± 0.96^{b}
Carotenoids	4.3 ± 0.13^{a}	2.7 ± 0.18^{b}
Ascorbic acid	166.4 ± 8.6^{a}	101.1 ± 5.1^{b}

Dvorska and Surai, 2001
^{ab}Means in a row differ ($P<0.05$).

Toxicity mechanisms

The chemical structure of T-2 toxin makes it

fat soluble, which allows it to be incorporated into cell membranes, potentially changing membrane structural and functional properties. Lipid peroxidation caused by T-2 toxin in the liver has also been identified as an important underlying mechanism of T-2 toxin-induced cell injury and DNA damage (Surai, 2002).

Impact of T-2 toxin on immunity

Decreased leukocyte count and regression of the bursa of Fabricius are common immunosuppressive effects associated with T-2 toxicosis. Increasing the concentration of T-2 toxin derivatives reduces macrophage viability (Kidd *et al.*, 1997). Occurrence of aflatoxin and T-2 toxin in the same diet is the most immunosuppressive combination of mycotoxins for poultry (Raju and Devegowda, 2002).

DIACETOXYSCIRPENOL AND MONOACETOXYSCIRPENOL

Diacetoxyscirpenol (DAS) and monoacetoxyscirpenol (MAS) are naturally occurring mycotoxins produced by several species of fungi in the genus *Fusarium* that are found in many cereals and feeds. DAS and MAS have been reported to cause mouth lesions, reduced feed intake and feed refusal in chickens and ducklings (Chi and Mirocha, 1978; Ademoyero and Hamilton, 1989). T-2 toxin and DAS are highly toxic for poultry. The LD_{50} values for DAS, T-2 toxin and HT-2 toxin in day-old chicks are 2.0, 5.0 and 7.2 mg/kg, respectively.

Laying hens

Diaz *et al.* (1994) reported a significant decrease in egg production and feed intake in commercial layers fed a diet containing 2 mg/kg DAS for 24 days. DAS has also been reported to reduce the hatchability of fertile eggs as well as female fertility in white leghorn hens (Allen *et al.*, 1982).

DEOXYNIVALENOL

Deoxynivalenol (DON) is produced by several *Fusarium* species, primarily *F. graminearum,* and is present as a co-contaminant with other mycotoxins in many commodities. In Austria the maximum permissible level of DON is 1 mg/kg in poultry feeds. Poultry are relatively resistant to the effects of DON in comparison to pigs and pet animals. However, in field situations a low DON level is sometimes associated with reduced feed consumption in layers and broiler breeders. This suggests that DON may be an indicator that T-2 toxin or other unknown *Fusarium* mycotoxins are present.

Harvey *et al.* (1991) reported a decrease in serum antibody titer to Newcastle disease virus (NDV) after feeding a DON-contaminated diet (18 mg/kg) to growing chickens for 18 wks. Moreover, a dose-response related decrease in serum NDV titers of broilers was observed after feeding diets containing no DON, 7.5 mg and 15 mg DON/kg. Such changes in immune response might occur without affecting performance, but should be viewed as indicating adverse DON effects on the health of the birds.

Residues in eggs

Deoxynivalenol has been shown to appear in body fat, muscle and eggs, with the egg concentration of DON being maximum within 24-96 hrs of consuming contaminated feed. In a study where hens were given radiolabeled DON, only 10% of the activity was recovered in egg yolk, suggesting that the balance was in the form of the many DON metabolites (Prelusky *et al.*, 1987).

FUSARIC ACID

Fusaric acid is primarily produced by *F. moniliforme*. Despite its wide occurrence and possible interactions with trichothecenes, fusaric acid is rarely analysed in experimental diets involving contaminated grains or culture materials. The lack of information about dietary fusaric acid content might explain the contradicting results observed in the earlier studies involving DON. Recent trials (Swamy *et al.*, 2002) indicated that broiler chickens were susceptible when grains naturally contaminated with a combination of DON and fusaric acid were fed.

An increase in redness of breast meat was observed in broiler chickens (Smith *et al.*, 2001) and in turkey poults fed *Fusarium* culture material (Wu *et al.*, 1994). The discolouration may be due to the effect of fusaric acid, which acts as a hypotensive agent.

FUMONISINS

Fumonisins are a relatively recently discovered group of mycotoxins, primarily produced by *F. verticillioides* (*moniliforme*) and *F. proliferatum*. The most toxic of this group is fumonisin B_1 (FB_1). Fumonisins are found in a wide range of commodities from cereal grains to bananas and are a major concern to producers of both human and animal foods since they directly affect health. Fumonisins are found worldwide, in tropical, subtropical and temperate regions; and they also have been found as co-contaminates with many other mycotoxins including aflatoxin, DON, ZEA and OTA. A high incidence of human oesophageal cancer observed in some countries (South Africa, China) has been attributed to chronic exposure to high levels of fumonisins (Smith *et al.*, 2001).

Increased liver sphinganine: sphingosine (SA:SO) ratios (Table 17), an early biomarker for fumonisin toxicosis, was observed in both broilers and turkeys fed diets that supplied 25 or 50 mg FB_1/kg (Broomhead *et al* ., 1998).

Spiking mortality or **toxic feed syndrome** in poultry has been attributed to fumonisin toxicity. The symptoms of spiking mortality include paralysis, extended legs and neck, wobbly gait, gasping and poor growth.

Table 17. Effects of fumonisin B_1 on liver SA:SO ratio of broiler chicks and turkey poults.

FB_1	*SA:SO*	
(mg/kg) diet	*Broiler*	*Turkey*
0	0.15[b]	0.18[c]
25	0.52[a]	0.62[b]
50	0.59[a]	1.21[a]

Broomhead *et al.*, 1998
[abc]Means in a column differ ($P<0.05$).

Broilers and turkeys are relatively resistant to the acute toxic effects of FB_1. In broiler chicks fed high levels of FB_1, acute toxic effects included poor performance, increased organ weights, and multifocal hepatic necrosis. In turkey poults fed high levels of FB_1, acute toxic effects include poor performance, increased organ weights, and hepatocellular hyperplasia (Weibking *et al.*, 1994).

Experiments with turkey poults using culture material with FB_1 concentrations of 100 or 200 mg/kg gave results similar to those for broiler chicks. Turkey poults fed a diet containing 200 mg/kg FB_1 were 25% less efficient than the controls (Ledoux *et al.*, 1992).

Under field conditions dietary FB_1 at 20 mg/kg or more is associated with elevated mortalities at 10 and 30 days, as well as feed refusal in commercial broiler flocks. In another experiment, birds fed the fumonisin toxin for four weeks responded poorly to an inactivated Newcastle vaccine compared to controls.

Investigation of the effects of fumonisin in poultry has been limited to dietary inclusion of culture material contaminated with fumonisin. Experiments with *F.* (*verticillioides*) *moniliforme* contaminated culture material can lead to confusing results and erroneous conclusions due to the high probability of other toxic metabolites being present. Nevertheless, such recent experiments indicate that feed contaminated with fumonisin-producing fungi can cause some adverse responses in poultry (Ledoux *et al.*, 1992).

Residues

Due to poor absorption and rapid excretion following metabolism, fumonisin residues are of no major public health significance. Though well absorbed, trichothecenes are rapidly metabolised and excreted without considerable accumulation in animal tissues and thus a trichothecene residue in foods of animal origin is of no public health significance.

MONILIFORMIN

Moniliformin is a mycotoxin produced by several *Fusarium* species including *F. proliferatum*. Moniliformin is highly toxic to chicks with a reported lethal dose of 4 mg/kg BW in day-old chicks. Exposure to moniliformin ranging from 144 to 575 mg/kg feed can lead to very high mortality rates (80-100%) in chicks, poults and ducklings (Engelhardt *et al.*, 1989). Cardiac injury and alterations in cardiac electrical conductance are shown to cause the mortality in affected birds (Nagaraj *et al.*, 1996). Even low levels of moniliformin can cause poor growth performance, increased serum pyruvate levels, enlarged hearts and cardiac lesions (Ledoux *et al.*, 1993), and poor immune response (Li *et al.*, 2000).

FUSAROCHROMANONE

Fusarochromanone, produced by *F. equiseti*, is a mycotoxin associated with tibial dyschondroplasia of growing broiler chicks and turkeys. Auburn University (US) publications indicate fusarochromanone to be the most likely cause of tibial dyschondroplasia. At a dietary level of 75 mg/kg, fusarochromanone can cause tibial dyschondroplasia in 100% of the chicks exposed to contaminated feed. In breeder hens, fusarochromanone is transferred into eggs resulting in increased embryonic mortality.

ZEARALENONE

Zearalenone (ZEA) is produced primarily by *F. graminearum* and *F. roseum*. Zearalenone is responsible for reproductive disorders due to its estrogenic effect at high concentrations.

Poultry are relatively resistant to ZEA in comparison to pigs and cattle. At high dietary concentrations, however, vent enlargement and enhanced secondary sex characteristics are seen in turkeys and chickens.

Zearalenone, which often occurs with DON in naturally contaminated cereals, is even less toxic to laying hens than DON. Allen *et al.* (1981) found no detrimental effects of increasing dietary ZEA concentration up to 800 mg/kg on layer performance or on growth of progeny. However, feeding ZEA-contaminated feed to laying hens might cause residues of the toxin in egg yolk as indicated by an experiment by Dailey *et al.* (1980), who found radioactivity in egg yolks after an oral bolus of ^{14}C-labeled ZEA. This could have implications for human health because of the estrogenic and genotoxic properties of ZEA. Moreover, a possible carryover of DON and ZEA to eggs could adversely affect reproductive performance of hens as indicated by the investigations of Bergsj *et al.* (1993), who reported an increased incidence of chick developmental anomalies when hens were fed these mycotoxins.

AUROFUSARIN

Aurofusarin, a metabolite of *F. graminearum*, was first described in 1968, but only recently has been identified in feed ingredients in the Ukraine, in particular in contaminated cereal grains (Kotyk and Trufanova, 1998). This mycotoxin was shown to have a profound effect on egg quality with no clinical signs of toxicosis. On the other hand aurofusarin compromised the immune system of laying chickens and quail with decreased fertility and hatchability (Sakhatsky, 1999). At the biochemical level aurofusarin was responsible for significant changes in fatty acid and antioxidant profiles of quail eggs. Furthermore the antioxidant system of the developing quail embryo was shown to be compromised as a result of aurofusarin inclusion in the maternal diet.

Aurofusarin pigment is transferred from feed to egg yolk, and due to a pH-dependent colour change, can cause colour change from yellow via red and pink to violet (Dvorska and Surai, 2003). Aurofusarin consumption is also associated with the deterioration of chicken meat quality.

LD_{50} values of mycotoxins

The LD_{50} values of some mycotoxins in poultry are presented in Table 18. At much lower levels, and in combination, mycotoxins can cause significant damage to poultry in terms of health, productivity and economics.

Table 18. The LD_{50} values of some mycotoxins.

Mycotoxins	*LD_{50}[1] (μg/kg)*
Aflatoxin	6.5
Ochratoxin	3.6
T-2 toxin	4.9 –5.2
Moniliformin	5.4
Diacetoxyscirpenol	3.8 –5.9
Citrinin	95
Deoxynivalenol	140
Fumonisin B_1	300*
Cyclopiazonic acid	100

[1] LD_{50} = Dose at which 50% of the test animals die.
*Not LD_{50}. Birds fed this concentration had severe growth depression (Reduced weight gain by 19% and increased liver weight by 30%)

Co-contamination of feeds by mycotoxins

As previously indicated, contaminated feeds or ingredients typically contain more than one toxin. The toxicity responses and clinical signs observed in poultry when more than one mycotoxin is present in feed are complex and diverse. Co-contamination of mycotoxins appears to exert greater negative impact on health and productivity than do single toxins. For this reason, symptoms typical of mycotoxicosis are often seen in poultry despite analysis of the feed indicating only very low concentrations of individual mycotoxins. Toxicity may be due to interactions between different mycotoxins that exaggerate the toxicity symptom.

Brazilian researchers demonstrated the symptoms and economic losses caused by combinations of mycotoxins. They observed that birds fed damaged corn estimated to contain 15.3 μg/kg aflatoxin B_1, 11.6 μg/kg ochratoxin A, 1200 μg/kg ZEA and 177 μg/kg T-2 toxin produced higher incidence of breast and thigh haemorrhage, liver steatosis and gizzard erosions in broilers (Santin and Fagundes, personal communication).

In a survey of sunflower oil cake in India, 8% of the samples were found to contain T-2 toxin and ochratoxin simultaneously (Chandrasekaran, 1996). Similarly Ravindran *et al.* (1996) reported the co-occurrence of aflatoxin, fumonisin and ZEA in Australian maize.

Over several years a considerable amount of research has been conducted into the co-occurrence of mycotoxins in feeds. Mycotoxin co-contamination can occur in many combinations, some of which are listed in Table 19. Since the fungi that produce mycotoxins do not exist in nature as pure cultures, the number of possible combinations of mycotoxins that can appear is vast.

Table 19. Co-contamination of mycotoxins in poultry.

- Aflatoxins and ochratoxin
- Aflatoxins and T-2 toxin
- T-2 toxin and diacetoxyscirpenol
- Ochratoxin and citrinin
- Deoxynivalenol and fusaric acid

Devegowda *et al.*, 2003

The microenvironment of feed is very complex and consists of many types of fungi, all competing for available nutrients. A feed sample positive for a particular mycotoxin not only confirms the presence of that mycotoxin, but also signifies the fact that at some point in time conditions favouring fungal growth and toxin production were present in an ingredient used to produce that feed. It is pertinent to check that sample for the presence of other mycotoxins as well; and recent findings strongly support this approach. The mycotoxin interactions scientifically examined are presented in Table 20.

AFLATOXINS AND OCHRATOXIN

Both aflatoxin and ochratoxin are extremely toxic to poultry and they interact synergistically (Campbell *et al.*, 1983; Raju and Devegowda, 2000), meaning that the toxicity resulting from dual exposure to aflatoxin and ochratoxin is much greater than the sum of the individual toxicities of these mycotoxins (Table 21). This may be one reason why very low levels of aflatoxin in the field are sometimes associated with significant effects in poultry compared to laboratory studies where only aflatoxin impact is evaluated.

Table 21. Effect of dietary aflatoxin and ochratoxin A on body weight of three week old broiler chickens.

Aflatoxin	*Ochratoxin*	*Body weight*
(μg toxin/g diet)		*(g)*
0	0	625.8 ± 9.7[b]
0	2.0	549.1 ± 11.7[a]
2.5	0	551.1 ± 10.7[a]
2.5	2.0	380.4 ± 5.5[c]

Huff and Doerr, 1981
[abc]Within columns means differ ($P<0.05$). Values represent the mean ± SEM of 6 groups of 10 birds.
*Significant ($P<0.05$) interaction between aflatoxin and OTA.

The symptoms of this co-mycotoxicosis in poultry are totally different than those observed with only aflatoxin. Ochratoxin, when present in combination with aflatoxin, suppresses the symptomatic manifestation of

Table 20. Impact of co-contamination by mycotoxins in poultry.

Mycotoxins	*Species tested*	*Production effects*	*References*
Aflatoxin & ochratoxin	Broiler chicks	Synergism	Huff *et al.* 1992
Aflatoxin & DAS	Broiler chicks	Synergism	Kubena *et al.* 1993
Aflatoxin & DON	Broiler chicks	Additive	Huff *et al.* 1986
Aflatoxin & T-2 toxin	Broiler chicks	Synergism	Aravind *et al.*, 2003, Girish and Devegowda, 2004b
Aflatoxin & ochratoxin	Broiler chicks	Synergism	Kubena *et al.* 1989, Raju and Devegowda, 2000
Ochratoxin & T-2 toxin	Broiler chicks	Synergism	Raju and Devegowda, 2000
DON & moniliformin	Turkey poults	Moniliformin effect only	Morris *et al.*, 1999
DON & fusaric acid	Broiler chicks	Synergism	Smith *et al.*, 2000, Swamy *et al.*, 2002
Fumonisin B_1 & moniliformin	Turkey poults	Moniliformin effect only	Li *et al.*, 2000
Fumonisin B_1 & T-2 toxin	Turkey poults	Additive	Kubena *et al.*, 1995
Fumonisin B_1 & DAS	Turkey poults	Additive	Kubena *et al.*, 1997
Ochratoxin & citrinin	Broiler chicks	Antagonism	Manning *et al.*, 1985

aflatoxin-mediated liver damage (fatty, yellow, enlarged, and friable liver), making diagnosis of aflatoxicosis difficult (Huff and Doerr, 1981) (Table 22). The target organ of the interaction between aflatoxin and ochratoxin in poultry appears to be the kidney.

Table 22. The individual and combined effects of dietary aflatoxin and ochratoxin A on liver lipid of broiler chickens.

Aflatoxin	*Ochratoxin*	*Liver lipid*
(μg toxin/g diet)		*(% dry wt)*
0	0	16.49 ± 0.51[b]
0	2.0	14.59 ± 0.20[b]
2.5	0	24.48 ± 1.58[a]
2.5	2.0	17.25 ± 0.53[b]

Huff and Doerr, 1981
[ab]Within columns means differ ($P<0.05$). Values represent the mean ± SEM of 6 groups of 10 birds.
*Significant ($P<0.05$) interaction between aflatoxin and OTA.

AFLATOXINS AND T-2 TOXIN

Both aflatoxins and T-2 toxin are important to the poultry industry because of their toxicity and co-occurrence in feeds. These two mycotoxins have significant synergistic interactions on body weight, relative organ weights (Figures 7 and 8) and antibody titres in broiler chickens (Huff *et al.*, 1988; Raju and Devegowda, 2002; Girish and Devegowda, 2004b).

T-2 TOXIN AND DIACETOXYSCIRPENOL

The co-occurrence of T-2 and DAS was reported in mixed diets and grains with concentrations ranging from 0.2 to 5.8 mg/kg for T-2 toxin and 0.5 to 2.1 mg/kg for DAS. A survey in Canada showed concentrations of 0.11–2.27 mg DAS /kg occurring in less than 10% of grain samples analysed, while T-2 toxin concentration ranged from 0.16 to 0.31 mg/kg in the same samples (Stratton *et al.*, 1993). T-2 and DAS are more toxic than aflatoxin and comparable in toxicity to ochratoxin in chickens (Huff *et al.*, 1974). Additive lethal toxicity has been reported in chickens given DAS and T-2 toxin in combination (Hoerr *et al.*, 1981). The effects of T-2 and DAS were additive in laying hens for feed intake, oral lesions, mild changes in plasma enzyme activities and reduced egg production (Diaz *et al.*, 1994) and in broilers for body weight gain and feed intake (Hoerr *et al.*, 1981).

OCHRATOXIN AND CITRININ

Citrinin is a nephrotoxin similar to ochratoxin,

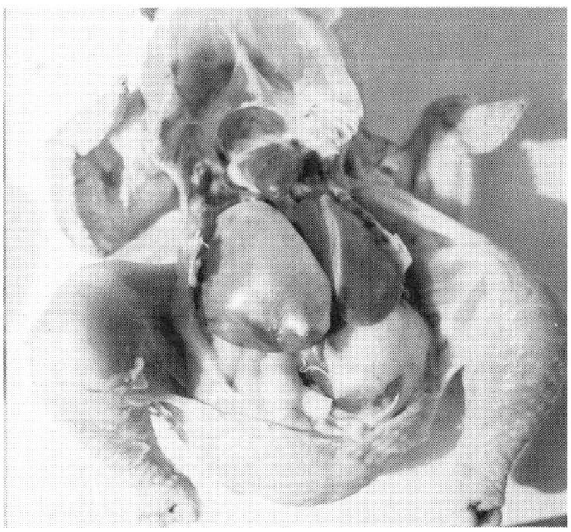

Figure 7. Pale, enlarged liver of broiler chickens: response to exposure to aflatoxin plus T-2 toxin (Girish and Devegowda, 2004).

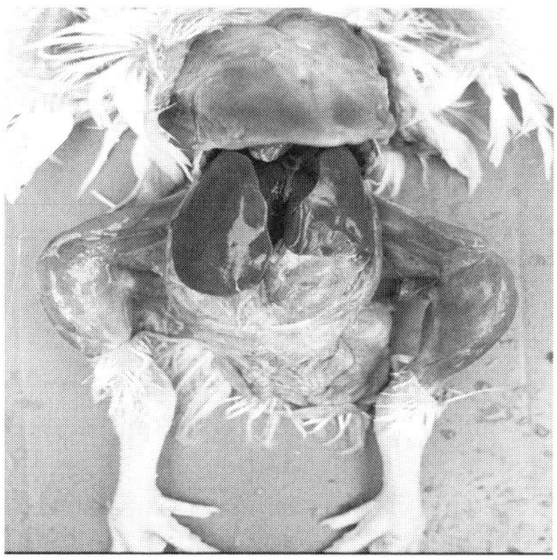

Figure 8. Enlarged liver and haemorrhagic thigh muscle of broiler chickens: response to exposure to aflatoxin plus T-2 toxin (Girish and Devegowda, 2004).

and some of the fungi that produce ochratoxin also produce citrinin. This is an important mycotoxin interaction to understand since it is very likely that when one is found the other mycotoxin will be present. The interaction between these two mycotoxins is synergistic, however the interaction of these two toxins can also be characterized as antagonistic. One of the symptoms of citrinin toxicity in poultry is a dramatic increase in the amount of water consumed and excreted, which is of significant diagnostic value. However, ochratoxin prevents this very important diagnostic index of citrinin, making the diagnosis in case of co-contamination difficult.

DEOXYNIVALENOL AND FUSARIC ACID

Feeding grains contaminated with DON and fusaric acid synergistically depressed broiler growth rate and feed intake. There was a significant increase in breast meat colour. In laying hens, feeding contaminated grains significantly reduced feed consumption, egg production and alleviated blood uric concentration (Smith *et al.*, 2003).

Safe/permissible levels of mycotoxins in poultry feeds

What is a safe level? Can a contaminated grain source be safely fed to other animals if not poultry? What will be the economic impact of a given level of contamination? These are some of the questions commonly asked by people involved in poultry and livestock farming.

Strictly speaking **there is no *safe level.*** With reference to mycotoxins the risk directly depends on the level of the major mycotoxins in the feed and also on the co-occurrence and levels of other mycotoxins. A mycotoxin level considered safe on one farm may not be safe on another farm because of differences in management and disease prevalence. Additionally, factors like interaction of mycotoxins with pathogens, animal genetics, environment (high temperature, humidity, ammonia etc.), sex and nutritional status, influence the severity of the manifestation of mycotoxicosis.

In order to reduce the toxic and economic impact of mycotoxins, several countries regulate the levels of certain mycotoxins in foods and feeds. Worldwide food and feed legislation safeguards the health of consumers and the economic interests of animal producers and traders. Virtually all countries with fully developed market economies have regulations with the exception of some African countries.

Control of the adverse effects of mycotoxins

Mycotoxins are produced at only certain stages of mould growth. Mycotoxin molecules are generally quite resistant to chemical or thermal treatments capable of destroying mould spores. Therefore, the presence or absence of visible mould spores does not serve as a reliable guide to the presence or levels of toxins.

Practical mycotoxin control must be accomplished in two stages: prevention of fungal growth and toxin production, and decontamination of existing mycotoxins in feeds. Several strategies, to be adopted in a coordinated manner, have been proposed to reduce the impact of mycotoxins. These include plant breeding for mould resistance, harvesting and storage practices that minimise contamination, and decontamination procedures to neutralise mycotoxins in contaminated commodities (Devegowda and Aravind, 2002).

The most effective method of neutralising mycotoxins present in feed is by inclusion of inert adsorbents that prevent absorption of the toxin from the intestine. The ideal features of a good mycotoxin adsorbent for animal feeds are:

- Ability to adsorb a wide range of mycotoxins
- Low effective inclusion rate
- Rapid and uniform dispersion in the feed during mixing
- Heat stability during pelleting, extrusion, and during storage
- No affinity for vitamins, minerals or other nutrients
- High stability over a wide pH range
- Biodegradability after excretion
- Safety to animals and humans
- Palatability

NUTRITIONAL MODIFICATIONS

Nutritional approaches for protection against mycotoxins include increased levels of methionine, selenium and vitamin supplementation of affected diets and some plant and herbal compounds including chlorophyll derivatives, aspartame etc. Mycotoxins, upon being absorbed, are detoxified in the liver utilizing the glutathione system, which contains cystine (derivative of methionine). Hence the metabolic level of methionine is depleted, leading to poor growth and feed efficiency. Higher levels of methionine, selenium and vitamin supplementation in feed, therefore, have been found to be beneficial.

CHEMICAL DETOXIFICATION

Among the chemicals tested for their ability to detoxify/inactivate mycotoxins, ammonia, sodium bisulfite, peroxide acids, bases and gases are most effective. However, most of the chemical methods are impractical and do not generally meet the requirements for ideal detoxifying agents, especially as regards safety and palatability.

APPLICATION OF MINERAL CLAYS

Many clays have been tested for ability to counteract mycotoxins, but amongst the most successful, very few are actually used commercially. These include bentonites, zeolites and aluminosilicates. Several studies have demonstrated that sodium aluminosilicate, hydrated sodium calcium aluminosilicate (HSCAS) , and sodium bentonites can adsorb aflatoxins (Phillips *et al.*, 1988; Doerr, 1989; Kubena *et al.*, 1990; Araba and Wyatt, 1991; Schiedeler, 1993; Santurio *et al.*, 1994; Mahesh and Devegowda, 1996). The chemisorption of aflatoxin to HSCAS involves formation of a complex between the ß-keto-lactone or bilactone system of aflatoxin with uncoordinated metal ions in HSCAS (Phillips *et al.*, 1990). The degree of adsorption depends on the chemical structure of the clay material with the adsorption capacities varying between 0 to 87% with various mineral clay materials (Schiedeler, 1993; Santurio *et al.*, 1994).

The high inclusion rates and narrow range of adsorptive efficacy are the main disadvantages of mineral clays. Clays are mainly effective against aflatoxins (Kubena *et al.*, 1990; Huff *et al.*, 1992; Ledoux *et al.*, 1999; Girish and Devegowda 2004a) and appear to have little or no beneficial effect against:

- Zearalenone (Volkl and Karlovsky, 1998)
- T-2 toxin (Dvorska and Surai 2001; Girish and Devegowda 2004a)
- Ochratoxin (Santin *et al.*, 2002)
- Diacetoxyscirpenol (Kubena *et al.*, 1993)
- Fescue toxin (Chestnut *et al.*, 1992).

Despite research demonstrating that mineral clays do not adsorb many of the common mycotoxins *in vitro* or *in vivo*, they are widely used in the poultry industry where mycotoxins are known or suspected. Furthermore, it has been demonstrated that mineral clays reduce the utilisation of minerals including manganese, zinc and magnesium (Chestnut *et al.*, 1992), chloride (Schiedeler, 1993), copper and sodium (Kramer, 1993).

YEAST CELL WALL-BASED ADSORBENTS

In the early 1990s, a commercial yeast culture preparation of *Saccharomyces cerevisiae1026*, which was earlier noted as a digestive aid and a growth promoter, was found to improve hatchability (McDaniel, 1991) and broiler body weights (Stanley *et al.*, 1993). Further investigations lead to the establishment of the yeast culture preparation's ability to adsorb aflatoxins in poultry feeds (Devegowda *et al.*, 1994; 1995) and ultimately to the development of yeast cell wall-derived modified glucomannan in the form of a product called Mycosorb® by Alltech Inc.

These studies signaled the beginning of a new strategy that utilized yeast-derived products to overcome problems associated with mycotoxin-contaminated feeds. Several researchers have confirmed that modified glucomannan, a derivative of the yeast cell wall (Figure 9), is able to adsorb higher levels of several important mycotoxins at lower inclusion rates than inorganic binders (Mahesh and Devegowda, 1996; Volkl and Karlovsky, 1998; Trenholm *et al.*, 1996; Dawson *et al.*, 2001). Some recent research findings are summarised below.

Modified glucomannan supplementation was found to be beneficial in reducing the individual and combined adverse effects of aflatoxin, ochratoxin and T-2 toxin in broilers. Apart from improving body weights and antibody titres (Figure 10) suppressed by combined dietary mycotoxins, modified glucomannan also improved serum biochemical and haematological parameters (Raju and Devegowda, 2000).

In broiler chickens given feeds naturally contaminated with different mycotoxins, supplementation with modified glucomannan in Mycosorb® has been demonstrated to improve general performance and health in several studies (Smith *et al.*, 2001; Swamy *et al.*, 2002; Aravind *et al.*, 2003). Feed containing wheat and barley contaminated with T-2 and DON reduced body weight gain by 5.8%. The addition of modified glucomannan protected the broilers from the

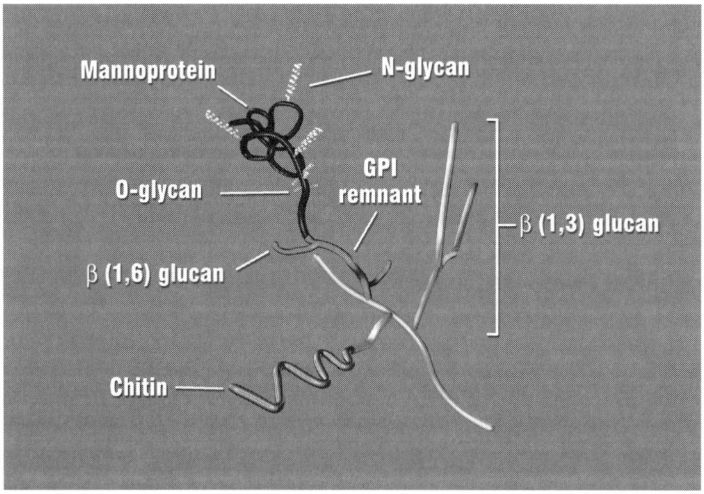

Figure 9. Relationships among components of *S. cerevisiae* cell wall (adapted from Lipke and Ovalle, 1998).

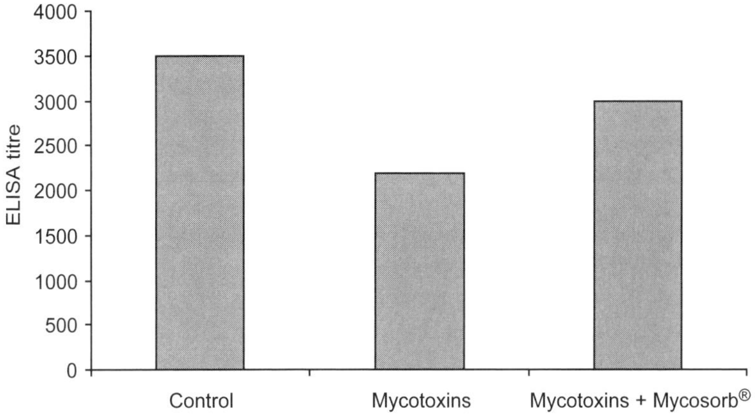

Figure 10. Antibody titres to infectious bursal disease in birds exposed to mycotoxin-contaminated diets with or without modified glucomannan (Mycosorb®) supplementation.

performance depressing effects of the mycotoxins (Smith *et al.*, 2001). Modified glucomannan also counteracted the toxic effects of DAS on body weight gain of broilers (Pavicic *et al.*, 2001) (Table 23).

Table 23. Modified glucomannan (Mycosorb®) and diacetoxyscirpenol (DAS) in broiler chicken diets.

DAS (mg/kg)	*Mycosorb® (%)*	*Body weight (g)*
0	0	1681
1	0	1374
0	0.1	1794
1	0.1	1678

Pavicic et al., 2001

Modified glucomannan was also shown to be beneficial in broiler diets contaminated with sorghum ergot alkaloid (SEA) with alleviation of body weight gain depression (Deo *et al.*, 1999).

Murthy and Devegowda (2004) demonstrated the rapid adsorption of aflatoxin and T-2 toxin (Reddy and Devegowda, 2004) by modified glucomannan in the digestive tract at different time intervals after ingestion of feed by broiler chickens. Similar findings were reported by Reddy and Devegowda (2004) with regard to T-2 toxin (Figure 11).

Diets containing aflatoxin (2 mg/kg) and T-2 toxin (1 mg/kg), significantly reduced broiler feed intake, weight gain and feed efficiency, weights of thymus and bursa of Fabricius, and antibody titres against Newcastle disease and infectious bursal disease. Modified glucomannan (1 kg/ton of feed) significantly improved body weight, feed intake, antibody titres, weights of thymus and bursa of Fabricius, suggesting action against both aflatoxin and T-2 toxin. HSCAS (10 kg/ton of feed) showed improvement against aflatoxin with little protection against T-2 toxin (Girish and Devegowda, 2004b), (Figure 12).

With dietary treatments of four concentrations of T-2 toxin (0, 0.5, 1.0, and 2.0 mg/kg) and two concentrations (0 and 0.1%) of Mycosorb®, Manoj and Devegowda (2001) observed that modified glucomannan reversed the suppression of egg production by T-2 toxin. Even at the highest concentration of T-2 toxin (2 mg/kg), wherein egg production was reduced to less than 90% of the control (Figure 13), modified glucomannan restored production to more than 95% of the control values.

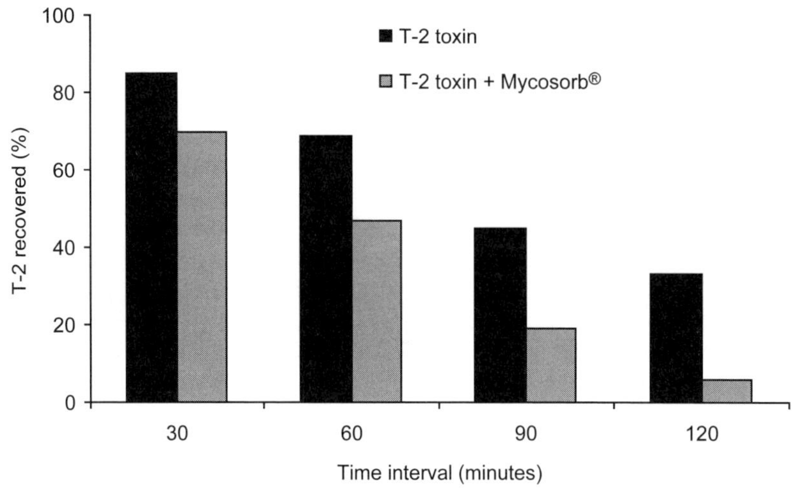

Figure 11. Recovery of T-2 toxin from the digestive tract of broilers at different time intervals (from Reddy and Devegowda, 2004).

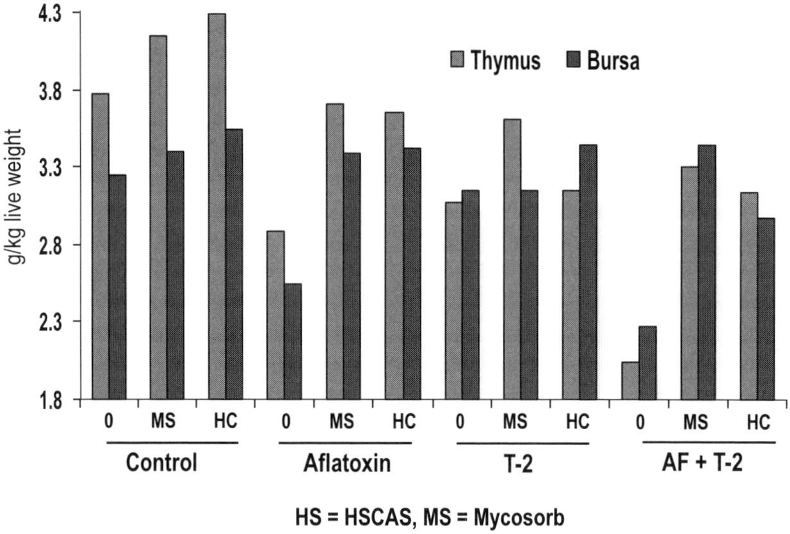

Figure 12. Efficacy of modified glucomannan and HSCAS to alleviate the individual and combined toxicity of aflatoxin and T-2 toxin on lymphoid organs of broilers at 5 weeks of age (Girish and Devegowda, 2004).

In ducklings, modified glucomannan supplementation can reverse the growth suppression induced by dietary aflatoxin. Modified glucomannan inclusion can also improve feather score, incidence of leg deformity, and web toe haemorrhage scores that are adversely affected by higher levels of dietary aflatoxin (Khajarern and Khajarern, 1999).

T-2 toxin is known to decrease the levels of innate antioxidants in birds and increase lipid peroxidation in the liver. Recent studies have revealed that modified glucomannan supplementation helps maintain the levels of systemic antioxidants and prevents hepatic lipid peroxidation in growing quail while zeolite supplementation may not be

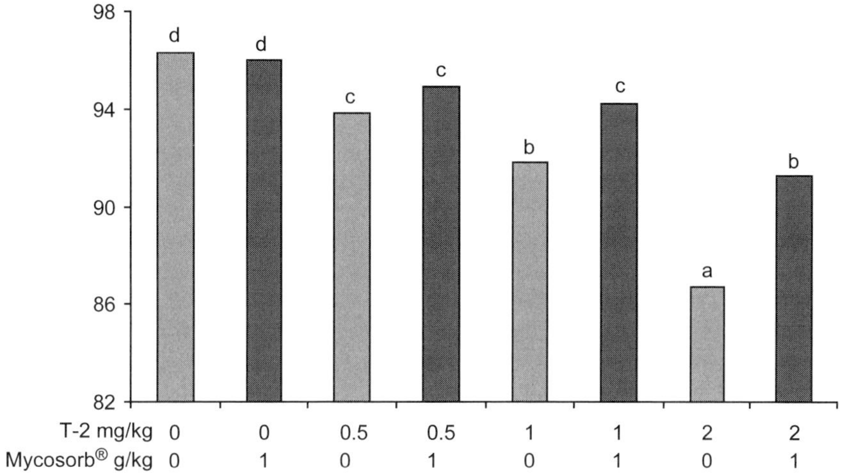

Figure 13. Effect of dietary T-2 toxin and modified glucomannan (Mycosorb®) on egg production (%).

beneficial (Dvorska and Surai, 2001) (Figure 14).

Conclusions

Under practical conditions, no poultry feed is completely free of mycotoxins. Additionally, no feed can be expected to contain only one mycotoxin. The adverse effects of mycotoxins on poultry are manifold indicating a clear and persistent danger.

A strong recommendation is made to view the problem as it naturally occurs, i.e., co-contamination of feeds and feedstuffs by mycotoxins, rather than a narrow approach of dealing with one mycotoxin at a time. It is therefore essential to take all necessary precautions:

- Prevention of fungal growth on crops in the field, at harvest, during storage of feedstuffs and processing of feed.

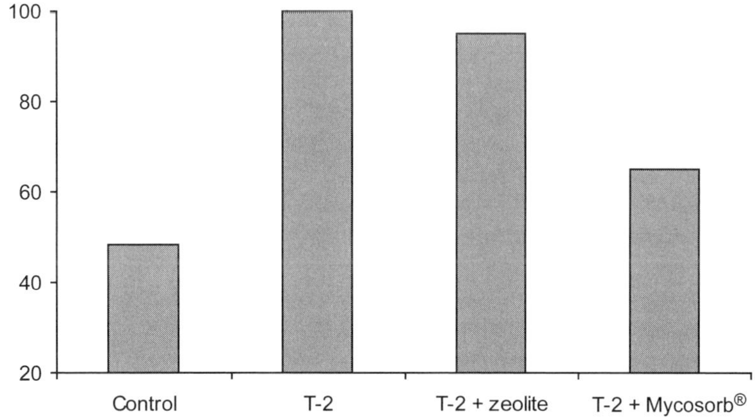

Figure 14. Effects of T-2 toxin, zeolite and modified glucomannan (Mycosorb®) on lipid peroxidation in the liver (mg MDA/g fresh tissue).

- The time to act is not after the birds are affected, but before the contaminated feed is consumed.
- Application of an appropriate mycotoxin adsorbent (which is safe to birds, for the environment and consumers) in order to achieve good productivity and profitability.

Acknowledgments

We would like to express our sincere gratitude to Dr. C.K. Girish and Dr. Vittal Kumar, working with Prof. G. Devegowda as research fellows in the area of mycotoxins, Dr. K.L. Aravind, Product manager, Alltech Biotechnology Pvt. Ltd, India and to Dr. Ravi Kiran, Ph.D scholar, for their significant contributions to this chapter.

References

Ademoyero, A.A.. and P.B. Hamilton. 1989. Research note: Influence of degree of acetylation of scirpenol mycotoxins on feed refusal by chickens. Poult. Sci. 68:854-856.

Afzali, N. and G. Devegowda. 1999. Ability of modified mannan-oligosaccharide to counteract aflatoxicosis in broiler breeder hens. Poult. Sci. 78(Suppl. 1):228. (Abstr.).

Allen, N.K., C.J. Mirocha, S. Aakhus-Allen, J.J. Bitgood, G. Weaver, and F. Bates. 1981. Effect of dietary zearalenone on reproduction of chickens. Poult. Sci. 60:1165-1174.

Allen, N.K., R.L. Jevne, C.J. Mirocha, and Y.W. Lee. 1982. The effect of *Fusarium roseum* culture and diacetoxyscirpenol on reproduction of white leghorn females. Poult. Sci. 61:2172-2175.

Araba, M. and R.D. Wyatt. 1991. Effects of sodium bentonite, hydrated sodium calcium aluminosilicate (Novasil) and Ethacal on aflatoxicosis in broiler chickens. Poult. Sci. 70:S6 (Abstr).

Aravind, K.L., V.S. Patil, G. Devegowda, B. Umakantha, and S.P. Ganpule. 2003. Efficacy of modified glucomannan to counteract mycotoxicosis in naturally contaminated feed on performance, serum biochemical and hematological parameters in broilers. Poult. Sci. 82:570-576.

Aravind. K.L. and G. Devegowda. 2003. Use of glucomannan to reduce the effects of ochratoxin A in broilers. In: *Proc. of 2nd World Mycotoxin Forum*, The Netherlands, pp. 122.

Bauer, J. 1998. Krankheit and Leistungs-deppression in der schweine haltung durch Mycotoxine. Tierarstliche Praxis Suppl. 3:40-47.

Bergsj, B., O. Herstad, and I. Nafstad. 1993. Effects of feeding deoxynivalenol-contaminated oats on reproduction performance in white leghorn hens. Br. Poult. Sci. 34:147-159.

Broomhead, J.N., D.R. Ledoux, A.J. Bermudez, and G.E. Rottinghaus. 1998. Effects of fumonisin B_1 in broilers and turkeys fed dietary treatments to market age. UMC Anim. Sci. Dept. Rep. 107-110.

Bryden, W.L. 1991. Occurrence and biological effects of cyclopiazonic acid. In: *Emerging Problem Resulting from Microbial Contamination* (K. Mixe and J.L. Richard, eds). National Institute of Hygienic Science, Tokyo, pp. 127-147.

Campbell, M.L. Jr., J.D. May, W.E. Huff, and J.A. Doerr. 1983. Evaluation of immunity of young broiler chickens during simultaneous aflatoxicosis and ochratoxicosis. Poult. Sci. 62:2138-2144.

Chestnut, A.B., P.D. Anderson, M.A. Cochran, H.A. Fribourg, and K.D. Twinn. 1992. Effects of hydrated sodium calcium aluminosilicate on fescue toxicosis and mineral absorption. J. Anim. Sci. 70:2838-2846.

Chi, M. S. and C. J. Mirocha. 1978. Necrotic oral lesions in chickens fed diacetoxy-scirpenol, T-2 toxin and crocotin. Poult. Sci. 57:807-808.

Chandrasekaran, D. 1996. Survey on the presence of T-2 toxin and ochratoxin in feed/ingredients in Namakkal area. In: *Proc. 20th World's Poultry Congress 4:268*, New Delhi, India.

Council for Agricultural Science and Technology (CAST). 2003. Mycotoxins: Risks in plant, animal, and human systems. Ames, Iowa, USA.

Cole, R.J. 1986. Etiology of turkey "X" disease in retrospect: A case for the involvement of cyclopiazonic acid. Mycotox. Res. 2:3-7.

Dawson, K.A., J. Evans, and M. Kudupoje. 2001. Understanding the adsorption characteristics of yeast cell wall preparations associated with mycotoxin binding. In: *Biotechnology in the Feed Industry: Proceedings of Alltech's 17th Annual Symposium* (T.P. Lyons and K.A. Jacques, eds). Nottingham University Press, UK, pp. 169-181.

Dailey, R.E., R.E. Reese, and E.A. Brouwer. 1980. Metabolism of ^{14}C zearalenone in laying hens. J. Agric. Food Chem. 28:286-291.

Deo, P., B.J. Blaney and J.G. Dingle. 1999. Effects of mineral and organic adsorbents in meat chicken diets contaminated with sorghum ergot alkaloid. In: *Proc. of Queensland Poult. Sci. Symp. 8*, Australia.

Devegowda, G., B.I.R. Arvind, K. Rajendra, M.G. Morton, A. Baburathna, and C. Sudarshan. 1994. A biological approach to counteract aflatoxicosis in broiler chickens and ducklings by the use of *Saccharomyces cerevisiae* cultures added to feed. In: *Biotechnology in the Feed Industry: Proceedings of Alltech's 10th Annual Symposium* (T.P. Lyons and K.A. Jacques, eds). Nottingham University Press, UK, pp. 235-245.

Devegowda, G., B.I.R. Aravind, M.G. Morton and K. Rajendra. 1995. A biotechnological approach to counteract aflatoxicosis in broiler chickens and ducklings by the use of *Saccharomyces cerevisiae*. In: *Proc. Feed Ingredients Asia '95*. Singapore, pp. 161-171.

Devegowda, G., M.V.L.N. Raju, N. Afzali, and H.V.L.N. Swamy. 1998a. Mycotoxin picture worldwide: Novel solutions for their counteraction. Feed Compounder 18(6):22-27.

Devegowda, G., M.V.L.N. Raju, N. Afzali, and H.V.L.N. Swamy. 1998b. Mycotoxin picture worldwide: Novel solutions for their counteraction. In: *Biotechnology in the Feed Industry: Proceedings of Alltech's 14th Annual Symposium* (T.P. Lyons and K.A. Jacques, eds). Nottingham University Press, Nottingham, UK, pp. 241-255.

Devegowda, G., M.V.L.N. Raju, and H.V.L.N. Swamy. 1998c. Mycotoxins: Novel solutions for their counteraction. Feedstuffs 70(50):12-17.

Devegowda, G. 2001. Mycotoxins: Economic risks and their control. In: *Handbook of Poultry Nutrition* (V.R. Reddy and D.T. Bhosale, eds). Published by American Soybean Association, pp. 246-260.

Devegowda, G. and K.L. Aravind. 2002. Global mycotoxin problem: Biological solutions. In: *Proc. 150th Year Celebration*. Institute Superior de Agronomia., Lisbon, Portugal, pp. 30-45.

Devegowda, G. 2002. Mycotoxins: A worldwide threat. Feeding Times 7:2-3.

Devegowda, G., K.L. Aravind, T.N.K. Murthy, and B.N. Reddy. 2003. Mycotoxins: Economic risk and control – The biological solutions. In: *3rd International Poultry Show and Seminar, World's Poultry Science Association*. Bangladesh, India, pp. 94-98.

Diaz, G.J., E.J. Squires, R.J. Julian, and H.J. Boermans. 1994. Individual and combined effects of T-2 toxin and DAS in laying hens. Br. Poult. Sci. 35:393-405.

Doerr, J.A. 1989. Effect of an aluminosilicate on broiler chickens during aflatoxicosis. Poult. Sci. 68:S 45 (Abstr).

Dorner, J.W., R.J. Cole, D.J. Erlington, S. Susksupath, G.H. McDowell, and W.L. Bryden. 1994. Cyclopiazonic acid residues in milk and eggs. J. Agric. Food Chem. 42:1516-1518.

Dvorska, J.E. and P.F. Surai. 2001. Effect of T-2 toxin, zeolite and Mycosorb on antioxidant systems of growing quail. Asian-Aust. J. Anim. Sci. 14:1752-1757.

Dvorska, J.E. and P.F. Surai. 2003. Aurofusarin, a *Fusarium graminearum* pigment: oxidative stress and protective effect of modified glucomannan. In: *Biotechnology in the Feed Industry: Proceedings of Alltech's 19th Annual Symposium* (T.P. Lyons and K.A. Jacques, eds). Nottingham University Press, UK, pp. 27-32.

Engelhardt, J.A., W.W. Carton, and J.F. Tuite. 1989. Toxicity of *Fusarium moniliforme var. subglutinans* for chicks, ducklings and turkey poults. Avian Dis. 33:357-360.

Fazekas, B., J. Tanyi, G. Salyi, A. Bistyak, and L. Kerekes. 1993. Effect of T-2 toxin on the hatching ability of turkey eggs and viability of day old turkey poults. Magy. Allatorv, Lapja. 48:221-225.

Gedek, B. 1998. Pilzifte lauern in Futter-getxeide. Tierzuchter 40:26-28.

Girish, C.K. and G. Devegowda. 2004a. Evaluation of modified glucomannan (Mycosorb) and hydrated sodium calcium aluminosilicate to ameliorate the individual and combined toxicity of aflatoxin and T-2 toxin in broiler chickens. Aust. Poult. Sci. Symp. 16:126-129. Sydney, Australia.

Girish, C.K. and G. Devegowda. 2004b. Efficacy of modified glucomannan (Mycosorb) and clay (HSCAS) to alleviate the individual and combined toxicity of aflatoxin and T-2 toxin in broilers. In: *22nd World's Poultry Congress*. Istanbul, Turkey, p. 591.

Hamilton, P.B., W.E. Huff, J.R. Harris, and R.D. Wyatt. 1982. Natural occurrences of ochratoxicosis in poultry. Poult. Sci. 61:1832-1841.

Hamilton, P.B.1984. Determining safe levels of mycotoxins. J. Food. Prot. 47:570-575.

Hamilton, P.B. 1987. Why the animal industry worries about mycotoxins. Presented at a Symposium on Recent Developments in the study of Mycotoxins. Kaiser Chemicals, Illinois.

Harvey, R.B., L.F. Kubena, W.E. Huff, M.H. Elissalde, and T.D. Phillips. 1991. Hematologic and immunologic toxicity of deoxynivalenol (DON)-contaminated diets to growing chickens. Bull. Environ. Contam. Toxicol. 46:410-416.

Hess, J.B. and S.F. Biligili. 1997. Mycotoxins: Hidden culprit in bruising. Broiler Industry, Watt Publications, Mt. Morris, Ill. 60(11):20-21.

Hoerr, F.J., W.W. Carlton, and B. Yagen. 1981. Mycotoxicosis caused by a single dose of T-2 toxin or diacetoxscirpenol in broiler chickens. Vet. Path. 18:652-664.

Hohler, D. 1999. A brief survey on important mycotoxins and possible detoxification methods. World Poult. Misset 15(11):13-15.

Hussein, H. S. and J.M. Brase. 2001. Toxicity, metabolism and impact of mycotoxins on humans and animals. Toxicology 167:101-134.

Huff, W.E., R.D. Wyatt, T.L. Tucker, and P.B. Hamilton. 1974. Ochratoxicosis in broiler chickens. Poult. Sci. 53:1585.

Huff, W.E. and J.A. Doerr. 1981. Synergism between aflatoxin and ochratoxin A in broiler chickens. Poult. Sci. 60:550.

Huff, W.E., L.F. Kubena, R.B. Harvey, W.M. Hagler, Jr, S.P. Swanson, T.D. Phillips, and C.R. Creger. 1986. Individual and combined effects of aflatoxin and deoxynivalenol (DON, vomitoxin) in broiler chickens. Poult. Sci. 64:1291-1298.

Huff, W.E., R.B. Harvey, L.F. Kubena, and G.E. Rottinghaus. 1988. Toxic synergism between aflatoxin and T-2 toxin in broiler chickens. Poult. Sci. 67:1418–1423.

Huff, W.E., L.F. Kubena, R.B. Harvey, and T.D. Phillips. 1992. Efficacy of hydrated sodium calcium aluminosilicate to reduce the individual and combined toxicity of aflatoxin and ochratoxin A. Poult. Sci. 71:64-69.

Jacobson, W.C. and H.G. Wiseman. 1974. Transmission of aflatoxin B_1 into eggs. Poult. Sci. 53:1743.

Jelinek, C.F., A.E. Pohland, and G.E. Wood. 1989. Worldwide occurrence of mycotoxins in foods and feeds - an update. J. Assoc. Off. Anal. Chem. 72:223-230.

Jones, F.T., W.M. Hagler, and P.B. Hamilton. 1982. Association of low levels of aflatoxins in feed with productivity losses in commercial broiler operations. Poult. Sci. 61:861-868.

Khajarern, J. and S. Khajarern. 1999. Protective effects of Mycosorb against aflatoxicosis in ducklings and broilers. Poster presentation of Alltech's 15th Annual Symposium on Biotechnology in the Feed Industry, Lexington, KY.

Kidd, M.T., M.A. Quershi, W.M. Hagler, Jr., and R. Ali. 1997. T-2 tetraol is cytotoxic to a chicken macrophage cell line. Poult. Sci. 76:311-313.

Kotyk, A.N. and V.A. Trufanova. 1998. Detection of naphthoquinone fusariotoxin aurofusarin in wheat. Mycology and Phytopathology 32:58-61.

Kramer, K., M. Anke, M. Glei, and M. Muller. 1993. Effects of oral bentonite on macro and trace element incorporation of different parts of the body and the milk. In: *Trace Elements in Man and Animals* (M. Anke, D. Meissner and C.F. Mills, eds), pp. 466-469.

Kubena, L.F., R.B. Harvey, W.E. Huff, D.E. Corrier, T.D. Phillips, and G.E. Rottinghaus. 1989. Influence of ochratoxin A and T-2 toxin singly and in combination on broiler chickens. Poult. Sci. 68:867 872.

Kubena, L.F., R.B. Harvey, W.E. Huff, D.E. Corrier, T.D. Phillips, and G.E. Rottinghaus. 1990. Efficacy of a hydrated sodium calcium aluminosilicate to reduce the toxicity of aflatoxin and T-2 toxin. Poult. Sci. 69:1078-1086.

Kubena, L.F., R.B. Harvey, W.E. Huff, A.G. Yersin, M.H. Elissalde, and D.A. Witzel. 1993. Efficacy of a hydrated sodium calcium aluminosilicate to reduce the toxicity of aflatoxin and diacetoxyscirpenol. Poult. Sci. 72(1):51-59.

Kubena, L.F., T.S. Edrington, C. Kamps-Holtzapple, R.B. Harvey, M.H. Elissalde, and G.E. Rottinghaus. 1995. Effects of feeding fumonisin B_1 present in *Fusarium moniliforme* culture material and aflatoxin singly and in combination to turkey poults. Poult. Sci. 74:1295-1303.

Kubena, L.F., R.B. Harvey, S.A. Buckley, T.S. Edrington, and G.E. Rottinghaus. 1997. Individual and combined effects of moniliformin present in *Fusarium fujikuroi* culture material and aflatoxin in broiler chicks. Poult. Sci. 76:265-270.

Kudupoje, M. and G. Devegowda. 2000. Efficacy of esterified glucomannan to ameliorate the toxic effects of T-2 toxin in laying hens. Poult. Sci. 79(Suppl. 1):62 (Abstr).

Kudupoje, M. and G. Devegowda. 2001. Use of esterified glucomannan to reduce the effects of T-2 toxin in laying hens. In: *Proc. of The World Mycotoxin Forum*, The Netherlands, pp. 71.

Ledoux, D.R., T.S. Weibking, and G.E. Rottinghaus. 1992. Effects of *Fusarium moniliforme* culture material containing known levels of fumonisin B_1 on turkey poults. Poult. Sci. 71(Suppl.1):162.

Ledoux, D.R., A.J. Bermudez, and G.E. Rottinghaus. 1993. Effects of *Fusarium fujikuroi* culture material containing known levels of moniliformin on turkey poult. Poult. Sci. 72(Suppl. 1):68 (Abstr).

Ledoux, D.R., G.E. Rottinghaus, A.J. Bermudez, and M. Alonso Debolt. 1999. Efficacy of a hydrated sodium calcium aluminosilicate to ameliorate the toxic effects of aflatoxin in broiler chicks. Poults. Sci. 78:204-210.

Leibetseder, J. 1995. The European perspective on mycotoxins. In: *Bio-technology in the Feed Industry: Proceedings of Alltech's 11th Annual Symposium* (T.P. Lyons and K.A. Jacques, eds). Nottingham University Press, UK, pp. 65-74.

Li, Y.C., D.R. Ledoux, A.J. Bermudez, K.L. Fritsche, and G.E. Rottinghaus. 2000. The individual and combined effects of fumonisin B_1 and moniliformin on performance and selected immune parameters in turkey poults. Poult. Sci 79(16):253-260.

Lipke, P.N. and R. Ovalle. 1998. Cell wall architecture in yeast: New structure and new challenges. J. Bacteriol. 180(15): 3735-3740.

Mahesh, B.K. and G. Devegowda. 1996. Ability of aflatoxin binders to bind aflatoxin in contaminated poultry feeds - an *in vitro* study. In: *Proc. XX World's Poultry Congress* 4:296. New Delhi, India.

Manning, R.O., T.P. Brown, R.D. Wyatt, and O.J. Fletcher. 1985. The individual and combined effects of citrinin and ochratoxin A in broiler chicks. Avian Dis. 29:986-997.

McDaniel, G. 1991. Effect of Yea-Sacc[1026] on reproductive performance of broiler breeder males and females. In: *Biotechnology in the Feed Industry, Proc. Alltech's 7th Annual Symposium* (T.P. Lyons, ed). Nottingham University Press, UK, pp. 413:415.

Miller, B.L. and R.D. Wyatt. 1985. Effect of dietary aflatoxin on the uptake and elimination of chlortetracycline in broiler chicks. Poult. Sci. 64:1637-1643.

Morris, C.M., Y.C. Li, D.R. Ledoux, A.J. Bermudez, and G.E. Rottinghaus. 1999. The individual and combined effects of feeding moniliformin, supplied by *Fusarium fujikuroi* culture material, and deoxynivalenol in young turkey poults. Poult. Sci. 78:1110–1115.

Muller, R.D., C.W. Carlson, G. Semeniuk, and G.S. Harshfield. 1970. The response of chicks, ducklings, goslings, pheasants and poults to graded levels of aflatoxins. Poult. Sci. 49:1346-1350.

Murthy, T.N.K. and G. Devegowda. 2004. Efficacy of modified glucomannan (Mycosorb) to adsorb aflatoxin B_1 in gut conditions of broiler chickens. In: *22nd World's Poultry Congress*. Istanbul, Turkey, pp. 471.

Nagaraj, R.Y., W. Wu, J.A. Will, and R.F. Vesonder. 1996. Acute cardio toxicity of moniliformin in broiler chickens as measured by electrocardiography. Avian Dis. 40:223-227.

Nascimento, J.A.F.B., V.A. Nunes, R.M.C. Guedes, and M.A. Rachid. 2001. T-2 toxin and disturbed endochondral bone growth in broiler chickens. Arq. Bras. Med. Vet. Zootec. 53:332-341.

Pavicic, P., P. Spring, N. Fuchs, and A. Nemanic. 2001. Efficacy of esterified glucomannan to reduce the toxicity of diacetoxyscirpenol in broiler chickens. In: *13th European Symposium Poult. Nutr., World's Poultry Science Association*. Blankenberge, Belgium.

Phillips, T.D., L.F. Kubena, R.B. Harvery, D.R. Taylor, and N.D. Heidelbaugh. 1988. Hydrated sodium calcium aluminosilicate: A high affinity sorbent for aflatoxin. Poult. Sci. 67:243-247.

Phillips, T.D., A.B. Sarr, B.A. Clement, L.F. Kubena, and R.B. Harvery. 1990. Prevention of aflatoxicosis in farm animals via selective chemisorption of aflatoxin. In: *Pennington Center Nutrition Series Volume 1, Mycotoxins, Cancer, and Health* (G.A. Bray and D.H. Ryan, eds). Louisiana State University Press, Baton Rouge, LA, pp. 223-237.

Prelusky, D.B. Trenholm, H.L., Hamilton, R.M.G., and Miller J.D. 1987. Transfer of [^{14}C] deoxynivalenol to eggs following oral administration to laying hens. J. Agric. Food Chem. 35:182-185.

Qureshi, M.A., J. Brake, P.B. Hamilton, W.M. Hagler, and S. Nesheim. 1998. Dietary exposure of broiler breeders to aflatoxin results in immune dysfunction in progeny chicks. Poult. Sci. 77:812-819.

Ravindran, P., R.J. Gill, and W.L. Bryden. 1996. Aflatoxin, fumonisin and zearalenone contamination of Australian maize. Proc. 20th World's Poultry Congress. New Delhi, India. 4:273.

Rafai, P., H. Pettersson, A. Bata, Z. Papa, R. Glavits, S. Tubolys, A. Vanyi, and P. Sooss. 2000. Effect of dietary T-2 fusariotoxin concentrations on the health and production of white Pekin duck broilers. Poult. Sci. 79:1548-1556.

Raju, M.V.L.N. and G. Devegowda. 2000. Influence of modified glucomannan on performance and organ morphology, serum biochemistry and hematology in broilers exposed to individual and combined mycotoxicosis (aflatoxin, ochratoxin and T-2 toxin). Br. Poult. Sci. 41:640-650.

Raju, M.V.L.N. and G. Devegowda. 2002. Esterified glucomannan in broiler chicken diets contaminated with aflatoxin, ochratoxin and T-2 toxin: evaluation of its binding ability (*in vitro*) and efficacy as immunomodulatory. Asian-Aust. J. Anim. Sci. 15(7):1051-1056.

Reddy, N.B. and G. Devegowda. 2004. Ability of modified glucomannan (Mycosorb) to sequester T-2 toxin in the gastrointestinal tract of chickens. Asian-Aust. J. Anim. Sci. 17(1) (accepted for publication).

Sakhatsky, I.M. 1999. Effect of *Fusarium graminearum* mycotoxins on reproduction and immunity of laying hens. Vet. Med. Ukraine. 9:34-35.

Salyi, G. and R. Glavits. 1995. A barbari Kasca Kiserletesen eloidezett T-2 toxikozisa (Experimental T-2 toxicosis in Moscovy ducks). Magy. Allatorv. Lapja 50:202-208.

Santurio, J.M., C.A. Mallmann, M.A. Balidisera, C. Ewald, and A. Heer. 1994. Niveis de adsorcao de Aflatoxina B_1 *in vitro* de aluminosilicatoes e bentonitas comercializados no Brasil. Proceedings of the 1st Congresso Latino-Americano de Micotoxicologia. Rio de Janeio, pp. 10-11.

Santin, E., A. Maiorka, E.L. Krabbe, and A.C. Alessi. 2002. Effect of hydrated sodium calcium aluminosilicate on the prevention of the toxic effects of ochratoxin. J. Appl. Poult. Res. 11:22-28.

Schiedeler, S.E. 1993. Effects of various types of aluminosilicates and aflatoxin B_1 on aflatoxin toxicity, chick performance, and mineral status. Poult. Sci. 72:282-288.

Shlosberg, A.S., Y. Klinger, and M.H. Malkinson. 1986. Muscovy ducklings, a particularly sensitive avian bioassay for T-2 toxin and diacetoxyscirpenol. Avian Dis. 30:820-824.

Shirley, H.V. and S.H. Tohala. 1983. Ochratoxicosis in laying hens.1982. Annual Science Progress Report 83-08. University of Tennessee Agriculture Experiment Station.

Smith, E.E., L.F. Kubena, C.E. Braithwaite, R.B. Harvey, T.D. Phillips, and A.H. Reine. 1992. Toxicological evaluation of aflatoxin and cyclopiazonic acid in broiler chickens. Poult. Sci. 71:1136-1144.

Smith, T.K., E.J. MacDonald, and S. Haladi. 2001. Mycosorb: Growth and neurological response of pigs given fusariotoxin contaminated diets. Poster presented at Alltech's 17th Annual Symposium, Lexington, KY, USA.

Smith, T.K., M. Modirsanei, and E.J. Macdonald. 2000. The use of binding agents and amino acid supplements for dietary treatment of Fusarium mycotoxicosis. In: *Biotechnology in the Feed Industry, Proc. of Alltech's 16th Annual Symposium*, (T.P. Lyons and K.A. Jacques, eds). Nottingham University Press, UK, pp. 383-390.

Smith, T.K., H.V.L.N. Swamy, S.R. Chowdhary, and S.L. Raymond. 2003. *In vivo* demonstrations of the efficacy of Mycosorb™ in prevention of *Fusarium* mycotoxicoses. The Second World Mycotoxin Forum, 17-18 February, pp. 63-64.

Stanley, V.G., R. Ojo, S. Woldesenbet, and D.H. Hutchinson. 1993. The use of *Saccharomyces cerevisiae* to suppress the effects of aflatoxicosis in broiler chicks. Poult. Sci. 72:1867-1872.

Stratton, G.W., A.R. Robinson, H.C. Smith, L. Kittilsen, and M. Barbour. 1993. Levels of five mycotoxins in grains harvested in Atlantic Canada as measured by high performance liquid. Archives of Environmental Contamination and Toxicology. 24: 399-409.

Surai, P.F. 2002. *Natural Antioxidants in Avian Nutrition and Reproduction*. Nottingham University Press, UK.

Swamy, H.V.L.N. and G. Devegowda. 1998. Ability of Mycosorb to counteract aflatoxicosis in commercial broilers. Indian J. Poult. Sci. 33:273-278.

Swamy H.V.L.N., T.K. Smith, P.F. Cotter, H.J. Boermans, and A.E. Sefton. 2002. Effects of feeding blends of grains naturally contaminated with *Fusarium* mycotoxins on production and metabolism in broilers. Poult. Sci. 81:966-975.

Tung, H.T., J.W. Smith, and P.B. Hamilton. 1971. Aflatoxicosis and bruising in chickens. Poult. Sci. 50:795-800.

Trenholm, H.L., L.L. Charmley, and D.B Prelusky. 1996. Mycotoxin binding agents: An update on what we know. In: *Biotechnology in the Feed Industry, Proc. of Alltech's 12th Annual Symposium* (T.P. Lyons and K.A. Jacques, eds). Nottingham University Press, UK, pp. 327-349.

Urano, J., M.W. Trucksess, R.W. Beaver, D.M. Wilson, J.W. Dorner, and F.E. Dowell. 1992. Co-occurrence of cyclopiazonic acid and aflatoxin in corn and peanuts. J. AOAC International 75:838-841.

Van Egmond, H.P. and G.J.A. Spijers. 1994. Survey of data on the incidence and levels of ochratoxin A in food and animal feed world wide. J. Natural Toxin. 3:125-144.

Vanyi, A., R. Glavits, A. Batta, and F. Kovacs. 1994. Pathomorphological changes caused by trichothecene fusariotoxin in geese. Acta Vet. Sci. Hung. 42:447-457.

Veldman, B. 2003. Effects of dietary mycotoxins on the quality of animal products and on human health. Proc. of the Second World Mycotoxin Forum, The Netherlands, pp. 52-54.

Volkl, A. and P. Karlovsky. 1998. Biological detoxification of fungal toxins and its use in plant breeding, feed and food production. Natural Toxins 7:1-23.

Wang, R., C. Miao, R. Chong, N. Sawatwiroj, K. Dawson, S. Fu, and D. Feng. 2003. Feed and feed ingredient mycotoxin contamination in China: A survey. In: *Proc. of Alltech's 17th Annual Asia Pacific Lecture Tour*.

Weibking, T.S., D.R. Ledoux, A.J. Bermudez, and G.E. Rottinghaus. 1994. Individual and combined effects of feeding *Fusarium moniliforme* culture material containing known levels of fumonisin B_1 and aflatoxin B_1 in the young turkey poult. Poult. Sci. 73:1517-1525.

World Health Organisation (WHO). 2001. Safety evaluation of certain mycotoxins in food. Geneva, Switzerland.

Wu, W., D. Jerome, and R. Nagaraj. 1994. Increased redness in turkey breast muscle induced by fusarial culture materials. Poult. Sci. 73:331-335.

Wyatt, R.D., P.B. Hamilton, and H.R. Burmeister. 1975. Altered feathering of chicks caused by T-2 toxin. Poult. Sci. 54:1172 (Abstr).

3

EFFECTS OF MYCOTOXINS IN HORSES

Kyle E. Newman[1] and Susan L. Raymond[2]
[1]*Venture Laboratories, Lexington, Kentucky, USA*
[2]*Equine Guelph, University of Guelph, Ontario, Canada*

Until recently, studies investigating the effects of mycotoxins on horses were limited to trials with few animals or depended on extrapolations based on data from other species (the pig or ruminant being the most obvious choices). However, a unique challenge is presented when attempting to use non-equine data effectively since the horse is comparable to the ruminant in that it is a forage-grazing animal but has a gastrointestinal tract more closely related to the pig with the addition of a hindgut fermentation process. The nature of the horse farm also makes the equine quite different from other livestock species. These other species are bred for growth and meat yield and have a relatively short lifespan while in most cases the horse is bred for athletic performance, conformation, temperament, beauty and/or durability.

Limiting values or safe concentrations of specific mycotoxins are, for the most part, unknown for the horse. Historically, mycotoxins were identified by their ability to produce severe, overt disease syndromes in animals as a result of relatively high intakes of mycotoxins, however low levels of exposure over long periods can elicit chronic or subchronic toxicological manifestations. A concern is that for horses, especially elite athletes or breeding stock, exposure to low levels of mycotoxins may affect performance or breeding ability without the appearance of overt signs of disease. The outcome from exposure to chronic low levels of toxin may include general unthriftiness, suppression of the immune system and increased risk of secondary infections. Unlike other livestock species, horses can have a long lifespan and are expected to be reproductively sound in their later years. For these reasons, probably more so than for other species, the safe amount of specific mycotoxins is unknown for the horse. In addition to the degree of exposure, incidence of disease can also be influenced by the presence of multiple mycotoxins. Factors influencing susceptibility to mycotoxins include disease, heat stress, marginal nutritional profile, drug interactions, presence of multiple toxins, crowding, age and reproductive status. The intent of this chapter is to inform the reader of available data on mycotoxins and horses or the possible implications of toxins for the horse in the absence of data.

Moulds and production of mycotoxins

The occurrence of mould and mycotoxins in food and animal feed is a problem of major concern internationally (Wood, 1992). Mycotoxin production can occur when favourable conditions allow fungi to grow on crops in the field, at harvest, in storage or during the processing of feed (Palmgren and Lee, 1986). Mycotoxins are the products of secondary fungal metabolism although not all fungi produce mycotoxins. Environmental and nutritional parameters and growth stage of the mould govern toxin production. Moulds that are present in the field before harvest can proliferate if moisture content remains high through inadequate processing and storage conditions. Mould and subsequent mycotoxin contamination of a feedstuff or forage can increase in extreme environmental conditions such as drought, excessive precipitation, or sudden frost. It can also increase due to physical damage to a crop, which allows fungal penetration of the plant tissue (Gregory *et al.*, 1963; Northolt *et al.*, 1976; Schindler, 1977). A large number of the predominant mycotoxins are produced by the *Fusarium* moulds, which thrive in temperate climates worldwide and are common contaminants of feed grains. A wet growing season followed by cool weather increases the likelihood that fungi, especially *Fusarium* and its mycotoxins, will be present in grains. The high moisture level in grain encourages fungal growth while the cool temperatures can increase the production of mycotoxins. These fungi require a moisture content of 25% concurrent with a relative humidity of more than 90% in order to grow. Such conditions can commonly occur when field-dried hay is not cured properly. Forage is the basis of most feeding programmes for the horse, with long-stemmed, field-dried hay being the traditional source. Any horseman can recognise the dust aerosol when opening a bale of hay that has a high mould count; and many individuals equate a high mould count with mycotoxin content. It is important to note that a forage or feed with a low level of measurable mould contamination may still contain significant amounts of mycotoxins. It has been shown that the subjective opinion of horse owners about their hay does not always correlate with objective analyses of mycotoxins, mould and actinomycete contamination (Raymond *et al.*, 2000). Mycotoxin production does not correlate to mould growth and conditions that support one may actually inhibit the other. Increased mycotoxin contamination and the potential for mixtures of mycotoxins to be co-contaminants in feeds are now even more likely due to improved global grain transportation systems and global trading of agricultural commodities.

CONDITIONS FAVOURING MYCOTOXIN PRODUCTION

Aspergillus, Fusarium and *Penicillium* are the three most important genera of fungi that have been extensively studied in relation to mycotoxicosis. The large surface area and vigorous activity of fungal mycelia ensure a close relationship of the mould with the environment. Extrinsic factors such as temperature, pH, water activity, and chemical composition have an influence on growth and mycotoxin biosynthesis. Other factors such as plant-fungal interactions, interactions among microorganisms, and the presence of certain chemical agents can influence mould growth and mycotoxin synthesis (Smith and Henderson, 1991). In the field situation specific fungal species may occur cyclically or seasonally. It is usually the case that the optimum conditions for mould growth do not coincide with those for mycotoxin production. The production of several different mycotoxins by the same species, or even the same strain, may not occur optimally under identical conditions. For example, the production of deoxynivalenol (DON) and zearalenone (ZEA) by a single strain of *Fusarium graminearum* responded differently to changes in temperature. ZEA production reached a maximum at 25°C, while the production of DON continued to increase with increasing temperature, and at 28°C the production was more than twice that produced at 25°C (Greenhalgh *et al.*, 1983). In field situations DON and ZEA have been produced simultaneously in infected corn and wheat by the above fungi (Cote *et al.*, 1985).

Aspergillus and aflatoxin

Aflatoxin is a mycotoxin produced primarily by *Aspergillus flavus* and *A. parasiticus*. These moulds prefer warm and humid conditions. Mycotoxin production on feedstuffs seems to be most prevalent when drought, insect infestation or other stresses occur late in the growing season of cottonseed, peanuts, corn, or other grain products. In documented outbreaks of aflatoxicosis on two separate horse farms, peanut meal and corn were found to be the sources of the toxin (Angsubhakorn *et al.*, 1981). In one case the elevated price of rice caused feed producers to use stored corn and peanut meal (comprising 70% of the feed mixture) in place of the rice in the feed. Representative samples of the mixed ration contained approximately 0.2 mg/kg aflatoxin B_1 (AFB_1) and 0.2 mg/kg aflatoxin B_2. A number of animals showed clinical signs of illness and 12 yearlings died. Clinical symptoms of aflatoxin intoxication include ataxia, tremor, elevated temperature, anorexia, loss of appetite, weight loss, icterus (yellowing of the eyes or skin), haemorrhages, bloody faeces, and brown urine (Angsubhakorn *et al.*, 1981; Bortell *et al.*, 1983; Cysewski *et al.*, 1982). The major target organ of aflatoxin is the liver, where the toxin induces centrilobular necrosis (Stoloff and Trucksess, 1979).

Clinical signs of toxicity were observed in ponies dosed with aflatoxin at levels ranging from 0.075 to 0.3 mg/kg BW/day (75 to 300 ppb) with all animals either dying or declining and euthanized during a 39 day trial (Cysewski *et al.*, 1982). A subsequent trial found that ponies given aflatoxin B_1 at dosages equal to or greater than 2 mg/kg BW died within 76 hrs (Bortell *et al.*, 1983). At the time of necropsy, ponies demonstrated petechia of the viscera, free blood in the small intestine, pale tan to yellow livers and icterus. From the data, it appears that the equine is more susceptible to aflatoxin B_1 than calves. Calves had a reduced rate of gain with 0.2 mg B_1/kg/day but died after 14 days given 0.5 mg B_1/kg/day (Pier *et al.*, 1976). In ponies, 0.3 mg B_1/kg/day caused death between 12 and 16 days and clinical symptoms and biochemical changes were observed even at 0.075 mg B_1/kg/day (Cysewski *et al.*, 1982). It should be noted that other studies in ponies have observed apparent resistance at levels of 0.2 mg B_1/kg/day (Walter *et al.*, 1981). Cysewski *et al.* hypothesised that previous exposure to aflatoxin of the ponies used in the Walter study may have influenced this resistance. The fact that Walter and co-workers dosed their ponies for 5 days and Cysewski *et al.* dosed for up to 39 days may also have played a role in the different responses observed in these two trials. A separate case study of an Arkansas farm where three horses died of severe hepatic necrosis and others became ill found concentrations of 130 ppb total aflatoxin (114 ppb AFB_1, 10 ppb AFB_2 and 6 ppb AFM_1) in the suspected corn (Vesonder *et al.*, 1991). A summary of equine aflatoxicosis cases indicates a very large range from approximately 55 ppb to 6,500 ppb to be associated with clinical symptoms (Asquith, 1985) with duration of exposure and previous history of exposure playing a role in aflatoxicosis.

Recently, authors of a study in Sweden have implied a possible link between chronic obstructive pulmonary disease (COPD) and inhaled mycotoxins (Larsson *et al.*, 2003). Data from this study show that activation of AFB_1 occurs not only in the liver but also in the olfactory and respiratory tissues of the horse. AFB_1 requires activation by cytochrome P450 enzymes to the reactive intermediate AFB_1-8,9-epoxide, which then binds to nucleic acids among other molecules and leads to the carcinogenic and cytotoxic properties normally associated with this toxin. Glutathione (GSH) may also conjugate with AFB_1-8,9-epoxide, however this binding essentially detoxifies the compound (Quinn *et al.*, 1990).

COPD is a disease of the equine lung that is similar to asthma in humans and caused by an allergic response. It is also called 'heaves', since horses typically are observed 'heaving' to push air out of the lungs at the end of an exhale. COPD is most often observed in older or stabled horses. It is rarely seen in situations where horses are pastured year round. As with asthma in humans, aerosolised particles of pollen, dander, microorganisms or other agents inhaled by a COPD horse elicit an allergic response. The moulds *A. fumigatus* and *Micropolyspora faeni* are two heavily implicated organisms that may evoke COPD in horses (Halliwell *et al.*, 1993).

Ochratoxin

Ochratoxin was first detected by van der Merwe *et al.* (1965) in laboratory tests to identify toxic feedstuffs in South Africa. Ochratoxin is a potent nephrotoxin and teratogen. Adverse effects have been noted in pigs and poultry at levels at or near 2 ppm. Effects observed in cattle include decreased performance, reduced milk production, kidney failure and death (at levels greater than 800 ppm). The outcome of this toxin on horses is not known but assumed to be similar to pigs due to the site of ochratoxin absorption. The fungus responsible for the production of ochratoxin can invade starchy cereal grains such as corn and wheat with a moisture content of 15.5-16%. Ochratoxin A (OTA) is only slightly soluble in water and when consumed is absorbed in the upper sections of the gastrointestinal tract. Maximal absorption occurs in the proximal jejunum in a passive manner in the non-ionized form and is then subjected to secretion and re-absorption via enterohepatic recycling (Kumagai and Aibara, 1982; Leeson *et al.*, 1995). Absorption is faster in areas of the GIT with a low pH. In mammals, OTA absorption through the lungs into the systemic circulation has also been documented (Di Paolo *et al.*, 1993). Toxicity of OTA in ruminants is thought to be relatively low due to the rumen microflora. It has been postulated that levels up to 12 mg of OTA/kg of contaminated feed are tolerated by ruminants (Hult *et al.*, 1976). However, Hohler and co-workers found that even at dosages of 2 mg/kg of concentrate feed, substantial amounts of OTA were detected in the serum of cattle fed the toxin (Hohler *et al.*, 1999).

Ochratoxin A enters circulation through the portal vein and the lymphatic vessels bound to plasma proteins, especially albumin. Due to the relatively high affinity of OTA for albumin a number of studies have examined increasing dietary protein levels to help alleviate the deleterious effects of this toxin. As with most mycotoxins, ochratoxin is not present in isolation and is commonly found with citrinin, penicillic acid and other mycotoxins in field situations (Ribelin *et al.*, 1978).

Fusarium mycotoxins

The genus *Fusarium* contains a large number of species of fungi, which produce various mycotoxins. There are more than 300 known *Fusarium* mycotoxins, but very little is known about the relative toxicity of most of these compounds. *Fusarium* mycotoxins of significant public health and agro-economic importance include trichothecenes, ZEA, fumonisins, moniliformin and fusaric acid (FA). The trichothecene mycotoxins are a group of over 180 compounds produced by *Fusarium, Stachybotrys, Myrothecium* and other fungal genera. The name trichothecene was derived from the fungus *Trichothecium* and these compounds are chemically related sesquiterpenoids that all possess a tetracyclic 12,13-epoxy-trichothec-9-ene ring system. Deoxynivalenol is considered the most common trichothecene but also one of the least toxic. Other prominent trichothecenes include T-2 toxin (considered one of the least prevalent but most toxic), diacetoxyscirpenol (DAS), fusarenon-X and satratoxins. ZEA, fumonisins, moniliformin and FA are classified as non-trichothecene *Fusarium* mycotoxins. Most trichothecenes are found in very small quantities.

The trichothecenes are known as feed refusal toxins due to loss of appetite being one of the first observable symptoms

(Trenholm *et al*., 1994). These compounds have been shown to alter brain neurochemistry by increasing tryptophan and serotonin levels through inhibition of hepatic protein synthesis resulting in hyperamino-acidemia (Meloche and Smith, 1995; Wannemacher and Dinterman, 1983). Excess brain serotonin, a neurotransmitter synthesized from tryptophan, can contribute to loss of appetite, lethargy, sleepiness and loss of muscle co-ordination. Studies in laboratory animals and pigs have shown enhanced serotonergic activity in the brain as one of the possible mechanisms for DON and FA induced feed refusal, vomiting and other behavioral changes (Smith *et al*., 1997).

Immunosuppression and the increased susceptibility to infectious diseases are the main features of chronic exposure of animals to low levels of trichothecene mycotoxins (Rotter *et al*., 1994). It has been shown that impairment of immunity against bacterial and parasitic infection increases after mycotoxin consumption as well as susceptibility to disease (Venturini *et al*., 1996). Immunosuppressant effects of ingested mycotoxins can also induce secondary mycotoxic diseases. In mice, ingestion of DON may also cause glomerulonephritis (kidney failure) from the overproduction in the intestines of IgA immunoglobulins, which accumulate in the kidneys. Johnson *et al*. (1997) fed barley containing 36-44 ppm DON (on an as-fed basis) to horses while at pasture. No detectable effects were found specifically on feed intake or serum IgA concentrations during the 40 days of the trial using non-pregnant mares and geldings. It was suggested that the horse may not be susceptible to the adverse effects of DON and the horse's gastric microflora may be able to detoxify it. Raymond *et al*. (2003a) fed a diet in which the concentrate averaged 15 ppm DON. The concentrate also contained on average 0.8 ppm 15-acetyldeoxynivalenol, 9.7 ppm FA and 0.2 ppm ZEA. The horses were consuming 2.8 kg concentrate and 5 kg of mixed hay per day. The contaminated grains used in this study were a mixture of naturally contaminated wheat and corn. Naturally contaminated mixed sources are very important as opposed to pure forms or single contaminated sources. The former mimics more closely what is likely to be found in industry where an animal is not likely to be exposed to only one mycotoxin. The grains used most commonly in horse feeds are corn, oats and barley. It was concluded that consumption of the contaminated diet reduced feed intake. It was also found that serum activities of γ-glutamyltransferase (GGT) were significantly higher in horses consuming contaminated grain sampled on day 7 and 14 of supplementation but not on day 21 (Table 1). An increase in GGT levels indicates a possible effect on liver function without obvious signs of damage since GGT is a hepatic membrane-associated enzyme. A similar study was conducted by the same authors incorporating an exercise programme (Raymond *et al*, 2003b). Results were comparable with the addition of significant weight loss observed over the 21 day supplementation period in horses fed contaminated grains as compared to control animals.

Table 1. Effect of feeding blends of grains naturally contaminated with *Fusarium* mycotoxins on serum γ-glutamyltransferase (GGT) activity in horses.

Diet	*serum GGT (U/L)*		
	7 days	*14 days*	*21 days*
Control[a]	11.44[c]	12.23	11.49
Mycotoxin[b]	22.98	23.86	15.27

[a]Diet with control corn and wheat.
[b]Mycotoxin-contaminated concentrate containing 36% contaminated wheat and 53% contaminated corn.
[c]Values are least square means; n = 6 horses.
Based on data from Raymond *et al*., 2003a.

Toxicological synergism has also been reported among *Fusarium* mycotoxins (Smith *et al*., 1997) and may help to explain the response differences noted between the studies of Johnson and Raymond. Studies have indicated that the presence of fusaric (5-butylpicolinic) acid, a compound synthesised from tryptophan by *Fusarium* moulds, will increase the growth depression seen when low levels of DON are fed to starter pigs (Smith *et al.*, 1997). Although fusaric acid was chemically characterised years ago, it has not been considered to be a significant factor in *Fusarium* mycotoxicosis because of its relatively low toxicity. Fusaric acid is pharmacologically active, however, and alters brain neurochemistry in a wide range of animal species. The physiological effect of fusaric acid is a drop in blood pressure. Such an effect could be of significance for the horse as one of its production parameters is athletic performance. Like DON, fusaric acid is associated with increased brain serotonin levels but through a different mechanism. Trichothecenes elevate brain serotonin levels by elevating total blood tryptophan. Fusaric acid increases the availability of this tryptophan to the brain. Fusaric acid is a tryptophan analogue and competes with tryptophan for binding to blood albumin, therefore an increase in free tryptophan in the blood occurs (Chaouloff *et al.*, 1986). It has been demonstrated that fusaric acid and DON act synergistically to reduce feed consumption and trigger loss of muscle coordination and lethargy (Smith *et al.*, 1997).

Horses could come in contact with DON and/or fusaric acid from the consumption of corn, wheat, or wheat by-products in grain concentrates and/or from the consumption of wheat straw bedding. An often overlooked potential source of equine exposure to mycotoxin contamination is bedding, with straw being a common source. Contamination of bedding material represents a risk through ingestion, inhalation and dermal contact. An unplanned DON concentration of 1.2 ppm in the straw bedding in the previously described feeding trial is an illustration of the potential mycotoxin contamination (Raymond *et al.*, 2003a). The amount of DON in straw depends primarily on the presence of contaminated grain and chaff, two sources known to have more concentrated levels of mycotoxins. Horses fed a grain ration containing significant levels of DON may also consume their bedding. If the bedding also contains a significant level of toxin, the total DON consumption could exceed a tolerance level. Additional exposure could also come from dried or pasture forage. Conversely the practice of including a large forage component as compared to concentrates in the traditional equine diet may provide some protection from mycotoxin exposure as a high concentration of plant fibre has been shown to reduce some of the deleterious effects of mycotoxins (Carson and Smith, 1983; James and Smith, 1982).

A published case report cited weight loss and elevated hepatic enzymes in horses with the probable cause being straw contaminated with DON. It was reported that in 2002, approximately half of 104 warmblood-type riding horses stabled in Germany suddenly lost weight. Further examination of nine of the affected horses revealed marked elevation of liver enzyme activity of both GDH and GGT. Analysis of the feed, hay and bedding revealed DON concentrations ranging from 0.5 to 2.7 ppm in the straw. Once the horses were removed from the contaminated bedding they slowly gained weight and their general condition progressively improved (Zeyner *et al.*, 2002). FDA levels of concern for DON are 2 ppm for wheat entering the milling process for humans and 1 ppm for the finished product for humans. The concern level for DON in wheat for livestock is 4 ppm (Wood, 1992). Since these papers represent a large range in potential toxic levels, it is suggested that until more precise research is completed, the maximum tolerable level for DON in the total feed for horses be comparable to levels allowed for humans.

Zearalenone is an oestrogenic, non-trichothecene *Fusarium* mycotoxin. ZEA binds to oestrogen binding sites and causes enlargement of the uterus and rectal and vaginal prolapse, abortions and infertility with pre-pubertal animals appearing to be more sensitive than older animals and swine being the most susceptible livestock species (D'Mello *et al.*, 1999; Kalleta and Ettala, 1981). ZEA toxicity is more readily recognised than trichothecene toxicity because these symptoms are very specific.

Juhasz *et al.* (2001) reported that ZEA was administered to mares for 10 days at 7 mg of purified ZEA *per os* daily to mimic a 1 ppm concentration in the feed. It had no adverse effect on the reproductive parameters of cyclic mares but skin lesions were observed around the mouth in 3 of 6 horses. Gimeno and Quintanilla (1983) reported a case of natural outbreak of ZEA mycotoxicosis on a commercial horse farm. Symptoms in the mares included feed refusal prolapse uterine and internal haemorrhage. Severe flaccidity of the genitals was observed in the males. It was found that the corn screenings fed to the horses contained on average 2.7 ppm ZEA. No other common mycotoxins were found. Corn screenings fed to a herd of horses on a commercial farm produced strong estrogenic symptoms after a feeding period of 30 days. Reports of feeding ZEA-contaminated diets to pigs indicated that 1 ppm is the minimum concentration required to produce hyperestrogenism (James and Smith, 1982). Although traditionally considered more resistant to the effects of ZEA, abortions in cattle have been attributed to ZEA-contaminated hay (Kalleta and Ettala, 1981).

Fumonisins are produced primarily by *F. moniliforme* and *F. proliferatum,* the *Fusarium* that invade and are prevalent in corn. Fumonisins are recently discovered mycotoxins that can impair immune function, cause kidney and liver damage, decrease animal performance and cause death. In pigs, fumonisin has been linked with **porcine pulmonary oedema (PPE)** (Harrison *et al.*, 1990). Fumonisin in horses can cause **equine leukoencephalomalacia (ELEM)**, which is typified by staggers, stupor, lameness, seizure (due to brain necrosis) and death. Fumonisin levels associated with PPE and ELEM ranged from 1 to 330 ppm fumonisin B_1 (FB_1) (PPE) and 1 to 126 ppm (ELEM) (Ross *et al.*, 1991). Additional studies (Ross *et al.*, 1992) suggested that fumonisin B_1 concentrations greater than 10 ppm in horse feeds made those animals consuming the feed likely candidates for ELEM. Although naturally contaminated feed containing fumonisin has been shown to elevate brain serotonin levels, it appears that fumonisin is not the source of this activity. Smith and MacDonald (1991) demonstrated fusaric acid would increase tryptophan, serotonin and 5-hydroxyindoleacetic acid in pigs. Previous work by Porter and coworkers (1990) examined corn associated with ELEM on tryptophan levels in rats and found elevated levels, however, these investigators did not measure fusaric acid levels in the corn. Subsequent work by Porter did not show similar effects with purified FB_1 (Porter *et al.*, 1993). It seems probable that fumonisin does not affect brain serotonin levels; other toxins produced along with fumonisin appear to be responsible for this activity. Fumonisins interfere with sphingolipid metabolism, disrupting endothelial cell walls and basal membranes (van der Westhuizen *et al.*, 2001). Sphingolipids are one of the primary groups of lipids in cell membranes. They have multiple functions, including signal transduction, cell growth, cell-to-cell communication, immunorecognition, and aid in defining the physical nature of lipoproteins.

There are three common fumonisin mycotoxins designated as FB_1, FB_2 and FB_3. FB_1 is well documented as the primary cause of leukoencephalomalacia in horses. Fumonisin B_1 has also been documented to alter cardiovascular and hepatic function and elevate serum cholesterol (Smith *et al.*, 2002; Haschek *et al.*, 2001). FB_2 has also been implicated as a possible cause of ELEM, however, the only trial examining this toxin (75 ppm FB_2) in the equine also had levels of FB_1 of 3 ppm (Ross *et al.*, 1994). Two of three ponies consuming this diet for 136 days demonstrated symptoms of ELEM. Ponies receiving a diet containing 75 ppm FB_3 and less than 1 ppm of both FB_1 and FB_2 for 56 days were clinically normal with no differences from control ponies at the time of necropsy. From this trial it is clear that FB_3 is not as toxic as FB_1 and FB_2. The effects of FB_2 are not as obvious due to the somewhat high concentrations of FB_1 in the diet containing 75 ppm FB_2.

In the equine, the consumption of fumonisins produces a brain neurological disease identified by multi-focal liquefactive necrosis of the white matter affecting multiple horses in a herd. Once clinical signs appear, the majority of affected horses die. Horses that survive typically have some degree of permanent neurological disorder. Corn screenings can be heavily contaminated with fumonisins and should never be fed to horses. Concentrations of FB_1, FB_2 and FB_3 in equine feedstuffs should not exceed 5 ppm and such feedstuffs should not exceed 20% of the diet on a dry matter basis.

Studies on the effects of T-2 toxin on the horse are few in number, with only one report involving the feeding of T-2 toxin. Feeding 7 mg of purified T-2 toxin to equines *per os* daily, to mimic a 1 ppm concentration in feed, had no effect on the ovarian activity of mares (Juhasz *et al.*, 1997). In other species, oral and intestinal lesions are frequently observed in animals consuming T-2 or diacetoxyscirpenol (DAS) at levels as low as 100 ppb for a period of 25 days (Sklan *et al.*, 2001).

Mycotoxins in forages

STACHYBOTRYOTOXICOSIS

One of the earliest recognized mycotoxin diseases was **stachybotryotoxicosis**. In eastern Europe and South Africa the mould *Stachybotrys* has been isolated from forage sources and straw and implicated in stachybotryotocosis in horses, cattle, swine and sheep. *Stachybotrys* has also been incriminated in domestic housing in the US as a source of 'sick house syndrome'. This fungus has been shown to produce macrocyclic trichothecene mycotoxins such as satratoxins F, G and H, verrucarin J and roridin E (Eppley, 1977; Harrach *et al.*, 1981) (Figure 1). *Stachybotrys chartarum* (formerly *S. atra*) is a fungus found in the soil and in moist dark indoor environments. This fungus produces black, soot-like spores in hay and straw and has been found to grow extensively on moist drywall in houses.

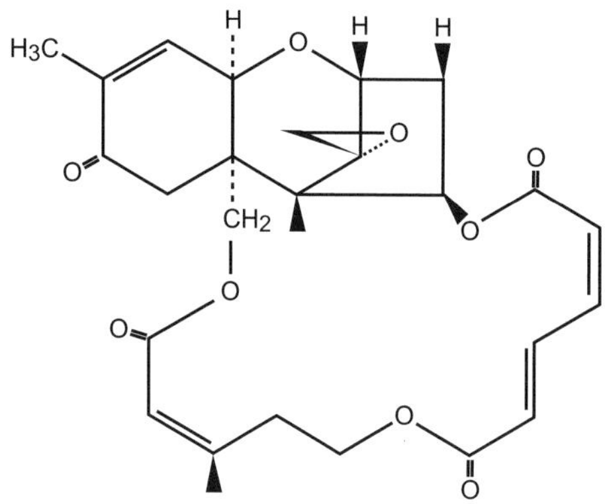

Verrucarin J

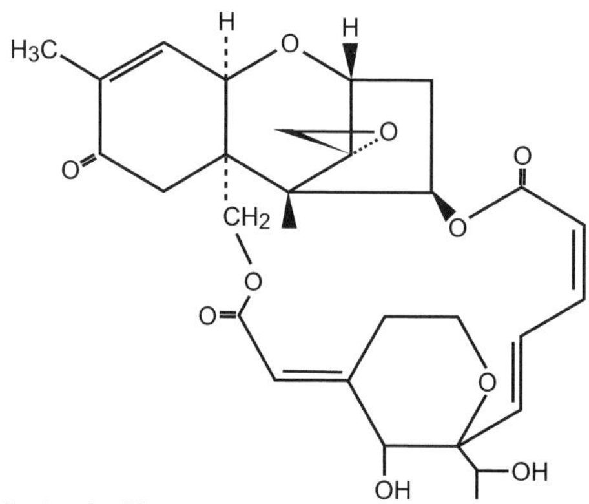

Satratoxin H

Figure 1. Structures of two macrocyclic trichothecene mycotoxins produced by *Stachybotrys* spp.

In the early 1930s, the horse was vital to the economy of Russia. In the Ukraine in 1931 thousands of horses were afflicted with an unknown ailment and a team of scientists was sent to investigate. By 1938 it was discovered that the disease was caused by *Stachybotrys alternans* (also now called *S. chartarum*) growing on the hay fed to the animals (Rodricks and Eppley, 1974). In the summer of 1983 and spring of 1984 three stables in Hungary experienced an outbreak of fungal toxicosis where nearly all of the 300 horses at these stables presented with drying and cracking of the facial mucosae, purulent nasal discharge and epistaxis. Postmortem examination of two of the affected animals showed extensive petechiation of the internal tissue. Bedding from these stables was found to contain satratoxin-G (20 to 35-μg/kg), satratoxin-H (75 to 107 μg/kg), and verrucarin-J (35 to 61 μg/kg). Samples of oats fed to horses during the 1984 outbreak were found to contain T-2 toxin, suggesting a synergism among toxins contributing to the intensity of the outbreak (Harrach *et al.*, 1987). When low levels of toxic feed are consumed, fissures at the corner of the mouth are followed by deeper lesions. Traditionally, the first cases of this disease appear in the autumn when animals are stabled and inhale or consume fodder containing the toxin. There is a subsequent increase in affected animals as the winter months continue into April when the toxicosis begins to decline as horses are returned to pasture and stall confinement is diminished (Forgacs, 1972). The toxins are nonantigenic therefore no immunity seems to develop to stachybotryotoxicosis and the same horse can become affected several times in the same year. Foals nursing affected mares do not show signs of illness, indicating the toxins are not excreted in milk.

Stachybotryotoxicosis may present itself in either of two forms; the typical form that most commonly occurs in nature and the atypical (shocking) form. The typical form develops progressively with three stages of onset.

Stage 1 begins with the aforementioned stomatitis within the first 2 to 10 days of exposure. The fissures of the mouth become necrotic. Edema of the underlying tissues leads to swelling of the lower and upper lips and perhaps the entire head in what has been described as a "hippopotamus-like" appearance (Forgacs, 1972). In many cases salivation, conjunctivitis, rhinitis, and enlarged lymph nodes develop in the next 120 hrs. In other cases these symptoms are absent and the only specific abnormality may be an increase in body temperature of 1-1.5°C accompanied by neutrophilia for a brief period. Typically, Stage 1 lasts from 8-12 days but may last for as long as a month.

Stage 2 is characterised by changes in blood profiles. The total leukocyte count may drop to 1000-3000 cells/mm^3 of whole blood with a concomitant decrease in the relative neutrophil count. Blood clotting time also increases or clotting fails to occur. Body temperature, respiration, and cardiac function remain in the normal range, but in rare cases it may be possible for the clinician to detect atony of the intestinal tract. The second stage duration varies from 5 to 20 days, but may last as long as 50 days.

In Stage 3 thrombocytopenia and leucopenia is exacerbated. Leukocyte counts may drop as low as 100/mm^3 and a complete loss of blood coagulation is observed. Hyperthermia is also a distinct characteristic of Stage 3 with the temperature rising to between 40 and 41.5°C. Affected horses may show signs of depression, poor or no appetite, problems swallowing and colic. This stage is usually quite short (1-6 days) and almost always is fatal.

The atypical form of stachybotrytoxicosis develops after ingestion of a large amount of toxin. In this form nervous disorders are the main visible symptoms. The animal will avoid movement and stand with legs in a wide gait or may cross the front legs or lean against objects for support. Swallowing is quite difficult, the pulse becomes weak and arrhythmic (80-100 beats/min), pulmonary edema and cyanosis of the mucous membranes is observed. The animal may become blind and death is due to respiratory failure usually within 72 hrs of ingestion.

FESCUE TOXINS

In 1993, Hoveland estimated that approximately 688,000 horses in the US graze tall fescue. Bacon and coworkers (1977) first reported the presence of the endophytic fungus now called *Neotyphodium coenophialum* in tall fescue. The presence of this fungus in consumed grasses has been associated with reduced weight gain, rough hair coats, higher body temperatures, reproductive problems and necrosis of the foot, tail and ears of horses and cattle. Increases in the number of cycles for conception and early embryonic death have also been associated with endophyte-infected fescue (Brendemuehl *et al.*, 1994). In addition, increased length of gestation, retained and so-called 'red bag' placentas, higher numbers of weak or stillborn foals, and agalactic mares were associated with endophyte-infected fescue consumption compared to mares consuming endophyte-free fescue (Monroe *et al.*, 1988).

Ergovaline is one of the primary ergot alkaloids produced by *N. coenophialum* - which exerts its activity by acting as a dopamine agonist and suppressing prolactin secretion (Strickland *et al*, 1994). Decreased serum prolactin is one of the most consistently reported symptoms of fescue toxicosis (Neal and Schmidt, 1985; Elsasser and Bolt, 1987; Redmond *et al.*, 1994). Ergotamine and ergonovine have been shown to be arterio- and vasoconstrictive agents that play a role in thermoregulation since vasoconstriction prevents dissipation of heat by slowing the transfer of internal body heat to the skin or respiratory tract (Abney *et al.*, 1993). This is of primary importance in warm climates and in the equine athlete. Trials examining the effects of endophyte-infected fescue on exercising horses found that recovery heart and respiratory rates (post-exercise) were significantly greater than for horses grazing non-infected pasture. Water consumption post-exercise was also found to be greater for horses consuming endophyte (9.45 gallons vs 5.59 gallons, $P<0.05$; Vivrette *et al.*, 2001). The minimum amount of exposure required to cause these effects and the reversibility is not clear. Symptoms in this trial were observed following 14 days of grazing endophyte-infected fescue. In cattle, the adverse effects were reversed following 8 days of grazing endophyte-free pastures (Rhodes *et al.*, 1991). Although endophyte-free fescue is currently available, it is less hardy, and less resistant to overgrazing, insect damage and drought than infected grass. New fescue products with so-called 'friendly endophyte' show some promising results, but the overall effect on the horse has not yet been defined.

RYEGRASS TOXINS

Perennial ryegrass intoxication or **ryegrass staggers** is a neurotoxic syndrome characterized by ataxia, lack of coordination, head shaking, and collapse. Animals ingesting this toxin appear normal until disturbed. The neurological effects are temporary, but lack of coordination can lead to drowning, and running into and through barns and fences. The causative agents are compounds called **tremorgens**. A number of tremorgens have been identified with the most important being lolitrem B, which is produced by the endophytic fungus *Neotyphodium (Acremonium) lolii*. Concentrations of lolitrem B are lowest in the leaf blades and highest in the leaf sheath, which tends to make ryegrass staggers most often noted in over-grazed pastures (DiMenna *et al.*, 1992). Symptoms of ryegrass staggers appear when the lolitrem B concentrations exceed 1.5-2.5 ppm. As is seen with fescue, the endophyte improves the vigor of the ryegrass.

SLAFRAMINE

The ubiquitous soil fungus *Rhizoctonia leguminicola* can infect certain legumes following heavy rainfall or in situations of high humidity. Although associated mostly with red clover (*Trifolium pratense*), white clover, alsike clover, kudzu, alfalfa, Korean lespedeza, cow pea, soybean, black medic and blue lupine have also been reported to be susceptible to infection by this fungus (Hagler and Behlow, 1981). Examination of the leaves of the legumes will reveal bronze to black spots or concentric rings ranging in size from pinhead to pencil eraser in diameter. *R. leguminicola* produces slaframine, an alkaloid that, following ingestion is activated by the liver yielding a cholinergic parasympathomimetic agent (Figure 2) (Guengerich and Aust, 1977). Toxicosis is manifested by profuse salivation ('slobbers') accompanied by increased water consumption. Bloating, piloerection, diarrhoea, agalactia, feed refusal, abortions and death in severe cases are also occasionally observed (Sockett *et al.*, 1982). Slobber-response tests in guinea pigs indicate that levels of 50-100 ppm in clover will elicit a response. The alkaloid is somewhat stable in that it may still be present in baled hay from infected fields but will degrade over time. Studies have shown a decrease in slaframine concentrations from 50-100 ppm to approximately 7 ppm following 10 months of storage (Hagler and Behlow, 1981). Removal of animals from affected pastures to non-infected areas or feeding grass hay will return the animals to normal in 3 to 4 days.

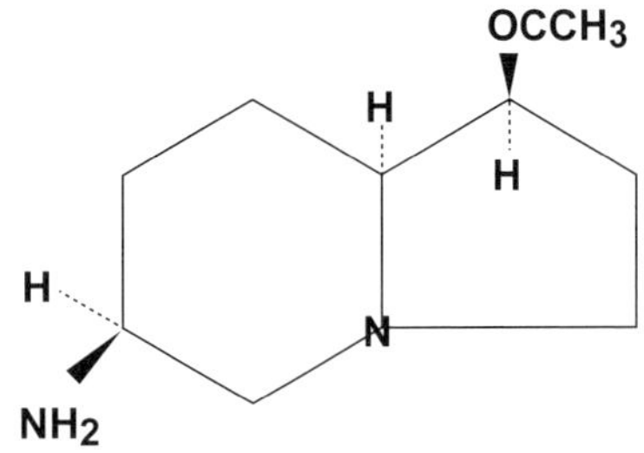

Figure 2. Slaframine.

OTHER TOXINS FROM FORAGE SOURCES

PR toxin produced by *P. roquefortii* has been found in forage sources and was the suspected toxic agent in a case study with symptoms of abortion in cattle (Figure 3) (Still *et al.*, 1972). In addition, fumigaclavine produced by *Aspergillus fumigatus*, and *Alternaria* spp. have been implicated with abortion and other health problems (Cole *et al.*, 1977; Pitt and Hocking, 1997).

As yet undiscovered mycotoxins may also hold the key to understanding mycotoxin effects on animal performance. It is well documented that the toxic effects of many mycotoxins are enhanced by the presence of more than one toxin. As more mycotoxins

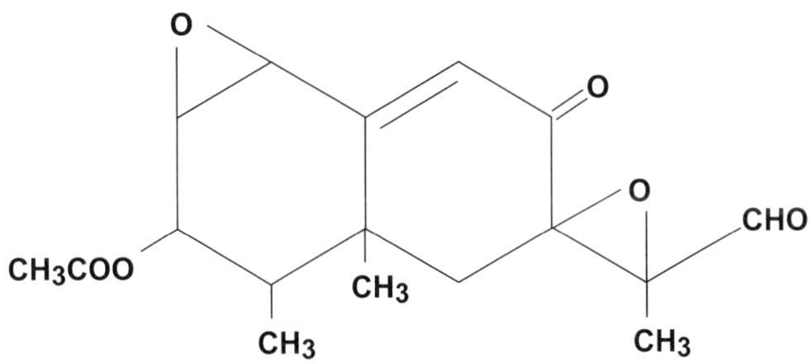

Figure 3. PR toxin.

are discovered, it may come to pass that the above mentioned toxins only serve as markers or indicators of other unknown toxins.

Mycotoxins and colic

Barnett *et al.* (1995) studied the relationship between mycotoxins and equine colic. They analysed feed samples from farms experiencing possible feed-related colics (n = 16) and control farms (n = 10) as indicated by the veterinarian. They found DON in the concentrate of 100% of colic cases (range 0.20 to 8.3 ppm) and 70% of the concentrate fed on control farms (n = 10) (range 0-2.5 ppm). T-2 toxin at levels greater than 0.5 ppm and zearalenone greater than 0.7 ppm were present in 31% and 44% of the colic concentrate samples, respectively; while neither were found in control samples. Forage samples were not provided by all farms. Hence the causal-effect relationship of DON and colic in horses is not clear.

Raymond *et al.* (2001) reported that higher levels of DON, ZEA and T-2 toxin were found in forages as compared to concentrates during a field study of performance horse farms in Ontario, Canada (Figures 4 and 5). Increased levels of mycotoxins in forages and feed were associated with increased mold contamination in forages and feed. Mycotoxins, like mould, are not uniformly distributed throughout a batch of feed or grains but occur as pockets of contamination. The concentrates sampled did not seem to contain significant levels. It is important to emphasise that in these studies commercial Enzyme Linked Immunosorbent Assays (ELISA) test kits were used for the analysis. These kits are not validated for the complex matrices found in forage samples. ELISA procedures are prone to false positive results on non-validated matrices and should be confirmed using HPLC and/or TLC testing.

Mycotoxins and mare reproductive loss syndrome

In late April and throughout the spring of 2001 what appears to have been a series of catastrophic events resulted in abortions, and heart and eye problems to the horse industry of Kentucky, West Virginia and southern

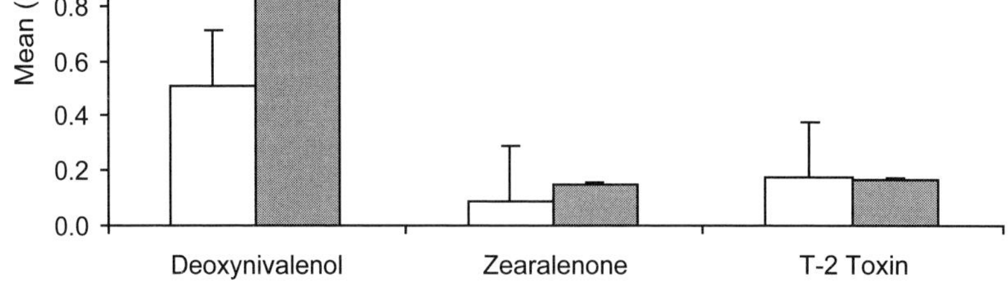

Figure 4. Mycotoxin levels found in forage and concentrate samples (from Raymond *et al.*, 2001).

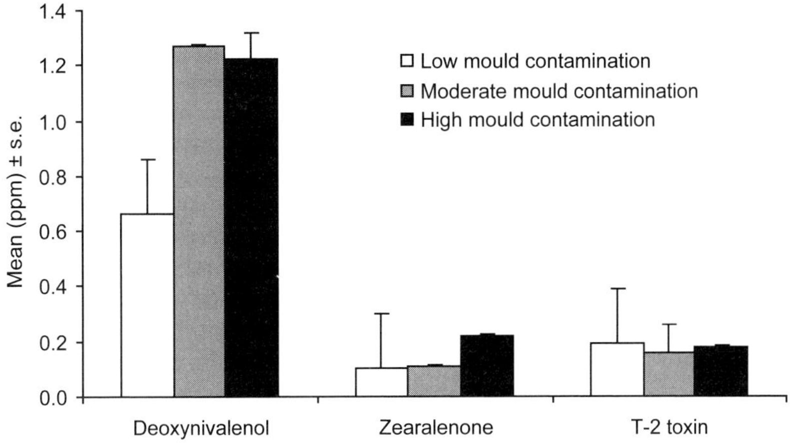

Figure 5. Mycotoxin levels found in forage and feed samples with varying levels of mould contamination.

Ohio. The episode was termed **'Mare Reproductive Loss Syndrome' (MRLS)** due to the number of early and late-term aborted foetuses. Estimates of losses from this syndrome range as high as 14,000 foals. A University of Louisville study estimates that about 31% of the 2002 thoroughbred crop and 20% of the standardbred crop for the following year were lost from this syndrome although all breeds of horses were affected. The economic impact of MRLS has been speculated to be over $336 million.

The signs observed in MRLS were quite unique and led primarily to drawing similarities to other known processes causing abortion and looking for common factors among horses exhibiting problems. Although reports indicated that abortion was by far the most consistent symptom, in the case of MRLS the aborted early foetuses were observed almost exclusively in mares 40 to 100 days into pregnancy. This may have been an artifact of the dominant stage of mare pregnancy at the time. In these mares, abortion was preceded by changes in the foetal fluid, then foetal death followed by expulsion of the dead foetus. According to veterinarians, abnormal echogenic fluid (cloudy and flocculent), was prevalent with a mucoid discharge from the mare associated with the retained dead foetus and foetal membranes. In some cases lethargy and loss of appetite were noted with no manure passage for 18 hrs. Late foetal losses were noticed in mares that were near term – about three weeks prior to expected parturition date. Gross and histopathological lesions in late foetal loss indicated a relatively consistent pattern in the foetus and amniotic sac of the placenta. The lungs of the foetus showed evidence of intrauterine insult and inflammation, as did the amniotic sac and umbilical cord. Foals born alive were quite weak and frequently required intensive veterinary care and in many cases suffered from respiratory distress. Toxicological results on foetuses were negative for nitrates and nitrites and serum levels of copper, iron, zinc and selenium in the mare were normal. Virological tests for equine herpesvirus, equine arteritis virus and adenovirus as well as tests to discover an as yet unknown virus also proved to be negative.

At first, the entire abortion episode seemed to be quite clear. Mares were aborting and a number of forage samples were submitted for analysis and found to contain the mycotoxin zearalenone. As was mentioned earlier, zearalenone has oestrogenic properties although it is not chemically an oestrogen. For this reason, females are more susceptible than males. In pigs, a swollen, red vulva, which may lead to rectal and vaginal prolapse, is normally only seen in prepubertal gilts. In MRLS, maiden and previously barren mares seemed to be more susceptible to the syndrome. A number of studies in other animal species have demonstrated the effects of zearalenone over extended periods of time, as compared to a more probable scenario under field conditions (and in examining MRLS) of dosing of toxin over a short period of time. Studies with 16 gilts given pure ZEA (108 mg on post-mating days 2 to 6, 7 to 10 or 11 to 15) showed that gilts given ZEA on post-mating days 7-10 had reduced embryonic survival compared to control or other treatment groups (1/4 pregnant in the 7-10 day group - 4/4 pregnant in control and other treatment groups). FSH and estradiol-17ß concentrations were unaffected by ZEA consumption. Serum concentrations of prolactin in the 7-10 and 11-15 day group were lower than controls or the 2-6 post-mating group (Long and Diekman, 1986). Because of the critical effects of ZEA on stage of pregnancy, it was thought that this toxin may have played a critical role in the events described in the spring of 2001. Indeed, the initial theory for the cause of MRLS was that the unusual weather patterns of the spring of 2001 led to an imbalance of the soil and forage ecosystem that led to a bloom of mould and the sudden frost caused mycotoxin production (Schindler, 1977; Newman, 2002a). The positive results for ZEA in the forage were obtained by commercial ELISA kits and could not be confirmed by HPLC or other methods. Indeed, an examination of split samples tested by ELISA and HPLC determined that ZEA levels as high as 1.5 ppm detected by ELISA could not be confirmed by HPLC.

As diligent work by the task force assembled to examine this syndrome continued; it was observed that a high percentage of affected fields or farms had black cherry trees (*Prunas serotina*) and these trees and surrounding areas were heavily infested with eastern tent caterpillars (ETC). Recent studies indicate that the abortions associated with MRLS are not linked to mycotoxins, but rather are associated with setae (hairs) from the caterpillar (unpublished data). However, the concurrent problems of pericarditis (fluid accumulation in the sac surrounding the heart), and uveitis (blindness in one eye) may still prove to be associated with mycotoxicosis. The heavy infestation of ETC led to a large amount of ETC frass (excrement) in fields where caterpillars were found and studies have shown that this frass supports luxuriant growth of *Aspergillus, Fusarium*, and *Penicillium* spp. (Newman, 2002b).

Human health hazards

An important consideration is also the human health risk associated with handling mycotoxin-contaminated materials. The allergic condition known as 'farmer's lung' has been well established to be caused by inhaling dust from mouldy hay and straw (Gregory and Lacey, 1963; Blyth, 1973). Airborne particles are generated as grains, forages and beddings are harvested, transported, processed, stored and used. Agricultural workers can be exposed to potential hazards as a result of inhalation of mycotoxin-contaminated dust. The human toxic effects associated with *Fusarium* mould and mycotoxins may include allergies, skin irritation, poor appetite, headache, vomiting, gastric and intestinal irritation, haemorrhaging, increased sensitivity to disease and reproductive problems. As a result, workers are advised to take safety precautions to reduce dust exposure and dermal contact when handling agricultural products.

Summary

Mycotoxins are best known for their ability to cause acute, severe disease in horses. However, the impact of mycotoxins is much wider. Subclinical problems such as loss of performance and decreased reproductive capacity are non-specific disorders that can be linked with mycotoxin ingestion. Mycotoxins may be unavoidable contaminants in feeds and forages due to the unpredictability of pre-harvest contamination of susceptible crops by fungi. Various genera of fungi can produce mycotoxins whenever optimal conditions of temperature, humidity and suitable substrate prevail in a given area. It can be argued that given the complex interactions of mycotoxins with feed ingredients, varying exposure situations and the state of the individual horse, safe levels cannot be identified and that no levels of mycotoxins can be demonstrated to be safe in a field situation. Levels can be reduced with the testing of feeds and forages, increased awareness, proper management and possibly the use of compounds that have been shown to have adsorptive capabilities.

Possible adverse effects in the horse due to mycotoxin exposure:

- Decreased feed intake
- Immunosuppression (decreased antibody titre values)
- Increased susceptibility to disease
- Damage to organs (liver, kidney, reproductive organs, etc.)
- Poor reproductive performance
- Excessive salivation
- Incoordination

Safe levels for horses and reducing risk:

- The safest amount of mycotoxins in horse feeds is zero, but realistically many feeds contain small amounts of various toxins. Low levels of mycotoxins can be found in numerous feed ingredients.
- Whether the amount harms the horse depends on the type of mycotoxin, what other types are present, the feeding programme and health status of the horse and how long the horse was exposed to the contaminated feeds.
- One of the first signs that a feed may contain mycotoxins may be reduced feed consumption.
- Feed, forage and bedding products known to be contaminated by mould should not be used for horses.
- During questionable growing seasons (i.e. damp/cold) feeds should be chosen carefully and tested if suspect.
- Mycotoxin contamination in feeds, forages and bedding are not always obvious by visual inspection.
- Moisture content is the single most important factor in determining if and to what extent mould will grow in feeds. This moisture could come from the feed ingredient, harvesting or processing methods or time of harvest and the storage environment. All sources of moisture must be controlled to reduce risk.

References

Abney, L.K., J.W. Oliver, and C.R. Reinemeyer. 1993. Vasoconstrictive effects of tall fescue alkaloids on equine vasculature. J. Equine Vet. Sci. 13:334-340.

Angsubhakorn, S., P. Poomvises, K. Romruen and P.M. Newberne. 1981. Aflatoxicosis in horses. J. Am. Vet. Med. Assoc. 178:274-278.

Asquith, R.L. 1985. Equine aflatoxicosis: A brief review. In: *Trichothecenes and Other Mycotoxins* (J.L. Lacey, ed). John Wiley and Sons. New York. pp. 487-494.

Bacon, C.W., J.K. Porter, J.D. Robbins, and E.S. Luttrell. 1977. *Epichloe typhina* from toxic tall fescue grasses. Appl. Enivron. Microbiol. 34:576-581.

Barnett, D.T., R.A. Mowery, W.M. Hagler, D.G. Bristol, and R.A. Mansmann. 1995. The correlation of selected mycotoxins to the incidence of colic in horses. In: *Proc. 14th Equine Nutrition and Physiology Symposium*. 242-247.

Blyth, W. 1973. Farmer's Lung Disease. In: *Actinomycetales: Characteristics and Practical Importance* (G. Sykes and F.A.

Skinner, eds). Society for Applied Bacteriology Symposium Series No. 2; Academic Press, London. 261-276.

Bortell, R., R.L. Asquith, G.T. Edds, C.F. Simpson, and W.W. Aller. 1983. Acute experimentally induced aflatoxicosis in the weanling pony. Am. J. Vet. Res. 44:2110-2114.

Brendemuehl, J.P., T.R. Boosinger, D.G. Pugh and R.A. Shelby. 1994. Influence of endophyte-infected tall fescue on cyclicity, pregnancy rate and early embryonic loss in the mare. Theriogen. 42:489-500.

Carson, M.S. and T.K. Smith. 1983. Effect of feeding alfalfa and refined plant fibers on the toxicity and metabolism of T-2 toxin in rats. J. Nutr. 113:304-313.

Chaouloff, F.D., D. Laude, D. Merino, B. Serrurrier, and J.L. Elghoji. 1986. Peripheral and central short-term effects of fusaric acid, a DBH inhibitor, on tryptophan and serotonin metabolism in the rat. J. Neural Transm. 65:219:232.

Cole, R.J., J.W. Kirksley, J.W. Dorner, D.M. Wilson, J.C. Johnson, A.N. Johnson, D.M. Bedell, J.P. Springer, K.K. Chexal, J.C. Clardy, and R.H. Cox. 1977. Mycotoxins produced by *Aspergillus fumigatus* species isolated from molded silage. J. Agric. Food Chem. 25:826-830.

Cote, L.M., V.R. Beasley, P.M. Bratich, S.P. Swanson, H.L. Shivaprasad, and W.B. Buck. 1985. Sex-related reduced weight gains in growing swine fed diets containing deoxynivalenol. J. Anim. Sci. 61:942-950.

Cysewski, S.J., A.C. Pier, A.L. Baetz, and N.F. Cheville. 1982. Experimental equine aflatoxicosis. Tox. and Appl. Pharmacol. 65:354-365.

DiMenna, M.E., P.H. Mortimer, R.A. Prestige, A.D. Hawkes, and J.M. Sprosen. 1992. Lolitrem B concentrations, counts of *Acremonium lolii* hyphae, and the incidence of ryegrass staggers in lambs on plots of *A. lolii* infected perennial ryegrass. N.Z.J. Agric. Res. 35:211-220.

Di Paolo, N., A. Guarnieri, F. Loi, G. Sacchi, A.M. Mangiarotti and M. Di Paolo. 1993. Acute renal failure from inhalation of mycotoxins. Nephron. 64:621-625.

D'Mello, J.P.F., C.M. Placinta, and A.M.C. Macdonald. 1999. *Fusarium* mycotoxins: a review of global implications for animal health, welfare and productivity. Anim. Feed Sci. Technol. 80:183-205.

Elsasser, T.H. and D.H. Bolt. 1987. Dopaminergic-like activity in toxic fescue alters prolactin but not growth hormone or thyroid stimulating hormone in ewes. Dom. Anim. Endocrinol. 4:259-269.

Eppley, R.M. 1977. Chemistry of stachybotryotoxicosis. In: *Mycotoxins in Human and Animal Health* (J.V. Rodricks, W. Hesseltine and M.A. Mehlman, eds). Pathotox Publishers, Inc. Park Forest South, Ill.

Forgacs, J. 1972. Stachybotryotoxicosis. In: *Microbial Toxins*, Vol. VIII (S.K. Adis, A. Ciegler, and S.J. Aji, eds). Academic Press. New York. Pp. 95-126.

Gimeno, A. and J.A. Quintanilla. 1983. Analytical and mycological study of a natural outbreak of zearalenone mycotoxicosis in horses. Proc. Int. Symp. Mycotoxins: 387-392.

Greenhalgh, R., G.A. Neish and J.D. Miller. 1983. Deoxynivalenol, acetyldeoxynivalenol, and zearalenone formation by Canadian isolates of *Fusarium graminearum* on solid substrates. Appl. Environ. Microbiol. 46: 625.

Gregory, P.H., M.E. Lacey, G.N. Festenstein, and F.A. Skinner. 1963. Microbial and biochemical changes during the molding of hay. J. Gen. Microbiol. 33: 147-174.

Guengerich, F.P. and S.O. Aust. 1977. Activation of the parasympathomimetic alkaloid slaframine by microsomal and phytochemical oxidation. Mol. Pharmacol. 13:185-195.

Hagler, W.M. and R.F. Behlow. 1981. Salivary syndrome in horses: identification of slaframine in red clover hay. Appl. Environ. Microbiol. 42:1067-1073.

Halliwell, R.E.W., B.C. McGorum, P. Irving,

and P.M. Dixon. 1993. Local and systemic antibody production in horses affected with chronic obstructive pulmonary disease. Vet. Immunol. Immunopath. 38:201-215.

Harrach, B., C.J. Mirocha, S.V. Pathre, and M. Palyusik. 1981. Macrocyclic trichothecene toxins produced by a strain of *Stachybotrys atra* from Hungary. Appl. Environ. Microbiol. 41:1428-1432.

Harrach, B., A. Batta, G. Sandor, and A. Vanyi. 1987. Isolation of macrocyclic and non-macrocyclic trichothecenes (*Stachybotrys* and *Fusarium* toxins) from the environment of 200 ill sport horses. Mycotoxin Res. 3:65-68.

Harrison, L.R., B.B. Colvin, J. Todd Greene, L. E. Newman, and J. R. Cole. 1990. Pulmonary edema and hydrothorax in swine produced by fumonisin B1, a toxic metabolite of *Fusarium moniliforme*. J. Vet. Diagn. Invest. 2:217-221.

Haschek, W.M., L.A. Gumprecht, G. Smith, M.E. Tumbleson, and P.D. Constable. 2001. Fumonisin toxicosis in swine: An overview of porcine pulmonary edema and current perspectives. Eviron. Health Perspect. 109:251-257.

Hohler, D., K.h. Sudekum, S. Wolffram, A.A. Frohlich, and R.R. Marquardt. 1999. Metabolism and excretion of ochratoxin A fed to sheep. J. Anim. Sci. 77:1217-1223.

Hoveland, C.S. 1993. Importance and economic significance of the *Acremonium* endophytes to performance of animals and grass plants. Agric. Ecosystems Environ. 44:3-12.

Hult, K., A. Teiling, and S. Gatenbeck. 1976. Degradation of ochratoxin A by a ruminant. Appl. Environ. Microbiol. 32:443-444.

James, L. J. and T. K. Smith. 1982. Effect of dietary alfalfa on zearalenone toxicity and metabolism in rats and swine. J. Anim. Sci. 55:110-118.

Johnson, P.J., S.W. Casteel, and N.T. Messer. 1997. Effect of feeding deoxynivalenol (vomitoxin)-contaminated barley to horses. J. Vet. Diagn. Invest. 9:219-221.

Juhasz, J., P. Nagy, G. Huszenicza, G. Szigeti, J. Reiczigel, and M. Kulesar. 1997. Long term exposure to T-2 fusarium mycotoxin fails to alter luteal function, follicular activity and embryo recovery in mares. Equine Vet. J. Suppl. 25:17-21.

Juhasz, J., P. Nagy, M. Kulesar, G. Szigeti, J. Reiczigel, and G. Huszenicza. 2001. Effect of low-dose zearalenone exposure on luteal function, follicular activity and uterine oedema in cycling mares. Acta Veterinaria Hungarica. 49(2):211-222.

Kalleta, K. and E. Ettala. 1981. The oestrogenic *Fusarium* toxin (zearalenone) in hay as a cause of early abortions in the cow. Nord. Vet. Med. 36:305-309.

Kumagai, S. and K. Aibara. 1982. Intestinal absorption and secretion of ochratoxin A in the rat. Toxicol. Appl. Pharmacol. 64:94-102.

Larsson, P., E. Persson, E. Tyden and H. Tjalve. 2003. Cell-specific activation of aflatoxin B1 correlates with presence of some cytochrome P450 enzymes in olfactory and respiratory tissues in horse. Res. Vet. Sci. 74:227-233.

Leeson, S., G. Diaz, and J.D. Summers. 1995. Ochratoxins. In: *Poultry Metabolic Disorders and Mycotoxins*. University Books. Ontario, Canada.

Long, G.G. amd M.A. Diekman. 1986. Characterization of effects of zearalenone in swine during early pregnancy. Am. J. Vet. Res. 47:184-187.

Meloche, J.L., and T.K. Smith. 1995. Altered tissue amino acid metabolism in acute T-2 toxicosis. Proc. Soc. Exp. Biol. Med. 210:260-266.

Monroe, J.L., D.L. Cross, L.W. Hudson, D.M. Hendricks, S.W. Kennedy, and C.W. Bridges, Jr. 1988. Effect of selenium and endophyte-contaminated fescue on performance and reproduction in mares. J. Equine Vet. Sci. 8:148-153.

Neal, W.D. and S.P. Schmidt. 1985. Effects of feeding Kentucky 31 tall fescue seed infected with *Acremonium coenophialum* to laboratory rats. J. Anim. Sci. 61:603-611.

Newman, K.E. 2002a. Searching for the cause of Mare Reproductive Loss Syndrome: Kentucky's $300 million disaster. In: *Nutritional Biotechnology in the Food and Feed Industries*. Proceedings of the 18th Annual Symposium (T.P. Lyons and K.A. Jacques, eds.) Nottingham University Press, UK. pp. 351-357.

Newman, K.E. 2002b. Review of mycotoxins as a possible cause of mare reproductive loss syndrome. Proceedings of the First Workshop on Mare Reproductive Loss Syndrome (D.G. Powell, A. Troppman and T. Tobin, eds). Kentucky Agricultural Experiment Station, University of Kentucky, Lexington, KY. Pp. 66-68.

Northolt, M.D., C.A.H. Verhulsdonk, and P.S.S. Soentoro. 1976. Effect of water activity and temperature on aflatoxin production by *Aspergillus parasiticus*. J. Milk Food Technol. 39:170-174.

Palmgren, M.S. and L.S. Lee. 1986. Separation of mycotoxin-containing sources in grain dust and determination of their mycotoxin potential. Environ. Health Perspect. 66:105-108.

Pier, A.C., S.J. Cysewski, J.L. Richard, A.L. Baetz, and L. Mitchell. 1976. Experimental mycotoxicoses in calves with aflatoxin, ochratoxin, rubratoxin and T-2 toxin. In: Proc. U.S. Amer. Health Assoc., 80th Annual Meeting, pp.130-148.

Pitt, J.L. and A.D. Hocking. 1997. *Fungi and Food Spoilage* (2nd Ed). Blackie Academic and Professional. New York.

Porter, J.K., K.A. Voss, C.W. Bacon, and W.P. Norred. 1990. Effects of *Fusarium moniliforme* and corn associated with equine leukoencephalomalacia on rat neurotransmitters and metabolites. Proc. Soc. Exper. Biol. Med. 194:265-269.

Porter, J.K., K.A. Voss, W. J. Chamberlain , C.W. Bacon, and W.P. Norred. 1993. Neurotransmitters in rats fed fumonisin B1. Proc. Soc. Exper. Biol. Med. 202:360-364.

Quinn, B.A., T.L. Crane T.E. Kocal, S.J. Best, R.G. Cameron, T.H. Rushmore, E. Farber, and M.A. Hayes. 1990. Protective activity of different hepatic cytosolic glutathione-S-transferases against DNA-binding metabolites of aflatoxin B1. Toxicol. Appl. Pharmacol. 105:351-363.

Raymond, S.L., M. Heiskanen, T.K. Smith, M. Reiman, S. Laitinen, and A.F. Clarke. 2000. An investigation of the concentrations of selected *Fusarium* mycotoxins and the degree of mold contamination of field-dried hay. J. Equine Vet. Sci. 20(10):616-621.

Raymond, S.L., T.K. Smith, and H.V.L.N. Swamy. 2003a. Effects of feeding a blend of grains naturally contaminated with *Fusarium* mycotoxins on feed intake, serum chemistry, and hematology of horses, and the efficacy of a polymeric glucomannan mycotoxin adsorbent. J. Anim. Sci. 81:2123-2130.

Raymond, S.L., T.K. Smith, and H.V.L.N. Swamy. 2003b. Effects of feeding a blend of grains naturally contaminated with *Fusarium* mycotoxins on feed intake and indices of athletic performance of horses. J. Anim. Sci. 81(Suppl. 1):263-264.

Raymond S.L., Smith T.K., Curtis E.F. and A.F. Clarke. 2001. An investigation of the levels of selected *Fusarium* mycotoxins and the degree of mold contamination found in the diets of performance horses in Ontario. In: *Equine Nutrition and Physiology Society Symposium*, Lexington, KY, USA, May 30-June 2, 86-92.

Redmond, L.M., D.L. Cross, J.R. Strickland, and S.W. Kennedy. 1994. Efficacy of domperidone and sulpiride for fescue toxicosis in horses. Am. J. Vet. Res. 55:722-729.

Ribelin, W.E., K. Fukushima, and P.E. Still. 1978. The toxicity of ochratoxin to ruminants. Can. J. Comp. Med. 42:172-176.

Rhodes, M.T., J.A. Patterson, M.S. Karley, H.E. Garner, and M.H. Laughlin. 1991. Reduced blood flow to peripheral and core body tissues in sheep and cattle induced by endophyte-infected tall fescue. J. Anim. Sci. 69:2033-2043.

Rodricks, J.V. and R.M. Eppley. 1974. *Stachybotrys* and stachybotryotoxicosis. In: *Mycotoxins*. (I.F. Purchase, ed). Elsevier, NY.

Ross, P.F., L.G. Rice, R.D. Plattner, G.D. Osweiler, T.M. Wilson, D.L. Owens, H.A. Nelson, and J.L. Richard. 1991. Concentrations of fumonisin B1 in feeds associated with animal health problems. Mycopathologia. 114:129-135.

Ross, P.F., L.G. Rice, G.D. Osweiller, P.E. Nelson, J.L. Richard, and T.M. Wilson. 1992. A review and update of animal toxicosis associated with fumonisin-contaminated feeds and production of fumonisins by *Fusarium* isolates. Mycopathologia. 117:109-114.

Ross, P.F., P.E. Nelson, D.L. Owens, L.G. Rice, H.A. Nelson, and T.M. Wilson. 1994. Fumonisin B2 in cultured *Fusarium proliferatum*, M-6104, causes equine leukoencephalomalacia. J. Vet. Diagn. Invest. 6:263-265.

Rotter, B.A., B.K. Thompson, M. Lessard, H.L. Trenholm, and H. Tryponas. 1994. Influence of low-level exposure to *Fusarium* mycotoxins on selected immunological and hematological parameters in young swine. Fundam. Appl. Toxicol. 23:117-124.

Schindler, A.F. 1977. Temperature limits for production of aflatoxin by 25 isolates of *Aspergillus flavus* and *Aspergillus parasiticus*. J. Food Protect. 40:39-40.

Sklan, D., E. Klipper, and A. Friedman. 2001. The effect of chronic feeding of DAS, T-2 toxin and aflatoxin on performance, health and antibody production in chicks. J. Appl. Poultry Research 10:79-85.

Smith, G.W., P.D. Constable, J.H. Foreman, R.M. Eppley, A.L. Waggoner, M.E. Tumbleson, and W.M. Haschek. 2002. Cardiovascular changes associated with intravenous administration of fumonisin B1 in horses. Am. J. Vet. Res. 63:538-545.

Smith, J.E. and R.S. Henderson. 1991. *Mycotoxins and animal foods*. CRC Press, Inc., Boca Raton, Florida.

Smith, T.K. and E.J. MacDonald. 1991. Effect of fusaric acid on brain regional neurochemistry and vomiting behavior in swine. J. Anim. Sci. 69:2044-2049.

Smith, T.K., E.G. McMillan, and J.B. Castillo. 1997. Effect of feeding blends of *Fusarium* mycotoxin-contaminated grains containing deoxynivalenol and fusaric acid on growth and feed consumption of immature swine. J. Anim. Sci. 75:2184-2191.

Sockett, D.C., J.C. Baker, and C.M. Stowe. 1982. Slaframine (*Rhizoctonia leguminicola*) intoxication in horses. 181:606.

Still, P.E., R.D. Wei, E.B. Smalley, and F.M. Strong. 1972. A mycotoxin from *Penicillium roqueforti* isolated from toxic cattle feed. Fedn. Proc. Fedn. Am. Socs. Exp. Biol. 31:733.

Strickland, J.R., D.L. Cross, G.P. Birrenkott, and L.W. Grimes. 1994. Effect of ergovaline, loline, and dopamine antagonists on rat pituitary cell prolactin release *in vitro*. Am. J. Vet. Res. 55:716-721.

Stoloff, L. and M.W. Trucksess. 1979. Distribution of aflatoxin B1 and M1 in contaminated calf and pig livers. J. Assoc. Off. Anal. Chem. 62:1361-1362.

Trenholm, H.L., B.C. Foster, L.L. Charmley, B.K., Thompson, K.E., Hartin, R.W. Coppock, and M.A. Albassam. 1994. Effects of feeding diets containing *Fusarium* (naturally) contaminated wheat or pure deoxynivalenol (DON) in growing pigs. Can. J. Anim. Sci. 74:361-369.

Van der Westhuizen, L., G.S. Shephard, and D.J. van Schalkwyk. 2001. The effect of a single gavage dose of fumonisin B1 on the sphinganine and sphingosine levels in vervet monkeys. Toxicon. 39:273-281.

Van der Merwe, K.J., P.S. Steyn, and L. Fourie, D.B. Scott, and J.J. Theron. 1965. Ochratoxin A, a toxic metabolite produced by *Aspergillus ochraceus*. Nature. 205:1112-1115.

Venturini M.C., M.A. Quiroga, M.A. Risso, C. Di Lorenzo, Y. Omata, L. Venturini, and H. Godoy. 1996. Mycotoxin T-2 and aflatoxin B_1 as immunosuppressors in mice chronically infected with *Toxoplasma gondii*. J. Comp. Path. 115: 229-237, 1996.

Vesonder, R., J. Haliburton, R. Stubblefield, W. Gilmore, and S. Peterson. 1991. *Aspergillus flavus* and aflatoxins B1, B2, and M1 in corn associated with equine death. Arch. Environ. Contam. Toxicol. 20:151-153.

Vivrette, S. M.E. Stebbins, O.Martin, K. Dooley, and D. Cross. 2001. Cardio-respiratory and thermoregulatory effects of endophyte-infected fescue in exercising horses. J. Eq. Vet. Sci. 21:65-67.

Walter, W., M.S. Allen, G.T. Edds, and R.L. Asquith. 1981. Effects of aflatoxins in young ponies. Amer. J. Vet. Res. 42:2162-2164.

Wannemacher, R.W. and R.E. Dinterman. 1983. Plasma amino acid changes in guinea pigs injected with T-2 mycotoxin. Fed. Proc. 42:625 (abstract).

Wood, G.E. 1992. Mycotoxins in foods and feeds in the United States. J. Anim. Sci. 70:3941-3949.

Zeyner, A., U. Fischer, and A. Lindner. 2002. Distinct weight loss and elevated hepatic enzymes in horses caused by straw contaminated with deoxynivalenol (case report). Proceedings of the Comparative Nutrition Society. Antwerp, Belgium.

4

EFFECTS OF MYCOTOXINS ON DOMESTIC PET SPECIES

Josef Böhm and Ebrahim Razzazi-Fazeli
Institute of Nutrition, Department of Veterinary Public Health, University of Veterinary Medicine, Vienna, Austria

Introduction

Domestic pets include all animals commonly cared for in households, such as cats, dogs, pet birds, rabbits, guinea pigs, and mink. Over the last decade, the importance of domestic pets in developed countries has grown dramatically. As an example from the EU, more than 28% of households in Austria keep cats, followed by dogs in 16%, caged birds in 5%, aquarium fish in 4%, rabbits in 2.5%, guinea pigs in 2%, and hamsters and turtles are kept in 1% of households. The adverse effects of fungal toxins have been well documented in many species, particularly farm animals. However, only a few papers address mycotoxicoses in pets. Puschner (2002) has reviewed the clinical and pathological effects of some mycotoxins on pet animals. In addition, Devegowda (2000) and Bird (2000) have reported on the relevance of mycotoxins in pet foods. This general lack of information makes it difficult to assess the relevance of mycotoxins to pet health.

In the cases of aflatoxins, penitrem A, roquefortine, and ochratoxins in dogs, descriptions of feed-induced natural mycotoxicoses have been reported. Experimentally, the effects of trichothecenes, patulin, penicillic acid, moniliformin, and cyclopiazonic acid have been studied in dogs, cats, rabbits, and mink. For the most toxic compound within the aflatoxin family, aflatoxin B_1, maximum regulatory limits in the EU have been defined as 10 μg/kg in complete feedstuffs (Council Directive 1999/29/EC). The US Food and Drug Administration has established only an action level (FDA Regulatory Guidance for Toxins and Contaminants) of 20 ppb for corn, peanut products, cottonseed meal, and other animal feeds and feed ingredients.

This chapter provides an overview of the toxicological effects of mycotoxins on domestic pets, as well as on the occurrence of mycotoxins in pet foods. Different mycotoxins and their relevance, as well as clinical signs of their mycotoxicosis, are described.

Aflatoxins

Aflatoxins are toxic secondary metabolites of *Aspergillus* spp., such as *A. parasiticus* and *A. flavus.* Aflatoxins have been shown to be the cause of toxicities in livestock, domestic animals, and humans throughout the world. Due to their high toxicity and carcinogenicity, aflatoxins have received greater attention than other mycotoxins. In the early 1960s, when more than 100,000 young turkeys in England died as the result of an apparently new disease, later named **Turkey X disease**, aflatoxicosis was soon found to be the cause of death.

In the 1950s, before Turkey X disease was reported, epizootics of fatal hepatitis in dogs were diagnosed provisionally as hepatitis X (Seibold and Bailey, 1952; Seibold, 1953; Newberne *et al.*, 1955). At that time, the etiology was unknown; however, the lesions were similar to those observed in mouldy corn poisoning of swine (Bailey and Groth, 1959), and the symptoms were very similar to those of aflatoxicosis when it was diagnosed. After isolation and chemical identification of the toxins produced by *A. flavus*, a close connection was made between aflatoxicosis and hepatitis X.

Since then, experimental studies have proven that the dog is susceptible to hepatic damage due to aflatoxins (Newberne *et al.*, 1966). In addition to hepatitis and sudden death, symptoms of acute aflatoxicosis in both dogs and cats include vomiting, depression, polydipsia, and polyuria. The pet animals most sensitive to aflatoxins are rabbits, for which the LD_{50} of aflatoxin is 0.4 mg/kg body weight (BW). In experiments with dogs, death usually occurs within 3 days of exposure, with LD_{50} levels ranging from 0.5 to 1.0 mg/kg BW. In cats, LD_{50} levels are from 0.3 to 0.6 mg/kg BW depending on the age of the animal (Patterson, 1977). Pathological examinations of such animals revealed enlarged livers, disseminated intravascular coagulation, and internal haemorrhaging. In subacute aflatoxicosis (0.5–1 mg/kg pet food) administered over 2-3 weeks, dogs and cats became lethargic, anorexic, and jaundiced. This was followed occasionally by disseminated intravascular coagulation and death. In chronic aflatoxicosis, as defined by doses of 0.05 – 0.3 mg/kg of pet food administered over 6-8 weeks, dogs and cats showed clinical signs similar to those for the subacute phase, but jaundice was the predominant manifestation. Histopathology of animals with chronic aflatoxicosis revealed shrunken livers with extensive fibrosis, which correlates with the hepatotoxic effects of aflatoxins observed in many animal species, including dogs.

Frequently observed clinical signs of experimentally induced aflatoxicosis in dogs include anorexia, followed by dehydration, somnolence, and jaundice. Gross pathological changes include subserosal and submucosal petechial haemorrhages throughout the thoracic and peritoneal cavities and yellow mottled livers. Necrosis and petechial haemorrhages in the gall-bladder, pancreas, kidneys, thymus, heart, and adrenal glands have also been described. Microscopically, acute hepatitis characterised by fatty degeneration and hepatic cell necrosis, as well as bile duct proliferation, fibrosis, and hepatic cell degeneration have also been observed (Edds, 1973).

NATURAL OUTBREAKS OF AFLATOXICOSIS

Ketterer *et al.* (1975) reviewed a case report where three dogs died after consuming a ration of mouldy bread containing *Aspergillus flavus* – the bread also contained 6.7 ppm aflatoxin. Analysis of two samples of vomit revealed 100 and 40 ppm aflatoxin, respectively.

Greene *et al.* (1977) reported an outbreak of chronic liver disease in a kennel of dogs. Anorexia, depression, polyuria, polydipsia, icterus and a terminal haemorrhagic diathesis were noted. Thrombocytopenia, hypofibrinogenemia, elevated fibrinogen degradation products, prolonged activated partial thromboplastin times, and one-stage prothrombin times were associated with haemorrhagic crises. Aflatoxicosis was confirmed by the presence of significant levels of aflatoxin in the commercial dog food being fed. In addition, a subacute hepatitis was found on necropsy.

Liggett *et al.* (1986) published case descriptions of high morbidity and mortality in hunting dogs. Clinical signs included icterus, anorexia, and listlessness; and additional laboratory examinations revealed toxic hepatitis, bilirubinuria, and anemia. The diagnosis was confirmed by the analysis of aflatoxin B_1 in tissues. The authors reported a second case of a dead Beagle that had been fed a ration containing cornmeal contaminated with 511 ng aflatoxin B_1 and B_2/g of feed.

Bastianello *et al.* (1987) reported a natural outbreak of aflatoxicosis in 10 dogs in the Republic of South Africa. The cases were classified as acute (1 case), subacute (7 cases), and chronic (2 cases) on the basis of the nature, degree, and extent of the following histopathological features: hepatocellular fatty degeneration, necrosis or regeneration, proliferation of bile ductules, accumulation of bile within the canaliculi. Fibroplasias, and mucoid degeneration, necrosis or segmental atrophy of the larger intrahepatic bile ducts. Fatty degeneration was noted grossly in the livers of all 10 cases, and bile stasis was noted in four of them. Varying degrees of fibrosis were present depending on the stage of the disease. In the two chronic cases, in which nodular regeneration also was observed, fibrosis was pronounced. Other macroscopic findings included icterus, anaemia, ascites, hydrothorax, hydropericardium, anasarca, pulmonary oedema, gastro-enterorrhagia, and nephrosis.

The high morbidity and mortality rate associated with outbreaks of aflatoxicosis and the failure of dogs to respond to symptomatic treatment often prompts practitioners to refer dogs or submit specimens to veterinary hospitals and diagnostic laboratories. An immediate ration change is indicated in most cases of toxic hepatitis pending analysis of the feed. In particular, mouldy dog food based on peanut meal is likely to be contaminated with aflatoxins and can result in toxic hepatitis if eaten.

In contrast, no natural outbreaks of aflatoxicosis have been reported in cats. This may be due to the fact that cat foods contain mainly meat and meat products (approximately 70%) and lower amounts of cereals (approximately 30%).

The introduction of strict quality control and quality assurance of incoming material in the pet food industry, and the development of accurate screening methods for animal food in recent years has led to a vast reduction in the amounts of aflatoxin B_1 in pet food. Consequently, there have been only occasional reports of aflatoxicosis in dogs in recent years.

AFLATOXIN OCCURRENCE IN PET FOODS

Sharma and Marquez (2001) recently investigated 35 samples of dog and cat foods in Mexico using immunoaffinity columns and HPLC. Maize was identified as the main ingredient in all contaminated samples. Also, aflatoxin B_1 was the mycotoxin found with the highest frequency and was present at particularly high levels (72 und 60 ng/g) in six samples (17%) of both dog and cat foods.

Bueno *et al.* (2001) investigated the mycoflora present in commercial dry pet foods from Argentina. Ten fungal genera and one fungus classified as *Mycelia sterilia* were identified. The predominant genera were *Aspergillus* (62%), *Rhizopus* (48%), and *Mucor* (38%). The most prevalent among *Aspergillus* was *A. flavus* followed by *A. niger*, and *A. terreus*. *Mucor racemosus* was followed in frequency of occurrence by *M. plumbus* and *M. globosus*. Some genera and species isolated and identified from these foods are potential mycotoxin producers and represent a risk for companion animal health. A particularly important factor in minimizing fungal growth is maintaining the moisture content below 10%.

Recently, the occurrence of aflatoxins in some Brazilian pet foods (dog, cat and bird foods) was investigated by Maia and Bastos de Siqueira (2002). They found aflatoxin in 12% of 100 samples investigated. The levels detected were between 15 and 374 μg/kg. All samples containing peanuts scored positive for aflatoxin B_1. Therefore, the authors concluded that the levels of aflatoxins present were of consequence, especially in bird foods, and that the consumption of such foods might be a risk for domestic animal health.

Henke *et al.* (2001) reported a survey of aflatoxin concentrations in wild birdseed purchased in Texas. Bags of wild birdseed (n = 142) were purchased from grain cooperatives, grocery stores, and pet shops located in Texas during the spring and summer of 1999. Aflatoxin concentrations ranged from non-detectable to 2780 μg/kg, with 17% of samples having aflatoxin concentrations greater than 100 μg/kg. The samples with greater than 100 μg/kg aflatoxin contained corn as the main ingredient. As a result, granivorous songbirds in Texas are exposed to aflatoxin at backyard feeders, which may contribute significantly to their morbidity and mortality.

Ochratoxin A

Ochratoxin A (OTA) is an important mycotoxin because of its high toxicity. It is a potent nephrotoxic fungal secondary metabolite produced by various moulds including *Aspergillus ochraceus* and *Penicillium verrucosum*. Ochratoxin A is widely distributed in Europe and North America and has been identified as a natural contaminant of agricultural commodities, apparently causing porcine nephropathy following degradation of the proximal tubules and interstitial fibroses (Blank *et al.*, 1999; Krogh, 1992).

Young Beagle dogs have reacted very sensitively to OTA. A daily oral dose of 0.2 and 0.3 mg/kg BW represents a deadly amount within 10 to 14 days of administration (Szczech *et al.*, 1973a,b; 1974). The following clinical signs were observed at this dose level: loss of appetite, vomiting, tenesmus, increased body temperature, tonsillitis, bloody viscous faeces, polydipsia, polyuria, dehydration, paralysis, and death. The pathological disturbances were associated with kidney damage. Gross dissections revealed pale brownish kidneys, hyperemic tonsils, and viscous bloody enteritis of the distal ileum, cecum, colon, and rectum. Additional oedemas and hyperemic and partial necrotic lymph nodes were also noted. Necrosis and desquamation of epithelial cells from the proximal renal tubules were observed. Ultrastructural changes such as proliferation and dilatation of the endoplasmic reticulum of the proximal epithelial tubuli cells were noticed.

In another study with dogs, Kitchen *et al.* (1977a,b,c) observed pacing and vomiting at an OTA dose of 0.2 mg/kg. At doses between 0.2 and 3.0 mg/kg, symptoms of intoxication in dogs included anorexia, polydipsia, polyuria, anxiety, prostration, and death. Necropsy findings included epithelial degeneration (proximal tubules), mucohaemorrhagic enteritis (cecum, colon, and rectum), and necrosis of the lymphoid tissues (spleen, tonsil, thymus, and peripheral lymph nodes). Ochratoxin A, dosed alone or in combination with citrinin, was administered to young Beagle dogs for two weeks. The effects observed were similar to those described above, but in combination-treated subjects the mortality rate was increased and clinical signs were more dramatic. That is, a synergistic toxic effect was observed when OTA and citrinin were administered together.

Gareis *et al.* (1987) have reported a case of fading puppy syndrome in Afghan puppies following consumption of OTA-contaminated milk powder. Clinical signs appeared a few hours after the first feeding of the milk powder. Pathological and histological findings suggested canine herpesvirus as the most probably cause of the syndrome. An association between the intake of the low OTA-contaminated feed, a possible dysbiosis and the manifestation of the multicausal viral infection is discussed.

OCCURRENCE OF OCHRATOXIN A IN PET FOODS

Gareis and Scheuer (1999) investigated meat and meat products in Germany and found high and frequent OTA contamination especially in pork, liver and blood sausages. Razzazi-Fazeli *et al.* (2001) investigated 26 canned and 17 dry pet foods for cats and dogs; and 26 feline kidney samples with and without pathological changes were surveyed for OTA residues. The separation and detection of OTA was carried out by an isocratic HPLC system based on reverse phase with fluorescence detection. After homogenization and extraction steps, immunoaffinity columns were utilised for sample cleanup. Ochratoxin A could be detected in 47% of the pet food samples. Those that were positive generally contained low amounts of OTA (0.1 – 0.8 μg/kg pet food). Higher levels were only detected in two pet food samples (3.2 and 13.1 μg/kg). Low concentrations of OTA could also be found in cat kidneys, 16 of which were positive. The concentrations found were between 0.35 and 1.5 μg/kg. No correlation between pathological findings and OTA levels in feline kidneys could be assessed.

Pühringer (2003) investigated 101 feline kidneys with and without pathological alterations and determined the OTA residue burden. Homogenisation, extraction and immunoaffinity columns were used for sample cleanup. An HPLC/fluorescence method was used to determine OTA. Thirty-nine kidneys were OTA positive. Minor OTA contents in the range of 0.11 to 0.3 μg/kg were identified in 23 cases, in 16 samples the range was 0.31 to 5.18 μg OTA/kg. No significant correlation was found between the occurrence of OTA in feline kidneys and pathological changes based on the analysed samples. Additionally, 55 foods for cats, 45 canned and 10 dry foods, were investigated. Fourteen pet foods showed positive levels of OTA in the range 0.11 to 2.17 μg/kg original substance. Seven of 45 canned pet foods and 7 of 10 dry pet foods tested positive for OTA. The detection limit of OTA both in feline kidneys and pet foods for cats was 0.1 μg/kg.

Fusariotoxins

TRICHOTHECENES

While aflatoxins are undoubtedly the most important mycotoxins, *Fusarium* toxins have recently been given more attention (Böhm, 2000a,b; Placinta *et al.*, 1999). In Europe, infection of agricultural crops with *Fusarium* spp. has been of major concern (Razzazi-Fazeli *et al.*, 2003). Many factors such as temperature, rainfall, and the type of host plant influence the distribution of *Fusarium* spp. in cereals. The importance of these toxigenic Fusaria is due to '*Fusarium* Head Blight' and '*Giberella* Ear Rot', two diseases with huge economic consequences. Moderate temperatures are required for *Fusarium* growth and toxin production.

Trichothecenes (TCT) are secondary metabolites of *Fusarium* moulds, one of the major contaminants of cereal crops. TCT are often found in cereals such as maize, barley, wheat, and oats grown in regions with temperate climates. Commercial dry pet foods could contain large amounts of TCT-contaminated grain and grain by-products. They can be divided into subclasses based on chemical structure. Ingestion of TCT by animals may lead to weight loss, decreased feed conversion, feed refusal and vomiting. Furthermore, TCT can inhibit protein synthesis and have immuno-suppressive effects. A-type trichothecenes are more toxic than B-type trichothecenes, since they have a dermatotoxic effect, and lead to necrosis and haemorrhage of the intestinal mucosa.

The most important trichothecenes are deoxynivalenol (DON) or vomitoxin, nivalenol (NIV), T-2 toxin, and diacetoxyscirpenol (DAS). B-type trichothecenes such as deoxynivalenol (DON) mainly lead to feed refusal and vomiting (Canady *et al.*, 2002). Toxin stability is of particular concern because DON is stable under normal food processing conditions such as sterilisation at 120°C (Hughes *et al.*, 1999) autoclaving, or extrusion cooking (Wolf-Hall *et al.*, 1999).

Type A-TCT (T-2 toxin, DAS) have extremely toxic effects on skin and mucous surfaces and can induce lesions on the mucosa of the mouth and oesophageal region of poultry and pigs. The most toxic trichothecene is T-2 toxin, which has been reported to cause dermatitis of the nose and buccal commissures of pigs. The immunotoxicity of type A-TCT, especially that of T-2 toxin, has been investigated and type A-TCT were found to be less immunotoxic than the type B-TCT (Hussein and Brasel, 2001).

Lutsky and Mor (1981) administered T-2 toxin to cats *per os* to evaluate the suitability of this species as a model for the human disease, alimentary toxic aleukia. The resulting chronic lethal intoxication was characterised by pancytopenia, haemorrhagic diatheses, bone marrow aplasia, diminished haemostasis, severe lymphatic tissue alterations, and histopathologic changes in proliferative tissues. Clinical signs included vomiting, bloody faeces, weakness, lassitude, ataxia, dyspnea, dehydration, loss of weight, and pre-terminal anorexia. The clinical course, haematologic picture, and both the gross and microscopic tissue changes seen in the experimentally produced disease in cats were similar to alimentary toxic aleukia, a frequently fatal mycotoxin-induced disease of man.

Coppock *et al.* (1989) have administered diacetoxyscirpenol (DAS) intravenously to dogs at concentrations of 0 and 0.5 mg/kg BW. Haemograms were conducted at 0.5 hr intervals for 8 hrs. The animals were euthanised 8 hrs after treatment, and bone marrow samples were taken and examined by light microscopy. Moderate to severe necrosis of bone marrow haematopoietic elements were observed in dogs given DAS. The sequential increase in the type and number of abnormal cells in the blood was noted, which indicated a successive destruction of haematopoietic elements. A marked left shift in the neutrophil population was found in animals given DAS. Metarubricytes and large platelets were found in blood, and lymphocytes were replaced with immature cells. The authors suggest that the order of species sensitivity to DAS was: pigs greater than dogs much greater than cattle. Residue studies of DAS and rapid recovery after exposure suggest that DAS is quickly metabolised and excreted.

T-2 toxin in rabbits

Fekete and Huszenicza (1993) investigated the effects of T-2 toxin on the ovarian activity and some metabolic variables in rabbits. Eight female, sexually mature New Zealand White rabbits were treated with 100 g of 284 ppb T2 toxin-containing wheat and 50 g alfalfa pellets for 32 days of exposure. Two rabbits died and these events were discussed in relation to the immunotoxic effects of T-2 toxin. Serum enzyme activities, such as alanine transaminase and the blood constituent creatinine, were increased and cholinesterase activity was decreased. Slight liver cell damage was discussed as the cause of all these effects. The morphologic examination of the ovaries showed no difference from the T-2 toxin-free control group, but there were alterations in the progesterone profile whereby three out of five GnRh-treated rabbits did not respond by an increase in plasma progesterone. Therefore, it would appear prudent to monitor feed ingredients regularly for possible mycotoxin contamination before preparation of batch rabbit diets.

Deoxynivalenol in cats and dogs

DON and other trichothecenes have been found to be highly toxic to animals if present in feed. These mycotoxins can result in sublethal toxicosis in lower doses, causing clinical effects such as reduced weight gain, emesis, and diarrhoea. The mechanisms of toxicity have been shown to be due to inhibition of protein synthesis, which results in cell death. Gastrointestinal haemorrhage and vomiting are acute signs; weight loss and loss of appetite are chronic sequelae. Clinical reports of natural intoxications with TCT have not been reported as yet (Puschner, 2002).

There are no specific gross or histological lesions associated with TCT mycotoxicosis. Exposure is confirmed by the detection of TCT mycotoxins in suspect feed. Clinical signs of DON toxicosis subside rapidly once the contaminated feed is removed.

Hughes *et al.* (1999) have conducted experimental feeding trials with DON in dogs and cats to determine the dietary amounts of DON required to produce overt signs of toxicity (e.g., vomiting or reduced food intake). The authors conducted toxicity and metabolism studies with Beagles and Brittany Spaniels. Food intake was significantly reduced at DON concentrations of 4.5 mg/kg food. Naturally contaminated wheat containing 37 mg of DON/kg and 15-acetylDON at 1 mg/kg was used to manufacture pet foods with levels of 0, 1, 2, 4, 6, 8, and 10 mg of DON/kg. No other *Fusarium* toxins were detected in these diets. In this study DON was found to be stable during conventional extrusion processing. Dogs previously fed DON-contaminated food were able to select preferentially the uncontaminated food. Food intake was significantly reduced by DON concentrations greater than 4.5 mg/kg. Vomiting by dogs was observed commonly at levels of 8 and 10 mg DON/kg food. In another study the same contaminated wheat source was fed for 14 days to groups of 2-7 mature American shorthair cats that were 1-9 years old. Cat food intake was significantly reduced when it contained DON exceeding 7.7 ± 1.1 mg/kg (equivalent to 0.38 mg/kg BW per day). For cats DON concentrations greater than 7.7 mg/kg reduced food intake. Vomiting was common at the highest concentrations of 8 and 10 mg of DON/kg.

Effects of deoxynivalenol on mink

Gibson *et al.* (1993) gave mink a choice between uncontaminated feed and DON-contaminated feed. They displayed a preference for the uncontaminated feed at a concentration of DON as low as 0.28 mg/kg. However, when no choice was available, the mink readily consumed feed containing DON at concentrations up to 1.2 mg/kg feed (equal to 0.20 mg/kg BW per day) with no apparent ill effects over a 28 day period. The study suggested that the sensitivity of mink to DON is close to that of pigs and greater than that of rats and chickens.

ZEARALENONE

Zearalenone (ZEA) is a secondary fungal metabolite of *Fusarium* species and has been shown to have oestrogenic activities. Pigs are especially sensitive to this mycotoxin and the ingestion of ZEA causes hyperoestrogenism associated with severe reproductive problems.

Oestrogenic effects of ZEA also have been shown in guinea pigs and rabbits (Mirocha *et al.*, 1968). In dogs, a deficient number of corpora lutea was observed following a 13 week diet containing ZEA (1 mg/kg BW per day) (Hidy *et al.*, 1977). Zöllner *et al.* (2002) investigated the residues of ZEA in liver and muscle of pigs fed diets containing 1.3 mg/kg feed. The authors found 13.3 μg/kg zearanol as a metabolite of ZEA in muscle tissue. However, the relevance of these levels in dog and cat foods is questionable.

Zearalenone in mink

Yamini *et al.* (1997) have investigated the effects of ZEA in mink. Hyperoestrogenic effects of ZEA were assessed by examination of pathological changes to the reproductive organs of female mink (*Mustela vison*) after feeding diets containing 20 mg ZEA/kg. All female mink fed ZEA mated, but only 25% whelped. Postmortem examination revealed moderate to severely distended uteri filled with caseated necrotic substances in all groups. Histological examination revealed a range of changes from mild to severe endometrial hyperplasia to uterine atrophy, endometritis, metritis and pyometra. Ovarian follicles were atrophied and degenerated.

EFFECTS OF MONILIFORMIN ON MINK

Morgan *et al.* (1999) have investigated the toxicological effects of moniliformin (MON), a member of the *Fusarium* toxin family. They studied acute and subacute effects of intraperitoneal injection of MON in mink (*Mustela vison*). MON was extracted from *Fusarium fujikuroi* culture material containing 9 mg/kg MON. An acute LD_{50} between 2.2 and 2.8 mg/kg BW was determined for 9-month-old female mink. Subacute exposure to 1.5 to 3.2 mg total MON/kg BW resulted in dilated (right side) hearts that were rounded in appearance. Electron microscopic examination of the right ventricular free heart wall of mink receiving acute or subacute doses of MON revealed ultrastructural damage to myofibres, mitochondria, nuclei, Z-and M-lines and sarcoplasmic reticula, and increased extracellular collagen deposition. The authors suggest that mink are among the most sensitive mammals to MON and that this mycotoxin specifically targets and damages the mink heart.

Tremorgenic mycotoxins (penitrem A, roquefortine)

Tremorgenic mycotoxins have an indole alkaloid moiety and are produced by different genera of *Penicillium*, *Aspergillus*, and *Claviceps* (Cole, 1981; Cole and Dorner, 1985). The main toxic effects have been associated with changes to the central nervous system (CNS). Penitrem A and roquefortine, two of the most important tremorgenic mycotoxins, have been shown to be associated with natural disease outbreaks in small animals. Different *Penicillium* spp., such as *P. crustosum, P. roqueforti*, and *P. cyclopium*, have been reported as potential producers of tremorgenic mycotoxins. These species have been isolated mainly from mouldy cream cheese and other moldy foodstuffs (Richard and Arp, 1979; Hocking *et al.*, 1988, Richard *et al.*, 1981). One of the most important producers of roquefortine, *Penicillium roqueforti,* is found primarily in blue cheese but has also been reported in other organic materials (Auerbach *et al.*, 1998; Ware *et al.*, 1980).

Hayes *et al.* (1976) have reported on the effects of an intraperitoneally administered dose of 0.5 mg penitrem A/kg, which resulted in the onset of acute tremors and death in dogs. Dogs developed tremors within 30 minutes when 0.13 mg of penitrem A was administered, however the animals recovered completely. Also, severe muscle tremors in dogs were observed after an oral dose of penitrem A of approximately 0.18 mg/kg BW.

Hocking *et al.* (1988) described a case of a 1-year-old Siberian Husky with severe muscle tremors after ingestion of a mouldy hamburger bun. *P. crustosum* and penitrem A were isolated from the remaining portion of the hamburger bun. When grown in pure culture, the isolate of *P. crustosum* produced large amounts of penitrem A, along with other penitrem compounds. This was the first reported Australian case of toxicosis by naturally occurring penitrem A. Penitrem A and roquefortine are absorbed in the gut and excreted through the bile (Wilson and Payne, 1994). However, enterohepatic recirculation and continued reabsorption may contribute to the prolonged recovery period.

Penitrem A is a lipophilic molecule, therefore it can cross the blood-brain barrier, and influence acetylcholine release (Wilson *et al.*, 1972). Cole and Cox (1981) have shown that penitrem A can antagonise glycine production. Selala *et al.* (1989) have found that penitrem A can substitute for γ-aminobutyric acid. Penitrem A toxicosis in dogs often appears within half an hour of toxin ingestion. The clinical symptoms have been described by Hocking *et al.* (1988) and Lowes *et al.* (1992) and include weakness, muscle tremors, increased irritability, rigidity, hyperactivity, and occasionally fever. Exhaustion may follow the increased muscle activity; elevated serum aspartate aminotransferase, increased creatine kinase, hyperthermia, rhabdomyolysis, and dehydration also can be seen. Often vomiting is observed before the onset of neurologic signs.

A differential diagnosis can be established after the exclusion of other tremorgenic agents such as strychnine, metaldehyde, bromethaline, organophosphorus compounds, carbamates, methylxanthines, pyrethroids, and chlorinated hydrocarbons. Analysis of penitrem A or roquefortine should be positive in samples such as vomitus, gastric lavage washings, stomach contents, or bile (Braselton and Rumler, 1996; Lowes *et al.,* 1992). The following treatments are recommended: Emesis should be provoked and gastric lavage should be performed under anesthesia. As with other intoxications, activated charcoal can be administered orally. Additionally, sodium bicarbonate administration is recommended if an acid-base imbalance has occurred. Diazepam is the treatment of choice followed by barbiturates (Lowes *et al.,* 1992; Richard *et al.,* 1981).

Puls and Ladyman (1988) described intoxication of an 8-year-old Fox Terrier/ Pekingese cross male dog with shivering and display of moderate tetanic spasms and mild opisthotonus. A tentative diagnosis of strychnine poisoning was made. Toxicological analysis revealed the presence of roquefortine in mouldy blue cheese, which the dog had eaten. No *Penicillium* spp. were found in any of the samples analysed.

CASE STUDY

Arp and Richard (1979) referred to intoxication of two Australian Shepherd dogs that had eaten mouldy cream cheese. The cream cheese was covered with the dark blue-green *Penicillium crustosum*. Experimental mice and a Beagle dog fed the sample showed similar tremorgenic symptoms as the two affected dogs. Penitrem A was identified as the causative agent. Acute muscle tremors, seizures, and prostration are typical clinical symptoms for this mycotoxin. Several hypothetical mechanisms of action have been proposed for penitrem A toxicosis. The severity of the clinical and pathologic features of penitrem A mycotoxicosis in dogs is dose dependent: mildly affected dogs display transitory muscle tremors and ataxia lasting 2-4 hrs, whereas higher doses may cause death during seizures or a progressive return to normal neurologic function after one or two days. Visceral petechial haemorrhages, liver necrosis, and hyperthermia have also been reported in dogs.

Although a specific antidote is not available for penitrem A toxicosis, pentobarbital has been used to control seizures and excessive motor activity. Fluid therapy and short-duration corticosteroids may counteract shock and stabilise cell membranes. Hyperthermia associated with the acute neurologic syndrome can be controlled by sedation and cold water baths. *Penicillium* and *Aspergillus* species are the fungi most frequently isolated from refrigerators and moldy foodstuffs in the home.

Cyclopiazonic acid

Another mycotoxin isolated from cultures of *Aspergillus flavus,* known as cyclopiazonic acid (CPA), may adversely affect dogs if ingested. The subacute toxic effects of orally administered CPA were characterized in dogs by Nuehring *et al.* (1985). Four groups of dogs were given CPA orally in gelatin capsules for 90 days at the following dosages: 0.05, 0.25, 0.5 and 1.0 mg/kg of BW. All dogs in the high dose groups died. Clinical signs of intoxication appeared 2 to 44 days after the start of dosing and consisted of anorexia, vomiting, diarrhoea, pyrexia, dehydration, weight loss, and CNS depression. Grossly, the entire alimentary tract exhibited diffuse hyperemia with focal areas of haemorrhage and ulceration. Other lesions identified included renal infarcts, necrotizing epididymitis, and ulcerative dermatitis. The authors also reported microscopic lesions, including ulceration, necrosis, vasculitis, lymphoid necrosis, karyomegaly in several organs, and decreased mitotic activity of the intestinal crypt epithelium. Ulcerative and necrotic lesions were usually associated with vascular lesions. Clinical pathologic changes included leukocytosis, neutrophilia, lymphopenia, monocytosis, and increased serum alkaline phosphatase activity.

Patulin and penicillic acid

Patulin and penicillic acid are produced by *Penicillium* and *Aspergillus* spp., which are often found in mouldy feeds and cheeses. Patulin was analysed in rotten fruits, apple juice, and mouldy bread. The toxicity of patulin and the interaction between patulin and penicillic acid in dogs have been studied experimentally by Reddy *et al.* (1979). The most sensitive organs were the lungs and gastrointestinal tract. Clinical signs of toxicosis included haematemesis, diarrhoea, lethargy, pulmonary haemorrhage and pulmonary oedema when patulin was administered intravenously at doses greater than 10 mg/kg of BW. Dogs that received 10 mg penicillic acid/kg BW alone showed slight inappetence but no other clinical signs. A synergistic interaction between patulin and penicillic acid was noted.

Multimycotoxin occurrence in pet foods

Scudamore *et al.* (1997) have analysed over 100 samples of pet foods for the presence of mycotoxins. These consisted of 35 samples each of cereal-based dog and cat foods, 15 samples of domestic birdseed, and 15 samples of wild bird food. All samples were examined for aflatoxins B_1, B_2, G_1, G_2, and ochratoxin A. Twenty samples of these were also examined for fumonisins B_1 and B_2. Of the 100 samples, 84% contained no measurable concentrations of mycotoxins. A low level of aflatoxin B_1 was found in one sample of cat food, and a concentration of 370 μg/kg aflatoxin B_1 was found in one sample of peanuts marketed for feeding to wild birds. Ochratoxin A was detected in 10% of the samples, but in low concentrations; the highest concentration of 7 μg/kg occurred in a sample of bird food. Fumonisins were found in 30% of the 20 samples tested, with a maximum of 750 μg/kg total fumonisins found in a sample of cat food. The significance of fumonisin contaminant levels is uncertain at present. Also, these authors have examined five samples each of dog and cat foods for mould count and fungal species. After storage under controlled simulated damp conditions, visible moulds were noted in the wet samples. *Penicillium*, *Eurotium* and *Aspergillus* were the predominant species found. However no OTA or aflatoxins were detected in spoiled samples.

Böhm and Hochleithner (1991) analysed feeds given Celestial parrotlets, Gouldian finches and budgerigars suspected of causing mycotoxicosis. In such samples, rearing feed and millet seed cobs contained 20 and 15 ng/g aflatoxin B_1, respectively. In an insect-contaminated feed, trichothecenes (2 ppm nivalenol, 0.3 ppm DON and 0.1 ppm DAS) were identified by gas chromatography. In the budgerigar feed *Mucor* spp. were isolated at a level of more than 1 million/g. In all cases a change of feed was successful at reversing clinical signs of mycotoxicosis. Loss of appetite, the appearance of neurological symptoms, and a high and rapid mortality in young birds are the clinical symptoms of aflatoxicosis. Trichothecenes may exacerbate the symptoms.

Ergot alkaloids

The fungus *Claviceps purpurea* produces three important alkaloids, ergotine, ergotamine and ergotoxin. They may contaminate all grains, but especially triticale, rye and wheat. Toxicity is mediated via vasoconstriction and thrombosis in peripheral capillaries, and can result in severe damage to extremities. Chronic intoxication is more likely to produce neuronal symptoms.

Wolter and Jean (1997) have reported on dietary intoxications of ergot alkaloids in domestic carnivores and suggest that these alkaloids are important contributors to feline dysautonomia. Intoxication of these carnivores results in acute peripheral vasoconstriction extending to gangrene, disseminated coagulation, haemorrhage, and female reproductive disorders including agalactia. Chronic intoxications produce neurological symptoms, such as hyperexcitation and tremors combined with ataxia. The toxic effects can be observed when dietary concentrations exceed 1 g ergot/kg. Quantitative determination of ergot alkaloids in foods is a difficult analytical task; however, if ergot is separated from the grains, the task is easier according to Mellor (2003), who suggests that since ergot is still common in wild grains, pet birds might still be at risk.

Conclusion

The problem of mycotoxin contamination is generally complex. *Aspergillus* spp., *Fusarium* spp., and *Penicillium* spp. may produce several different types of mycotoxins, and contamination might occur anywhere in the production chain of cereals, soyabean meal, corn, other grains, and their by-products. Normally, high temperatures of extrusion processing or canning cannot be relied upon to destroy possible mycotoxin contamination in pet foods. As expected, uncleaned grain is more likely to be contaminated with mycotoxins than cleaned grain. Aflatoxins, ochratoxins and cyclopiazonic acid as well as fusariotoxins such as zearalenone, deoxynivalenol, moniliformin and fumonisins are common mycotoxins and all of them can be found in pet foods. Aflatoxins still dominate the list of possible mycotoxin contaminants that are considered high-risk substances in pet foods. Contamination can occur directly through *A. flavus*-spoiled cereal ingredients or indirectly through residues in inner organs such as livers or kidneys from slaughtered animals such as poultry, pigs or cattle. Consequently, aflatoxicosis continues to be a problem and might be considered in a differential diagnosis in sick dogs. Also, direct high aflatoxin contamination can be seen in peanuts or other kernels that are often the main ingredients of bird feed.

Since all mycotoxins affect animal health, they should be of concern to all pet owners. Dose and time of orally ingested mycotoxins relate to the severity of the diseases. Usually toxicosis is associated with a combination of stress and infection, which give rise to a variety of severe diseases classified as mycotoxicosis. Not unexpectedly, the symptoms of mycotoxicosis in pets are very similar to those seen in farm animals. Symptoms include jaundice, loss of appetite, and slight dizziness, to severe organ changes including hepatitis, blood coagulation, neuronal symptoms, and even death in advanced cases.

Different mycotoxins affect different organs and tissues. Thus, combinations of mycotoxins may produce undesirable synergistic effects on animal health, which have not yet been fully characterised, and thus merit further attention. One of the major sequelae of mycotoxins, and of greatest concern long-term, is their possible carcinogenicity (aflatoxins, fumonisins). Due to the relatively long life span of companion animals, long-term exposure to mycotoxins may succeed in inducing cancer. This aside, although acute mycotoxicosis rarely occurs, chronic mycotoxicosis is more common and leads to increased susceptibility to infectious diseases, damage to the liver, kidney, and reproductive organs, poor coat condition, anorexia, feed refusal, vomiting, and lethargy.

Fortunately, because strict quality control standards are practiced at most pet food manufacturing plants, including careful screening of incoming ingredients and thorough cleaning of equipment, mycotoxin contamination of pet food is rare. Additionally, the widespread use of screening techniques and proper storage of feedstuffs has reduced the incidence of outbreaks associated with commercial feed consumption.

The simplest precaution to prevent mycotoxicosis in pets remains avoiding mould-spoiled foodstuffs. Keep them out of reach of companion animals.

References

Arp, L.H. and J.L. Richard. 1979. Intoxication of dogs with the mycotoxin penitrem A. JAVMA 175:565-566.

Auerbach, H., E. Oldenburg and F. Weissbach. 1998. Incidence of *Penicillium roqueforti* and roquefortine C in silages. J. Sci. Food Agric. 76:565-572.

Bailey, W.S. and A.H. Groth. 1959. The relationship of hepatitis X of dogs and moldy corn poisoning of swine. JAVMA 134:514-516.

Bastianello, S.S., J.W. Nesbit, M.C. Williams and A.L. Lange. 1987. Pathological findings in a natural outbreak of aflatoxicosis in dogs. Onderstepoort J. Vet. Res. 54:635-640.

Bird, C. 2000. Detecting and controlling mycotoxins in petfoods. In: *Technical Symposium on Mycotoxins*. Alltech Inc., Nicholasville, KY.

Blank, R., D. Höhler and S. Wolffram. 1999. Ochratoxin A in der Nahrungskette Vorkommen Toxizität und Dekontamination. Übers. Tierernährung. 27: 123-163.

Böhm, J. 2000a. Fusariotoxins in animal nutrition. Habilitation Vetmed. Univ. Vienna.

Böhm, J. 2000b. Fusariotoxins and their importance in animal nutrition. Übers. Tierernährg. 28:95-132.

Böhm, J. and M. Hochleithner. 1991. Mycotoxicosis in pet birds – selected cases. 1st Conference of the European Committee of the Association of Avian Veterinarians. Vienna, March 13-16, pp. 255-257.

Braselton, W.E. and P.C. Rumler. 1996. MS/MS screen for the tremorgenic mycotoxin roquefortine and penitrem A. J. Vet. Diagn. Invest. 8:515-518.

Bueno, D.J., J.O. Silva and G. Oliver. 2001. Mycoflora in commercial pet foods. J. Food Protect. 64:741-743.

Canady, R.A., R.D. Coker, S.K. Egan, R. Krska, T. Kuiper-Goodman, M. Olsen, J. Pestka, S. Resnik and J. Schlatter. 2002. The joint FAO/WHO expert committee on food additives (JECFA) Deoxynivalenol. *http://www.inchem.org/documents/jecfa/jecmono/v47je05.htm, last access: 2003-12-9.*

Cole, R.J. 1981. Fungal tremorgens. J. Food Prot. 44:715-722.

Cole, R.J. and J.W. Dorner. 1985. Role of fungal tremorgens in animal disease. In: *Mycotoxins and Phycotoxins* (P.S. Steyn and R. Vleggar, eds). Elsevier Science Publishers, Amsterdam, pp. 501-511.

Cole, R.J. and R.H. Cox. 1981. Tremorgen group. In: *Handbook of Toxic Fungal Metabolites*. New York, Academic Press, pp. 355-509.

Coppock, R.W., W.E. Hoffmann, H.B. Gelberg, D. Bass and W.B. Buck. 1989. Hematologic changes induced by intravenous administration of diacetoxyscirpenol in pigs, dogs and calves. Am. J. Vet. Res. 50:411-415.

Devegowda, G. 2000. Mycotoxins: hidden killers in pet foods. Is there a biological solution? In: *Technical Symposium on Mycotoxins*. Alltech Inc., Nicholasville, KY.

Edds, G.T. 1973. Acute aflatoxicosis: a review. JAVMA 162:304-309.

Fekete, S. and G. Huszenicza. 1993. Effects of T-2 toxin on ovarian activity and some metabolic variables of rabbits. Lab. Anim. Sci. 43:646-649.

Gareis, M. and R. Scheuer. 1999. Ochratoxin A in Fleisch und Fleischerzeugnissen. Proceedings 21, Mycotoxin Workshop, p. 148.

Gareis, M., G. Reubel, T. Kröning and R. Porzig. 1987. Ein Fall von infektiösem Welpensterben bei Afghanen in Verbindung mit der Verfütterung von Ochratoxin A - haltigem Milchpulver. Tierärztl. Umschau 42:77-80.

Gibson, M.K., S.J. Bursian and R.J. Aulerich. 1993. Effects of deoxynivalenol on feed consumption and body weight gains in mink (*Mustela vison*). Bull. Environ. Contam. Toxicol. 51:6-11.

Greene, C.E., J.A. Barsanti and B.D. Jones. 1977. Disseminated intravascular coagulation complicating aflatoxicosis in dogs. Cornell Vet. 67:29-49.

Hayes, A.W., D.B. Presley and J.A. Neville. 1976. Acute toxicity of penitrem A in dogs. Toxicol. Appl. Pharmacol. 35:311-320.

Henke, S.E., V.C. Gallardo, B. Martinez and R. Balley. 2001. Survey of aflatoxin concentrations in wild bird seed purchased in Texas. J. Wildl. Dis. 37:831-835.

Hidy, P.H., R.S. Baldwin, R.L. Greasham, C.L. Keith and J.R. McMullen. 1977. Zearalenone and some derivates: Production and biological activities. In: *Adv. Appl. Microbiol.* (D. Perlman, ed.) 22:59-82, Acad. Press, New York.

Hocking, A.D., K. Holds and N.F. Tobin. 1988. Intoxication by tremorgenic mycotoxin (penitrem A) in a dog. Aust. Vet. J. 65:82-85.

Hughes, D.M., M.J. Gahl, C.H. Graham and S.L. Grieb. 1999. Overt signs of toxicity to dogs and cats of dietary deoxynivalenol. J. Anim. Sci. 77:693-700.

Hussein, S.H. and J.M. Brasel. 2001. Toxicity, metabolism and impact of mycotoxins on humans and animals. Toxicol. 167:101-134.

Ketterer, P.J., E.S. Williams, B.J. Blaney and M.D. Connole. 1975. Canine aflatoxicosis. Aust. Vet. J. 51:355-357.

Kitchen, D.N., W.W. Carlton and E.J. Hinsman. 1977c. Ochratoxin A and citrinin induced nephrosis in Beagle dogs. III. Terminal renal ultrastructural alterations. Vet. Path. 14:392-406.

Kitchen, D.N., W.W. Carlton and J. Tuite. 1977a. Ochratoxin A and citrinin induced nephrosis in Beagle dogs. I. Clinical and clinicopathological features. Vet. Path. 14:154-172.

Kitchen, D.N., W.W. Carlton and J. Tuite. 1977b. Ochratoxin A and citrinin induced nephrosis in Beagle dogs. II. Pathology. Vet. Path. 14:261-272.

Krogh, P. 1992. Role of ochratoxin in disease causation. Fd. Chem. Toxicol. 30:213-244.

Liggett, A.D., B.M. Colvin, R.W. Beaver and D.M. Wilson. 1986. Canine aflatoxicoses: a continuing problem. Vet. Hum. Toxicol. 28:428-430.

Lowes, N.R., R.A. Smith and B.E. Beck. 1992. Roquefortine in the stomach contents of dogs suspected of strychnine poisoning in Alberta. Can. Vet. J. 33:5819-5828.

Lutsky, I. and N. Mor. 1981. Experimental alimentary toxic aleukia in cats. Lab. Anim. Sci. 31:43-47.

Maia, P.P. and M.E.P. Bastos de Sequeira. 2002. Occurrence of aflatoxins B_1, B_2, G_1 and G_2 in some Brazilian pet foods. Food Add. Contam. 19:1180-1183.

Mellor, S. 2003. Keep petfood mycotoxins on the leash. Feed Mix 11:23-24.

Mirocha, C.J., C.M. Christensen and G.H. Nelson. 1968. Physiologic activity of some fungal estrogens produced by Fusarium. Cancer Res. 28:2319-2322.

Morgan, M.K., S.D. Fitzgerald, G.E. Rottinghaus, S.J. Bursian. and R.J. Aulerich. 1999. Toxic Effects to mink of moniliformin extracted from *Fusarium fujikuroi* culture material. Vet. Human Toxicol. 41:1-5.

Newberne, J.W., W.S. Bailey and H.R. Seibold. 1955. Notes on a recent outbreak and experimental reproduction of hepatitis in dogs. JAVMA 127:59-62.

Newberne, P.M., R. Russo and G.N. Wogan. 1966. Acute toxicity of aflatoxin B_1 in the

dog. Path. Vet. 3:331-340.
Nuehring, L.P., G.N. Rowland, L.R. Harrison, R.J. Cole and J.W. Dorner. 1985. Cyclopiazonic acid mycotoxicosis in the dog. Am. J. Vet. Res. 46:1670-1676.
Patterson, D.S.P. 1977. Toxin-producing fungi and susceptible animal species. In: *Mycotoxic Fungi, Mycotoxins, Mycotoxicoses Vol. 1* (T.D. Wyllie and L.G. Morehouse, eds). Marcel Dekker Inc., pp. 156-158.
Placinta, C.M., J.P.F. D'Mello and A.M.C. Macdonald. 1999. A review of worldwide contamination of cereal grains and animal feed with Fusarium mycotoxins. Anim. Feed Sci. Tech. 78:21-37.
Pühringer, S. 2003: Untersuchungen zum Vorkommen von Ochratoxin A in Katzennieren und Katzenfuttermitteln. Diss. Vetmed. Univ. Wien.
Puls, R. and E. Ladyman. 1988. Roquefortine toxicity in a dog. Can. Vet. J. 29:569.
Puschner, B. 2002. Mycotoxins. Vet. Clin. Small Anim. 32:409-419.
Razzazi-Fazeli, E., J. Böhm, J. Grajewski, K. Szczepaniak, A. Kübber-Heiss and C. Iben. 2001. Residues of ochratoxin A in pet foods, canine and feline kidneys. J. Anim. Physiol. A. Anim. Nutr. 85:212-216.
Razzazi-Fazeli, E., J. Böhm, A. Adler and J. Zentek. 2003. Fusarientoxine und ihre Bedeutung in der Nutztierfütterung, eine Übersicht. Wiener Tierärztliche Monatsschrift 90:202-210.
Reddy, C.S., P.K. Chan, A.W. Hayes and W.L. Williams. 1979. Acute toxicity of patulin and its interaction with penicillic acid in dogs. Food Cosmet. Toxicol. 17:605-609.
Richard, J.L. and L.H. Arp. 1979. Natural occurrence of the mycotoxin penitrem A in moldy cream cheese. Mycopathologica 67:107-109.
Richard, J.L., P. Baccetti and L.H. Arp. 1981. Moldy walnut toxicosis in a dog, caused by the mycotoxin penitrem A. Mycopathologica 76:55-58.
Scudamore, K.A., M.T. Hetmanski, S. Nawaz, J. Naylor and S. Rainbird. 1997. Determination of mycotoxins in pet foods sold for domestic pets and wild birds using linked-column immunoassay clean-up and HPLC. Food Add. Contam. 14:175-186.
Seibold, H.R. 1953. Hepatitis X in dogs. Vet. Med. 48:242-243.
Seibold, H.R. and W.S. Bailey. 1952: An epizootic of hepatitis in dogs. JAVMA 121:201-206.
Selala, M.I., F. Daelemans and P.J. Schepens. 1989. Fungal tremorgens: The mechanism of action of single nitrogen containing toxins - a hypothesis. Drug Chem. Toxicol. 17:605-609.
Sharma, M. and C. Marquez. 2001. Determination of aflatoxins in domestic pet foods (dog and cat) using immunoaffinity column and HPLC. Anim. Feed Sci. Tech. 93:109-114.
Szczech, G.M., W.W. Carlton and E.J. Hinsman. 1974. Ochratoxicosis in Beagle dogs. III. Terminal renal ultrastructural alterations. Vet. Path. 11:385-406.
Szczech, G.M., W.W. Carlton and J. Tuite. 1973a. Ochratoxicosis in Beagle dogs. I. Clinical and clinicopathological features. Vet. Path. 10:135-154.
Szczech, G.M., W.W. Carlton and J. Tuite. 1973b. Ochratoxicosis in Beagle dogs. II. Pathology. Vet. Path. 10:219-231.
Ware, G.M., C.W. Thorpe and A.E. Pohland. 1980. Determination of roquefortine in blue cheese dressing by high pressure liquid chromatography with ultraviolet and electrochemical detectors. JAOAC 63:637-641.
Wilson, B.J., T. Hoekman and W.D. Dettbarn. 1972. Effects of a fungus tremorgenic toxin (penitrem A) on transmission in rat phrenic nerve-diaphragm preparations. Brain Res. 40:540-544.
Wilson, D.M. and G.A. Payne. 1994. Factors affecting *Aspergillus flavus* group infection and aflatoxin contamination of crops. In: *The Toxicology of Aflatoxins: Human Health, Veterinary and Agricultural Significance* (D.L. Eaton and J.D.

Groopman, eds). San Diego, Academic Press, pp. 309-326.

Wolf-Hall, C.E., M.A. Hanna and L.B. Bullerman. 1999. Stability of deoxynivalenol in heat-treated foods. J. Food Prot. 62:962-964.

Wolter, R. and C. Jean. 1997. Intoxications alimentaires: principales mycotoxicoses chez les carnivores domestiques. Prat. Med. Chir. Anim. Comp. 32:157-162.

Yamini, B., S.J. Bursian and R.J. Aulerich. 1997. Pathological effects of dietary zearalenone and/or tamoxifen on female mink reproductive organs. Vet. Human Toxicol. 39:74-78.

Zöllner, P., J. Jodlbauer, M. Kleinova, H. Kahlbacher, T. Kuhn, W. Hochsteiner and W. Lindner. 2002. Concentration levels of zearalenone and its metabolites in urine, muscle tissue and liver samples of pigs fed with mycotoxin-contaminated oats. J. Agric. Food Chem. 50:2494-501.

5

EFFECTS OF MYCOTOXINS ON ANTIOXIDANT STATUS AND IMMUNITY

[1*]Peter F. Surai and [2]Julia. E. Dvorska
[1]*Avian Science Research Centre, Scottish Agricultural College, Auchincruive, Ayr, Scotland, UK and* [2]*Sumy National Agrarian University, Sumy, Ukraine*

Effects of mycotoxins on antioxidant systems

A delicate balance between antioxidants and pro-oxidants in the body in general and specifically in the cell is responsible for regulation of various metabolic pathways leading to maintenance of immunocompetence, growth and development and protection against stress conditions associated with commercial poultry production (Surai and Dvorska, 2001). This balance can be regulated by dietary antioxidants, including vitamin E (Surai *et al.*, 1999), carotenoids (Surai and Speake, 1998; Surai *et al.*, 2001) and selenium (Se) (Surai, 2000). On the other hand, nutritional stress factors have a negative impact on this antioxidant/pro-oxidant balance. In this respect mycotoxins are considered to be among the most important feed-borne stress factors.

It is not clear at present whether mycotoxins stimulate lipid peroxidation directly by enhancing free radical production or if the increased tissue susceptibility to lipid peroxidation is a result of a compromised antioxidant system. It seems likely that both processes are at work. In most cases lipid peroxidation in tissues caused by mycotoxins was associated with decreased concentrations of natural antioxidants. For example, in an experiment with quail, levels of the primary liver antioxidants (α–tocopherol, γ–tocopherol, carotenoids and ascorbic acid) were significantly decreased as a result of T-2 toxin consumption (Dvorska and Surai, 2001; Figure 1).

Similarly, the presence of T-2 toxin in the diet decreased the concentration of α–tocopherol in the chicken liver (Hoehler and Marquardt, 1996). T-2 toxin consistently depressed concentrations of vitamin E in chicken plasma (Coffin and Combs, 1981). Addition of micelle-promoting compounds (taurocholic, monoolein, and oleic acids) alleviated depression in plasma vitamin E, indicating interference of T-2 toxin with micelle formation during vitamin E absorption. Similarly, aflatoxin B_1 (AFB_1) in the feed interfered with the accumulation of carotenoids in chicken tissues (Schaeffer *et al.*, 1988) inducing pale bird syndrome in birds. In fact, AFB_1 caused a significant depression of lutein in the toe web, liver, serum and mucosa (Schaeffer *et al.*, 1988a). Pigment restoration was accomplished by feeding the same diet supplemented with lutein (70 mg/kg). In young chickens AFB_1 reduced the lutein content of jejunal mucosa up to 35% while serum lutein was reduced up to 70% (Tyczkowski and Hamilton, 1987), suggesting that AFB_1 interfered with the absorbtion, transport and deposition of carotenoids. More precisely, AFB_1 impaired lutein absorption in chickens (Tyczkowski and Hamilton, 1987a). In similar fashion, ochratoxin A (OTA) was shown to affect

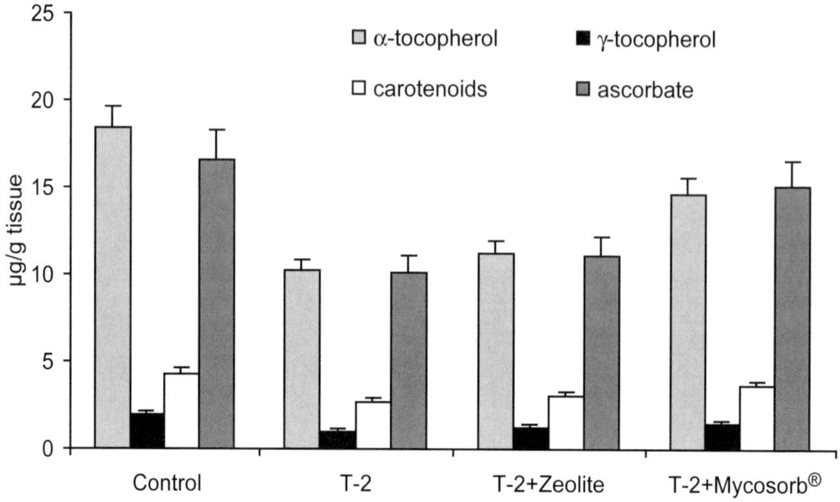

Figure 1. Effect of T-2 toxin with and without supplemental toxin adsorbents on antioxidant concentrations in quail liver (adapted from Dvorska and Surai, 2001).

carotenoid assimilation in chickens. Again, depression in uptake of carotenoids by intestinal mucosa and depressed transport in serum were considered to be important mechanisms of AFB_1 action on carotenoid metabolism (Schaeffer *et al.*, 1988; Huff and Hamilton, 1975).

In general, **malabsorption syndrome** is considered a common result of mycotoxicoses. For example, aflatoxicosis, ochratoxicosis and T-2 toxicosis were produced by feeding diets containing graded concentrations of the appropriate toxin to broiler chickens from hatching until 3 weeks of age (Osborne *et al.*, 1982). In this experiment AFB_1, even at levels lower than needed for growth inhibition, produced a malabsorption syndrome characterized by steatorrhea, hypocarotenoidemia, and decreased concentrations of bile salts and pancreatic enzymes. T-2 toxin also produced malabsorption, but at concentrations higher than required to inhibit growth. Ochratoxicosis produced mainly hypocarotenoidemia (Osborne *et al.*, 1982). It is postulated that the decreased level of vitamin A in the quail liver as a result of T-2 toxin consumption (Dvorska and Surai, 2001) is also a reflection of the decreased intestinal absorption of fat soluble nutrients.

The presence of OTA in the diet significantly decreased the concentration of α–tocopherol in the chicken liver (Hoehler and Marquardt, 1996). Furthermore, aflatoxin-treated barrows had decreased serum tocopherol and retinol concentrations compared with control and pre-test values, and decreased tocopherol concentration in cardiac tissue (Harvey *et al.*, 1994). Aurofusarin decreased vitamin E and carotenoid concentration in quail egg yolk (Dvorska *et al.*, 2001a), in the liver, yolk sac membrane and other tissues of newly hatched quail (Dvorska *et al.*, 2002; Dvorska *et al.*, 2003; Dvorska and Surai, 2004). It is interesting to note that the carotenoid and vitamin A concentrations in the quail liver were also decreased when the maternal diet was supplemented with aurofusarin.

A pro-oxidant effect of mycotoxins in many cases could be mediated through changes in reduced glutathione (GSH) concentration. For example, Rizzo *et al.* (1994) demonstrated that T-2 toxin decreased GSH in rat liver. Treatment of fasted mice with a single dose of T-2 toxin (1.8 or 2.8

mg/kg BW) by oral gavage led to a marked decrease in hepatic GSH (Atroshi *et al.*, 1997). In male broiler chicks, hepatic GSH concentration decreased after 7 days of treatment (1.5 mg T-2 toxin/kg BW/day) (Leal *et al.*, 1999). Acute exposure of mice to T-2 toxin (4 mg/kg, s.c.) resulted in a progressive decrease in hepatic GSH, reaching a minimum 6-8 hrs after toxin administration (Fricke and Jorge, 1991). Intraperitoneal administration of AFB_1 to rats (2 mg/kg) was also associated with decreased GSH in the liver. In contrast, in 3-week-old male chickens daily aflatoxin gavage (2 mg/kg BW, in corn oil) for 5 and 10 days elevated hepatic GSH; and renal GSH was elevated after 10 days (Beers *et al.*, 1992a). Similarly, hepatic GSH increased 2 and 8 hrs following a single AFB_1 dose and continued to increase through five daily doses of AFB_1 (Beers *et al.*, 1992b).

There was GSH depletion in cultured rat hepatocytes as a result of AFB_1 toxicosis (Liu *et al.*, 1999). Similarly, consumption of OTA for two weeks was associated with a depletion of GSH from the mouse liver (Atroshi *et al.*, 2000). The mycotoxin patulin also decreased GSH concentration in rat hepatocytes (Busbee *et al.*, 1999). Considering that glutathione is responsible for the maintenance of redox status of the cell (Sies, 1999) and therefore participates in regulation of gene expression (Arrigo, 1999), changes in GSH status could be detrimental.

One of the most important mycotoxin actions is their effect on antioxidant enzymes. Depending on experimental conditions (species, doses, route and duration of administration, concentrations of other antioxidants etc.), antioxidant enzyme activities can be increased in response to oxidative stress or decreased by direct or indirect action of mycotoxins. For example, treatment of pig kidney cells with 50 mM fumonisin B_1 (FB_1) for 24 hrs significantly decreased cellular GSH and increased the activities of glutathione reductase (Kang and Alexander, 1996). The activities of glutathione peroxidase (GSH-Px), catalase, and Cu/Zn-superoxide dismutase (SOD) were not changed by this treatment. Oral administration of T-2 mycotoxin to rats (1.25 mg/kg) for 5 days decreased the activity of liver glutathione-S-transferase (Ahmed and Ram, 1986). In contrast, feeding a single dose of T-2 toxin (2 mg/kg BW) to rats increased activities of GSH-shuttle enzymes including GSH-Px, glutathione reductase and glucose-6-phosphate dehydrogenase (Suneja *et al.*, 1989), probably reflecting an adaptive response to oxidative stress. On the other hand, when male rats were given a diet deficient in vitamins C and E and Se and were administered orally a single dose of deoxynivalenol (DON) or T-2 toxin, there was a significant decrease in activities of GSH-Px, catalase, SOD and glutathione reductase (Rizzo *et al.*, 1994). Activity of GSH-Px in rat blood was decreased due to consumption of AFB_1 (Choi *et al.*, 1995). Administration of AFB_1 to rats (2 mg/kg intraperitoneally) caused a significant decrease in the activities of SOD, catalase, GSH-Px, glutathione-S-transferase and glutathione reductase in liver (Rastogi *et al.*, 2001). A significant increase in SOD activity occurred in the liver following AFB_1 exposure of the ducks (Barraud *et al.*, 2001).

When considering detrimental effects of mycotoxins on antioxidant systems it is necessary to be aware that combinations of several mycotoxins can be more toxic than individual mycotoxins, as was shown for a combination of T-2 toxin and OTA (Kubena *et al.*, 1989).

One of the most important targets for mycotoxins is embryonic development. Since chicken embryo tissues contain high levels of PUFA, they are vulnerable to peroxidation and oxidative stress caused by mycotoxins could be lethal. As mentioned previously, aurofusarin increased late mortality of quail embryos (Dvorska *et al.*, 2001a). Furthermore, contamination of the diet with T-2 toxin markedly decreased egg production and impaired hatchability (Tobias *et al.*, 1992). Confirmation of a possible association of this effect with oxidative stress came from data indicating that increased dietary vitamin E during the first week of the experiment significantly decreased the number of infertile eggs and significantly improved the hatching percentage (Tobias *et al.*, 1992).

Increased lipid peroxidation as a consequence of mycotoxicoses

As illustrated in Table 1, OTA has a stimulating effect on lipid peroxidation. In most of cases, accumulation of thiobarbituric acid reactive substances (TBARS) was used as a measurement of lipid peroxidation. Ethane exhalation, EPR-registered free radicals, hydroxyl radical formation, single-strand cleavage DNA, DNA adduct formation as well as LDH release were also used to confirm pro-oxidant properties of OTA. Various *in vitro* and *in vivo* systems were also used including liver microsomes, phospholipid vesicles, primary cell cultures, whole organ and whole body measurements.

T-2 toxin was also shown to have pro-oxidant properties (Table 2). Those properties were confirmed with rat, mouse and quail liver tissue and yeast. TBARS accumulation was the method used in most of the studies, however, conjugate diene formation and DNA fragmentation were also demonstrated. Effect of AFB_1 on lipid peroxidation has been studied in rat liver and kidney as well as in cultured hepatocytes and in an *in vitro* model

Table 1. Stimulation of lipid peroxidation by ochratoxin A.

Mycotoxin	*Tissue*	*Lipid peroxidation measurement*	*Protective effect of antioxidants*	*References*
OTA *in vitro*	Rat liver microsomes	TBARS ↑	-	Rahimtula *et al.*, 1988, Gautier *et al.*, 2001; Khan *et al.*, 1989
OTA in feed	Rats	Ethane exhalation ↑	-	Rahmitula *et al.*, 1988
OTA in feed	Rats	MDA in serum, liver, kidney ↑	-	Meki and Hussein, 2001
OTA *in vitro*	Phospholipid vesicles	TBARS ↑, oxygen uptake ↑	-	Omar *et al.*, 1990
OTA and its analogs	*Bacillus brevis* bacteria	Free radical generation (EPR) ↑	Vitamin E	Hoehler *et al.*, 1996
OTA in feed	Chicken liver	TBARS ↑	-	Hoehler and Marquardt, 1996; Hoehler *et al.*, 1997
OTA *in vitro*	Vero cells in culture	TBARS ↑	SOD, catalase aspartame	Baudrimont *et al.*, 1997a; 1997b
OTA in diet	Rat liver	TBARS ↑	-	Hoehler *et al.*, 1997
OTA *in vitro*	A model oxidation system	OH• ↑, single-strand cleavage DNA ↑	-	Gillman *et al.*, 1999
OTA *in vitro*	Astrocytes and neurons	TBARS ↑, LDH release ↑	-	Belmadani *et al.*, 1999
OTA in feed	Mouse and rat kidney	DNA adduct formation ↑	Retinol, ascorbic acid, vitamin E	Grosse *et al.*, 1997
OTA in feed	Rats	DNA adduct formation ↑	Aspartame	Creppy *et al.*, 1998
OTA *in vitro*	Rat kidney microsomes	TBARS ↑	-	Gautier *et al.*, 2001

Table 2. Stimulation of lipid peroxidation by T-2 toxin.

Mycotoxin	*Tissue*	*Lipid peroxidation measurement*	*Protective effect of antioxidants*	*References*
T-2 in feed	Rat liver	TBARS ↑	Se, ascorbic acid, vitamin E	Tsuchida *et al.*, 1984; Suneja *et al.*, 1989; Schuster *et al.*, 1987; Rizzo *et al.*, 1994
T-2 in feed	Rat liver nuclei	TBARS ↑	-	Ahmed and Ram, 1986
T-2 in feed	Mouse liver	TBARS ↑	-	Karppanen *et al.*, 1989
T-2 in feed	Mouse	MDA in liver after 24-48 hrs after supplementation↑		Vila *et al.*, 2002
T-2 in feed	Mouse liver	DNA fragmentation ↑	CoQ_{10} and vitamin E	Atroshi *et al.*, 1997
T-2 in feed	Quail liver	TBARS ↑	-	Dvorska and Surai, 2001
T-2 in feed	Rat tissues	Conjugate diene ↑	-	Chang and Mar, 1988
T-2 in feed	Chicken	Liver MDA ↑	Lycopene	Leal *et al.*, 1999
T-2 in feed	Chicken, duck, goose	Liver MDA ↑		Mezes *et al.*, 1999

system (Table 3). Similar to the examples above, TBARS accumulation was substantially increased as well as conjugate diene production. At the same time GSH concentration and activities of antioxidant enzymes significantly declined as a result of AFB_1 action.

Fumonisin B_1 also stimulated lipid peroxidation in rat liver, rat liver nuclei fraction, primary rat hepatocytes, Vero cells in culture and phosphatidyl choline (PC) bilayers. In those systems TBARS accumulation and DNA strand breaks were increased (Table 4). DON increased TBARS formation in rat and mouse liver and decreased GSH in rat brain and spleen. There are also data indicating pro-oxidant properties of zearalenone (Karagezyan *et al.*, 1995; Ghedira-Chekir *et al.*, 1999) and citrinin (Ribeiro *et al.*, 1997). Aurofusarin has been shown to decrease antioxidant defences and stimulate lipid peroxidation in quail egg and tissues of newly hatched quail (Dvorska *et al.*, 2002; 2003; Dvorska and Surai, 2004).

It is clear from these data that mycotoxins strongly promote lipid peroxidation in various *in vitro* and *in vivo* systems. This effect was obvious no matter which measurement was used to assess the process of lipid peroxidation.

Mycotoxins and apoptosis

The maintenance of tissue homeostasis involves the removal of superfluous and damaged cells. This process is often referred to as **programmed cell death** or **apoptosis**, since it is thought that cells activate an intrinsic death program contributing to their own demise (Sastre *et al.*, 1996). Several processes, such as initiation of death signals at the plasma membrane, expression of pro-apoptotic oncoproteins, activation of death proteases, endonucleases etc., ultimately coalesce to a common irreversible execution phase leading to cell demise. A balance between cell death and cell survival factors plays a major role in the decision process as to whether a cell should live or die (Ray *et al.*, 2000).

Apoptosis is distinguishable from necrosis. When cell death is induced by osmotic, physical or chemical damage, early disruption of external and internal membranes takes place with subsequent liberation of denatured

Table 3. Stimulation of lipid peroxidation by aflatoxin.

Mycotoxin	*Tissue*	*Lipid peroxidation measurement*	*Protective effect of antioxidants*	*References*
In feed	Rat liver	TBARS ↑, conjugated dienes ↑	Se , vitamin E	Shen *et al.*, 1994
In feed	Rat liver	MDA ↑, NO↑	melatonin - loaded chitosan	El-Gibaly *et al.*, 2003
In feed	Rat liver	MDA ↑, NO ↑	melatonin	Meki *et al.*, 2001
In feed	Rat testis	MDA ↑	Vitamin E	Verma and Nair, 2001
In vitro	Cultured rat hepatocytes	TBARS ↑, LDH release ↑	SOD, catalase	Shen *et al.*, 1995
In feed	Rat liver	TBARS ↑	*Semecarpus anacardium* nut extract	Premalatha *et al.*, 1997
In vitro	Cultured primary rat hepatocytes	TBARS ↑ ROS formation ↑	*Silvia miltorrhiza* extract	Liu *et al.*, 1999
Intraperitoneally	Rat liver	TBARS ↑	Vitamin E, ternatin	Souza *et al.*, 1999
In feed	Rat liver	GSH-Px ↓	Se, vitamin E	Choi *et al.*, 1995
Intraperitoneally	Rat liver and kidney	TBARS ↑	Picroliv	Rastogi *et al.*, 2001
Intraperitoneally	Rat liver	GSH, SOD, Catalase, GSH-Px ↓	-	Rastogi *et al.*, 2001b
In vitro	Primary rat hepatocytes	TBARS ↑, GSH ↓, ROS generation ↑	-	Yang *et al.*, 2000

proteins into the cellular space and induction of an inflammatory response in the vicinity of the dying cell (Sastre *et al.*, 1996). In contrast, apoptosis is characterised by cell shrinkage, nuclear pyknosis, chromatin condensation, DNA cleavage into fragments of regular sizes and activation of proteases called caspases (Dare *et al.*, 2001).

Reactive oxygen species (ROS) are thought to play a major role in apoptosis (Hockenbery *et al.*, 1993; Ratan *et al.*, 1994), being involved in both initiation and execution of apoptosis (Herdener *et al.*, 2000). GSH depletion increases the percentage of apoptotic cells in a given population; and increased GSH concentration is shown to decrease the percentage of apoptosis in fibroblasts (Sastre *et al.*, 1996). In fact, GSH depletion sensitises cells for intracellular induction of apoptosis (Zucker and Bauer, 1997). Therefore, a decrease in GSH, or an increase in GSSG or perhaps a change in the ratio of the two constitutes a trigger for apoptosis (Beaver and Waring, 1995). For example ROS from mitochondria can cause apoptosis after GSH depletion (Zucker *et al.*, 1997). Therefore, apoptosis is induced by oxidative damage either directly from oxygen free radicals or hydrogen peroxide or from their generation in cells by injurious agents. For example, hydrogen peroxide is considered a common mediator for the apoptosis induced by various anticancer drugs (Simizu *et al.*, 1998). In line with those findings there are data showing protective effects of catalase and SOD from different inducers of apoptosis (Sandstrom and Buttke, 1993; Herdener *et al.*, 2000). Indeed, intracellular induction of apoptosis depends on ROS production and can be efficiently blocked by antioxidants (Langer *et al.*, 1996; Schaefer *et al.*, 1995).

As previously mentioned, in many cases mycotoxins decreased cellular GSH, which

Table 4. Stimulation of lipid peroxidation by fumonisin, deoxynivalenol, aurofusarin and citrinin.

Mycotoxin	*Tissue*	*Lipid peroxidation measurement*	*Protective effect of antioxidants*	*References*
Fumonisin	Rat liver nuclei	TBARS ↑, DNA strand breaks ↑	Catalase, mannitol	Sahu *et al.*, 1998
Fumonisin B_1	Primary rat hepatocytes	TBARS ↑	Vitamin E	Abel and Gelderblom, 1998
Fumonisin B_1	Rat liver	TBARS ↑	-	Abel and Gelderblom, 1998; Lemmer *et al.*, 1999
Fumonisin B_1	Phosphatidyl choline bilayers	Rate of peroxidation ↑, free radical Formation ↑, Acceleration of chain formation ↑	-	Yin *et al.*, 1998
Fumonisin B_1	Vero cells	TBARS ↑	-	Abado-Becognee *et al.*, 1998
Fumonisin B_1	Glioma cells, Mouse fibroblasts	TBARS ↑, DNA fragmentation	-	Mobio *et al.*, 2003
Fumonisin B_1	Macrophage cell line	MDA↑	-	Ferrante *et al.*, 2002
DON in feed	Rat liver	TBARS ↑	Se, ascorbic acid, vitamin E	Rizzo *et al.*, 1994
DON + 3-AcDON in feed	Mouse liver	TBARS ↑	-	Karppanen *et al.*, 1989
Aurofusarin in feed	Quail egg yolk	TBARS ↑	-	Dvorska *et al.*, 2001b, Dvorska, 2001
Aurofusarin in feed	Liver of newly hatched quail	TBARS ↑	-	Dvorska *et al.*, 2002 Dvorska and Surai, 2004
Zearalenone injected IV	Rat liver	TBARS ↑	-	Karagezyan *et al.*, 1995
Zearalenone *in vitro*	Vero cells	TBARS ↑	Vitamin E	Ghedira-Chekir *et al.*, 1999
Citrinin in feed	Rat liver	TBARS ↑	-	Ribeiro *et al.*, 1997

can trigger apoptosis. In general, T-2 toxin is the most potent apoptotic agent among mycotoxins. However, there are also reports indicating apoptosis caused by FB_1, OTA and AFB_1. For example, based on the DNA fragmentation profile in gel electrophoresis and the morphological changes seen in electron microscopy, the induction of apoptotic nuclear changes by various mycotoxins was investigated in HL-60 human promyelotic leukemia cells (Ueno *et al.*, 1995). The results showed that T-2 toxin, nivalenol (NIV), DON, OTA, citrinin, AFB_1 and some other mycotoxins induced DNA fragmentation. Morphological evidence of apoptosis was related to the toxicity of the mycotoxins and more toxic NIV and DON resulted in more late stage apoptotic events than FB_1. The results suggested that DNA damage and apoptosis rather than plasma membrane damage and necrosis may be responsible for the observed cytotoxicity.

Fumonisin B_1, T-2 toxin, fusarenon-X and DON could induce apoptosis of liver cells, kidney cells, gastrointestinal epithelial cells and immunological cells as well as several cell lines. The possible mechanisms of apoptosis induction are not well understood. It seems likely that sphingol, p21 gene, protein kinase and intracellular Ca^{2+} level might be involved in the cellular apoptosis induction by *Fusarium* mycotoxins (Wang and Zhang, 2000).

T-2 TOXIN

The rank order of the potency of trichothecene mycotoxins to induce internucleosomal DNA fragmentation was found to be T-2, satratoxin G, roridin A > > diacetoxyscirpenol > baccharin B-5 > > nivalenol, deoxynivalenol, 3-acetyldeoxynivalenol, fusarenon-X, baccharin B-4 vehicle control (Nagase *et al.*, 2001). The authors showed that T-2-induced apoptosis involved activation of caspase-3 through cytosolic accumulation of cytochrome c along with caspase-9 activation. Both T-2 and HT-2 induced apoptosis after 24 hrs in HL-60 cells, with T-2 being somewhat more potent than HT-2 and the apoptotic process was almost completely blocked in the presence of a caspase inhibitor (Holme *et al.*, 2003). Development of apoptosis and changes in lymphocyte subsets were examined in thymus, mesenteric lymph nodes and Peyer's patches of mice up to 24 hrs after oral inoculation with T-2 toxin (10 mg/kg). The degree of lymphocyte apoptosis was prominent in the thymus, moderate in the Peyer's patches, and somewhat mild in the mesenteric lymph nodes (Nagata *et al.*, 2001). Cytocentrifugation and light microscopy of leukocyte-enriched cell samples from the pronephros (the primary hematopoietic compartment in the fish) demonstrated T-2-related increases in apoptotic bodies and the apoptotic probes confirmed apoptosis in fish (Gogal *et al.*, 2000).

Female mice were treated orally with T-2 toxin (10 mg/kg BW) and thymus and spleen were examined to detect apoptotic changes (Shinozuka *et al.*, 1997). The number of apoptotic lymphocytes in the thymus increased dramatically from 9 to 24 hrs after treatment and began to increase in the spleen at 12 hrs after treatment indicating that thymus was more sensitive to apoptosis than spleen. In another experiment, mice were given 5 mg/kg BW of T-2 toxin and were killed 12 hrs later (Ihara *et al.*, 1997). Massive cellular destruction was observed in the thymus and spleen and electron microscopy revealed the presence of apoptotic bodies. Similarly, in the liver of mice given 2.5 mg/kg T-2 toxin and killed 2 hrs later, the induction of apoptotic cellular lesions was observed. Electron microscopic characteristics of damaged lymphocytes were shrinkage of the cell body, nuclear chromatin condensation and fragmentation (Li *et al.*, 1997). Of nine trichothecene mycotoxins tested, T-2 toxin was the most potent agent to induce apoptosis in the thymus in young female mice (Islam *et al.*, 1998; 1998a). It is interesting that both the acetyl group at the C-4 position and the isovaleryl or 3′-hydroxy isovaleryl group at the C-8 position of the T-2 toxin molecule are important for inducing apoptotic changes in the thymus (Islam *et al.*, 1998). T-2 toxin was shown to induce apoptosis in the hematopoietic (bone marrow) tissues of female mice (Shinozuka *et al.*, 1998). It should be noted that the gastrointestinal tract is also sensitive to trichothecene-induced apoptosis, since such changes were observed in the gastric mucosa, gastric glandular epithelium and intestinal crypt cell epithelium (for review see Bondy and Pestka, 2000). These changes could impair nutrient assimilation as well as decrease defences against enteric pathogens and endotoxin.

To examine the effect of T-2 toxin on the developing embryo, pregnant mice were dosed orally with T-2 toxin (3 mg/kg BW) at 11 days of gestation (Ishigami *et al.*, 1999). Ultrastructural changes characteristic of apoptosis were observed. For example, in some layers of the central nervous system moderate pyknosis or karyorrhexis was seen. Using DNA fragmentation and fluorescence microscopy assays Yang *et al.* (2000a) demonstrated that rank order of trichothecene-mediated apoptosis is similar to their cytotoxicity. Therefore, cytotoxicity of trichothecene mycotoxins is mediated through an apoptotic process. In particular, the embryotoxic effect of T-2 toxin on the mouse fetus is mediated through apoptosis. T-2 toxin (2 mg/kg BW) was given orally to pregnant mice at various gestational days and the fetuses were examined 24 hrs later. T-2 toxin-induced apoptosis was detected. Since T-2 toxin readily crosses the placenta, T-2 toxin could directly cause fetal apoptosis (Ishigami *et al.*, 2001).

It was shown that in human promyelotic cell line HL-60, T-2 toxin induced apoptosis at 10 ng/ml levels within 2-6 hrs (Ueno *et al.*, 1995). Apoptosis in HL-60 cells induced by T-2 toxin was dose dependent when cells were treated with concentrations of 5-100 ng/ml for more than 2 hrs (Yoshino *et al.*, 1996). The authors suggested that the Ca^{2+} signal triggered by T-2 toxin is transduced by the activation of endonuclease and protease, and ultimately evokes apoptosis.

Molecular mechanisms of T-2 stimulation of apoptosis are not clear at present. However, recently it has been shown that trichothecene mycotoxins trigger a ribotoxic stress response that activates the stress-activated kinases c-Jun N-terminal kinase and/or p38 mitogen-activated kinase and induces apoptosis (Shifrin and Anderson, 1999). Such activation may signal cell survival or induce apoptosis in various cell types depending on the conditions, e.g. length of signal. However, the pathway leading from the activation of these kinases to the activation of caspases and apoptosis has not been elucidated and needs further research. Albarenque *et al.* (2001) suggested that the elevation of TNF-α mRNA expression may play an important role in T-2 toxin-induced epidermal cell apoptosis. The induction of c-fos and perhaps of c-jun mRNAs may be associated with T-2 toxin-induced epidermal cell apoptosis (Albarenque *et al.*, 2001).

FUMONISIN B_1

The effects of FB_1 on apoptosis in various *in vitro* and *in vivo* studies have received substantial attention over the last few years. For example, FB_1-treated keratinocytes developed morphological features consistent with apoptosis. In particular, they released nucleosomal DNA fragments into the medium 2-3 days after exposure to 0.1 mM FB_1 and showed increased DNA strand breaks (Tolleson *et al.*, 1996). FB_1 induced a significant and dose-related increase of modifications of DNA bases (8-OH-dG) and DNA fragmentation in both C6 glioma and mouse embryonic fibroblasts cells (Mobio *et al.*, 2003). In the same study, apoptotic C6 glioma cells were also observed after FB_1 incubation. Recently genes that inhibit FB_1-induced apoptosis in African green monkey kidney fibronlasts (CV-1 cells) and two mouse embryo fibroblasts have been identified (Jones *et al.*, 2001). Primary mouse embryo fibroblasts underwent apoptosis following FB_1 treatment (Ciacci-Zanella *et al.*, 1999). The authors also demonstrated that the tumour necrosis factor (TNF) pathway and caspases play an important role in FB_1-induced apoptosis. Effects of exposure of human fibroblasts to FB_1 were investigated. After 72 hrs of treatment, FB_1 (50 and 100 mM) induced DNA damage, an enhancement of caspase-3-activity and cleavage of poly(ADP-ribose)polymerase compared to controls (Galvano *et al.*, 2002).

The cytotoxicity and genotoxicity of FB_1 in rabbit kidney RK13 cells were examined. Exposure to FB_1 caused a significant increase in micronucleus frequency in a concentration- and in a time-dependent manner and an increased number of the cells were dying by the process of apoptosis (Rumora *et al.*, 2002). The author suggested that nanomolar concentrations of FB_1 induced apoptosis, which subsequently may proceed to secondary necrosis. Treatment of monkey kidney cells (CV-1 cells) with FB_1 led to cell cycle arrest and apoptosis (Zhang *et al.*, 2001). It was suggested that the ability of FB_1 to alter gene expression and signal transduction pathways may be necessary for its carcinogenic and toxic effects.

In general, FB_1 induces apoptosis of hepatocytes and of proximal tubule epithelial cells. More advanced lesions in both organs were characterised by simultaneous cell loss (apoptosis and necrosis) and proliferation (Voss *et al.*, 2001). A dose-dependent increase in TNF-α–induced apoptosis was observed in porcine renal epithelial cells pretreated with FB_1. In particular, cells treated with FB_1 showed increased DNA fragmentation and terminal uridine nucleotide end labeling in response to TNF-α treatment (Johnson *et al.*, 2003). In the same study, FB_1 also increased DNA synthesis and resulted in cell cycle arrest in the G2/M phase of the cell cycle.

FB_1 toxicity caused induction of cytokine networks in liver with involvement of the TNF-α signaling pathway (Bhandari and Sharma, 2002). The authors suggested that increased expression of caspase 8 involved in the TNF-α signaling pathway may contribute to the apoptosis, whereas IL-1Ra induction could contribute to the proliferating effects observed in FB_1 toxicity. Similarly, FB_1 caused an activation of the cytokine network in liver, particularly the TNF-α signaling pathway, suggesting its involvement in hepatotoxic mechanisms (Bhandari *et al.*, 2002). Induction of apoptosis was suggested to be a consequence of ceramide synthase inhibition and disruption of sphingolipid metabolism by FB_1 (Dragan *et al.*, 2001). It seems likely that the elevated endogenous sphinganine acts as a contributing factor to the fumonisin-induced cell death (Yu *et al.*, 2001). Moreover recent results of Mobio *et al.* (2000) showed that cytotoxic concentrations of FB_1 induce cellular cycle arrest in phase G(2)/M in rat C6 glioma cells. Furthermore, under experimental conditions (9 or 18 mM FB_1 in the medium) there was an induced DNA fragmentation and laddering and many apoptotic bodies in glioma cells.

As mentioned above, FB_1 caused morphological changes (i.e., cell shrinkage, membrane blebbing) and time-dependent increases in DNA fragmentation characteristic for apoptosis. For example, Kim *et al.* (2001) demonstrated that FB_1 kills LLC-PK(1) kidney cells by inducing apoptosis. This included initial disruption of sphingolipid metabolism and accumulation of sphinganine (or a metabolite), which, in turn, induced expression of calmodulin. FB_1, but not AFB_1, induced the apoptosis of swine alveolar macrophages (AM) with evidence of DNA laddering and nuclear fragmentation. However, both FB_1 and AFB_1 exposure induced the expression of apoptosis-related heat shock protein 72 in AM (Liu *et al.*, 2002).

FB_1 has been shown to cause apoptosis in a variety of cell types and tissues, but the apoptotic potential of other fumonisins and fumonisin metabolites could substantially vary. For example, Seefelder *et al.* (2003) exposed human proximal tubule-derived cells (IHKE cells) to FB_1, FB_2, FB_3, hydrolyzed FB_1 and N-palmitoyl-hydrolyzed FB_1 and investigated caspase-3 activation, chromatin condensation and DNA fragmentation. The results demonstrated that all compounds led to increased sphinganine levels in IHKE cells, but only FB_1 was able to induce apoptosis. In the short-term studies, increases in tissue concentration of the ceramide synthase substrate sphinganine and the sphinganine:sphingosine ratio were correlated with apoptosis (Voss *et al.*, 2002).

Although it has been demonstrated that FB_1 induces apoptosis in many cell lines, the precise mechanism of apoptosis is not fully understood. For example, FB_1-induced apoptosis involves the activation of caspase 3, which is associated with the repression of protein kinase C and possibly its downstream effectors, NF-κB and TNF-α (Gopee *et al.*, 2003).

After intravenous injection of male rats with FB_1 (1.25 mg/kg), hepatic and renal GSH concentrations were depressed and proliferation and apoptosis were observed in the outer medulla of the kidney cell (Lim *et al.*, 1996). When male mice were injected with FB_1 subcutaneously at doses of 0, 0.25, 0.75, 2.25 and 6.75 mg/kg BW daily for 5 days, a dose-dependent increase in apoptosis in liver and kidney was observed (Tsunoda *et al.*, 1998). Similar observations were described by Sharma *et al.* (1997). Male mice were injected subcutaneously with vehicle or 2.25 mg/kg/day of FB_1 for 5 days and sampled 1 day after the last treatment. FB_1 increased apoptotic cells in liver (Bhandari *et al.*, 2002). Similarly, histopathology revealed the occurrence of apoptosis in the liver of rats exposed to FB_1 (Pozzi *et al.*, 2001).

Multiple daily doses of FB_1 after surgery elevated the number of apoptotic hepatocytes in rats (Li *et al.*, 2000). Including FB_1 in the diets of rats increased hepatocellular and renal tubule epithelial cell apoptosis (Howard *et al.*, 2001). It is interesting that female rats demonstrated more sensitivity than male rats in the induction of hepatocellular apoptosis and mitosis. Mild toxic effects, including apoptosis, proliferation of bile duct epithelial cells, and early signs of fibrosis were noticed in the liver of rats fed FB_1 (25 mg/FB_1/kg diet; Gelderblom *et al.*, 2001). Six hours after administration of FB_1, marked morphologic changes of rat hepatocytes included the appearance of small vacuoles along the margin of the cell membrane. Electron microscopic analysis revealed margination of nuclear chromatin and swollen mitochondria with amorphous matrical deposit (Moon *et al.*, 2000). The authors suggested that FB_1-induced alteration of hepatocyte membrane lipid composition is an early key event in the FB_1 apoptotic effect. Therefore, the induction of tubular epithelial cell apoptosis was the primary response of the kidneys to dietary FB_1. Apoptosis was present at all doses (99-484 ppm for 28 days) in the kidney of male rats and occurred in females at doses of 153-484 ppm FB_1 (Tolleson *et al.*, 1996). Wistar rats were fed diets containing 0 (control) or 100 ppm FB_1 for 12 weeks. Necrosis and apoptosis of tubular epithelial cells in the kidney were observed and increased mitotic figures and lymphocytic infiltrate in the small intestine were found (Theumer *et al.*, 2002). Increased hepatocellular apoptosis was also detected in the mice consuming diets containing 72 and 143 mmol/kg FM_1 (Howard *et al.*, 2002).

FB_1 can induce renal injury and organ sphingolipid alterations in cattle. For example, renal lesions in FB_1-treated calves consisted of vacuolar change, apoptosis, caryomegaly, and proliferation of proximal renal tubular cells, as well as dilation of proximal renal tubules, which contained cellular debris and protein (Mathur *et al.*, 2001). Probably there is species-specificity in susceptibility to FB_1. For example, FB_1 was investigated in four groups of growing ducks, each receiving 0, 5, 15 or 45 mg/kg FB_1 by daily oral administration over 12 days. No sign of apoptosis was present in the liver or in peripheral blood lymphocytes and only moderate oxidative damage was noted (Bailly *et al.*, 2001).

The apoptosis-stimulating effect of FB_1 has been clearly demonstrated; however molecular mechanisms of this stimulation are not clear at present. Nevertheless, stimulation of lipid peroxidation by FB_1 and decreased antioxidant concentrations including GSH in tissues could lead to changes in redox status of the cell and trigger a cascade of apoptotic changes.

OCHRATOXIN A

Lipid peroxidation due to OTA consumption may be implicated in DNA damage. When a specially designed oxidation system was used, OTA was found to facilitate single-strand cleavage of supercoiled plasma DNA through ROS production (Gillman *et al.*, 1999). In general, OTA is implicated in DNA-adduct formation (Grosse *et al.*, 1997). OTA treatment of rats resulted in a 5-fold increase in the expression of the protein haem oxygenase-1 generated during oxidative stress (Gautier *et al.*, 2001). Changes in redox status of the cell due to the pro-oxidant action of OTA could trigger cell apoptosis. For example, OTA at a dose of 20 mg/kg caused nuclear changes characteristic of apoptosis in hamster kidney (HaK) and HeLa cells (Seegers *et al.*, 1994). Furthermore, the results indicated that OTA might activate different cellular processes involved in the degradation of DNA in HaK and HeLa cells. Exposure of human proximal tubule-derived cells to 30 nmol/L or more OTA led to DNA fragmentation and chromatin condensation (Schwerdt *et al.*, 1999). It was concluded that exposure to low OTA concentrations can lead to direct or indirect caspase-3 activation and subsequently to apoptosis in cultured human proximal tubule cells and in other renal epithelial cell lines of different origins.

Administration of OTA twice a week for one or two weeks resulted in apoptosis in the liver of mice (Atroshi *et al.*, 2000). In particular, the presence of intracellular apoptosis bodies was detected two weeks after toxin treatment. Light microscopic examination demonstrated the presence of eosinophilic globules, often containing apoptotic bodies. They were found within the cytoplasm of intact hepatic cells. The number of apoptotic bodies was further enhanced at two weeks, resulting in an 8-fold increase in liver over the control values (Atroshi *et al.*, 2000). Wistar rats were treated with a low dose (5 mM and 1 mg/kg, respectively) or a high dose (12.5 mM and 10 mg/kg, respectively) of OTA for 24 or 72 hrs. Marked treatment-specific transcriptional changes were detected for genes involved in DNA damage response and apoptosis (Luhe *et al.*, 2003). Because of OTA consumption, there was a striking increase in the counts of eosinophils and of apoptotic phagocytes in weaner pigs in a dose-dependent manner (Muller *et al.*, 1999).

Nanomolar concentrations of OTA promote apoptosis in a cell-type specific fashion. For example, OTA, at noncytotoxic doses, was able to detach collagen- and fibronectin-adherent cells from immobilized substratum causing apoptosis as measured by caspase-3 activation (Scibelli *et al.*, 2003). Relatively low OTA concentrations (547.2, 752.5 and 930.3 ng OTA/g kidney tissue) have activated apoptotic processes and oxidative damage in kidney cells (Petrik *et al.*, 2003). In human monocyte/macrophage line THP-1 metabolic activity, cell proliferation, cell membrane integrity, cell differentiation, phagocytic behaviour, NO synthesis and cell surface markers were largely suppressed by OTA at concentrations between 10 and 1000 ng/ml (Muller *et al.*, 2003). OTA greatly induced apoptosis in human myc-transfected cell line (Horvath *et al.*, 2002).

Exposure to low OTA concentrations (5-30 nmol/L) can lead to time- and concentration-dependent direct or indirect caspase-3 activation and subsequently to apoptosis in cultured human proximal tubule cells and in other renal epithelial cell lines of different origins (Schwerdt *et al.*, 1999). Additional evidence was provided indicating that OTA interacts in a cell type-specific way with distinct members of the mitogen-activated protein kinase family providing induction of apoptosis via the c-jun amino terminal kinase pathway (Gekle *et al.*, 2000).

OTHER MYCOTOXINS

When the *in vitro* effect of deoxynivalenol (DON) on apoptosis in specific T- and B-cell subsets within thymus, spleen and Peyer's patch cultures was studied (Pestka *et al.*, 1994) it was shown that depending on lymphocyte subset, tissue source and glucocorticoid induction, DON could either enhance or inhibit apoptosis. Combined treatment of mice with DON and lipopolysaccharide significantly increased the amount of apoptotic thymic and splenic tissue (Zhou *et al.*, 2000). It was suggested that inhibition of protein synthesis and induction of apoptosis are the main mechanisms of DON toxicity in intestinal cells (Maresca *et al.*, 2002).

It seems likely that DON can interact with other inducers of apoptosis to increase apoptotic action. As demonstrated by DNA fragmentation and flow cytometric analysis, apoptosis in thymus, Peyer's patches, and bone marrow was marked in mice 12 hrs after administering *Escherichia coli* LPS (0.1 mg/kg BW i.p.) concurrently with DON (12.5 mg/kg BW p.o.), whereas apoptosis in control mice or mice treated with either toxin alone was minimal (Islam *et al.*, 2002). Therefore, lipopolysaccharide can interact with DON in mice to induce the glucocorticoid-driven apoptotic loss of immature thymocytes and cytotoxic T-lymphocytes in thymus, mature-B-lymphocytes in Peyer's patch, and pro/pre-B-lymphocytes and mature B-lymphocytes in bone marrow in mice (Islam *et al.*, 2003).

Dead lymphocytes in the thymus, spleen and Peyer's patches showed ultrastructural characteristics of apoptosis. DON and aflatoxin G_1 (AFG_1) could induce and accelerate apoptosis in human peripheral blood lymphocytes. For example, a typical sub-diploid apoptosis peak was demonstrated in lymphocytes treated with DON and AFG_1 (Wang *et al.*, 1999). In particular, a significant dose response effect and time effect correlation was found between apoptosis rates and mycotoxin concentrations and the treated time.

In an experiment with male rats, AFB_1 was administered i.p. or intratracheally at 2-8 mg/kg BW and animals were killed after 26 hrs (Raj *et al.*, 2001). Dose-dependent induction of micronuclei in bone marrow and lung cells was observed. Furthermore, apoptotic bodies were found in lung cells of rats given AFB_1, which was correlated with micronuclei induction. Clearly, AFB_1 induced micronuclei and apoptosis in lung and bone marrow cells. This stimulation could be associated with the stimulation of the phosphatidylinositol (PI) cycle and activation of PI kinase by AFB_1 (Pasupathy *et al.*, 1999). Apoptotic rates in rat liver were significantly reduced when melatonin was co-administered with AFB_1 (El-Gibaly *et al.*, 2003). Rats were treated with AFB_1 (50 mg/kg BW) for 8 weeks. The levels of caspase-3 activity in the AFB_1 group were significantly higher than in controls and the apoptosis was associated with degenerative and necrotic changes in the hepatocytes (Meki *et al.*, 2001a). In the study, caspase-3 activity was positively correlated with MDA while negatively correlated with GSH, GSH-Px and GSH-reductase in rat livers treated with AFB_1. It is interesting that melatonin treatment of rats could enhance hepatic antioxidant/detoxification system, which consequently reduced the apoptotic rate and the necrobiotic changes in the liver.

It seems likely that apoptosis is the principal mechanism contributing to germ cell depletion and testicular atrophy following zearalenone (ZEA) exposure. In particular, Kim *et al.* (2003) demonstrated that a single dose of ZEA (5 mg/kg) induced testicular germ cell apoptosis in a time-dependent and stage-specific pattern. In another study, ZEA (10, 20 and 40 mM) induced concentration-dependent DNA fragmentation in three cell lines (Vero, Caco-2 and DOK) resulting in DNA laddering patterns on agarose gel electrophoresis. This observation was consistent with apoptosis, which was confirmed by observations of formation of apoptotic bodies. Furthermore, ZEA induced cell cycle arrest in the three cell lines characterised by an increase in the number of cells in the G2/M phase of the cell cycle (Abid-Essefi *et al.*, 2003). It is interesting that vitamin E (25 mM) added simultaneously with ZEA partially reduced DNA fragmentation and apoptotic body formation after 24 hrs incubation.

Mycotoxins and the immune system

All animals protect themselves from invasion by microorganisms, parasites, fungi, viruses and any foreign molecules. This protective capacity requires an effective immune system, which is an important determinant of animal health and well being. In that sense, the remarkable ability of the immune system to distinguish between self and non-self is a great achievement of evolution. Commercial animal production is based on balanced feed, providing nutrient requirements and optimised environmental conditions. However, it is very difficult to avoid various nutritional or environmental stresses, which are responsible for immunosuppression and increased susceptibility to various diseases and consequently decreased productive and reproductive performance of farm animals. In this respect mycotoxins are one of the most immunosupressive factors in animal diets.

There are two major types of immune function: natural and acquired immunity (Figure 2). Natural immunity, called the innate immune system, includes physical barriers (e.g. skin, mucus layer of the gastrointestinal tract), specific molecules such as agglutinins, precipitins, acute phase proteins and lysozyme, phagocytic destruction of pathogens by macrophages and neutrophils, and lysing activity by lymphocytes called natural killer (NK) cells.

Macrophages perform a range of functions, including phagocytosis of foreign particles, destruction of bacterial or tumor cells, secretion of prostaglandins and cytokines, and as a result regulating activity of lymphocytes and other macrophages (Qureshi, 1998). In fact, phagocytosis is the major mechanism by which microbes are removed from the body and is especially important for defence against extracellular microbes. As a result of stimulation (for example by microbes) monocytes differentiate into macrophages that are more powerful in mediating host defence (Klasing, 1998a).

Macrophage activation and phagocytosis of foreign particles are regularly accompanied by a **respiratory burst**, an increase in the production of reactive oxygen species (ROS), exerted by the enzyme complex NADPH oxidase. Therefore macrophages as well as

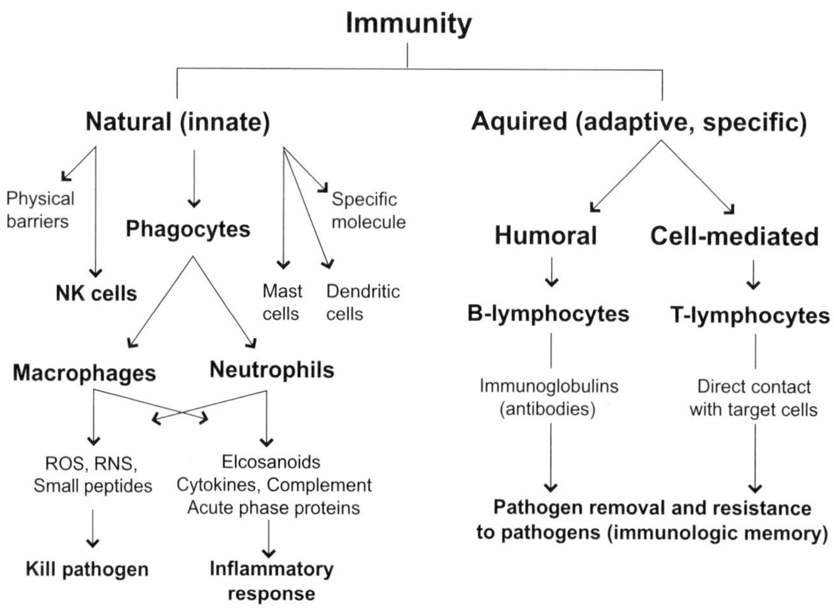

Figure 2. General overview of the immune system.

other phagocytic leukocytes (e.g., neutrophils, monocytes and eosinophils) can synthesize toxic oxygen metabolites such as superoxide anion (O_2^-), hydroxyl radical (OH•), singlet oxygen ($^1O_2$), hydrogen peroxide (H_2O_2), nitric oxide (NO) and peroxinitrite ($ONOO^-$) during the respiratory burst (Zhao *et al.*, 1998). For example, a bacterium coming into contact with the plasma membrane is enclosed in a plasma membrane vesicle containing NADPH oxidase, and exposed to an intensive flow of superoxide radical (Giller and Sigler, 1995). Superoxide radical can dismutate to H_2O_2, which penetrates the bacterium with production of hydroxyl radical, which is ultimately deadly to biological molecules. In general, the production of ROS and reactive nitrogen species (RNS) is a characteristic of both mammalian and avian macrophages (Qureshi *et al.*, 1998a).

In general ROS, RNS and eicosanoids (e.g. leukotrienes and prostaglandins) have recently received substantial attention as major metabolites produced by macrophages (Dietert and Golemboski, 1998). Because of this powerful weapon, macrophages bind, internalize, and degrade foreign antigens (e.g. bacteria) quite quickly. It takes only 15 minutes for chicken macrophages to kill more than 80% of the internalized *Salmonella* (Qureshi *et al.*, 1998a). Therefore natural immunity works rapidly, and gives rise to the acute inflammatory response. Macrophages contain various substances (including enzymes producing free radicals and small peptides with antibiotic activity) involved in microbial killing. They also have receptors for chemoattractive factors released from microbes. In addition to ROS, RNS and eicosanoids mentioned above, macrophages also synthesize and secrete a great number of cellular communication molecules such as cytokines, including the pro-inflammatory cytokines interleukin 1 (IL-1), interleukin 6 (IL-6), and TNF-α. They also produce cytokine inhibitors, hormones and neurotransmitters (Klasing, 1998a), which regulate specific immunity (initiating and directing the immune and inflammatory responses) and many other related physiological responses. Therefore macrophages are important amplifiers of the immune response, both by cytokine production and by serving to present parasite derived peptides to T-cells.

Acquired or specific immunity consists of humoral and cell-mediated immunity. There are two major types of lymphocytes, B-cells and T-cells. Humoral immunity is mediated by antibodies that are released by B-cells into the bloodstream. The bursa of Fabricius is the site of B-lymphocyte development and differentiation in birds. This immunity is based on the production of immunoglobulins. They are responsible for recognition and elimination of specific antigens: they bind and remove from the host invading organisms/substances.

Cell-mediated immunity is based on specific antigen recognition by thymus-derived T-lymphocytes. Due to this immunity cells infected with a foreign agent, for example virus, are destroyed via direct contact between an activated T-cell and the target (infected) cell (Qureshi *et al.*, 1998). Cell-mediated immunity is responsible for delayed-type hypersensitivity (DTH) reactions, foreign graft rejections, resistance to many pathogenic microorganisms and tumor immunosurveillance (Wu and Maydani, 1998).

In birds, both T-cell and B-cell precursors originate in the bone marrow. Actual development of T-cells takes place in the thymus and B-cells develop in the bursa of Fabricius in birds and in bone marrow of mammals (Lahtala, 1998). Interactions between T- and B-cells, as well as antigen presenting cells, are responsible for the development of specific immunity. The defence mechanisms of specific immunity are induced or stimulated by exposure to foreign substances, are specific for distinct macromolecules and increase in magnitude with each successive exposure to a particular macromolecule (Miles and Calder, 1998). In comparison to natural immunity, specific immunity takes longer to develop, but is highly specific for antigens and has memory. These two parts of the immune system work together through direct cell contact and through interactions involving such chemical mediators as cytokines and chemokines. Therefore, the animal immune system requires the co-operation of macrophages, bursa-derived B-lymphocytes and thymus-derived T-lymphocytes with various other types of cells.

The immune response involves cellular proliferation (T-lymphocytes), enhanced protein synthesis (including immunoglobulin synthesis by B-lymphocytes and acute phase protein synthesis in the liver) and inflammatory mediator production. Physiological changes resulting from stimulation of the immune system include fever, anorexia and loss of tissue components (Grimble, 1997).

When analysing data related to immunomodulation careful attention must be paid to methods used to assess immunological functions. For example, *in vivo* methods of immune function assessment are based on two main approaches: antibody response to vaccine or delayed-type hypersensitivity (DTH) reactions. In the first method, immunisation with appropriate antigens (viral or bacterial) can elicit serum antibodies. The haemagglutination (HA) assay measures serum antibody concentration (titre) against antigens, which are often sheep red blood cells (SRBC). This assay provides information about humoral immunity (B-cell responsiveness) and its association with cell-mediated immunity (T-cell co-operation). The second (DTH) method is used to assess cell-mediated immune function.

In vitro indices of immune function include (Wu and Meydani, 1998):

- *Lymphocyte proliferation assay*. This assay provides information about cell-mediated immune response and consists of measuring the number of cells in culture with or without addition of a stimulatory agent (mitogen). In this assay, isolated lymphocytes are incubated with mitogens, which activate division of either T-or B-lymphocytes. Various mitogens are used in such assays, but most often they include concanavalin A (con A, a T-cell mitogen), phytohemagglutin (PHA, a T-cell mitogen), lipopolysaccharide (LPS, a B-cell mitogen) and pokeweed mitogen (PWM, T- and B-cell mitogen) (Hayek *et al.*, 1996). Decreased proliferation may indicate impaired cell-mediated immunity.
- *Cytokine production*. T-cells produce a range of protein mediators called cytokines, which regulate cell activation, growth, differentiation, inflammation and immunity.
- *Cytotoxicity assay*. This assay assesses activity of cytotoxic T-lymphocytes (a group of T-cells that kill other cells by recognising their cell-surface antigens) and NK cells (a group of non-T and non-B lymphocytes that kill virus-infected and tumour cells).
- *Flow cytometric analysis*. This assay identifies cells with different surface markers. The results can be used for understanding the cellular basis of immune response.
- *Plaque-forming cell (PFC) test*. This shows the number of antibody-producing cells.

The impact of the avian immune system in modern production cannot be overestimated. The banning of feed grade antibiotics in Europe has made immunocompetence the major factor determining efficiency of poultry production. Molecular immunology is developing very quickly; and mechanisms and factors affecting avian immunocompetence have recently received substantial attention (McCorkle, 1998; Saif and Swayne, 1998) and nutritional modulation of resistance to infectious diseases in poultry (Klasing, 1998) is a frontline for future research. It must be emphasised that cellular integrity is very important for receiving and responding to the messages needed to co-ordinate an immune response (Latshaw, 1991). Therefore, the antioxidant/pro-oxidant balance of the host is a critical consideration in optimal immune system function.

The clinical toxicologic syndromes caused by ingestion of mycotoxins have been characterized in domestic animals, poultry and laboratory animals and range from acute mortality to decreased performance and immunosuppression. In fact, consumption of certain mycotoxins, at levels that do not cause overt clinical mycotoxicosis, suppresses immune function and decrease resistance to infectious disease (Corrier, 1991). The effects of several mycotoxins on immune responses have been investigated (Tables 5-9); however, most data were obtained with laboratory animals. In some instances, farm animals and cells derived from livestock species have been employed to evaluate the immunotoxicity of mycotoxins (Sharma, 1993; Bondy and Pestka, 2000).

Immunosupressive potency of various mycotoxins differs substantially. Effects of DON, 3-acetyldeoxynivalenol, fusarenon-X, T-2 toxin, ZEA, α–zearalenol, ß-zearalenol and NIV on T- and B-cells were studied using a proliferation assay and antibody-dependent cellular cytotoxic NK cell activity on human peripheral blood mononuclear cells (Berek *et al.,* 2001). The mycotoxin concentrations used in the experiments were comparable to those found in normal human peripheral blood (0.2-1800 ng/ml). Among the mycotoxins tested, T-2 toxin, fusarenon-X, NIV and DON showed the highest immunosuppressing effects *in vitro*. In particular, mycotoxin-induced immunosuppression was related to depressed T- or B-lymphocyte activity. Furthermore, they also inhibited NK cell activity (Berek *et al.,* 2001).

As can be seem from Table 5, immunosuppression caused by AFB_1 has been demonstrated in various livestock species (e.g., chickens, turkeys, pigs and lambs) and also in laboratory animals (mice and rats) and in various *in vitro* systems. Aflatoxin is an immunomodulating agent that acts primarily on cell-mediated immunity and phagocytic cell function. Therefore, AFB_1 mainly decreases lymphocyte activity and may affect macrophages assisting lymphocyte function. For example, in macrophages exposed to AFB_1, NO production stimulated by LPS significantly decreased (Moon *et al.,* 1998). Furthermore, AFB_1 decreased pro-inflammatory and increased anti-inflammatory cytokine mRNA expression in weanling piglets (Marin *et al.,* 2002). AFB_1 can be transferred from chicken to the egg and further to the developing embryo. Therefore, the progeny of hens consuming an AFB_1-contaminated diet may be increasingly susceptible to disease owing to suppression of humoral and cellular immunity. For example, anti-*Brucella abortus* antibody production was compromised and ROS production by macrophages from AFB_1 progeny decreased (Qureshi *et al.,* 1998). Long-term immune depression of macrophage-mediated functions can occur following embryonic exposure to AFB_1 (Neldon-Ortiz and Qureshi, 1992). Furthermore, depressed complement and interferon production could also result from AFB_1 exposure. Therefore, acquired immunity from vaccination programs may be substantially suppressed; and in such cases the signs of disease observed are those of the infectious process rather than those of the aflatoxin that predisposed the animal to infection. Mixtures of aflatoxin with other mycotoxins can result in greatly augmented biological responses in terms of weight gain depression, lethality, and immune reactivity (Pier, 1992).

The immunomodulatory effects of ochratoxins have also been considered for many years and a summary is shown in Table 6. OTA has been shown to affect mainly humoral immune function at the level of antibody synthesis in chickens, rats and mice. However, number and phagocytic activity of macrophages were also decresed in growing gilts receiving OTA for 35 days (Harvey *et al.,* 1992). IL production was also compromised. It has been demonstrated that exposure of purified

Table 5. Effects of aflatoxin B_1 on immunity.

Species	*Dose*	*Effects on immune system*	*Reference*
Broiler chicks	1 mg/kg, 7-49 days	Titres to NK ↓	Shivachandra *et al.*, 2003
Broiler chicks	2.5 mg/kg diet for 21 days	Peripheral T-lymphocyte counts ↓, Splenic plasma cell counts ↓	Celik *et al.*, 2000
Broiler chicks	2.5 mg/kg diet for 21 days	Percentage/mean phagocytic activity ↓; Immune response measured by HI test ↓	Ibrahim *et al.*, 2000
Broiler breeder hens	0.2; 1; 5 or 10 mg/kg	In progeny, anti-*Brucella abortus* titres ↓; Phagocytosis of SRBC and ROS production by macrophages ↓ in 5 mg AFB_1/kg	Qureshi *et al.*, 1998
Broiler chicks	5 mg/kg	Secondary antibodies against IBD↑	Okotie-Eboh *et al.*, 1997
Broiler chicks	400 μg/kg AFB_1 + AFB_2 for 5 weeks	Cell-mediated immunity measured by DTH skin test ↓; Humoral immunity unchanged; Acquired immunity to ND or fowl cholera vaccination unchanged	Giambrone *et al.*, 1985a
Broiler chickens	2.5 μg/g of feed for 42 days	Total complement activity ↓	Stewart *et al.*, 1985
Chickens	2.5 μg/g diet from hatching to 4 weeks of age	Cell-mediated immunity (graft-vs. host reaction) ↓; DTH reactions to tuberculin ↓; Humoral immunity (to rabbit RBC) unchanged; Serum IgG and IgA↓; IgM unchanged	Giambrone *et al.*, 1978
Chickens	0.3 mg/kg from 0 to 6 weeks	Cell mediated immune response (DTH reaction) suppressed at 30, 45 and 60 days of age	Kadian *et al.*, 1988
14-day-old turkeys	200 μg/kg	Cell-mediated immunity (DTH skin test) ↓; Humoral immunity unchanged; Acquired immunity to ND or fowl cholera vaccination unchanged	Giambrone *et al.*, 1985b
4-month-old wild turkeys	100-400 μg/kg feed for 2 wk	Lymphoblast transformation ↓	Quist *et al.*, 2000
Chick embryo	1.09-17.4 μg/g embryo at day 18 of incubation	Mitotic index B-cells ↓; Mitotic index T-cells unchanged; Sister chromatid exchanges in B-lymphocytes ↑; Sister chromatid exchanges in T-lymphocytes ↑	Potchinsky and Bloom, 1993
6-day chick embryos	0.1, 0.5, and 1 μg/per embryo	Macrophages recruited in the peritoneal cavity after i.p. Sephadex elicitation ↓; Substrate adherence potential of peritoneal exudate cells ↓ at 1 μg AFB_1; Macrophage phagocytic potential ↓ at 0.5-1.0 μg AFB_1	Neldon-Ortiz and Qureshi, 1992

Table 5. Continued.

Species	*Dose*	*Effects on immune system*	*Reference*
6-day chick embryos	0.1 μg	Incidence of sister chromatid exchanges in blood cells ↑ (5-fold); Cell-mediated immunity (graft vs host and cutaneous basophil hypersensitivity reactions) ↓	Dietert *et al.*, 1985
Weanling piglets	140-280 μg/kg, 4 wks	AFB_1 decreased proinflammatory (IL-1ß, TNF-α) and increased anti-inflammatory (IL-10) cytokine mRNA expression	Marin *et al.*, 2002
Weaned pigs	140, or 280 μg/kg, for 3 wks	Skin thickness (PHA test) ↓ linearly with ↑ dietary AFB_1	van Heugten *et al.*, 1994
Pigs	300 or 500 mg/kg of feed	Humoral immune response to *Erysipelothrix rhusiopathiae*, measured by ELISA unchanged; Proliferative responses to mitogens unchanged; Complement titers↓; Serum IgG and M values↑	Panangala *et al.*, 1986
Female lambs	2 mg/kg for 37 d	Response to intradermal injection of PHA↓	Fernandez *et al.*, 2000
White-tailed deer	0.8 mg/kg for 8 wks	Lymphocyte proliferation and DTH reactions unchanged	Quist *et al.*, 1997
Weanling rats	60, 300 and 600 μg/kg BW	Cell mediated immunity, measured by DTH response assay, ↓ at the 300 and 600 μg dose levels	Raisuddin *et al.*, 1993
Weanling rats	350 and 700 μg/kgBW	Population and phagocytic capacity of macrophages ↓	Raisuddin *et al.*, 1990
Rats and mice	Aerosol inhalation or intratracheal instillation to AFB_1	Alveolar macrophage phagocytosis ↓ at 16.8 μg/kg with the effect persisting for ~2 weeks	Jakab *et al.*, 1994
Mice	0.03, 0.145 or 0.70 mg/kg orally every other day for 2 weeks	[^{3}H]thymidine uptake in lymphocyte cultures↓; Synthesis of DNA in mixed lymphocyte ↓; Primary antibody production by splenic cells from animals challenged with SRBC; DTH reaction to hemocyanin ↓	Reddy *et al.*, 1987
Murine peritoneal macrophages	10 or 50 mM	NO production stimulated by LPS ↓, mediated by the reduction of iNOS activity, mRNA, and protein	Moon *et al.*, 1998
Human monocytes	0.1-1 pg/ml	Phagocytosis and microbicidal activity ↓, Intrinsic antiviral activity or superoxide production unchanged	Cusumano *et al.*, 1996
Rat Kupffer cells *in vitro*	0.01 pg/ml	Phagocytosis ↓, intracellular killing of *Candida albicans* ↓, intrinsic anti-Herpes virus activity ↓	Cusumano *et al.*, 1995

Table 6. Effect of Ochratoxin A on immunity.

Species	*Dose*	*Effects on immune system*	*Reference*
Broiler chicks	2 mg/kg	Vaccination titres to NK ↓, Mitotic cells in the bursa ↓	Santin *et al.*, 2002
Chicks	5 mg/kg	Weight of lymphoid organs ↓, Humoral immune response ↓	Stoev *et al.*, 2002
Broiler chicks	130, 305 and 790 mg/kg	HI titers to ND vaccination ↓	Stoev *et al.*, 2000
Broiler chicks	Up to 4 mg/kg for 20 days from hatch	Ig-containing cells in lymphoid organs ↓; Total Ig levels ↓ at 2-4 ppm	Dwivedi and Burns, 1984
Broiler chickens	0.5-8 μg/g from day-old to 3 weeks of age	Total circulating lymphocytes ↓; Monocytes ↓ at 2.0 μg/g; Circulating heterophils unchanged	Chang *et al.*, 1979
Salmonella-challenged broiler chicks	3.0 mg/kg	PHA- and ConA-stimulated blastogenesis ↓; *S. typhimurium* alone had no effect on the variables measured	Elissalde *et al.*, 1994
Growing gilts	2.5 mg/kg of feed for 35 days	Cutaneous basophil hyper-sensitivity response to PHA ↓, DTH to tuberculin ↓, Stimulation index for lymphoblastogenesis ↓, Il-2 production when lymphocytes were stimulated with ConA ↓, Number and phagocytic activity of macrophages ↓	Harvey *et al.*, 1992
Rats	Gavage with 0, 0.07, 0.34 or 1.68 mg OTA/kg BW for 4 wks	Thickening of the basement membrane and reduction in splenic T-cell fraction. IgG ↓ at 0.34-1.68 mg/kg OTA	Dortant *et al.*, 2001
Rats	Dam's exposure to a single dose of OTA (0, 10, 50 or 250 mg/kg BW) on day 11 of lactation	The proliferative response of splenocytes to ConA ↑ in pups from dams given 10 or 50 μg OTA/kg BW; Proliferation of thymocytes in response to ConA ↑ in pups from dams exposed to 50 μg OTA/kg BW	Thuvander *et al.*, 1996b
Mice	Dams, OTA, 200 μg/kg before and during gestation	In offspring on days 14 and 28 postpartum percentages of splenic CD4+ and CD8+ cells ↓	Thuvander *et al.*, 1996
Mice	Dam's exposure to OTA (500 μg/kg BW) on day 16 of gestation	Proliferation of splenic and thymic lymphocytes in response to mitogens in pups at 15 days of age ↓; percentages of mature CD4+ cells ↓ and percentages of immature, double-positive (CD4/CD8+) cells in the exposed pups ↑	Thuvander *et al.*, 1996a
Mice	Dam's exposure to OTA (500 μg/kg BW) on day 10 postpartum	Proliferative responsiveness of lymphocytes in the offspring when stimulated with B- or T-cell mitogens 3 days after exposure ↑	Thuvander *et al.*, 1996a

Table 6. Continued.

Species	*Dose*	*Effects on immune system*	*Reference*
Mice	250 or 2600 mg/kg diet, 28 days	After 28 days, AB production and plaque-forming cells ↓; Thymocyte cell counts ↓ in the 250-μg/kg group; After 90 days, proportion of mature CD4+ or CD8+ cells ↓	Thuvander *et al.*, 1995
Mice	0.005 μg/kg BW	Immune response to SRBC ↓	Haubeck *et al.*, 1981

human lymphocyte populations and subpopulations to the toxin will abrogate the cell's ability to respond to activating stimuli *in vitro* (Lea *et al.*, 1989). Thus, both IL-2 production and IL-2 receptor expression of activated T-lymphocytes were severely impaired. The results strongly suggest that the toxin caused immunosuppression through interference with essential processes in cell metabolism (Lea *et al.*, 1989). In particular, OTA appears to suppress NK cell activity by inhibiting production of basal interferon (Luster *et al.*, 1987).

It is important to note that subchronic oral exposure to OTA affects certain immune functions in mice at exposure levels that may be found in contaminated food products (Thuvander *et al.*, 1995). In the mouse model OTA had a non-selective suppressive effect on various immune reactions. They include weight depression, lymphopenia, neutrophilia and eosinophilia. Furthermore, antibody-producing cells, antibody titres in blood serum and phagocytosis of *E. coli* by blood phagocytes become suppressed. Immunized animals also showed a lower survival rate after experimental infection with *Pasteurella multocida* as well as an increase in oxygen radicals in blood cells (Muller *et al.*, 1995). Similarly, in *Salmonella*-challenged broiler chicks PHA- and ConA-stimulated blastogenesis was depressed as a result of OTA consumption, while *Salmonella* alone had no effect on these parameters (Elissalde *et al.*, 1994).

In weaner pigs subtoxic amounts of OTA produced immunomodulation in a dose-dependent mode (Muller *et al.*, 1999). It included increased counts of total leukocytes and neutrophils in the blood and reduced lymphocyte levels. There was also a substantial increase in the counts of eosinophils and of apoptotic phagocytes. Reduced phagocytosis and reduced expression of a swine-specific surface marker on lymphocytes were also observed.

It is interesting that the time of exposure significantly influences the immunotoxic effects of OTA on the developing immune system in rodents (Thuvander *et al.*, 1996a). Recent studies have examined immune function in the offspring of rats and mice exposed to OTA pre- and perinatally (Bondy and Pestka, 2000). It has been shown that the decrease in proliferation and antibody production in young animals resulted from prenatal modulation of the immune system. In particular, prenatal exposure to relatively low doses of OTA may induce immunosuppression in the progeny. In contrast, short-term exposure of suckling pups to OTA via the milk stimulated the proliferative responses of lymphocytes to polyclonal activation.

Fumonisin toxicity has been characterized relatively recently in comparison to aflatoxin and ochratoxin (Table 7), and fumonisin-induced immunotoxicity is an area of active research (Bondy and Pestka, 2000). In fact, FB_1 has diverse effects on the immune system, causing both stimulation and suppression of responses to foreign antigens, and apparently inducing an antigenic response to FB_1. For example, in chickens FB_1 caused a decrease in total immunoglobulins, in IgG and macrophage phagocytic activity (Qureshi

Table 7. Effects of Fumonisins on immunity.

Species	*Dose*	*Effects on immune system*	*Reference*
White Leghorn chicks	6.1, 10.5, and 42.7 mg/kg from day 7 to 42	Splenic, thymic, and liver weights ↓; Bursa of Fabricius unchanged; Total Ig and IgG; Macrophage phagocytic activity ↓; NK cell activity unchanged	Qureshi *et al.*, 1995
Turkey poults	75 mg/kg feed and AB_1, 200 µg/kg feed, from day 1 to 21	Primary immune response to SRBC ↑ (combination of FB_1 and AFB_1); PHA response unchanged	Weibking *et al.*, 1994
Weaned pigs	1-100 mg/animal/d	Titres unchanged; Lymphocyte stimulation by PHA, ConA, LPS unchanged	Tornyos *et al.*, 2003
Rats	Injected at 7.5 or 10 mg/kg BW/day for 4 days	Thymus weight ↓, Serum IgM ↑; Circulating phagocytic cell numbers ↑	Bondy *et al.*, 1995
Mice	50 (LD), FM1 + FM_2, 6 weeks	After challenge with *T. cruzi*, NO production by MG↑ after 14 days	Dresden *et al.*, 2002
Mice	Five daily s.c. injections of 2.25 mg/kg/d FB_1	Relative spleen and thymus weights in F ↓, No effect on organ weights in M; Splenic cellularity and the basal rate of lymphocyte proliferation in F ↓; PHA-induced T-lymphocyte and LPS-induced B-lymphocyte proliferation in F ↓; Expression of IL-2 mRNA in splenocytes in F ↓	Johnson and Sharma, 2001
Mice	I.p. with SRBC, at 5-100 µg	Number of plaque-forming cells produced against SRBC ↓	Martinova *et al.*, 1995
Mice	I.p daily, at 1 to 50 µg	4 to 12-fold increase in the number of of plaque-forming cells after SRBC injection	Martinova *et al.*, 1995
Macrophage cell line	1-10 µM	Membrane fluidity ↑; Membrane peroxidative damage ↑	Ferrante *et al.*, 2002
Murine macrophage cell line	1-100 µM	LPS-induced production of NO and PGE_2↑ at 0.1-10 µM	Meli *et al.*, 2000
Murine macrophage cell	1, 10, and 100 µM	LPS-induced NO production ↑	Rotter and Oh, 1996
Rat splenic macrophages and lymphocytes	1-100 µg/ml	NO production ↑; ConA-induced proliferation of splenic cells in the presence of NO synthase inhibitor ↑	Dombrink-Kurtzman *et al.*, 2000

et al., 1995). However, in turkey poults a combination of FB_1 and AFB_1 was responsible for increased primary immune response to SRBC (Weibking *et al.*, 1994). In weaned pigs, FB_1 (up to 100 mg/animal/day) did not affect immunity (Tornyos *et al.*, 2003). Effects of FB_1 on immunocompetence of laboratory animals also vary substantially (Table 7).

Immunomodulatory properties of FB_1 probably depend on its effect on lipid metabolism, antioxidant/pro-oxidant balance and interactions with other factors. For example, FB_1 decreased CD3 receptor expression on the surface of thymus cells *in vitro* and *in vivo*, which is consistent with the sharp decrease in the ceramide content in this organ (Martynova *et al.*, 1995).

Similar to FB_1, the trichothecenes can both suppress and stimulate immune function (Table 8). Most of the research on T-2 toxin effects on immunity was perfomed with laboratory animals and there are only a few publications addressing immunomodulating properties of T-2 toxin in farm animals. Futhermore, the molecular basis of immune modulation by mycotoxins remains mostly unknown and continues to be an active area of research (Bondy and Pestka, 2000). It appears that trichothecenes are potent immunosuppressive agents that directly affect immune cells and modify immune responses because of tissue damage elsewhere. For example, sheep and calves treated with *Fusarium* T-2 toxin develop leukopenia and decreased function of peripheral lymphocytes (Sharma, 1993). In fact, exposure of experimental animals and humans to T-2 toxin has been shown to have a variety of immunosuppressive effects, including altered parameters of humoral immunity. It is well established that T-2 toxin is cytotoxic to lymphocytic cells *in vitro*, however, limited information is presently available regarding the contribution of such a mechanism to immunosuppression *in vivo*, or to potential immune cell targets. It seems likely that lymphocyte progenitors, in contrast to thymocytes, represent highly sensitive targets of T-2 toxin exposure, responsible for thymic atrophy (Holladay *et al.*, 1993). For example, subchronic T-2 toxin treatment of timed-pregnant B6C3F1 mice resulted in significant and selective depletion of fetal liver cells expressing low levels of surface CD44 and CD45 antigens, suggestive of possible lymphoid progenitor cell sensitivity to this agent (Holladay *et al.*, 1995). The authors showed that the precursors of B-cells might represent highly sensitive targets of T-2 toxin exposure. Impaired resistance to pathogenic microorganisms occurs after exposure to the trichothecenes T-2 toxin and DON. This may predispose food animals to infectious disease and could result in decreased productivity as well as increased animal-to-human transmission of pathogens such as *Salmonella* and *Listeria* (Pestka and Bondy, 1990). For example, mice challenged with *S. typhimurium* and then treated with T-2 toxin every other day for 10 days had markedly larger and more bacteria-related lesions in the spleen, kidney, and liver than animals challenged with *S. typhimurium* alone (Tai and Pestka, 1990). In great contrast, in mice exposed to T-2 toxin at 3 mg/kg BW by gavage, virulence of both *E. coli* and *S. auerus* decreased. Apparently there are species-specific features in sensitivity to trichothecenes. For example, recently it has been shown that T-2 toxin (up to 1 ppm) in the poult diet did not affect antibody production stimulated by various antigens (Sklan *et al.*, 2003).

Immunomodulating properties of DON have been studied mainly with rodents (Table 9). The evidence is quickly accumulating to show that DON can be immunosuppressive or immunostimulatory, depending upon the dose and duration of exposure. While immunosuppression is probably related to the inhibition of translation, immunostimulation can be a result of interference with various cellular regulatory mechanisms (Rotter *et al.*, 1996). For example, *in vivo* DON suppresses normal immune response to pathogens and simultaneously induces autoimmune-like effects. Characteristic effects of DON also include superinduction of cytokine production by T-helper cells (*in vitro*) and activation of macrophages and T-cells to produce proinflammatory cytokines. Although these effects have been largely characterised in the mouse, several investigations with DON suggest that immunotoxic effects are

Table 8. Effects of T-2 toxin on immunity.

Species	*Dose*	*Effects on immune system*	*Reference*
Young poults	Up to 1 mg/kg for 32 days	Antibody titres unchanged	Sklan *et al.*, 2003
7-week-old pigs	0.5, 1.0, 2.0 or 3.0 mg/kg diet for 3 weeks	Leukocyte count ↓; Proportion of T-lymphocytes ↓; Antibody formation ↓; Blastogenic transformation of lymphocytes ↓	Rafai *et al.*, 1995
9-10-week-old pigs	Aerosol, 390 μg/L	Lymphocyte count ↓, Neutrophil count ↑; ConA, PHA, and PWM blastogenic responses and H1 titers to SRBC↓	Pang *et al.*, 1988
9-10-week-old pigs	15 mg/kg	Blastogenic responses to ConA, PHA and PWM ↓ at early (3 to 5 days) and late (20 to 28 days) postdosing intervals; HA titer to SRBC unchanged	Pang *et al.*, 1987
Calves	0.6 mg/kg/day	Lymphocyte responses to mitogens ↓; Chemotaxic migration of neutrophils ↓	Buening *et al.*, 1982
Calves	0.6 mg/kg per day for 43 days	IgM and IgA ↓; IgG→	Mann *et al.*, 1982
Monkeys	Gastric intubation at 100 μg/kg BW/ day for 4-5 wk	Leucocyte counts↓ at the end of wk 4; Bactericidal activity of neutrophils ↓, T-cell number ↓; Lymphocyte transformation ↓, B-cell number ↓; IgG and IgM levels ↓	Jagadeesan *et al.*, 1982
Rats	2 mg/kg	In progeny, the cortex was atrophic while the medulla was proliferative. In the spleen, both B- and T-cells were impaired and responsiveness to PHA and LPS decreased	Lafarge-Frayssinet *et al.*, 1990
Mice	3 mg/kg BW by gavage	Virulence of both *E. coli* and *S. aureus* ↓	Cooray and Jonsson, 1990
Mice challenged with *S. typhimurium*	1 mg/kg every other day for 10 d	Markedly larger and more bacteria-related lesions in the spleen, kidney, and liver than animals challenged with *S. typhimurium* alone	Tai and Pestka, 1990
Mice	20 ng	Immune response to sheep RBC ↓	Holt and DeLoach, 1988
Mice	1.5 or 3.0 ppm in the diet for 16 months	T-lymphocyte-dependent humoral immunity unaffected	Schiefer *et al.*, 1987
Mice exposed to *Listeria*	4.0 mg/ kg of BW	Growth of *Listeria* ↑; Mortality ↑	Corrier and Ziprin, 1986

Table 8. Continued.

Species	*Dose*	*Effects on immune system*	*Reference*
Mice	20 ppm for 1- 4 weeks	Spleen proliferative responses to ConA ↓;	Friend *et al.*, 1983
Human lymphocytes	1 -5 ng/ml	50% inhibition in cell proliferation	Meky *et al.*, 2001

Table 9. Effects of DON on immune system.

Species	*Dose*	*Effects on immune system*	*Reference*
Growing pigs	0.6, 1.8 and 4.7 mg/kg	Secondary antibody response to tetanus toxoid ↓ dose-dependently	Overnes *et al.*, 1997
Young pigs	0.75- 3.00 mg/kg diet for 28 days	α-globulin levels, and antibody titers to SRBC↓	Rotter *et al.*, 1994
Mice	5 or 25 mg/kg BW	mRNAs for IL-1ß, IL-6, and TNF-α; the T-helper 1 cytokines IFN-γ and IL-2; and the T helper 2 cytokines IL-4 and IL-10 ↑	Zhou *et al.*, 1997
Mice	25 mg/kg for 8 or 24 ws	Serum IgA and microhematuria index ↑ after 4 to 8 wks and continued to increase with further exposure	Dong and Pestka, 1993
Mice	25 ppm for 4 and 8 wks	Total IgA ↑; Total IgG and IgM ↓	Rasooly ands Pestka, 1992
Mice	10-100 mg/kg in the diet	Serum levels of anti-SRBC antibodies ↓; Stimulation of B- and T-cells by mitogens ↓	Robbana-Barnat *et al.*, 1988
Weanling mice	0.5- 25.0 mg/kg over 8 wks	Total circulating white blood cells ↓;PMN neutrophils ↑; Serum IgM ↓; IgA ↑	Forsell *et al.*, 1986
Weanling mice	0.75, 2.5 or 7.5 mg/kg BW by gavage	Serum IgM to SRBC ↓; Plaque-forming cell numbers ↓	Tryphonas *et al.*, 1984
Murine peritoneal macrophages	0.1 ng/ml-1 μg/ml	Phagocytosis ↓; Microbiocidal activity ↓; Superoxide anion production ↓ at 1 ng/ml; Phagosome-lysozome fusion ↓ at 100 ng/nl	Ayral *et al.*, 1992
Lymphocytes *in vitro*	0.005-100 ng/ml	PHA-induced lymphocyte proliferation ↑ at 0.005- 0.5 ng DON/ml; at 50-100 ng/ml lymphocyte proliferation ↓	Miller and Atkinson, 1986
Human lymphocytes	216 ng/ml	50% inhibition in cell proliferation	Meky *et al.*, 2001
Murine macrophage cell line	25-100 ng/ml	Production of H_2O_2 after LPS stumulation ↑, but ↓ at 250 ng/m; NO production ↓ at 25-250 ng/ml	Ji *et al.*, 1998

also likely in domestic animals (Rotter *et al.,* 1996). Therefore, DON can selectively and concurrently upregulate or downregulate critical functions associated with activated macrophages. For example, a single DON exposure rapidly induces gene expression *in vivo* for a wide range of cytokines with apparently complete recovery occurring after 24 hrs. Elevated cytokine expression may play an important role in the pathophysiological effects of DON and other trichothecenes (Zhou *et al.,* 1997). Males appeared more susceptible than female mice to DON-induced IgA dysregulation and IgA nephropathy in terms of latency, threshold dose, and severity (Greene *et al.,* 1994). It is interesting that experimental dysregulation of IgA production and IgA nephropathy persisted up to four months after a discrete period of dietary DON exposure (Dong and Pestka, 1993).

Mycotoxin-induced immunosuppression is related to both natural and adaptive immunity:

- depressed T- or B-lymphocyte activity
- depressed NK cell activity
- suppressed immunoglobulin and antibody production
- reduced complement or interferon activity
- impaired macrophage functions

The sensitivity of the immune system to mycotoxin-induced immunosuppression arises from the vulnerability of the continually proliferating and differentiating cells that participate in immunomediated activities and regulate the complex communication network between cellular and humoral components (Corrier, 1991). In fact, high levels of polyunsaturated fatty acids in the immune cells and presence of sensitive receptors on cell surfaces make them an important target for free radical attack (Surai, 2002). As mentioned above, oxidative stress caused by mycotoxin consumption could be responsible for breaking effective communication between immune cells, which ultimately would misregulate the immune system and result in immunosuppression. Although the molecular basis for many of the specific immunosuppressive effects of mycotoxins is presently unclear, inhibition of DNA, RNA and protein synthesis via a variety of different mechanisms appears to be directly or indirectly responsible for the immunosuppressive action of many mycotoxins (Corrier, 1991). Furthermore, detrimental effects on antioxidant defences, stimulation of lipid peroxidation and apoptosis and damage to immune cell receptors appear to be involved in the immunomodulation properties of mycotoxins. Consequences of oxidative stress for immunocompetence are shown in Figure 3.

Protective effect of antioxidants against mycotoxicoses

The wide range of mycotoxins that can contaminate animal feed and their differing chemical compositions makes protection against mycotoxin-related toxicity a difficult task. There are various approaches to control or combat mycotoxin problems. The simplest strategy is based on the prevention of mycotoxin formation in feeds by special management programmes including storage at low moisture levels and prevention of grain damage during processing (Dawson, 2001). However, modern agronomic technology cannot eliminate pre-harvest infection of susceptible crops by fungi (Wood, 1992). Therefore, this strategy can be only partially effective; and in countries with warm and humid conditions, this strategy can be quite costly.

Other strategies based on microbial or thermal inactivation of toxins, physical separation of contaminated feedstuffs, irradiation, ammoniation and ozone degradation have not been developed to an industrial level because they are either time-consuming or comparatively expensive (Dawson, 2001a). In recent years, nutritional manipulation has been actively used to

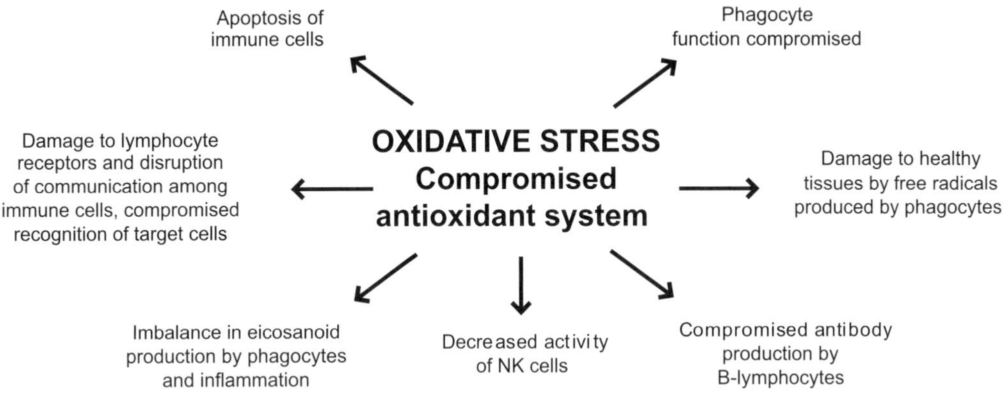

Figure 3. Oxidative stress and the immune system (Adapted from Surai, 2002).

improve animal defence against mycotoxins or to decrease the detrimental consequences of mycotoxin consumption.

Since lipid peroxidation plays an important role in mycotoxin toxicity, a protective effect of antioxidants is expected (Galvano *et al.*, 2001). Indeed, in several experiments with various animal species, protective effects of antioxidants against the toxic effects of mycotoxins were observed. For example, vitamin E supplementation ameliorated the pro-oxidative effects of OTA in the chicken (Hoehler and Maraquardt, 1996) and mouse (Grosse *et al.*, 1997). Trolox C, a water-soluble form of vitamin E, prevented the genotoxic effects of OTA in *E. coli* (Malavielle *et al.*, 1994). Protective effects of vitamin E were also obvious in aflatoxicosis (Shen *et al.*, 1994; Choi *et al.*, 1995; Souza *et al.*, 1999; Verma and Nair, 2001), DON or T-2 toxicoses (Segal *et al.*, 1983; Tsuchida *et al.*, 1984; Rizzo *et al.*, 1994) in rats; T-2 toxicosis in mice *in vivo* (Atroshi *et al.*, 1997) or in an *in vitro* system using cell lines (Shokri *et al.*, 2000). Vitamin E was also effective in protection against fumonisin B_1 (Abel and Gelderblom, 1998; Atroshi *et al.*, 1999) or ZEA (Grosse *et al.*, 1997; Ghedira-Chekir *et al.*, 1998; 1999). It also decreased the cytotoxic effect of T-2 toxin in cell culture (Shokri *et al.*, 2000). Pre-incubation of the glioma cells with vitamin E (25 mM) for 24 hrs before FB_1 (18 mM) significantly reduced DNA fragmentation and apoptotic bodies induced by FB_1 (Mobio *et al.*, 2000).

Other antioxidant compounds also have protective effects against various mycotoxins. Burguera *et al.* (1983) indicated that selenium (Se) has a protective effect against AFB_1 toxicity in turkey poults. To investigate the biochemical mechanism of the protective effect of dietary Se during aflatoxin exposure, hepatic metabolism of AFB_1 in turkey poults was examined at various dietary Se concentrations (0.2, 2.0 or 4.0 ppm). The experimental results provided clear evidence of Se-induced enhancement of aflatoxin detoxification (Gregory and Edds, 1984). It was suggested that the protective action of Se was not mediated by an increase in glutathione availability for aflatoxin conjugation or by effects on the activities of these enzymes as measured *in vitro*. Therefore dietary supplements such as Se are considered effective in the reduction of aflatoxicosis in poultry (Dalvi, 1986). Recently the protective antioxidant effect of organic Se in combination with vitamin E has been shown in chicks exposed to cold stress and aflatoxin-contaminated feed (Stanley, 1998).

Similar protective effects of Se against aflatoxicosis have been shown with mammalian species. Pigs fed a diet containing 2.5 mg Se/kg were protected from the toxic effects of AFB_1 (Davila *et al.*, 1983). The effects were measured by alteration of clinical responses and hematologic (prothrombin times), electrophoretic and clinical chemistry values. Effects of a single intramuscular injection of Se-vitamin E (5 mg of Se + 68 IU of α–tocopherol/60 kg of BW) as a pre-treatment 14 days before an oral dose of AFB_1 (1.0 mg/kg) were studied in 24 dairy calves. Although aflatoxin exposure caused a significant decrease in body weight and feed intake, Se was found to interact significantly with AFB_1 for feed intake, causing an improvement in this parameter (Brucato *et al.*, 1986).

Milks *et al.* (1985) showed that Se is able to protect against the hepatocarcinogenic effects of AFB_1 in the rat. Moreover, inhibitory activity of Se on mutagenesis induced by AFB_1 in the presence of a rat liver microsomal activation system has been shown in *S. typhimurium* tests (Francis *et al.*, 1988). The results of another experiment showed that Se could effectively protect cells from AFB_1 cytotoxicity in cultured cells but had no effect on AFB_1-DNA adduct formation or mutagenesis (Shi *et al.*, 1995). In a different study, the same authors reported that Se could effectively inhibit AFB_1-induced DNA damage (Shi *et al.*, 1994).

It has been revealed that Se can inhibit the formation of hyperplastic foci and enzyme-altered foci as well as hepatocarcinogenesis induced by AFB_1, but Se can neither prevent the enlargement nor accelerate the regression of the foci already developed after administration of carcinogens (Wang, 1990). Therefore Lei *et al.* (1990) concluded that Se had an inhibitory effect on the initiation and promotion stages of AFB_1-induced preneoplastic foci and nodules. Selenium also prevented progression of these nodules to hepatocellular carcinoma even after cessation of AFB_1 administration. Additional experiments were conducted to verify the effect of Se on the mutagenic activity of AFB_1. After 14 days of Se administration to Chinese hamsters, the incidence of chromosomal aberrations in bone marrow cells due to a single administration of AFB_1 (5 mg /kg BW) was significantly reduced (Petr *et al.*, 1990). A significant decrease in the frequency of aberrant cells, breaks and gaps was observed throughout the investigation.

A protective effect of Se is not restricted to aflatoxins, but is obvious with T-2 toxin as well. When male Wistar rats were fed diets supplemented with Se (0.5 and 2.5 mg/kg) for six weeks, signs of intoxication from T-2 toxin were less distinct and mortality caused by T-2 toxin was 2-fold (Kravchenko *et al.*, 1990) or 3-5-fold (Tutelyan *et al.*, 1990) lower compared to the unsupplemented group. The acute lethal toxicity of T-2 toxin was reduced by administration of sodium selenite (Yazdanpanah *et al.*, 1997).

In summary, Se is shown to have protective effects against T-2 toxicity (Rizzo *et al.*, 1994; Shokri *et al.*, 2000; Kravchenko *et al.*, 1990; Tutelyan *et al.*, 1990; Yazdanpanah *et al.*, 1997), to decrease damaging effects of AFB_1 (Shen *et al.*, 1994; Choi *et al.*, 1995; Shi *et al.*, 1994; Shi *et al.*, 1995) and to prevent DON toxicosis (Rizzo *et al.*, 1994; Atroshi *et al.*, 1997). A synthetic seleno-organic compound (ebselen) showed a potent protective effect against AFB_1-induced cytotoxicity (Yang *et al.*, 2000). Indeed, as can be seen from data presented in Tables 1-4, protective effects against lipid peroxidation caused by mycotoxins were attributed to various antioxidant compounds including vitamins A and E, ascorbic acid, CoQ10, selenium, antioxidant enzymes as well as synthetic antioxidants and various plant extracts (Surai, 2002).

Conclusions

Recent results show that in many cases membrane-active properties of various mycotoxins determine their toxicity. Indeed, incorporation of mycotoxins into membrane structures causes various detrimental changes. These changes are associated with alteration of fatty acid composition of the membrane structures and with peroxidation of long chain PUFAs inside membranes. This ultimately damages membrane receptors, causing alterations in second messenger systems; then to inactivation of a range of membrane-binding enzymes responsible for regulation of important pathways. Finally, this causes alterations in membrane permeability, flexibility and other important characteristics determining membrane function. Consequently, functional alterations in many biochemical pathways and changes in physiological functions including growth, development, reproduction etc. occur. An importance of lipid peroxidation in all these processes is confirmed by the protective effects of natural antioxidants against mycotoxin toxicity. However, molecular mechanisms of mycotoxin-antioxidant interactions *in vivo* await investigation.

Considering data reviewed, we can suggest a hypothetical scheme of mycotoxin-antioxidant interactions.

- Mycotoxins in the feed (at least OTA, T-2 toxin and AFB_1) cause malabsorption in the intestine, which in turn decreases concentrations of vitamin E, C, and carotenoids in tissues. Mycotoxins promote free radical formation (O_2^- and OH•) in the intestine, which cause antioxidant depletion, oxidative stress, enterocyte apoptosis and contribute to the development of malabsorption and decreased antioxidant absorption and accumulation.
- Mycotoxins and their active metabolites are absorbed from the intestine and accumulate in target tissues.
- Mycotoxins in tissues can generate free radicals, decreasing further antioxidant protection, causing lipid peroxidation and damage to other biological molecules including lipids, proteins and DNA. This could lead to antioxidant/pro-oxidant imbalance causing oxidative stress, which further leads to apoptosis and other cytotoxic effects of mycotoxins.
- Immunosupressive action of mycotoxins could be associated with downregulation of communications among various cells due to damage to receptors as well as downregulation of communicating molecule (cytokines, eicosanoids, etc.) production by macrophages.
- Increased antioxidant supplementation protects against toxic actions of mycotoxins by interfering with one or several steps described above, including gastrointestinal tract, plasma and tissue membranes.
- A combination of natural antioxidants with mycotoxin adsorbents could be a next step in combating mycotoxicoses in poultry production.

References

Abel, S. and W.C. Gelderblom. 1998. Oxidative damage and fumonisin B_1-induced toxicity in primary rat hepatocytes and rat liver *in vivo*. Toxicol. 131:121-131.

Abado-Becognee, K., T.A. Mobio, R. Ennamany, F. Fleurat-Lessard, W.T. Shier, F. Badria and E.E. Creppy. 1998. Cytotoxicity of fumonisin B_1: implication of lipid peroxidation and inhibition of protein and DNA synthesis. Archives of Toxicol. 72:233-236.

Abid-Essefi, S., I. Baudrimont, W. Hassen, Z. Ouanes, T.A. Mobio, R. Anane, E.E.Creppy and H. Bacha. 2003. DNA fragmentation, apoptosis and cell cycle arrest induced by zearalenone in cultured DOK, Vero and Caco-2 cells: prevention by vitamin E. Toxicol. 192:237-248.

Ahmed, N. and G.C. Ram. 1986. Nuclear lipid peroxidation induced in rat liver by T-2 mycotoxin. Toxicon. 24:947-949.

Albarenque, S.M., K. Suzuki, H. Nakayama and K. Doi. 2001. Kinetics of cytokine mRNA expression in the dorsal skin of WBN/ILA-Ht rats following topical application of T-2 toxin. Exp. Toxicol. Pathol. 53:271-274.

Arrigo, A.P. 1999. Gene expression and the thiol redox state. Free Rad. Bio. Med. 27:936-944.

Atroshi, F., A. Rizzo, I. Biese, P. Veijalainen,

E. Antila and T. Westermarck. 1997. T-2 toxin-induced DNA damage in mouse livers: the effect of pretreatment with coenzyme Q10 and α-tocopherol. Molecular Asp. Med. 18(Suppl. 1):S255-S258.

Atroshi, F., I. Biese, H. Saloniemi, T. Ali-Vehmas, S. Saari, A. Rizzo and P. Veijalainen. 2000. Significance of apoptosis and its relationship to antioxidants after ochratoxin A administration in mice. J. Pharmacy Pharmac. Sci. 3:281-291.

Atroshi, F., A. Rizzo, I. Biese, P. Veijalainen, H. Saloniemi, S. Sankari and K. Andersson. 1999. Fumonisin B_1-induced DNA damage in rat liver and spleen: effects of pretreatment with coenzyme Q10, L-carnitine, α-tocopherol and selenium. Pharmac. Res. 40:459-467.

Ayral, A.M., N. Dubech, J. Le Bars and L. Escoula. 1992. *In vitro* effect of diacetoxyscirpenol and deoxynivalenol on microbicidal activity of murine peritoneal macrophages. Mycopath. 120:121-127.

Bailly, J.D., G. Benard, J.Y. Jouglar, S. Durand and P. Guerre. 2001. Toxicity of *Fusarium moniliforme* culture material containing known levels of fumonisin B_1 in ducks. Toxicol. 163:11-22.

Barraud, L., T. Douki, S. Guerret, M. Chevallier, C. Jamard, C. Trepo, C.P. Wild, J. Cadet and L. Cova. 2001. The role of duck hepatitis B virus and aflatoxin B_1 in the induction of oxidative stress in the liver. Cancer Det. Prev. 25:192-201.

Baudrimont, I., R. Ahouandjivo and E.E. Creppy. 1997a. Prevention of lipid peroxidation induced by ochratoxin A in Vero cells in culture by several agents. Chemico-Biological Interactions 104:29-40.

Baudrimont, I., A.M. Betbeder and E.E. Creppy. 1997b. Reduction of the ochratoxin A-induced cytotoxicity in Vero cells by aspartame. Archives of Toxicol. 71:290-298.

Beaver, J.P. and P. Waring. 1995. A decrease in intracellular glutathione concentration precedes the onset of apoptosis in murine thymocytes. Euro. J. Cell Bio. 68:47-54.

Beers, K.W., H. Nejad and W.G. Bottje. 1992b. Aflatoxin and glutathione in domestic fowl (*Gallus domesticus*). I. Glutathione elevation and attenuation by high dietary methionine. Comp. Biochem. Physiol. 101C:239-244.

Beers, K.W., R.P. Glahn, W.G. Bottje and H.E. Huff. 1992a. Aflatoxin and glutathione in domestic fowl (*Gallus domesticus*). II. Effects on hepatic blood flow. Comp. Biochem. Physiol. 101C:463-467.

Belmadani, A., P.S. Steyn, G. Tramu, A.M. Betbeder, I. Baudrimont and E.E. Creppy. 1999. Selective toxicity of ochratoxin A in primary cultures from different brain regions. Arch. Toxicol. 73:108-114.

Berek, L., I.B. Petri, A. Mesterhazy, J. Teren and J. Molnar. 2001. Effects of mycotoxins on human immune functions *in vitro*. Toxicol. *in vitro* 15:25-30.

Bhandari, N., C.C. Brown and R.P. Sharma. 2002. Fumonisin B_1-induced localized activation of cytokine network in mouse liver. Food Chem. Toxicol. 40:1483-1491.

Bhandari, N. and R.P. Sharma. 2002. Fumonisin B(1)-induced alterations in cytokine expression and apoptosis signaling genes in mouse liver and kidney after an acute exposure. Toxicol. 172:81-92.

Bondy, G.S. and J.J. Pestka. 2000. Immunomodulation by fungal toxins. J. Toxicol. Environ. Health. Critical Reviews (Part B) 3:109-143.

Bondy, G., C. Suzuki, M. Barker, C. Armstrong, S. Fernie, L. Hierlihy, P. Rowsell and R. Mueller. 1995. Toxicity of fumonisin B1 administered intraperitoneally to male Sprague-Dawley rats. Food Chem. Toxicol. 33:653-665.

Buening, G.M., D.D. Mann, B. Hook and G.D. Osweiler. 1982. The effect of T-2 toxin on the bovine immune system: cellular factors. Vet. Immunol. Immunopathol. 3:411-417.

Brucato, M., S.F. Sundlof, J.U. Bell and G.T. Edds. 1986. Aflatoxin B1 toxicosis in dairy calves pretreated with selenium-vitamin E. Amer. J. Vet. Res. 47:179-183.

Busbee, D., R. Barhoumi, R.C. Burghardt, C. Gauntt, B. McAnalley and H.R. McDaniel. 1999. Protection from glutathione depletion by a glyconutritional mixture of saccharides. Age 22:159-165.

Burguera, J.A., G.T. Edds and O. Osuna. 1983. Influence of selenium on aflatoxin B1 or crotalaria toxicity in turkey poults. Amer. J. Vet. Res. 44:1714-1717.

Celik, I., H. Oguz, O. Demet, H.H. Donmez, M. Boydak and E. Sur. 2000. Efficacy of polyvinylpolypyrrolidone in reducing the immunotoxicity of aflatoxin in growing broilers. Brit. Poult. Sci. 41:430-439.

Chang, I.M. and W.C. Mar. 1988. Effect of T-2 toxin on lipid peroxidation in rats: elevation of conjugated diene formation. Toxicol. Letters 40:275-280.

Chang, C.F., W.E. Huff and P.B. Hamilton. 1979. A leucocytopenia induced in chickens by dietary ochratoxin A. Poult. Sci. 58:555-558.

Choi, Y.K., K.K. Jung, K.Y. Chae, I. Jang, B.D. Lee and K.H. Nahm. 1995. Effects of vitamin E and selenium supplementation to diets containing aflatoxin B_1 on the contents of liver lipids and various blood parameters in rats. Asian-Austr. J. Anim. Sci. 8:379-385.

Ciacci-Zanella, J.R. and C. Jones. 1999. Fumonisin B1, a mycotoxin contaminant of cereal grains, and inducer of apoptosis via the tumour necrosis factor pathway and caspase activation. Food Chem. Toxicol. 37:703-712.

Coffin, J.L. and G.F. Combs, Jr. 1981. Impaired vitamin E status of chicks fed T-2 toxin. Poult. Sci. 60:385-392.

Corrier, D.E. 1991. Mycotoxicosis: mechanisms of immunosuppression. Vet. Immunol. Immunopathol. 30:73-87.

Corrier, D.E. and R.L. Ziprin. 1986. Immunotoxic effects of T-2 toxin on cell-mediated immunity to listeriosis in mice: comparison with cyclophosphamide. Amer. J. Vet. Res. 47:1956-1960.

Cooray, R. and P. Jonsson. 1990. Modulation of resistance to mastitis pathogens by pretreatment of mice with T-2 toxin. Food Chem. Toxicol. 28:687-692.

Creppy, E.E., I. Baudrimont and A.M. Betbeder. 1998. How aspartame prevents the toxicity of ochratoxin A. J. Toxicol. Sci. 23(Suppl 2):165-172.

Cusumano, V., F. Rossano, R.A. Merendino, A. Arena, G.B. Costa, G. Mancuso, A. Baroni and E. Losi. 1996. Immunobiological activities of mould products: functional impairment of human monocytes exposed to aflatoxin B1. Res. Microbiol. 147:385-391.

Cusumano, V., G.B. Costa, R. Trifiletti, R.A. Merendino and G. Mancuso. 1995. Functional impairment of rat Kupffer cells induced by aflatoxin B1 and its metabolites. FEMS Immunol. Med. Microbiol. 10:151-155.

Dalvi, R.R. 1986. An overview of aflatoxicosis of poultry: its characteristics, prevention and reduction. Vet. Res. Commun. 10:429-443.

Dare, E., W. Li, B. Zhivotovsky, X. Yuan and S. Ceccatelli. 2001. Methylmercury and H_2O_2 provoke lysosomal damage in human astrocytoma D384 cells followed by apoptosis. Free Rad. Bio. Med. 30:1347-1356.

Davila, J.C., G.T. Edds, O. Osuna and C.F. Simpson. 1983. Modification of the effects of aflatoxin B1 and warfarin in young pigs given selenium. Amer. J. Vet. Res. 44:1877-1883.

Dawson, K.A. 2001. The application of yeast and yeast derivatives in the poultry industry. Proc. Austr. Poult. Sci. Symp. 13:100-105.

Dawson, K.A. 2001a. Understanding the absorption characteristics of yeast cell wall preparations associated with mycotoxin binding. In: *Biotechnology in the Feed Industry, Proceedings of Alltech's 17th*

Annual Symposium (T.P. Lyons and K.A. Jacques, eds). Nottingham University Press, UK, pp. 169-181.

Dietert, R.R. and K.A. Golemboski. 1998. Avian macrophage metabolism. Poult. Sci. 77:990-997.

Dietert, R.R., M.A. Qureshi, U.C. Nanna and S.E. Bloom. 1985. Embryonic exposure to aflatoxin-B1: mutagenicity and influence on development and immunity. Environ. Mutagen. 7:715-725.

Dombrink-Kurtzman, M.A., R. Gomez-Flores and R.J. Weber. 2000. Activation of rat splenic macrophage and lymphocyte functions by fumonisin B1. Immunopharm. 49:401-409.

Dong, W. and J.J. Pestka. 1993. Persistent dysregulation of IgA production and IgA nephropathy in the B6C3F1 mouse following withdrawal of dietary vomitoxin (deoxynivalenol). Fundam. Appl. Toxicol. 20:38-47.

Dortant, P.M., G.W. Peters-Volleberg, H. Van Loveren, R.R. Marquardt and G.J. Speijers. 2001. Age-related differences in the toxicity of ochratoxin A in female rats. Food Chem. Toxicol. 39:55-65.

Dragan, Y.P., W.R. Bidlack, S.M. Cohen, T.L. Goldsworthy, G.C. Hard, P.C. Howard, R.T. Riley and K.A. Voss. 2001. Implications of apoptosis for toxicity, carcinogenicity, and risk assessment: fumonisin B(1) as an example. Toxicol. Sci. 61:6-17.

Dresden Osborne, C., G. Pittman Noblet, E.N. Enongene, C.W. Bacon, R.T. Riley and K.A. Voss. 2002. Host resistance to *Trypanosoma cruzi* infection is enhanced in mice fed *Fusarium verticillioides* (*F. moniliforme*) culture material containing fumonisins. Food Chem. Toxicol. 40:1789-1798.

Dvorska, J.E. 2001. Effect of aurofusarin, a mycotoxin produced by *Fusarium graminearum* on Japanese quail. Abstracts of International Symposium on Bioactive Fungal metabolites – Impact and Exploitation. University of Wales Swansea, April 22-27, p. 32.

Dvorska, J.E. and P.F. Surai. 2004. Protective effect of modified glucomannans against changes in antioxidant systems of quail egg and embryo due to aurofusarin consumption. Asian-Aust. J. Anim. Sci. 17: 434-440 (In press).

Dvorska, J.E., P.F. Surai, B.K. Speake and N.H.C. Sparks. 2001a. Effect of the mycotoxin aurofusarin on the fatty acid profile and antioxidant composition of quail eggs. Brit. Poult. Sci. 42:643-649.

Dvorska, J.E., P.F. Surai, B.K. Speake and N.H.C. Sparks. 2003. Protective effect of modified glucomannans against aurofusarin-induced changes in quail egg and embryo. Comp. Biochem. Physiol. 135:337-343.

Dvorska, J.E. and P.F. Surai. 2001. Effects of T-2 toxin, zeolite and Mycosorb on antioxidant systems of growing quail. Asian-Aust. J. Anim. Sci. 14:1-6.

Dvorska, J.E., P.F. Surai, B.K. Speake and N.H.C. Sparks. 2002. Antioxidant systems of the developing quail embryo are compromised by the mycotoxin aurofusarin. Comp. Biochem. Physiol. 131C:197-205.

Dvorska, J.E., P.F. Surai. and N.H.C. Sparks. 2001b. Mycotoxin aurofusarin: effect on quail. Scientific programme and abstracts of 6th International Symposium on Poisonous Plants, Glasgow, UK, p. 86.

Dwivedi, P. and R.B. Burns. 1984. Effect of ochratoxin A on immunoglobulins in broiler chicks. Res. Vet. Sci. 36:117-121.

Elissalde, M.H., R.L. Ziprin, W.E. Huff, L.F. Kubena and R.B. Harvey. 1994. Effect of ochratoxin A on *Salmonella*-challenged broiler chicks. Poult. Sci. 73:1241-1248.

El-Gibaly, I., A.M. Meki. and S.K. Abdel-Ghaffar. 2003. Novel B melatonin-loaded chitosan microcapsules: *in vitro* characterization and antiapoptosis efficacy for aflatoxin B1-induced apoptosis in rat liver. Int. J. Pharm. 260:5-22.

Fernandez, A., M. Hernandez, M.T. Verde and M. Sanz. 2000. Effect of aflatoxin on performance, hematology and clinical immunology in lambs. Can. J. Vet. Res. 64:53-58.

Ferrante, M.C., R. Meli, G. Mattace Raso, E. Esposito, L. Severino, G. Di Carlo and A. Lucisano. 2002. Effect of fumonisin B1 on structure and function of macrophage plasma membrane. Toxicol. Lett. 129:181-187.

Forsell, J.H., M.F. Witt, J.H. Tai, R. Jensen and J.J. Pestka. 1986. Effects of 8-week exposure of the B6C3F1 mouse to dietary deoxynivalenol (vomitoxin) and zearalenone. Food Chem. Toxicol. 24:213-219.

Francis, A.R., T.K. Shetty and R.K. Bhattacharya. 1988. Modifying role of dietary factors on the mutagenicity of aflatoxin B1: *in vitro* effect of trace elements. Mutation Res. 199:85-93.

Fricke, R.F. and J. Jorge. 1991. Methylthiazolidine-4-carboxylate for treatment of acute T-2 toxin exposure. J. Appl. Toxicol. 11:135-140.

Friend, S.C., H.B. Schiefer and L.A. Babiuk. 1983. The effects of dietary T-2 toxin on acute herpes simplex virus type 1 infection in mice. Vet. Pathol. 20:737-760.

Galvano, F., A. Piva, A. Ritieni and G. Galvano. 2001. Dietary strategies to counteract the effects of mycotoxins: a review. J. Food Prot. 64:120-131.

Galvano, F., A. Russo, V. Cardile, G. Galvano, A. Vanella and M. Renis. 2002. DNA damage in human fibroblasts exposed to fumonisin B(1). Food Chem. Toxicol. 40:25-31.

Gautier, J.C., D. Holzhaeuser, J. Markovic, E. Gremaud, B. Schilter and R.J. Turesky. 2001. Oxidative damage and stress response from ochratoxin A exposure in rats. Free Rad. Bio. Med. 30:1089-1098.

Gekle, M., G. Schwerdt, R. Freudinger, S. Mildenberger, D. Wilflingseder, V. Pollack, M. Dander and H. Schramek. 2000. Ochratoxin A induces JNK activation and apoptosis in MDCK-C7 cells at nanomolar concentrations. J. Pharm. Exper. Therapy 293:837-844.

Gelderblom, W.C., D. Galendo, S. Abel, S. Swanevelder, W.F. Marasas and C.P. Wild. 2001. Cancer initiation by fumonisin B(1) in rat liver-role of cell proliferation. Cancer Letters 169:127-37.

Ghedira-Chekir, L., K. Maaroufi, A. Zakhama, F. Ellouz, S. Dhouib, E.E. Creppy and H. Bacha. 1998. Induction of a SOS repair system in lysogenic bacteria by zearalenone and its prevention by vitamin E. Chemico-Biological Interactions 113:15-25.

Ghedira-Chekir, L., K. Maaroufi, E.E. Creppy and H. Bacha. 1999. Cytotoxic and genotoxic effects of zearalenone: prevention by vitamin E. J. Toxicol. - Toxin Reviews 18:355-368.

Giambrone, J.J., U.L. Diener, N.D. Davis, V.S. Panangala and F.J. Hoerr. 1985a. Effects of purified aflatoxin on broiler chickens. Poult. Sci. 64:852-858.

Giambrone, J.J., M. Partadiredja, C.S. Eidson, S.H. Kleven and R.D. Wyatt. 1978. Interaction of aflatoxin with infectious bursal disease virus infection in young chickens. Avian Dis. 22:431-439.

Giambrone, J.J., U.L. Diener, N.D. Davis, V.S. Panangala and F.J. Hoerr. 1985b. Effects of aflatoxin on young turkeys and broiler chickens. Poult. Sci. 64:1678-1684.

Gille, G. and K. Sigler. 1995. Oxidative stress and living cells. Folia Microbiol. 40:131-152.

Gillman, I.G., T.N. Clark and R.A. Manderville. 1999. Oxidation of ochratoxin A by an Fe-porphyrin system: model for enzymatic activation and DNA cleavage. Chem. Res. Toxicol. 12:1066-1076.

Gogal, R.M. Jr., B.J. Smith, J. Kalnitsky and S.D. Holladay. 2000. Analysis of apoptosis of lymphoid cells in fish exposed to immunotoxic compounds. Cytometry 39:310-318.

Gopee, N.V., Q. He and R.P. Sharma. 2003. Fumonisin B1-induced apoptosis is associated with delayed inhibition of protein kinase C, nuclear factor-κB and tumor necrosis factor-α in LLC-PK1 cells. Chem. Biol. Interact. 146:131-145.

Greene, D.M., J.I. Azcona-Olivera and J.J. Pestka. 1994. Vomitoxin (deoxynivalenol)-induced IgA nephropathy in the B6C3F1 mouse: dose response and male predilection. Toxicol. 92:245-260.

Gregory, J.F. and G.T. Edds. 1984. Effect of dietary selenium on the metabolism of aflatoxin B1 in turkeys. Food Chem. Toxicol. 22:637-642.

Grimble, R.F. 1997. Effect of antioxidative vitamins on immune function with clinical applications. Intl. J. Vit. Nutr. Res. 67:312-320.

Grosse, Y., L. Chekir-Ghedira, A. Huc, S. Obrecht-Pflumio, G. Dirheimer, H. Bacha and A. Pfohl-Leszkowicz. 1997. Retinol, ascorbic acid and α-tocopherol prevent DNA adduct formation in mice treated with the mycotoxins ochratoxin A and zearalenone. Cancer Letters 114:225-229.

Harvey, R.B., L.F. Kubena and M.H. Elissalde. 1994. Influence of vitamin E on aflatoxicosis in growing swine. Amer. J. Vet. Res. 55:572-577.

Harvey, R.B., M.H. Elissalde, L.F. Kubena, E.A. Weaver, D.E. Corrier and B.A. Clement. 1992. Immunotoxicity of ochratoxin A to growing gilts. Amer. J. Vet. Res. 53:1966-1970.

Haubeck, H.D., G. Lorkowski, E. Kolsch and R. Roschenthaler. 1981. Immunosuppression by ochratoxin A and its prevention by phenylalanine. Appl. Environ. Microbiol. 41:1040-1042.

Hayek, M.G., G.A. Boissonneault and G.E. Mitchell, Jr. 1996. Influence of vitamin E on immune response of livestock and poultry. BASF Manual. pp. 185-190.

Herdener, M., S. Heigold, M. Saran and G. Bauer. 2000. Target cell-derived superoxide anions cause efficiency and selectivity of intercellular induction of apoptosis. Free Rad. Bio. Med. 29:1260-1271.

Hockenbery, D.M., Z.N. Oltvai, Y.M. Yin, C.L. Milliman and S.J. Korsmeyer. 1993. Bcl-2 functions in an antioxidant pathway to prevent apoptosis. Cell 75:241-251.

Hoehler, D. and R.R. Marquardt. 1996. Influence of vitamins E and C on the toxic effects of ochratoxin A and T-2 toxin in chicks. Poult. Sci. 75:1508-1515.

Hoehler, D., R.R. Marquardt, A.R. McIntosh, and H. Xiao. 1996. Free radical generation as induced by ochratoxin A and its analogs in bacteria (*Bacillus brevis*). J. Bio. Chem. 271:27388-27394.

Hoehler, D., R.R. Marquardt and A.A. Frohlich. 1997. Lipid peroxidation as one mode of action in ochratoxin A toxicity in rats and chicks. Can. J. Anim. Sci. 77:287-292.

Holladay, S.D., B.L. Blaylock, C.E. Comment, J.J. Heindel and M.I. Luster. 1993. Fetal thymic atrophy after exposure to T-2 toxin: selectivity for lymphoid progenitor cells. Toxicol. Appl. Pharmacol. 121:8-14.

Holladay, S.D., B.J. Smith and M.I. Luster. 1995. B-lymphocyte precursor cells represent sensitive targets of T-2 mycotoxin exposure. Toxicol. Appl. Pharmacol. 131:309-315.

Holme, J.A., E. Morrison, J.T. Samuelsen, R. Wiger, M. Lag, P.E. Schwarze, A. Bernhoft and M. Refsnes. 2003. Mechanisms involved in the induction of apoptosis by T-2 and HT-2 toxins in HL-60 human promyelocytic leukemia cells. Cell Biol. Toxicol. 19:53-68.

Holt, P.S. and J.R. DeLoach. 1988. *In vitro* effect of T-2 mycotoxin on the immune response of mice. Am. J. Vet. Res. 49:1480-1484.

Horvath, A., B.L. Upham, V. Ganev and J.E. Trosko. 2002. Determination of the epigenetic effects of ochratoxin in a human kidney and a rat liver epithelial cell line. Toxicon. 40:273-282.

Howard, P.C., L.H. Couch, R.E. Patton, R.M. Eppley, D.R. Doerge, M.I. Churchwell, M.M. Marques and C.V. Okerberg. 2002. Comparison of the toxicity of several fumonisin derivatives in a 28-day feeding study with female B6C3F(1) mice. Toxicol. Appl. Pharm. 85:153-165.

Howard, P.C., A. Warbritton, K.A. Voss, R.J. Lorentzen, J.D. Thurman, R.M. Kovach and T.J. Bucci. 2001. Compensatory regeneration as a mechanism for renal tubule carcinogenesis of fumonisin B1 in the F344/N/Nctr BR rat. Environ. Health Perspect. 109(Suppl. 2):309-314.

Huff, W.E. and P.B. Hamilton. 1975. Decreased plasma carotenoids during ochratoxicosis. Poult. Sci. 54:1308-1310.

Ibrahim, I.K., A.M. Shareef and K.M. Al-Joubory. 2000. Ameliorative effects of sodium bentonite on phagocytosis and Newcastle disease antibody formation in broiler chickens during aflatoxicosis. Res. Vet. Sci. 69:119-122.

Ihara, T., M. Sugamata, M. Sekijima, H. Okumura, N. Yoshino and Y. Ueno. 1997. Apoptotic cellular damage in mice after T-2 toxin-induced acute toxicosis. Natural Toxins 5:141-145.

Ishigami, N., J. Shinozuka, K. Katayama, K. Uetsuka, H. Nakayama and K. Doi. 1999. Apoptosis in the developing mouse embryos from T-2 toxin-inoculated dams. Hist. Histopathol. 14:729-733.

Ishigami, N., J. Shinozuka, K. Katayama, H. Nakayama and K. Doi. 2001. Apoptosis in mouse fetuses from dams exposed to T-2 toxin at different days of gestation. Exp. Toxicol. Pathol. 52:493-501.

Islam, Z., M. Nagase, A. Ota, S. Ueda, T. Yoshizawa and N. Sakato. 1998. Structure-function relationship of T-2 toxin and its metabolites in inducing thymic apoptosis *in vivo* in mice. Biosci. Biotech. Biochem. 62:1492-1497.

Islam, Z., M. Nagase, T. Yoshizawa, K. Yamauchi and N. Sakato. 1998a. T-2 toxin induces thymic apoptosis *in vivo* in mice. Toxicol. Appl. Pharm. 148:205-214.

Islam, Z., L.E. King, P.J. Fraker and J.J. Pestka. 2003. Differential induction of glucocorticoid-dependent apoptosis in murine lymphoid subpopulations *in vivo* following co-exposure to lipopolysaccharide and vomitoxin (deoxynivalenol). Toxicol. Appl. Pharmacol. 187:69-79.

Islam, Z., Y.S. Moon, H.R. Zhou, L.E. King, P.J. Fraker and J.J. Pestka. 2002. Endotoxin potentiation of trichothecene-induced lymphocyte apoptosis is mediated by up-regulation of glucocorticoids. Toxicol. Appl. Pharmacol. 180:43-55.

Jagadeesan, V., C. Rukmini, M. Vijayaraghavan and P.G. Tulpule. 1982. Immune studies with T-2 toxin: effect of feeding and withdrawal in monkeys. Food Chem. Toxicol. 20:83-87.

Jakab, G.J., R.R. Hmieleski, A. Zarba, D.R. Hemenway and J.D. Groopman. 1994. Respiratory aflatoxicosis: suppression of pulmonary and systemic host defenses in rats and mice. Toxicol. Appl. Pharmacol. 125:198-205.

Ji, G.E., S.Y. Park, S.S. Wong and J.J. Pestka. 1998. Modulation of nitric oxide, hydrogen peroxide and cytokine production in a clonal macrophage model by the trichothecene vomitoxin (deoxynivalenol). Toxicol. 125:203-214.

Johnson, V.J. and R.P. Sharma. 2001. Gender-dependent immunosuppression following subacute exposure to fumonisin B1. Int. Immunopharm. 1:2023-2034.

Johnson, V.J., Q. He, S.H. Kim, A. Kanti and R.P. Sharma. 2003. Increased susceptibility of renal epithelial cells to TNF-α-induced apoptosis following treatment with fumonisin B1. Chem. Biol. Interact. 145:297-309.

Jones, C., J.R. Ciacci-Zanella, Y. Zhang, G. Henderson and M. Dickman. 2001. Analysis of fumonisin B1-induced apoptosis. Environ. Health Persp. 109(Suppl. 2):315-320.

Kadian, S.K., D.P. Monga and M.C. Goel. 1988. Effect of aflatoxin B1 on the delayed type hypersensitivity and phagocytic activity of reticuloendothelial system in chickens. Mycopathologia 104:33-36.

Kang, Y.J. and J.M. Alexander. 1996. Alterations of the glutathione redox cycle status in fumonisin B1-treated pig kidney cells. J. Biochem. Toxicol. 11:121-126.

Karagezyan, M.K., L.M. Ovsepyan, S.S. Ovakimyan, A.S. Boyadzhyan, L.L. Osipyan and K.G. Karagezyan. 1995. Mycotoxin zearalenone is an activator of the free-radical lipid peroxidation process in the mitochondrial and microsomal fractions of white rat liver. Doklady, Biochem. 341:45-47.

Karppanen, E., A. Rizzo, L. Saari, S. Berg and H. Boström. 1989. Investigation on trichothecene-stimulated lipid peroxidation and toxic effects of trichothecenes in animals. Acta Vet. Scand. 30:391-399.

Khan, S., M. Martin, H. Bartsch and A.D. Rahimtula. 1989. Perturbation of liver microsomal calcium homeostasis by ochratoxin A. Biochem. Pharmacol. 38:67-72.

Kim, I.H., H.Y. Son, S.W. Cho, C.S. Ha and B.H. Kang. 2003. Zearalenone induces male germ cell apoptosis in rats. Toxicol. Lett. 138:185-192.

Kim, M.S., D.Y. Lee, T. Wang and J.J. Schroeder. 2001. Fumonisin B(1) induces apoptosis in LLC-PK(1) renal epithelial cells via a sphinganine- and calmodulin-dependent pathway. Toxicol. Appl. Pharmacol. 176:118-126.

Klasing, K.C. 1998. Nutritional modulation of resistance to infectious diseases. Poult. Sci. 77:1119-1125.

Klasing, K.C. 1998a. Avian macrophages: regulators of local and systemic immune responses. Poult. Sci. 77:983-989.

Kravchenko, L.V., E.E. Kuz'mina, L.I. Avren'eva and V.A. Tutel'ian. 1990. Protective effect of selenium in acute T-2 mycotoxicosis. Voprosy Medizinskoi Khimii 36:36-38.

Kubena, L.F., R.B. Harvey, W.E. Huff, D.E. Corrier, T.D. Philips and G.E. Rottinghaus. 1989. Influence of ochratoxin A and T-2 toxin singly and in combination on broiler chickens. Poult. Sci. 68:867-872.

Langer, C., J.M. Jurgensmeier and G. Bauer. 1996. Reactive oxygen species act at both TGF-ß-dependent and -independent steps during induction of apoptosis of transformed cells by normal cells. Exper. Cell Res. 222:117-124.

Lafarge-Frayssinet, C., K. Chakor, P. Lafont and C. Frayssinet. 1990. Transplacental transfer of T-2-toxin: pathological effect. J. Environ. Pathol. Toxicol. Oncol. 10:64-68.

Lahtala, M. 1998. Chicken CD4, CD8ab, and CD8aa T cell co-receptor molecules. Poult. Sci. 77:1858-1873.

Larsen, H.J. 1993. Relations between selenium and immunity. Norwegian J. Agric. Sci. Suppl. 11:105-119.

Latshaw, J.D. 1991. Nutrition—mechanisms of immunosuppression. Vet. Immun. Immunopath. 30:111-120.

Lea, T., K. Steien and F.C. Stormer. 1989. Mechanism of ochratoxin A-induced immunosuppression. Mycopathologia 107:153-159.

Leal, M., A. Shimada, F. Ruiz and E. Gonzalez de Mejia. 1999. Effect of lycopene on lipid peroxidation and glutathione-dependent enzymes induced by T-2 toxin *in vivo*. Toxicol. Letters 109:1-10.

Lei, D.N., L.Q. Wang, B.H. Ruebner, D.P. Hsieh, B.F. Wu, C.R. Zhu and M.J. Du. 1990. Effect of selenium on aflatoxin hepatocarcinogenesis in the rat. Biomed. Environ. Sci. 3:65-80.

Lemmer, E.R., W.C.A. Gelderblom, E.G. Shephard, S. Abel, B.L. Seymour, J.P. Cruse, R.E. Kirsch, W.F.O. Marasas and P. de la M. Hall. 1999. The effects of dietary iron overload on fumonisin B_1-induced cancer promotion in the rat liver. Cancer Letters 146:207-215.

Li, G., J. Shinozuka, K. Uetsuka, H. Nakayama and K. Doi. 1997. T-2 toxin-induced apoptosis in intestinal crypt epithelial cells of mice. Exper. Toxicol. Pathol. 49:447-450.

Li, W., R.T. Riley, K.A. Voss and W.P. Norred. 2000. Role of proliferation in the toxicity of fumonisin B1: enhanced hepatotoxic response in the partially hepatectomized rat. J. Toxicol. Environ.

Health A. 60:441-457.
Lim, C.W., H.M. Parker, R.F. Vesonder and W.M. Haschek. 1996. Intravenous fumonisin B1 induces cell proliferation and apoptosis in the rat. Natural Toxins 4:34-41.
Liu, J., C.F. Yang, B.L. Lee, H.M. Shen, S.G. Ang and C.N. Ong. 1999. Effect of *Salvia miltiorrhiza* on aflatoxin B1-induced oxidative stress in cultured rat hepatocytes. Free Rad. Res. 31:559-568.
Liu, B.H., F.Y. Yu, M.H. Chan and Y.L. Yang. 2002. The effects of mycotoxins, fumonisin B1 and aflatoxin B1, on primary swine alveolar macrophages. Toxicol. Appl. Pharmacol. 180:197-204.
Luhe, A., H. Hildebrand, U. Bach, T. Dingermann and H.J. Ahr. 2003. A new approach to studying ochratoxin A (OTA)-induced nephrotoxicity: expression profiling *in vivo* and *in vitro* employing cDNA microarrays. Toxicol. Sci. 73:315-328.
Luster, M.I., D.R. Germolec, G.R. Burleson, C.W. Jameson, M.F. Ackermann, K.R. Lamm and H.T. Hayes. 1987. Selective immunosuppression in mice of natural killer cell activity by ochratoxin A. Cancer Res. 47:2259-2263.
Lydyard, P.M., A. Whelan and M.W. Fanger. 2000. Instant notes in immunology. BIOS Scientific Publishers Ltd., UK.
Malaveille, C., G. Brun and H. Bartsch. 1994. Structure-activity studies in *E. coli* strains on ochratoxin A (OTA) and its analogues implicate a genotoxic free radical and a cytotoxic thiol derivative as reactive metabolites. Mutation Res. 307:141-147.
Mann, D.D., G.M. Buening, B.S. Hook and G.D. Osweiler. 1982. Effect of T-2 toxin on the bovine immune system: humoral factors. Infect. Immun. 36:1249-1252.
Maresca, M., R. Mahfoud, N. Garmy and J. Fantini. 2002. The mycotoxin deoxynivalenol affects nutrient absorption in human intestinal epithelial cells. J. Nutr. 132:2723-2731.
Marin, D.E., I. Taranu, R.P. Bunaciu, F. Pascale, D.S. Tudor, N. Avram, M. Sarca, I. Cureu, R.D. Criste, V. Suta and I.P. Oswald. 2002. Changes in performance, blood parameters, humoral and cellular immune responses in weanling piglets exposed to low doses of aflatoxin. J. Anim. Sci. 80:1250-1257.
Martinova, E.A. and A.H. Merrill, Jr. 1995. Fumonisin B1 alters sphingolipid metabolism and immune function in BALB/c mice: immunological responses to fumonisin B1. Mycopathologia 130:163-170.
Martynova, E.A., A.S. Solov'ev, A.V. Khrenov, T.N. Zabotina and A.V. Alesenko. 1995. Modulation of the level of sphingomyelin cycle products and expression of the CD3 receptor in immunocompetent organs by fumonisin B1. Biokhimiia 60:618-625.
Mathur, S., P.D. Constable, R.M. Eppley, A.L. Waggoner, M.E. Tumbleson and W.M. Haschek. 2001. Fumonisin B(1) is hepatotoxic and nephrotoxic in milk-fed calves. Toxicol. Sci. 60:385-396.
Meki, A.R., S.K. Abdel-Ghaffar and I. El-Gibaly. 2001. Aflatoxin B1 induces apoptosis in rat liver: protective effect of melatonin. Neuroendocrinol. Lett. 22:417-426.
Meki, A.R. and A.A. Hussein. 2001. Melatonin reduces oxidative stress induced by ochratoxin A in rat liver and kidney. Comp. Biochem. Physiol. C Toxicol. Pharmacol. 130:305-313.
Meky, F.A., L.J. Hardie, S.W. Evans and C.P. Wild. 2001. Deoxynivalenol-induced immunomodulation of human lymphocyte proliferation and cytokine production. Food Chem. Toxicol. 39:827-836.
Meli, R., M.C. Ferrante, G.M. Raso, M. Cavaliere, R. Di Carlo and A. Lucisano. 2000. Effect of fumonisin B1 on inducible nitric oxide synthase and cyclooxygenase-2 in LPS-stimulated J774A.1 cells. Life Sci. 67:2845-2853.
Mezes, M., M. Barta and G. Nagy. 1999. Comparative investigation on the effect

of T-2 mycotoxin on lipid peroxidation and antioxidant status in different poultry species. Res. Vet. Sci. 66:19-23.

McCorkle, F.M. 1998. Symposium: Nonlymphoid cells and their factors in immune response. Introduction to the symposium. Poult. Sci. 77:963.

Miles, E.A. and P.C. Calder. 1998. Modulation of immune function by dietary fatty acids. Proc. Nutr. Soc. 57:277-292.

Milks, M.M., S.R. Wilt, I.I. Ali and D. Couri. 1985. The effects of selenium on the emergence of aflatoxin B1-induced enzyme-altered foci in rat liver. Fund. Appl. Toxicol. 5:320-326.

Miller, K. and H.A. Atkinson. 1986. The *in vitro* effects of trichothecenes on the immune system. Food Chem. Toxicol. 24:545-549.

Minervini, F., F. Fornelli and K.M. Flynn. 2004. Toxicity and apoptosis induced by the mycotoxins nivalenol, deoxynivalenol and fumonisin B(1) in a human erythroleukemia cell line. Toxicol. *in vitro* 18:21-28.

Mobio, T.A., I. Baudrimont, A. Sanni, T.W. Shier, D. Saboureau, S.D. Dano, Y. Ueno, P.S. Steyn and E.E. Creppy. 2000. Prevention by vitamin E of DNA fragmentation and apoptosis induced by fumonisin B1 in C6 glioma cells. Arch. Toxicol. 74:112-119.

Mobio, T.A., E. Tavan, I. Baudrimont, R. Anane, M.R. Carratu, A. Sanni, M.F. Gbeassor, T.W. Shier, J.F. Narbonne and E.E. Creppy. 2003. Comparative study of the toxic effects of fumonisin B1 in rat C6 glioma cells and p53-null mouse embryo fibroblasts. Toxicol. 183:65-75.

Moon, W.S., J.H. Kim, M.J. Kang and D.G. Lee. 2000. Early ultrastructural changes of apoptosis induced by fumonisin B1 in rat liver. Yonsei Medical Journal 41:195-204.

Moon, E.Y., J.J. Han, D.K. Rhee and S. Pyo. 1998. Aflatoxin B1-induced suppression of nitric oxide production in murine peritoneal macrophages. J. Toxicol. Environ. Health A. 55:517-530.

Mueller, R. 1995. Toxicity of fumonisin B1 administered intraperitoneally to male Sprague-Dawley rats. Food Chem. Toxicol. 33:653-665.

Muller, G., H. Rosner, B. Rohrmann, W. Erler, G. Geschwend, U. Grafe, B. Burkert, U. Moller, R. Diller, K. Sachse and H. Kohler. 2003. Effects of the mycotoxin ochratoxin A and some of its metabolites on the human cell line THP-1. Toxicol. 184:69-82.

Muller, G., P. Kielstein, H. Rosner, A. Berndt, M. Heller and H. Kohler. 1999. Studies of the influence of ochratoxin A on immune and defence reactions in weaners. Mycoses 42:495-505.

Muller, G., P. Kielstein, H. Kohler, A. Berndt and H. Rosner. 1995. Studies of the influence of ochratoxin A on immune and defense reactions in the mouse model. Mycoses 38:85-91.

Nagase, M., M.M. Alam, A. Tsushima, T. Yoshizawa and N. Sakato. 2001. Apoptosis induction by T-2 toxin: activation of caspase-9, caspase-3, and DFF-40/CAD through cytosolic release of cytochrome c in HL-60 cells. Biosci. Biotechnol. Biochem. 65:1741-1747.

Nagata, T., H. Suzuki, N. Ishigami, J. Shinozuka, K. Uetsuka, H. Nakayama and K. Doi. 2001. Development of apoptosis and changes in lymphocyte subsets in thymus, mesenteric lymph nodes and Peyer's patches of mice orally inoculated with T-2 toxin. Exp. Toxicol. Pathol. 53:309-315.

Neldon-Ortiz, D.L. and M.A. Qureshi. 1992. Effects of AFB1 embryonic exposure on chicken mononuclear phagocytic cell functions. Dev. Comp. Immunol. 16:187-196.

Okotie-Eboh, G.O., L.F. Kubena, A.D. Chinnah and C.A. Bailey. 1997. Effects of ß-carotene and canthaxanthin on aflatoxicosis in broilers. Poult. Sci. 76:1337-1341.

Omar, R. F., B.B. Hasinoff, F. Mejilla and A.D. Rahimtula. 1990. Mechanism of ochratoxin A stimulated lipid peroxidation. Biochem. Pharm. 40:1183-1191.

Osborne, D.J., W.E. Huff, P.B. Hamilton and H.R. Burmeister. 1982. Comparison of ochratoxin, aflatoxin, and T-2 toxin for their effects on selected parameters related to digestion and evidence for specific metabolism of carotenoids in chickens. Poult. Sci. 61:1646-1652.

Overnes, G., T. Matre, T. Sivertsen, H.J. Larsen, W. Langseth, L.J. Reitan and J.H. Jansen. 1997. Effects of diets with graded levels of naturally deoxynivalenol-contaminated oats on immune response in growing pigs. Zentralbl Veterinarmed A. 44:539-550.

Panangala, V.S., J.J. Giambrone, U.L. Diener, N.D. Davis, F.J. Hoerr, A. Mitra, R.D. Schultz and G.R. Wilt. 1986. Effects of aflatoxin on the growth performance and immune responses of weanling swine. Am. J. Vet. Res. 47:2062-2067.

Pang, V.F., R.J. Lambert, P.J. Felsburg, V.R. Beasley, W.B. Buck and W.M. Haschek. 1988. Experimental T-2 toxicosis in swine following inhalation exposure: clinical signs and effects on hematology, serum biochemistry, and immune response. Fundam. Appl. Toxicol. 11:100-109.

Pang, V.F., P.J. Felsburg, V.R. Beasley, W.B. Buck and W.M. Haschek. 1987. The toxicity of T-2 toxin in swine following topical application. II. Effects on hematology, serum biochemistry and immune response. Fundam. Appl. Toxicol. 9:50-59.

Pasupathy, K., M. Krishna and R.K. Bhattacharya. 1999. Alterations of phosphatidylinositol signal transduction pathway in hepatic mitochondria following aflatoxin B1 administration. Indian J. Exper. Biol. 37:876-880.

Pestka, J.J. and G.S. Bondy. 1990. Alteration of immune function following dietary mycotoxin exposure. Can. J. Physiol. Pharmacol. 68:1009-1016.

Pestka, J.J., D. Yan and L.E. King. 1994. Flow cytometric analysis of the effects of *in vitro* exposure to vomitoxin (deoxynivalenol) on apoptosis in murine T, B and IgA+ cells. Food Chem. Toxicol. 32:1125-1136.

Petr, T., I. Barta and B. Turek. 1990. *In vivo* effect of selenium on the mutagenic activity of aflatoxin B1. J. Hyg. Epidemiol. Microbiol. Immunol. 34:123-128.

Petrik, J., T. Zanic-Grubisic, K. Barisic, S. Pepeljnjak, B. Radic, Z. Ferencic and I. Cepelak. 2003. Apoptosis and oxidative stress induced by ochratoxin A in rat kidney. Arch. Toxicol. 77:685-693.

Pier, A.C. 1992. Major biological consequences of aflatoxicosis in animal production. J. Anim. Sci. 70:3964-3967.

Potchinsky, M.B. and S.E. Bloom. 1993. Selective aflatoxin B1-induced sister chromatid exchanges and cytotoxicity in differentiating B and T lymphocytes *in vivo*. Environ. Mol. Mutagen. 21:87-94.

Pozzi, C.R., B. Correa, J.G. Xavier, G.M. Direito, R.B. Orsi and S.V. Matarazzo. 2001. Effects of prolonged oral administration of fumonisin B1 and aflatoxin B1 in rats. Mycopathologia 151:21-27.

Premalatha, B., V. Sujatha and P. Sachdanandam. 1997. Modulating effect of *Semecarpus anacardium Linn.* Nut extract on glucose metabolising enzymes in aflatoxin B-1-induced experimental hepatocellular carcinoma. Pharmacol. Res. 36:187-192.

Quist, C.F., D.I. Bounous, J.V. Kilburn, V.F. Nettles and R.D. Wyatt. 2000. The effect of dietary aflatoxin on wild turkey poults. J. Wildl. Dis. 36:436-444.

Quist, C.F., E.W. Howerth, J.R. Fischer, R.D. Wyatt, D.M. Miller and V.F. Nettles. 1997. Evaluation of low-level aflatoxin in the diet of white-tailed deer. J. Wildl. Dis. 33:112-121.

Qureshi, M.A. 1998. Role of macrophages in avian health and disease. Poult. Sci. 77:978-982.

Qureshi, M.A., I. Hussain and C.L. Heggen. 1998a. Understanding immunology in disease development and control. Poult. Sci. 77:1126-1129.

Qureshi, M.A., J. Brake, P.B. Hamilton, W.M. Hagler, Jr. and S. Nesheim. 1998. Dietary exposure of broiler breeders to aflatoxin results in immune dysfunction in progeny chicks. Poult. Sci. 77:812-819.

Qureshi, M.A., J.D. Garlich, W.M. Hagler, Jr. and D. Weinstock. 1995. *Fusarium proliferatum* culture material alters several production and immune performance parameters in White Leghorn chickens. Immunopharm. Immun. 17:791-804.

Rafai, P., S. Tuboly, A. Bata, P. Tilly, A. Vanyi, Z. Papp, L. Jakab and E. Tury. 1995. Effect of various levels of T-2 toxin in the immune system of growing pigs. Vet. Rec. 136:511-514.

Rahimtula, A.D., J.C. Bereziat, V. Bussacchini-Griot and H. Bartsch. 1988. Lipid peroxidation as a possible cause of ochratoxin A toxicity. Biochem. Pharm. 37:4469-4477.

Raisuddin, S., K.P. Singh, S.I. Zaidi, B.N. Paul and P.K. Ray. 1993. Immunosuppressive effects of aflatoxin in growing rats. Mycopathologia 124:189-194.

Raisuddin, S., K.P. Singh, S.I. Zaidi, A.K. Saxena and P.K. Ray. 1990. Effects of aflatoxin on lymphoid cells of weanling rats. J. Appl. Toxicol. 10:245-250.

Raj, H.G., E. Kohli, V. Rohil, B.S. Dwarakanath, V.S. Parmar, S. Malik, J.S. Adhikari, Y.K. Tyagi, S. Goel, K. Gupta, M. Bose and C.E. Olsen. 2001. Acetoxy-4-methylcoumarins confer differential protection from aflatoxin B(1)-induced micronuclei and apoptosis in lung and bone marrow cells. Mut. Res. 494:31-40.

Rasooly, L. and J.J. Pestka. 1992. Vomitoxin-induced dysregulation of serum IgA, IgM and IgG reactive with gut bacterial and self antigens. Food Chem. Toxicol. 30:499-504.

Rastogi, R., A.K. Srivastava and A.K. Rastogi. 2001a. Long term effect of aflatoxin B(1) on lipid peroxidation in rat liver and kidney: effect of picroliv and silymarin. Phytotherapy Res. 15:307-310.

Rastogi, R., A.K. Srivastava and A.K. Rastogi. 2001b. Biochemical changes induced in liver and serum of aflatoxin B1-treated male Wistar rats: preventive effect of picroliv. Pharmacol. Toxicol. 88:53-58.

Ratan, R.R., T.H. Murphy and J.M. Baraban. 1994. Oxidative stress induces apoptosis in embryonic cortical neurons. J. Neurochem. 62:376-379.

Ray, P.K., T. Das, G. Sa, A.K. Ghosh and S. Chattopadhyay. 2000. Protection of apoptotic cell death by protein A. Apoptosis 5:509-514.

Reddy, R.V., M.J. Taylor and R.P. Sharma. 1987. Studies of immune function of CD-1 mice exposed to aflatoxin B1. Toxicol. 43:123-132.

Ribeiro, S.M.R., G.M. Chagas, A.P. Campello and M.L.W. Kluppel. 1997. Mechanism of citrinin-induced dysfunction of mitochondria. 5. Effect on the homeostasis of the reactive oxygen species. Cell Biochem. Function 15:203-209.

Rizzo, A.F., F. Atroshi, M. Ahotupa, S. Sankari and E. Elovaara. 1994. Protective effect of antioxidants against free radical-mediated lipid peroxidation induced by DON or T-2 toxin. J. Vet. Med. A 41:81-90.

Robbana-Barnat, S., C. Lafarge-Frayssinet, H. Cohen, G.A. Neish and C. Frayssinet. 1988. Immunosuppressive properties of deoxynivalenol. Toxicol. 48:155-166.

Rotter, B.A. and Y.N. Oh. 1996. Mycotoxin fumonisin B1 stimulates nitric oxide production in a murine macrophage cell line. Nat. Toxins 4:291-294.

Rotter, B.A., B.K. Thompson, M. Lessard, H.L. Trenholm and H. Tryphonas. 1994. Influence of low-level exposure to *Fusarium* mycotoxins on selected immunological and hematological parameters in young swine. Fundam. Appl. Toxicol. 23:117-124.

Rumora, L., S. Kovacic, R. Rozgaj, I. Cepelak, S. Pepeljnjak and G.T. Zanic. 2002. Cytotoxic and genotoxic effects of fumonisin B1 on rabbit kidney RK13 cell line. Arch. Toxicol. 76:55-61.

Sandstrom, P.A. and T.M. Buttke. 1993. Autocrine production of extracellular catalase prevents apoptosis of the human CEM T-cell line in serum-free medium. Proc. Nat. Acad. Sci. 90:4708-4712.

Sahu, S. C., R.M. Eppley, S.W. Page, G.C. Gray, C.N. Barton and M.W. O'Donnell. 1998. Peroxidation of membrane lipids and oxidative DNA damage by fumonisin B_1 in isolated rat liver nuclei. Cancer Letters 125:117-121.

Saif, Y.M. and D.E. Swayne. 1998. Symposium: Infectious poultry diseases. Introduction. Poult. Sci. 77:1110.

Santin, E., A.C. Paulillo, P.C. Maiorka, A.C. Alessi, E.L. Krabbe and A. Maiorka. 2002. The effects of ochratoxin/aluminosilicate interaction on the tissues and humoral immune response of broilers. Avian Pathol. 31:73-79.

Sastre, J., F.V. Pallardo and J. Vina. 1996. Glutathione, oxidative stress and aging. Age 19:129-139.

Scibelli, A., S. Tafuri, M.C. Ferrante, E. Alimenti, B. Naso, A. Lucisano, N. Staiano and R. Della Morte. 2003. Ochratoxin A affects COS cell adhesion and signaling. Toxicol. Appl. Pharmacol. 192:222-230.

Schuster, A., G. Hunder, B. Fichtl and W. Forth. 1987. Role of lipid peroxidation in the toxicity of T-2 toxin. Toxicon. 25:1321-1328.

Schwerdt, G., R. Freudinger, S. Mildenberger, S. Silbernagl and M. Gekle. 1999. The nephrotoxin ochratoxin A induces apoptosis in cultured human proximal tubule cells. Cell Biology and Toxicol. 15:405-415.

Schaeffer, J.L., J.K. Tyczkowski, J.E. Riviere and P.B. Hamilton. 1988. Aflatoxin-impaired ability to accumulate oxycarotenoid pigments during restoration in young chickens. Poult. Sci. 67:619-625.

Schaeffer, J.L., J.K. Tyczkowski and P.B. Hamilton. 1988a. Depletion of oxycarotenoid pigments in chickens and the failure of aflatoxin to alter it. Poult. Sci. 67:1080-1088.

Schaefer, D., J.M. Jurgensmeier and G. Bauer. 1995. Catechol interferes with TGF-ß-induced elimination of transformed cells by normal cells: implications for the survival of transformed cells during carcinogenesis. Intl. J. Cancer 60:520-526.

Schiefer, H.B., C.G. Rousseaux, D.S. Hancock and B.R. Blakley. 1987. Effects of low-level long-term oral exposure to T-2 toxin in CD-1 mice. Food Chem. Toxicol. 25:593-601.

Seefelder, W., H.U. Humpf, G. Schwerdt, R. Freudinger and M. Gekle. 2003. Induction of apoptosis in cultured human proximal tubule cells by fumonisins and fumonisin metabolites. Toxicol. Appl. Pharmacol. 192:146-153.

Seegers, J.C., L.H. Bohmer, M.C. Kruger, M.L. Lottering and M. de Kock. 1994. A comparative study of ochratoxin A-induced apoptosis in hamster kidney and HeLa cells. Toxicol. Appl. Pharm. 129:1-11.

Segal, R., I. Milo-Goldzweig, A.Z. Joffe and B. Yagen. 1983. Trichothecene-induced hemolysis. I. The hemolytic activity of T-2 toxin. Toxicol. Appl. Pharm. 70:343-349.

Sharma, R.P. 1993. Immunotoxicity of mycotoxins. J. Dairy Sci. 76:892-897.

Sharma, R.P., R.R. Dugyala and K.A. Voss. 1997. Demonstration of *in situ* apoptosis in mouse liver and kidney after short-term repeated exposure to fumonisin B1. J. Comp. Path. 117:371-381.

Shen, H.M., C.Y. Shi, H.P. Lee and C.N. Ong. 1994. Aflatoxin B1-induced lipid peroxidation in rat liver. Toxicol. Appl. Pharm. 127:145-150.

Shen, H.M., C.N. Ong and C.Y. Shi. 1995. Involvement of reactive oxygen species in aflatoxin B1-induced cell injury in cultured rat hepatocytes. Toxicol. 99:115-123.

Shi, C.Y., Y.C. Hew and C.N. Ong. 1995. Inhibition of aflatoxin B1-induced cell injury by selenium: an *in vitro* study.

Human Exper. Toxicol. 14:55-60.

Shi, C.Y., S.C. Chua, H.P. Lee and C.N. Ong. 1994. Inhibition of aflatoxin B1-DNA binding and adduct formation by selenium in rats. Cancer Letters 82:203-208.

Shifrin, V.I. and P. Anderson. 1999. Trichothecene mycotoxins trigger a ribotoxic stress response that activates c-Jun N-terminal kinase and p38 mitogen-activated protein kinase and induces apoptosis. J. Biol. Chem. 274:13985-13992.

Shinozuka, J., L. Guanmin, K. Uetsuka, H. Nakayama and K. Doi. 1997. Process of the development of T-2 toxin-induced apoptosis in the lymphoid organs of mice. Exper. Anim. 46:117-126.

Shinozuka, J., M. Suzuki, N. Noguchi, T. Sugimoto, K. Uetsuka, H. Nakayama and K. Doi. 1998. T-2 toxin-induced apoptosis in hematopoietic tissues of mice. Toxicol. Path. 26:674-681.

Shivachandra, S.B., R.L. Sah, S.D. Singh, J.M. Kataria and K. Manimaran. 2003. Immunosuppression in broiler chicks fed aflatoxin and inoculated with fowl adenovirus serotype-4 (FAV-4) associated with hydropericardium syndrome. Vet. Res. Commun. 27:39-51.

Shokri, F., M. Heidari, S. Gharagozloo and M. Ghazi-Khansari. 2000. *In vitro* inhibitory effects of antioxidants on cytotoxicity of T-2 toxin. Toxicol. 146:171-176.

Sies, H. 1999. Glutathione and its role in cellular functions. Free Rad. Bio. Med. 27:916-921.

Simizu, S., M. Takada, K. Umezawa and M. Imoto. 1998. Requirement of caspase-3-like protease-mediated hydrogen peroxide production for apoptosis induced by various anticancer drugs. J. Biol. Chem. 273:26900-26907.

Sklan, D., M. Shelly, B. Makovsky, A. Geyra, E. Klipper and A. Friedman. 2003. The effect of chronic feeding of diacetoxy-scirpenol and T-2 toxin on performance, health, small intestinal physiology and antibody production in turkey poults. Brit. Poult. Sci. 44:46-52.

Souza, M.F., A.R. Tome and V.S. Rao. 1999. Inhibition by the bioflavonoid ternatin of aflatoxin B1-induced lipid peroxidation in rat liver. J. Pharm. Pharmac. 51:125-129.

Stanley, G.V., H. Chakwu and D. Thompson. 1998. Singly and combined effects of organic selenium (Se-yeast) and vitamin E on ascites reduction in broilers. Proc. Poult. Sci. Assoc. Conf., abst.111.

Stewart, R.G., J.K. Skeeles, R.D. Wyatt, J. Brown, R.K. Page, I.D. Russell and P.D. Lukert. 1985. The effect of aflatoxin on complement activity in broiler chickens. Poult. Sci. 64:616-619.

Stoev, S.D., D. Djuvinov, T. Mirtcheva, D. Pavlov and P. Mantle. 2002. Studies on some feed additives giving partial protection against ochratoxin A toxicity in chicks. Toxicol. Lett. 135:33-50.

Stoev, S.D., G. Anguelov, I. Ivanov and D. Pavlov. 2000. Influence of ochratoxin A and an extract of artichoke on the vaccinal immunity and health in broiler chicks. Exp. Toxicol. Pathol. 52:43-55.

Suneja, S.K., D.S. Wagle and G.C. Ram. 1989. Effect of oral administration of T-2 toxin on glutathione shuttle enzymes, microsomal reductases and lipid peroxidation in rat liver. Toxicon 27:995-1001.

Surai, P.F. 2002. In: *Natural Antioxidants in Avian Nutrition and Reproduction*. Nottingham University Press, UK.

Surai, P.F. 2000. Effect of the selenium and vitamin E content of the maternal diet on the antioxidant system of the yolk and the developing chick. Brit. Poult. Sci. 41:235-243.

Surai, P.F. and J.E. Dvorska. 2001. Is organic selenium better for animals than inorganic sources? Two different scenarios in stress conditions. Feed Mix 9:8-10.

Surai, P.F., R.C. Noble and B.K. Speake. 1999. Relationship between vitamin E content and susceptibility to lipid peroxidation in tissues of the newly hatched chick. Brit. Poult. Sci. 40:406-410.

Surai, P.F. and B.K. Speake 1998. Distribution of carotenoids from the yolk to the tissues of the chick embryo. J. Nutr. Biochem. 9:645-651.

Surai, P.F., B.K. Speake and N.H.C. Sparks. 2001. Carotenoids in avian nutrition and embryonic development. 2. Antioxidant properties and discrimination in embryonic tissues J. Poult. Sci. 38:117-145.

Tai, J.H. and J.J. Pestka. 1990. T-2 toxin impairment of murine response to *Salmonella typhimurium*: a histopathologic assessment. Mycopathologia 109:149-155.

Theumer, M.G., A.G. Lopez, D.T. Masih, S.N. Chulze and H.R. Rubinstein. 2002. Immunobiological effects of fumonisin B1 in experimental subchronic mycotoxicoses in rats. Clin. Diagn. Lab Immunol. 9:149-155.

Thuvander, A., A. Breitholtz-Emanuelsson, D. Brabencova and I. Gadhasson, I. 1996. Prenatal exposure of Balb/c mice to ochratoxin A: effects on the immune system in the offspring. Food Chem. Toxicol. 34:547-554.

Thuvander, A., P. Dahl, P. and A. Breitholtz-Emanuelsson. 1996a. Influence of perinatal ochratoxin A exposure on the immune system in mice. Nat. Toxins 4:174-180.

Thuvander, A., E. Funseth, A. Breitholtz-Emanuelsson, I.P. Hallen and A. Oskarsson. 1996b. Effects of ochratoxin A on the rat immune system after perinatal exposure. Nat. Toxins 4:141-147.

Thuvander, A., A. Breitholtz-Emanuelsson and M. Olsen. 1995. Effects of ochratoxin A on the mouse immune system after subchronic exposure. Food Chem. Toxicol. 33:1005-1011.

Tobias, S., I. Rajic and A. Vanyi. 1992. Effect of T-2 toxin on egg production and hatchability in laying hens. Acta Vet. Hungarica 40:47-54.

Tolleson, W.H., W.B. Melchior, Jr., S.M. Morris, L.J. McGarrity, O.E. Domon, L. Muskhelishvili, S.J. James and P.C. Howard. 1996. Apoptotic and anti-proliferative effects of fumonisin B1 in human keratinocytes, fibroblasts, esophageal epithelial cells and hepatoma cells. Carcinogenesis 17:239-249.

Tornyos, G., M. Kovacs, M. Rusvai, P. Horn, J. Fodor and F. Kovacs. 2003. Effect of dietary fumonisin B1 on certain immune parameters of weaned pigs. Acta Vet Hung. 51:171-179.

Tryphonas, H., L. O'Grady, D.L. Arnold, P.F. McGuire, K. Karpinski and R.F. Vesonder. 1984. Effect of deoxynivalenol (vomitoxin) on the humoral immunity of mice. Toxicol. Lett. 23:17-24.

Tsuchida, M., T. Miura, T. Shimizu and K. Aibara. 1984. Elevation of thiobarbituric acid values in the rat liver intoxicated by T-2 toxin. Biochem. Med. 31:147-166.

Tsunoda, M., R.P. Sharma and R.T. Riley. 1998. Early fumonisin B1 toxicity in relation to disrupted sphingolipid metabolism in male BALB/c mice. J. Biochem. Molecular Toxicol. 12:281-289.

Tutelyan, V.A., L.V. Kravchenko, E.E. Kuzmina, L.I. Avrenieva and J.T. Kumpulainen. 1990. Dietary selenium protects against acute toxicity of T-2 toxin in rats. Food Add. Contam. 7:821-827.

Tyczkowski, J.K. and P.B. Hamilton. 1987. Altered metabolism of carotenoids during aflatoxicosis in young chickens. Poult. Sci. 66:1184-1188.

Tyczkowski, J.K. and P.B. Hamilton. 1987a. Metabolism of lutein diester during aflatoxicosis in young chickens. Poult. Sci. 66:2011-2016.

Ueno, Y., K. Umemori, E. Niimi, S. Tanuma, S. Nagata, M. Sugamata, T. Ihara, M. Sekijima, K. Kawai and I. Ueno. 1995. Induction of apoptosis by T-2 toxin and other natural toxins in HL-60 human promyelotic leukemia cells. Natural Toxins 3:129-137.

Van Heugten, E., J.W. Spears, M.T. Coffey, E.B. Kegley and M.A. Qureshi. 1994. The effect of methionine and aflatoxin on immune function in weanling pigs. J.

Anim. Sci. 72:658-664.

Verma, R.J. and A. Nair. 2001. Ameliorative effect of vitamin E on aflatoxin-induced lipid peroxidation in the testis of mice. Asian J. Andrology 3:217-221.

Vila, B., Z.W. Jaradat, R.R. Marquardt and A.A. Frohlich. 2002. Effect of T-2 toxin on *in vivo* lipid peroxidation and vitamin E status in mice. Food Chem. Toxicol. 40:479-486.

Voss, K.A., R.T. Riley, W.P. Norred, C.W. Bacon, F.I. Meredith, P.C. Howard, R.D. Plattner, T.F. Collins, D.K. Hansen and J.K. Porter. 2001. An overview of rodent toxicities: liver and kidney effects of fumonisins and *Fusarium moniliforme*. Environ. Health Persp. 109(Suppl 2):259-266.

Wang, L.Q. 1990. The effect of selenium on hepatocarcinogenesis of rats induced by aflatoxin B1. Chung Hua Ping Li Hsueh Tsa Chih 19:46-49.

Wang, H. and X. Zhang. 2000. Recent progress in the study of common *Fusarium* mycotoxins on apoptosis of cells. Wei Sheng Yan Jiu. 29:181-183.

Wang, H., X. Sun, X. Zhang and L. Zuo. 1999. Effect of deoxynivalenol and aflatoxin G1 on apoptosis of human blood lymphocytes *in vitro*. Wei Sheng Yan Jiu. 28:102-104.

Weibking, T.S., D.R. Ledoux, A.J. Bermudez and G.E. Rottinghaus. 1994. Individual and combined effects of feeding *Fusarium moniliforme* culture material, containing known levels of fumonisin B1, and aflatoxin B1 in the young turkey poult. Poult. Sci. 73:1517-1525.

Wood, G.E. 1992. Mycotoxins in foods and feeds in the United States. J. Anim. Sci. 70:3941-3949.

Wu, D.O. and S.N. Meydani. 1998. Antioxidants and immune function. In: *Antioxidant Status, Diet, Nutrition and Health* (A.M. Papas, ed). CRC Press, Boca Raton, pp. 371-400.

Yang, C.F., J. Liu, H.M. Shen and C.N. Ong. 2000. Protective effect of ebselen on aflatoxin B1-induced cytotoxicity in primary rat hepatocytes. Pharm. Toxicol. 86:156-161.

Yang, G.H., B.B. Jarvis, Y.J. Chung and J.J. Pestka. 2000a. Apoptosis induction by the satratoxins and other trichothecene mycotoxins: relationship to ERK, p38 MAPK and SAPK/JNK activation. Toxicol. Appl. Pharm. 164:149-160.

Yazdanpanah, H, F. Roshanzamir, B. Shafaghi, M. Faizi, M. Elhami and H.R. Rasekh. 1997. Assessment of possible protective roles of selenium, zinc and cis-stilbene oxide against acute T-2 toxin poisoning: a preliminary report. Natural Toxins 5:133-135.

Yin, J.J., M.J. Smith, R.M. Eppley, S.W. Page and J.A. Sphon. 1998. Effects of fumonisin B1 on lipid peroxidation in membranes. Biochimica et Biophysica Acta 1371:134-142

Yoshino, N., M. Takizawa, H. Akiba, H. Okumura, F. Tashiro, M. Honda and Y. Ueno. 1996. Transient elevation of intracellular calcium ion levels as an early event in T-2 toxin-induced apoptosis in human promyelotic cell line HL-60. Natural Toxins 4:234-241.

Yu, C.H., Y.M. Lee, Y.P. Yun and H.S. Yoo. 2001. Differential effects of fumonisin B1 on cell death in cultured cells: the significance of the elevated sphinganine. Arch. Pharm. Res. 24:136-143.

Zhang, Y., C. Jones and M.B. Dickman. 2001. Identification of differentially expressed genes following treatment of monkey kidney cells with the mycotoxin fumonisin B(1). Food Chem. Toxicol. 39:45-53.

Zhao, W., Y. Han, B. Zhao, S. Hirota, J. Hou and W. Xin. 1998. Effect of carotenoids on the respiratory burst of rat peritoneal macrophages. Biochimica et Biophysica Acta 1381:77-88.

Zhou, H.R., J.R. Harkema, J.A. Hotchkiss, D. Yan, R.A. Roth and J.J. Pestka. 2000. Lipopolysaccharide and the trichothecene vomitoxin (deoxynivalenol) syner-gistically induce apoptosis in murine lymphoid organs. Toxicol. Sci. 53:253-263.

Zhou, H.R., D. Yan and J.J. Pestka. 1997. Differential cytokine mRNA expression in mice after oral exposure to the trichothecene vomitoxin (deoxynivalenol): dose response and time course. Toxicol. Appl. Pharmacol. 144:294-305.

Zucker, B. and G. Bauer. 1997. Intercellular induction of apoptosis of transformed cells is modulated by their intracellular glutathione concentration. Intl. J. Oncology 10:141-146.

Zucker, B., J. Hanusch and G. Bauer. 1997. Glutathione depletion in fibroblasts is the basis for apoptosis-induction by endogenous reactive oxygen species. Cell Death and Differentiation 4:388-395.

6

MYCOTOXINS IN AQUACULTURE

Bruce Manning
Thad Cochran National Warmwater Aquaculture Center, Mississippi State University, Stoneville, Mississippi, USA

Introduction

An understanding of how mycotoxins affect aquaculture becomes more important as fish nutritionists rely more heavily on plant ingredients to formulate fish diets. Commercial diets for warmwater fish, such as channel catfish, *Ictalurus punctatus*, have evolved over the past decade to include little or no animal protein. It has been common practice during the past 10 to15 years to include in the diets of channel catfish up to 10% fish meal or other sources of animal protein. More recently, nutritional research with salmonids is striving to develop practical diets that contain reduced levels of animal protein that are replaced with higher levels of plant protein and energy sources from oilseed meals and cereal grains. Therefore, with increasing use of plant sources of protein or energy, the likelihood of exposing cultured fish to mycotoxin-contaminated diets becomes eminent.

Concerns about mycotoxins in aquaculture began over 40 years ago when hepatocellular carcinomas of rainbow trout, *Onchorhyncus mykiss,* being cultured in commercial and governmental fish hatcheries in the US were determined to be caused by aflatoxin-contaminated cottonseed meal. Since that time, many of the recognised mycotoxins have been evaluated with coldwater fish like salmonids and warmwater species such as channel catfish and tilapia. Other mycotoxins that have been investigated with rainbow trout and more recently with warmwater aquaculture fish include the *Aspergillus* or *Penicillium* mycotoxins, cyclopiazonic acid and ochratoxin A (OTA), and the *Fusarium* trichothecene mycotoxins T-2 toxin and deoxynivalenol (DON). Other *Fusarium* mycotoxins like fumonisin B_1, (FB_1) and moniliformin have been studied recently.

Background

The occurrence of aflatoxicosis at fish hatcheries in the US producing rainbow trout and in turkey farms in the UK caused great concern in the early 1960s about the safety of the food supply for both domesticated animals and humans. Initially, the condition affecting turkeys was referred to as **turkey X disease** (Blount, 1961), but was subsequently identified as being caused by mouldy peanut meal imported from Brazil that was infected with *Aspergillus flavus.*

In the US, rainbow trout reared in government and commercial hatcheries were found to have developed liver nodules or hepatomas (Post, 1987). Much laboratory and research work went into understanding the cause of the hepatomas that appeared in rainbow trout. These efforts have been thoroughly described by Dr. J.E. Halver (1969). After numerous studies it was found that trout fed pelleted dry diets containing cottonseed meal that was infected with *A. flavus* developed hepatomas. These were later determined to be hepatocellular carcinomas (HCC).

The active agent causing HCC was identified by chromatographic techniques and named aflatoxin in reference to genus and species names of the fungal organism from which it arises. It was determined quickly that the presence of aflatoxin in foods for humans posed a threat to the health of people who might consume such food, since aflatoxin was determined to be a powerful carcinogen. Measures were taken through the regulatory authority of the United States Food and Drug Administration (FDA) to limit the concentration of aflatoxin that can be present in the feeds of domesticated animals and foods for humans (Meronek and Xie, 1999). At first, the maximum allowable concentrations of aflatoxin were set by the limits of analytical detection of available methods of chromatography. Subsequently, the 20 parts per billion (ppb) action level for total aflatoxins was established in 1969 and remains in effect today even though advances in chromatography allow for the detection of aflatoxin down to <1 ppb. Over the years, exceptions to the 20 ppb action level have been made in the US for feed ingredients that have been designated for use in certain classes of poultry and livestock as shown in Table 1 (Meronek and Xie, 1999). This was done, in part, to accommodate the normal commerce of animal feed ingredients, such as corn and cottonseed meal, that contained levels of aflatoxin higher than the 20 ppb limit. Additionally, research determined that only very low levels of aflatoxin were deposited in the edible portions of meat and eggs of agricultural animals that consumed feeds manufactured with corn or cottonseed meal containing levels of aflatoxin higher than 20 ppb, but not exceeding the levels shown in Table 1. Total aflatoxins include the four commonly detected mycotoxins with similar chemical structures. These include aflatoxin B_1 (AFB_1), the most toxic of the aflatoxins, aflatoxin B_2 (AFB_2), aflatoxin G_1 (AFG_1), and aflatoxin G_2 (AFG_2). The latter three have varying degrees of toxicity, but are usually considered to be less carcinogenic than AFB_1 (Coulombe, 1993).

Mycotoxins produced by *Aspergillus* and *Penicillium* moulds

When aflatoxin was determined to be the cause of HCC in rainbow trout, research was conducted to determine the levels of

Table 1. US Food and Drug Administration guidelines for acceptable levels of total aflatoxins in foods and feeds.

Action level (ppb)	*Commodity*	*Species*	*Comments*
0.5 (AFM_1)	Milk	Human	Metabolite of AFB_1
20	Any food except milk	Human	
20	Animal feeds	All species	
Exceptions to above guidelines			
300	Cottonseed meal used in feeds	All species	
300	Corn	Finishing beef cattle	Must be used for designated species
200	Corn	Finishing swine	Same as above
100	Corn	Breeding cattle, swine, or poultry	Same as above

aflatoxin contamination of trout diets required to produce liver carcinomas. One of these studies showed that rainbow trout receiving a diet containing only 0.4 ppb AFB_1 for 15 months had a 14% incidence of HCC (Lee *et al.*, 1968). Another later study reported by Schoenhard *et al.* (1981) showed that rainbow trout developed a higher incidence of HCC as the time period of feeding AFB_1 was extended. Trout fed a diet containing 20 ppb AFB_1 for 8 months had an incidence of 58% HCC while trout fed the same dietary level for 12 months had 83% incidence.

Species differences have been observed among salmonids regarding sensitivity to aflatoxin. Rainbow trout fed a diet containing 20 ppb AFB_1 had a 96% incidence of HCC. On the other hand, coho salmon, *Onchorhyncus kisutch*, fed the same dietary level of AFB_1 had no incidence of HCC (Coulombe *et al.*, 1984). The difference observed between the two species was ascribed to the greater ability of trout hepatocytes to convert AFB_1 to the 8,9-epoxide of AFB_1. The 8,9-epoxide of AFB_1 forms mutagenic adducts with guanine of DNA; these also appear to persist longer in trout than in coho salmon (Coulombe 1993). Other species differences have been observed among laboratory animals. For example, AFB_1 is more toxic to the rat than the mouse, because hepatic conversion of AFB_1 to the 8,9-epoxide occurs more readily in the rat than the mouse. Also, detoxification of aflatoxin is more efficient in the mouse, because of greater activity of the GSH-S-transferase system in the liver, which promotes the formation of glutathione-aflatoxin conjugates. Other conjugates of AFB_1 are formed in the liver with sulfate and glucuronic acid (Coulombe 1993).

These differences in hepatic detoxification systems affect the manner in which fishes other than rainbow trout process aflatoxin and render it less harmful to them. For instance, warmwater fish such as channel catfish and Nile tilapia, *Oreochromis nilotica*, are less sensitive to dietary aflatoxin than rainbow trout. Halver (1969) ranked sensitivity of fish to the toxic effects of aflatoxin from least sensitive to most sensitive as channel catfish, coho salmon, and rainbow trout. Jantrarotai and Lovell (1990a) found that channel catfish fed 10,000 ppb purified AFB_1 in a semipurified diet for 10 weeks had reduced growth and moderate internal lesions, but no HCC. They also reported that feeding 2,154 ppb AFB_1 in the same study produced no reductions in growth or any internal lesions. Nile tilapia showed significant reductions in growth rate when fed diets containing at least 1,880 ppb AFB_1, but not when fed 940 ppb for 25 days (Chavez-Sanchez *et al.*, 1994). Fish were maintained on the control diet (no aflatoxin) for an additional 50 days to monitor development of lesions. Precancerous liver lesions were observed in tilapia fed diets with 1,880 ppb AFB_1 or higher, but none developed into carcinomas during the time frame of the experiment. More recently, Nguyen *et al.* (2002) reported that juvenile Nile tilapia had significantly reduced weight gains, poorer feed conversion ratio (FCR), and reduced hematocrit (%) values when fed diets with 2.5 ppm AFB_1 or above supplied from an *A. parasiticus* culture material (CM), but not when given 0.25 ppm AFB_1. Histological examination revealed that tilapia fed diets with CM that supplied 10 or 100 ppm AFB_1 had evidence of excess liver lipofuscin and irregularly shaped hepatocellular nuclei. No histological lesions were observed in the spleen, heart, head kidney, stomach, or intestine of tilapia fed any of the five dietary levels of AFB_1.

Warmwater fish appear to have lower suceptibility to aflatoxin than rainbow trout. HCC was not demonstrated in any of the studies involving channel catfish or Nile tilapia that are reviewed here. Jantrarotai and Lovell (1990a) conjectured that the reason for the reduced sensitivity to aflatoxin in the case of channel catfish was that this species of fish has a more efficient detoxification system compared to rainbow trout. Strength *et al.* (1982) reported that channel catfish detoxify xenobiotics effectively because of the activity of liver glucuronyltransferase and sulfotransferase systems. Rainbow trout, on the other hand, have only the glucuronyltransferase system to detoxify aflatoxin (Loveland *et al.*, 1984).

Mycotoxins other than aflatoxin that are produced by *Aspergillus* and *Penicillium* fungal organisms include ochratoxin A (OTA), cyclopiazonic acid (CPA), sterigmatocystin, and versicolorin A. The latter two are biosynthetic precursors in the production of aflatoxin, but lack the toxicity of aflatoxin in rainbow trout. A study that compared the effects of aflatoxin and sterigmatocystin on rainbow trout embryos showed that 14-day embryos exposed to 0.5 ppm aqueous solution of sterigmatocystin by immersion for 1 hr had a 13% incidence of liver tumors 12 months later. Similar exposure to comparable concentrations of aflatoxin resulted in a 53% incidence of liver tumors. Exposure of 21-day trout embryos to a 5.0 ppm solution of versicolorin A resulted in an incidence of HCC of 42% at one year compared to 68% for embryos exposed to 0.5 ppm AFB_1 (Hendricks *et al.*, 1980). Jantrarotai and Lovell (1990b) evaluated the toxic effects of purified CPA on channel catfish and found its subchronic toxicity to be greater than that of purified AFB_1 when weight gain was applied as the criterion. Catfish receiving diets that contained 100 μg CPA/kg diet or higher had lower body weights than those fed the control diet. Feeding a 10,000 μg CPA/kg diet resulted in the presence of histological lesions of the trunk kidneys and stomach, but no lesions of the liver. No hematological changes were observed in catfish fed the various levels of CPA. Concerns about the contamination of aquaculture diets with CPA are valid, especially if it is found to be co-contaminating diets with aflatoxin. Cyclopiazonic acid is produced by *A. flavus* and may occur in combination with aflatoxin in feed grains infected with this fungal organism (Luk *et al.*, 1977). While the toxic effects of this co-contamination have not been evaluated in fish, CPA has been implicated as contributing to pathological changes observed in the turkey X disease epizootic described by Blount (1961) (Cole 1986).

Ochratoxin A is produced primarily by *A. ochraceus* or *P. verrucosum* on a variety of substrates, including cereal grains and oil seeds. In most animals studied, the target organ of OTA is the kidney. It causes necrosis of the proximal tubules. The two mould organisms appear to have different substrate and environmental requirements for optimum production of OTA. *A. ochraceus* grows mostly on oil seeds such as peanuts or soyabeans while *P. verrucosum* prefers cereal grains like corn, wheat, or barley for substrates. Mycotoxin production is optimised at higher temperatures (12-37°C) for *A. ochraceus*, which is the typical temperature range for oil seeds grown in warmer climates. *P. verrucosum*, on the other hand, has better mycotoxin production at the cooler temperatures (4-31°C) and thus is more likely to occur in grain growing regions of the northern latitudes (Marquardt and Frohlich, 1992). Both organisms are classified as storage fungi, which may produce additional amounts of OTA under poor storage conditions for commodities that provide optimum conditions e.g. increased moisture for enhanced fungal development and mycotoxin production.

Ochratoxicosis in swine (Krogh, 1991) and poultry (Huff *et al.*, 1974) has been previously described. Consumption of feeds contaminated with OTA results in nephropathy in both groups of animals. Of additional concern is the possibility of carryover of the mycotoxin in edible muscle portions of animals that consume OTA-contaminated feeds. Ochratoxin A has been linked to a fatal human disease referred to as **Balkan endemic nephropathy**, which is characterized by tubular degeneration and renal lesions (Marquardt and Frohlich, 1992). Until recently, evaluation of the toxic effects of OTA in fish had been limited to work with rainbow trout to determine the acute toxicity of OTA. Intraperitoneal injection of OTA revealed that the LD_{50} value was 4.67 mg/kg, resulting in necrosis of kidney tubular cells and liver (Doster *et al.*, 1972). A study was completed at the National Warmwater Aquaculture Center, Stoneville, Mississippi, USA with juvenile channel catfish (initial weight 6.1 g) fed semipurified diets amended with CM containing a known concentration of OTA (Manning *et al.*, 2003b). Results after 8 weeks showed that catfish fed 1.0 mg OTA/kg diet had significantly lower weight gain than catfish fed 0.5 mg OTA/kg and the control diet. FCR was significantly reduced for catfish fed 4.0 and 8.0 mg OTA/kg and hematocrit values were significantly reduced for catfish fed 8.0 mg OTA/kg diet. Histological examination revealed that, unlike other species studied, there was no necrosis of the renal tubules. Severe multifocal melanomacrophage centers (MMC) were observed in interstitium of renal tissue of fish fed 4.0 and 8.0 mg OTA/kg diet. The prominent histopathological lesion was necrosis of hepatopancreatic tissue, specifically in the exocrine pancreatic cells that surround the hepatic portal veins of catfish receiving 1.0 mg OTA/kg diet or higher. This pathology resulted in obliteration of normal exocrine pancreatic tissue and replacement with accumulated multifocal MMC.

Fusarium mycotoxins

FUMONISIN

Interest in *Fusarium* mycotoxins has been renewed in the past decade due to the discovery and chemical characterisation of fumonisin B_1 by Gelderblom and associates in South Africa in 1988 (Gelderblom *et al.*, 1988). The *Fusarium* feed-borne mycotoxins that could have an impact on aquaculture include fumonisin, moniliformin, deoxynivalenol, T-2 toxin, and zearalenone. These mycotoxins develop on corn, rice and other feed grains that are infected with various moulds of this genus. The major fumonisin-producing organism, *Fusarium verticillioides*, (formerly *Fusarium moniliforme*) infects most agricultural corn. The foremost fumonisins have been classified as fumonisin B_1 (FB_1), fumonisin B_2, fumonisin B_3, and fumonisin B_4. Fumonisin B_1 is usually found in greater abundance than the other members of this group of mycotoxins and it is also considered to be the most toxic of this group. Fumonisin B_1 has generated the greatest interest because of its severe pathological effects on certain animal species. All equines are extremely sensitive to FB_1. Consumption of feeds containing only 5-10 mg FB_1/kg of diet over several days results in severe pathology of the brain called **equine leukoencephalomalacia**, a fatal condition typified by liquefaction of the cerebellum (Ross *et al.* 1993). Swine fed diets with FB_1 develop **porcine pulmonary oedema**, a condition characterised by oedema of the pleural cavity (Colvin and Harrison, 1992). Consumption of corn contaminated with fumonisin has been linked to human oesophageal cancer in areas of South Africa and northern China where corn is a dietary staple (Norred, 1993).

Examination of the effects of exposure to fumonisin on channel catfish and Nile tilapia by Lovell and associates at Auburn University, Alabama, USA showed that both species of warmwater fish experienced reduced growth. Juvenile channel catfish with initial weight of 1.2 g/fish had significantly reduced weight gain after consuming a diet containing 20 mg FB_1/kg for 10 weeks. Two-year-old channel catfish (initial weight 31.0 g) experienced reduced weight gain after being fed 80 mg FB_1/kg diet for 14 weeks (Lumlertdacha *et al.*, 1995). In a similar feeding study, Li *et al.* (1994) found that juvenile channel catfish (initial weight 6.1 g), fed 40 mg FB_1/kg of diet for 12 weeks had reduced weight gain and feed efficiency. The results of these three studies indicate that smaller catfish are more susceptible to the toxic effects of FB_1. Nile tilapia with an initial weight of 2.7 g had reduced weight gains and poorer feed conversion when fed diets containing levels of FB_1 at 40 mg/kg or greater for 8 weeks (Nguyen *et al.*, 2003). Decreased hematocrit was observed for tilapia fed the highest level of FB_1 at 150 mg/kg of diet. Fumonisin B_1 in these studies was supplied from corn-based *F. verticillioides* CM containing a known concentration of FB_1. Additional research at the same laboratory compared the effects of feeding semipurified diets containing various concentrations of either crude FB_1 from CM or purified FB_1 on juvenile channel catfish. Catfish (initial weight 5.0 g) showed significantly reduced weight gains of 58.4% of control values when fed a diet containing 50 mg crude FB_1/kg versus 67.5% for fish fed 200 mg purified FB_1/kg (Manning, 1998). A consistent feature of fumonisin toxicosis in all animals is the accumulation of sphinganine in liver, kidney, and serum that occurs as a result of competitive inhibition of the enzyme ceramide synthetase. The consequence of tissue sphingosine accumulation is elevation of the sphinganine:sphingosine ratio, which is an accepted biomarker of fumonisin intoxication (Riley *et al.*, 1994). Consumption of fumonisin-containing diets by catfish and tilapia in these experiments resulted in significantly elevated sphinganine: sphingosine ratios (Goel *et al.*, 1994; Nguyen *et al.*, 2003; Manning, 1998).

MONILIFORMIN

The effects of another *Fusarium* mycotoxin, moniliformin, on channel catfish was studied at Auburn University and compared singly or in combination with FB_1 (Yildirim *et al.*, 2000). Both mycotoxins were supplied from CM, each with known concentrations of the respective mycotoxins. Moniliformin CM was prepared from a culture of *F. proliferatum* and FB_1 CM was prepared from *F. verticillioides*. Experimental diets consisted of semipurified diets containing five levels of moniliformin: 0, 20, 40, 60, or 120 mg/kg diet. Also included were two diets containing 20 or 40 mg FB_1/kg, and two combinations 20 and 40 mg/kg (20:40) or 40 and 40 mg/kg (40:40) of FB_1 and moniliformin, respectively. Results of the 10-week study showed that catfish (initial weight 1.5 g) fed diets with 20 mg moniliformin/kg or higher had significantly lower weight gains than the control-fed catfish. Likewise, catfish fed 20 or 40 mg FB_1/kg diet had significantly lower weight gains than controls. Feeding diets with combinations of the two mycotoxins resulted in reduced weight gains for both the 20:40 and 40:40 combinations. A synergistic effect on weight gain was detected for the 40:40 combination. Weight gain for the combination of the two mycotoxins was significantly lower than that of either mycotoxin fed singly at 40 mg/kg diet, and equivalent to the weight gain of catfish fed 60 mg moniliformin/kg. Hematocrit values were significantly lower than those of the controls for catfish fed moniliformin at 60 and 120 mg/kg diet, 40 mg FB_1/kg diet and the 40:40 combination.

The modes of action are different for these two mycotoxins. Moniliformin disrupts intermediary metabolism of the tricarboxylic acid cycle at the conversion of pyruvate to the starting intermediate, acetyl-CoA, resulting in increased levels of serum pyruvate. Fumonisin B_1, as mentioned previously, inhibits sphingolipid metabolism resulting in elevation of liver sphinganine. Evaluation of the effects of the two mycotoxins on these metabolites showed that catfish fed moniliformin had significantly higher serum pyruvate levels when fed 60 mg/kg diet. Feeding fish the 40:40 combination resulted in an enhanced serum pyruvate response; pyruvate levels were significantly greater than fish fed 40 mg moniliformin/kg and the same as catfish fed the 60 mg moniliformin mg/kg diet. However, combinations of the two mycotoxins did not significantly enhance the fumonisin effect on sphingolipid response;

there were no additional increases in the sphinganine:sphingosine ratio above that observed for catfish fed FB_1 alone. In another aspect of the experiment described earlier, Nguyen *et al.* (2003) found that Nile tilapia with an initial weight of 2.7 g had reduced weight gains, poorer feed conversion, and reduced hematocrit when fed diets containing moniliformin supplied from *F. proliferatum* CM for eight weeks. Dietary moniliformin concentrations 70 or 150 mg/kg were required to elicit these responses. On the other hand, significant elevations of serum pyruvate were observed for tilapia fed all levels of moniliformin: 10, 40, 70 or 150 mg/kg of diet.

Corn may contain levels of moniliformin and/or fumonisin that are toxic to channel catfish or tilapia, since the moulds that produce these mycotoxins commonly infect corn plants. Some *Fusarium* spp. produce more than one mycotoxin. For instance, *F. proliferatum* is also a producer of fumonisin. Additionally, corn lots destined for feed manufacture will probably be intermingled at some point during the course of normal commercial movement resulting in corn lots that contain mixtures of these two or other mycotoxins.

TRICHOTHECENES

Other *Fusarium* mycotoxins that have significant detrimental effects on cultured fish are the trichothecenes. These include T-2 toxin, deoxynivalenol (DON, vomitoxin), and diacetoxyscirpenol (DAS). All three mycotoxins are structurally similar chemical compounds characterised by having the tetracyclic 12,13-epoxy-trichothecene moiety of sesquiterpenoids. These mycotoxins have been studied extensively in farm animals. Exposure of cattle, poultry, and swine to the trichothecenes resulted in oral lesions and gastrointestinal disturbances such as vomiting, diarrhoea, and intestinal inflammation and haemorrhaging. Various changes in haematology have been observed in animals exposed to dietary trichothecene mycotoxins. These include anaemia and leukopenia (CAST, 2003). Feed refusal syndrome sometimes associated with vomiting has been described for poultry and swine that consumed feed grains contaminated with trichothecene mycotoxins. These alterations in feeding behaviour are the result of neurotransmitter imbalance in the appetite control centers of the brain (Smith, 1992). One mechanism of trichothecene mycotoxicosis is inhibition of protein synthesis in liver and other tissues, which causes hyperaminoacidemia and results in elevated levels of circulating amino acids. An amino acid of importance is tryptophan, the precursor of the neurotransmitter serotonin (Smith, 1992). Consequently, consumption of feeds containing trichothecenes may elevate brain tryptophan levels and cause increased levels of serotonin (Smith and Seddon, 1998). Increased levels of serotonin in the brain are known to influence appetite, specifically, to initiate the perception of satiety.

Studies to evaluate the effects of feeding diets containing T-2 toxin and DON have been conducted first with rainbow trout and most recently with channel catfish. Rainbow trout fingerlings, initial body weight of 1.0 g, were fed diets containing graded levels of T-2 toxin (0 to 15 mg/kg diet) for 16 weeks. Trout fed diets with 2.5 mg T-2 toxin/kg or higher had depressed growth, feed efficiency, hematocrit, blood haemoglobin, and diet acceptance. Force-feeding diets containing 15 mg T-2 toxin/kg to adult trout resulted in intestinal haemorrhaging and regurgitation (Poston *et al.*, 1982). In another study conducted with juvenile rainbow trout, progressively lower body weight gains were observed for fish fed various levels of 1.0 to 13.0 mg DON/kg diet for four weeks (Woodward *et al.*, 1983). In another aspect of the study, feed refusal syndrome was experienced in trout given feed containing 20 mg DON/kg. Feeding diets with DON did not elicit vomiting in either group of fish. Trout resumed normal feeding of uncontaminated diets after a recovery period. Results of this study illustrate that the sensitivity of rainbow trout with respect to feeding behaviour is similar to that of the pig (Woodward *et al.*, 1983).

Two aquarium-feeding studies were conducted at the National Warmwater Aquaculture Center to evaluate the effects of DON on channel catfish. In the first study, channel catfish (initial weight 6.8 g), were fed a cold extruded, semipurified diet containing 0, 1.25, 2.5, 5.0, and 10.0 mg purified DON/kg to apparent satiation for eight weeks. Feeding diets with DON to juvenile catfish had no effect on feed consumption, growth, hematocrit values, or liver weights. In the second trial, juvenile catfish (initial weight 5.0 g) were fed practical diets prepared with naturally-contaminated wheat containing 37.5 mg DON/kg. The diets were formulated to contain 0, 2.5, 5.0, 10.0, 15.0, and 17.5 mg DON/kg. After six weeks, catfish fed dietary concentrations of DON up to 10.0 mg/kg experienced no reductions in growth or feed consumption. Catfish fed dietary levels of 15.0 and 17.5 mg DON/kg experienced reductions in growth and poorer feed conversion. There was no effect on hematocrit as a result of feeding any of the diets containing wheat naturally contaminated with DON. These results are in contrast to those of dogs that were fed diets prepared by the cooker-extrusion process (Hughes *et al.*, 1999). The diets were formulated with the same naturally DON-contaminated wheat used in the catfish study. Dogs fed diets containing 3-4 mg DON/kg developed feed refusal syndrome and exhibited vomiting behaviour under continuous consumption of DON-contaminated diets. Also, it was determined that the cooker-extrusion process did not degrade DON. Because of structural similarities of trichothecene mycotoxins, one could expect the same lack of detoxification by extrusion or other heat producing feed manufacturing processes for ingredients contaminated with T-2 toxin or DAS.

In an attempt to continue the evaluation of the effects of trichothecene mycotoxins on fish, another feeding study was conducted with channel catfish (Manning *et al.*, 2003a). Juvenile catfish with an initial body weight of 8.9 g were fed semipurified diets containing 0, 0.625, 1.25, 2.5, or 5.0 mg purified T-2 toxin/kg. These five diets were fed twice daily to apparent satiation. To correctly evaluate any toxic responses, three additional pair-fed treatments of fish were limited to being fed the control diet (no mycotoxin) at the same daily levels as the treatments receiving the three highest levels of T-2 toxin. After eight weeks catfish fed diets containing any amount of T-2 toxin had significantly lower body weight gains than the control fish. Feed conversion ratios of catfish fed the diet with 5.0 mg T-2 toxin/kg were significantly poorer than the control fish, but FCR of the pair-fed comparative treatment was not significantly different from that of the control treatment. Body weight gain for the pair-fed treatment was significantly greater than its comparison treatment fed 5.0 mg T-2 toxin/kg of diet, but still lower than the control-fed fish because of limited daily feed allowance. Hematocrit values of catfish fed the three highest levels of T-2 toxin were significantly lower than the control-fed fish and their pair-fed counterparts. Pair-fed fish had hematocrit values that were not different than those of the control fish. Survival of catfish was 78.8 and 80.0% for fish fed 2.5 and 5.0 mg T-2 toxin/kg, respectively, which was significantly lower than those of fish fed diets with no toxin (including the pair-fed fish) or fed the diets with lower concentrations of T-2 toxin. Catfish fed diets containing the three highest levels of T-2 toxin were found upon histological inspection to have substantially greater incidence of gastritis than the control-fed fish. Furthermore, histological examination revealed greater depletion of hematopoietic tissue of the anterior kidney in catfish fed the three highest levels of T-2 toxin.

The trichothecene mycotoxins T-2 toxin and DON have detrimental effects when fed to both coldwater and warmwater cultured fish. Diacetoxyscirpenol, while not reviewed here, would probably elicit similar responses in fish as T-2 toxin, because it is considered to be more toxic than the other two mycotoxins based on research conducted with poultry (Wyatt, 1991). Although DON, either from purified or naturally contaminated sources, does seem to be toxic to channel catfish at levels that are likely to appear in commercial feeds, T-2 toxin appears to be toxic to both rainbow trout and catfish. More thorough understanding of T-2 toxicosis of channel catfish may provide clues about causes of some less understood diseases of catfish such as idiopathic anaemia. Consumption of diets containing T-2 toxin for even relatively short periods of time, such as described here, resulted in individual hematocrit values below 10%. It has been demonstrated in this and other studies with poultry that consumption of T-2 toxin causes a destructive pathology of hematopoietic tissue of the spleen of chickens and the anterior kidney of catfish resulting in reduced numbers of circulating erythrocytes.

ZEARALENONE

Zearalenone is a mycotoxin produced by *F. graminearum* and other *Fusarium* moulds using corn, wheat, barley, oats, and sorghum as substrates. Production of this mycotoxin is augmented by high humidity and cool temperatures. Zearalenone is a nonsteroidal compound that exhibits oestrogen-like activity in certain farm animals. Cattle, sheep, and pigs have been shown to respond to its presence in feeds (CAST, 2003). Pigs show the most striking responses to zearalenone at doses of 1 mg/kg diet. Affected gilts and sows display signs of reproductive activity including swollen vulva and mammary glands and enlarged uterus. Young male pigs develop testicular atrophy and swollen nipples (Coulombe, 1993). Zearalenone appears to function biologically by competing for and occupying oestrogenic receptors, thus preventing normal endogenous oestrogenic compounds from achieving normal physiological function.

FUSARIC ACID

Fusaric acid, 5-butylpicolinic acid, is another toxin produced by fungi of the *Fusarium* genus including *F. verticillioides*. Bacon *et al.* (1996) found that 75 strains of *Fusarium* moulds produce fusaric acid, many of which also produce other mycotoxins that should be of concern to livestock and fish producers. Samples of corn, wheat, and barley were found to contain an average of 12 mg fusaric acid/kg with a maximum level of 36 mg/kg found in corn (Smith and Sousadias, 1993). Fusaric acid is usually classified as a phytotoxin with limited toxicity towards animals. At toxic levels, it causes behavioural responses that are similar to those elicited by DON and other trichothecene mycotoxins such as loss of appetite and vomiting (Smith and MacDonald, 1991). Consumption of fusaric acid elevates brain levels of tryptophan and serotonin (Chaouloff *et al.*, 1986), but by a different mechanism than DON. Fusaric acid bears a chemical similarity to tryptophan, from which it is derived. Fusaric acid competes with tryptophan for binding sites on blood albumin, thus raising the levels of free tryptophan in blood (Chaouloff *et al.*, 1986). Because of these circumstances, there is increased uptake of tryptophan by the brain with consequential increased synthesis of serotonin.

Synergistic effects of mycotoxins

Interest in fusaric acid stems from its potential co-occurrence with other *Fusarium* mycotoxins such as fumonisin and DON and its potential to act synergistically with these mycotoxins. Bacon *et al.* (1995) demonstrated that non-toxic levels of fusaric acid and fumonisin injected concurrently were toxic to developing chicks *in ovo*. Injection of 5 μg FB_1/egg in combination with 5 μg fusaric acid/egg caused embryonic death in approximately 50% of embryos, compared to no deaths from an injection of 5 μg FB_1/egg singly. In a subsequent study, Voss *et al.* (1999) were unable to demonstrate a synergistic response to feeding laboratory rodents diets with combinations of fusaric acid and fumonisin. Rats were fed diets containing 3.4, 18.4, or 437 ppm total fumonisins (FB_1 + FB_2 + FB_3) with each inclusion level augmented with 0, 20, 100, or 400 ppm fusaric acid. Results indicated that there were no observed synergistic effects on body weight gains and serum chemistry measurements, which included liver enzymes, cholesterol, triglycerides and creatinine. The sphinganine:sphingosine ratio was elevated in rats consuming diets with fumonisins at two levels, 3.4 and 18.4 ppm, compared to the control rats fed two diets with no fumonisins or 400 ppm fusaric acid (no fumonisins). Addition of 400 ppm fusaric acid to the two dietary levels of fumonisins did not increase the sphingolipid ratios above those observed for rats fed diets with fumonisins only. An explanation for the lack of synergistic effect may be low sensitivity to fusaric acid in rats.

Nevertheless, synergistic toxic responses to mycotoxins by animals have been reported under experimental conditions. Synergistic responses of barrows fed a diet containing a combination of FB_1 and aflatoxins from CM were reported by Harvey *et al.* (1995). Yildirim *et al.* (2000), as previously mentioned, observed a synergistic response for reduced growth of channel catfish fed a diet containing equal concentrations of FB_1 and moniliformin. Synergistic responses may be observed when the two mycotoxins that were evaluated have similar biochemical effects on the exposed animal. For example, pigs exposed to fusaric acid require high levels of this fungal metabolite to elicit the typical symptoms of fusaric acid intoxication, i.e. feed refusal and vomiting. Smith and MacDonald (1991) orally administered to barrows a high dose of fusaric acid (200 mg/kg of body weight) which elicited vomiting and altered brain neurochemical levels with increased tryptophan, serotonin, and 5-hydroxyindoleacetic acid. In contrast, Smith *et al.* (1997) fed pigs practical diets comprised of blends of *Fusarium*-contaminated grains that contained various combined levels of DON and fusaric acid. Pigs fed a diet with only 1.1 μg of DON and 14.1 μg fusaric acid/g of diet had reduced feed intake (loss of appetite) and growth.

Mycotoxin effects on immune response

Mycotoxin exposure often decreases disease resistance. Mycotoxins that impair the immune system include aflatoxin, T-2 toxin, OTA, DON, and fumonisin. The effects of mycotoxins on immunological responses of terrestrial animals have been examined extensively. Aflatoxin causes impairment of the immune system by inhibition of protein synthesis as a result of the formation of adducts between the 8,9-epoxide of AFB_1 and DNA or RNA. The effect on the immune system is to reduce the production of certain cell components such as C4 complement and lymphokines e.g. interleukins, and T lymphocytes. Aflatoxin suppresses phagocytosis by macrophages, which alters subsequent processing and presentation of antigen to B lymphocytes with consequential reductions in disease resistance (Pier, 1991). Giambrone *et al.* (1978) found that broiler chickens had decreased immunoglobulin concentrations after administration of aflatoxins. In fish, Boonyaratpalin *et al.* (2001) found that administration of AFB_1 at 0-2,500 ppb in the diets of black tiger shrimp (*Panaeus monodon*) for 8 weeks resulted in decreased total haemocyte count for shrimp fed the diet with 2,500 ppb AFB_1. Haemocytes, in conjunction with fixed phagocytes, form the immunocompetent components of the shrimp immune system, and as such, declination in their numbers could result in reduced disease resistance. Exposure of Indian major carp, *Labeo rohita,* to AFB_1 by a single intraperitoneal injection resulted in decreased resistance as measured by increased mortality to challenges from *Edwardsiella tarda* and *Aeromonas hydrophila* (Sahoo and Mukherjee, 2002).

Trichothecene mycotoxins have detrimental effects on the competency of the immune systems of poultry (Wyatt, 1991) and other animals (Pier, 1991). Immune system impairment is reflected, in part, by reductions in the numbers of circulating phagocytes, which reduces phagocytic activity and chemotaxis by macrophages. These events reduce macrophage uptake and presentation of antigens to B cells resulting in lower production of immunoglobulins. Also, immunoglobulin production is impaired in animals that have consumed trichothecene mycotoxins by inhibiting protein synthesis. Wyatt *et al.* (1973) found that bursa of Fabricius weights were reduced in chickens fed 4 or 8 mg T-2 toxin/kg diet. Bursa of Fabricius is the discrete organ of avian species with B cell activity that results in the production of circulating antibodies. Ochratoxin A adversely affects immune system function by inhibition of protein synthesis by impeding a translation step through competitive inhibition of the enzyme phenylalanine-tRNA synthetase. Inhibition of protein synthesis at this point stops amino-acylation and peptide elongation. The result is a reduction in the levels of circulating antibodies IgG and IgM (Kuiper-Goodman and Scott, 1989). Also, OTA affects the structural components of the immune system. Decreases in lymphoid cell population was observed in chickens fed diets with 2-4 mg OTA/kg diet for 20 days (Dwivedi and Burns, 1984). Reduction of 33% in mouse thymus size was observed after four intraperitoneal injections of 20 mg OTA/kg body weight on alternative days (Boorman *et al.*, 1984). This dose regimen, which caused minimal nephrotoxicity, also caused decreased bone marrow cellularity, specifically reductions in the numbers of macrophage progenitors and hematopoietic stem cells.

A study was conducted in our laboratory to evaluate the effect of feeding a semipurified diet containing two levels of T-2 toxin or OTA on disease resistance in juvenile channel catfish. Fish were fed diets containing 1.0 or 2.0 mg T-2 toxin/kg and 2.0 or 4.0 mg OTA/kg and compared to catfish fed a control diet containing no mycotoxin for 4 weeks. The fish were challenged with a preparation of *Edwardsiella ictaluri*, an important pathogen of farm-raised channel catfish, containing 2.25×10^6 colony-forming units (CFU). Results showed that significant increases in mortality occurred for catfish receiving both levels of T-2 toxin and the higher level of OTA. Fish fed the higher level of T-2 toxin had mortality of 99.3% compared to 68.3% for control fish. The presence of *E. ictaluri* was confirmed by culturing kidney specimens of all mortality on blood agar plates. Results indicate that consumption of commercial diets containing either of these two mycotoxins could significantly exacerbate an epizootic of this catfish pathogen and may have similar ramifications for other aquaculture fish.

Fumonisin B_1 impairs immune system response of animals, in part, by inhibiting protein synthesis. Norred *et al.* (1990; 1992) showed that inhibition in rat hepatocytes occurred through interference with valine incorporation during protein synthesis. Additionally, FB_1, it should be remembered, disrupts sphingolipid metabolism, and as such, may interfere with cellular regulatory functions i.e. cell signalling pathways. Therefore, alterations of sphingolipid metabolism may be a factor in reducing the immune system response of fish that consume diets contaminated with FB_1 (Lumlertdacha and Lovell, 1995). Feeding a practical diet containing known levels of FB_1 derived from *F. verticillioides* CM for 14 weeks reduced the disease resistance of channel catfish challenged with *E. ictaluri*. Two-year-old catfish fed 80 mg FB_1/kg of diet had significantly increased mortality compared to control fish due to *E. ictaluri* infection (Lumlertdacha and Lovell, 1995). In the same study, antibody production of catfish fed 20 or 80 mg FB_1/kg of diet was significantly lower than that of control fish.

Manufacture and storage of aquaculture feeds

Storage of the feed grains and plant protein products under poor conditions can increase the occurrence of potentially toxigenic types of moulds. Storage of these ingredients at grain storage facilities and feed manufacturing mills should be in structures that provide adequate protection from the environment to minimise their exposure to moisture from rain and other forms of precipitation. Feedstuffs that are maintained at a moisture content of 12% or less are less likely to develop mould growth and subsequently produce mycotoxin. It is important that grain storage facilities and feed mills are vigilant about the potential presence of mycotoxin- contaminated grains and oilseed meals. All incoming new shipments of these commodities should be inspected for water damage, moisture content, and the presence of mould infection. New shipments of commodities that are found to have these deficiencies should not be accepted upon delivery. Development and detection of mould and mycotoxin contamination of feed ingredients on the premises of a feed mill dictates that the mycotoxin-contaminated ingredients not be used in the production of aquaculture feeds.

Production of nutritionally complete aquaculture feeds requires that these feeds are processed into compressed steam pellets or extruded pellets. The processes by which both types of pellets are manufactured require the introduction of steam (steam pellets) or heated water and steam (cooker-extrusion pellets) with an increase in moisture content of finished product. The usual processes used to decrease moisture content are cooling, which in the case of steam pellets results in the simultaneous loss of moisture as part of evaporative cooling such that pellets that contain 15-16% moisture as they are ejected from the pelleting machine are reduced to less than 12% moisture after 20-25 minutes exposure to evaporative cooling in a pellet cooler. For pellets produced by the cooker-extrusion method, uptake of added moisture can be as high as 12-14% above starting levels, so a different method is employed to bring moisture levels down to an acceptable 10-12%. Immediately after being expelled from the extruder the pellets are conveyed to a horizontal drier equipped with three to four counter-moving, continuous metal belts where the pellets are exposed to heated, forced air at 121°C for 15-20 minutes. Dried pellets are cooled by exposure to environmental air for several minutes. Pellets are then conveyed to bulk bins for load-out on bulk trucks or moved to a bagging station within the feed mill for bagging in paper or polypropylene feed bags.

Because aquaculture is usually carried on in a wet environment, care should be taken not to expose dry feed to water or high levels of environmental moisture. Fish farms are often located in tropical or semitropical areas where environmental conditions are such that feed is also exposed not only to higher levels of moisture, but also higher ambient temperatures. Such conditions can provide optimum conditions for the growth of mould on feed. The presence of mould on feed does not always mean that the feed is contaminated with mycotoxins, but further use of the feed should be discontinued until negative results have been verified through laboratory testing. The use of mouldy feed for animals is a questionable practice at best and even after negative results for the presence of mycotoxins serious consideration should be given to disposing of the feed by other potentially less harmful means.

Keeping feed grains and other fish feed

ingredients free of the damage incurred after invasion by rodents and insects is important to controlling mould development and eventual contamination by mycotoxins. These pests increase the possibility of fungal invasion because their activities in stored grains damage the natural protective outer seed covering of grains, which facilitates attack by moulds.

Testing feed grains and oilseed meals can be accomplished nowadays by relatively simple, straightforward procedures. These methods can quickly provide information about the presence of many of the common mycotoxins in these products. Usually these tests detect the presence of total aflatoxins, fumonisins, DON, and OTA. The enzyme-linked immunosorbent assay (ELISA) test format is generally employed. This is a desirable format because it requires no special training or laboratory equipment. The company selling the test provides it as a kit with all required reagents, instruments, and instructions included. The basic kit provides the user with qualitative information about the presence of the mycotoxin in the feed ingredient. But with only a slight modification in the ELISA test procedure one can determine specific quantities of the mycotoxin in the ingredient. Monoclonal antibody mini-columns are another approach to evaluating the presence of a mycotoxin in a feed ingredient or feed sample. This method enables one to easily isolate a mycotoxin and quantify it by techniques that use instruments such as a fluorometer or HPLC. The monoclonal antibody columns are single-use, disposable plastic columns that have been loaded with a resin coated with antibody that was prepared to bind a specific antigen, in this case, a mycotoxin. The mycotoxin is extracted from the sample using a solvent mixture, usually a mixture of methanol and water, and placed on the mini-column. After elution from the column, the mycotoxin is chemically derivatised and the concentration determined after processing in a fluorometer. Alternatively, the concentration of the eluted mycotoxin can be determined using an HPLC method.

Mitigation of mycotoxin contamination in aquaculture feeds

One area of interest has been to find ways to decrease the concentration of mycotoxin in feeds and feed ingredients and evaluate the effect on agricultural animals. Reasonable success to achieve this objective has been obtained through the use of special, adsorbent clays, which are referred to as hydrated sodium calcium aluminosilicates (HSCAS). These clays when added to complete feeds bind the aflatoxins in feed and prevent intestinal absorption by animals. There has been an on-going debate about the effectiveness of HSCAS or other clay preparations to bind mycotoxins other than aflatoxin. Researchers who have evaluated their effectiveness both in the laboratory and in animals generally find that their binding effectiveness is limited to aflatoxins (Taylor, 1999). This is because of the distinctive structural and chemical characteristics of the aflatoxin molecule; it possesses rigid coplanarity and a 1,3-diketone moiety, respectively. Assertions that clays or other adsorbent preparations bind mycotoxins other than aflatoxin are usually based on *in vitro* evaluation rather than *in vivo* assessment. Comparative testing of inorganic binding products with poultry showed that *in vitro* test results do not always agree with *in vivo* evaluation (Ledoux and Rottinghaus, 1999). Claims that mycotoxin binders that bind aflatoxin will also bind other mycotoxins do not always stand up to *in vivo* evaluation. Ledoux and Rottinghaus (1999) found that adsorbents tested with poultry did not offer protection against feed-borne FB_1, moniliformin, OTA, or DON. Commercial HSCAS compounds are considered to be effective in preventing aflatoxicosis in poultry, swine, and other animals (CAST,

2003). In addition to clays, other sequestering agents are commercially available; these include activated carbons, glucomannan derived from yeast cell walls, and absorptive resins such as cholestyramine (Huwig *et al.*, 2001). These compounds have not been evaluated in aquaculture fish feeds. In any event, all mycotoxin sequestering agents should be tested in the animal species of choice to be certain that they have effective mycotoxin binding properties and are safe (CAST, 2003). The reader is referred to a detailed presentation of mycotoxin sequestering agents in a later chapter in this book.

Other methods that have been used to reduce the effects of mycotoxins, mainly aflatoxin, on animals include ammoniation and the application of heat to feed grains such as corn. Studies have shown that aflatoxin concentrations in whole corn and whole cottonseed could be substantially reduced, up to 99%, by ammoniation with gaseous ammonia or ammonium hydroxide. The time required to complete detoxification of aflatoxin varies with the method used. One method that incorporates a combination of high temperature (80-120°C) with high pressure (3 bars) and another that uses ambient temperature with atmospheric pressure (1 bar). The former method would be used in a production plant setting and can be completed within 3 hrs, while the latter method would be used on-farm and takes 14-21 days to completely detoxify the product (Smith, 1997). Use of these ammonia-detoxified products has been limited to feeds for cattle, pigs, and poultry. Whether these products are of any value in commercial fish feeds remains to be determined. Another method that was developed involved roasting aflatoxin-contaminated corn in commercial corn roasters at 145°C and 165°C with reductions in aflatoxin levels of 69% and 77%, respectively (Conway *et al.*, 1978).

Buser and Abbas (2001) found that extrusion of cottonseed contaminated with aflatoxin reduced the concentration of aflatoxin by 50% without decreasing the nutritional value of the cottonseed. Experimental catfish feeds containing aflatoxin from mouldy corn were processed by cooker-extrusion technology. The conditions of this type of feed processing subject the feed mixture to a cooker phase with exposure to temperatures of 120°C for up to 3-4 minutes and subsequent higher temperatures from steam in the extrusion barrel. The extruded pellets are exposed to more heat (120°C) for approximately 20-25 minutes in the drying process. Our analysis of the mouldy corn in catfish feeds showed that levels of aflatoxin declined over 60% after completing the cooker-extrusion-drying process. Since many feeds for aquaculture are manufactured by the cooker-extrusion method, use of this method might permit the successful use of aflatoxin contaminated feed grains. Research needs to be conducted to verify that the degradation products of this process are not toxic. Attempts to detoxify other mycotoxins have met with limited success. As mentioned beforehand, the toxic effects of fumonisin, ochratoxin A, DON and T-2 toxin are not mitigated by most sequestering agents. There is limited breakdown of these mycotoxins after exposure to heat treatment.

Conclusion

Understanding the effect that mycotoxins have on the numerous species of fish that are raised in the aquaculture systems around the world is important. Fish comprise an important part of the daily diet of people in many parts of world. Even though problems with mycotoxins are not always present, conditions conducive to the proliferation of mould growth on feedstuffs and finished feeds occur from time to time. Natural conditions such as plant stress associated with drought and insect damage can increase the invasion of moulds and development of mycotoxin contamination. Ubiquitous moulds can develop on feed ingredients and complete feeds that are not stored properly. As more aquaculture fish are fed diets that contain increasingly higher proportions of cereal grains and plant protein products, it becomes important to have the knowledge of not only how to control mycotoxin contamination, but in the event of exposure of fish to the various mycotoxins to understand how they are affected.

References

Bacon, C.W., J.K. Porter and W.P. Norred. 1995. Toxic interaction of fumonisin B_1 and fusaric acid measured by injection into fertile chicken egg. Mycopathologia 129: 29-35.

Bacon C.W., J.K. Porter, W.P. Norred and U. Leslie. 1996. Production of fusaric acid by *Fusarium* species. Applied Environmental Microbiology 62:4039-4043.

Blount, W.P. 1961. Turkey "X" disease. Journal of British Turkey Federation 9 (2): 52:55-58.

Boonyaratpalin, M., K. Supamattaya, V. Verakunpiriya, and D. Suprasert. 2001. Effects of aflatoxin B_1 on growth performance, blood components, immune function and histopathological changes in black tiger shrimp (*Panaeus monodon* Fabricius). Aquaculture Research (Supplement. 1) 32:388-398.

Boorman, G.A., H.L. Hong, M.P. Dieter, H.T. Hates, and A.E. Pohland. 1984. Myelotoxicity and macrophage alteration in mice exposed to ochratoxin A. Toxicology and Applied Pharmacology 72:304-312.

Buser, M.D. and H.K. Abbas. 2001. Mechanically processing cottonseed to reduce gossypol and aflatoxin levels. Journal of Toxicology-Toxin Reviews 20(3&4):179-208.

CAST. 2003. Mycotoxins: risks in plant, animal, and human systems. Council for Agricultural Science and Technology, Report No. 139. Ames, Iowa.

Chaouloff, F., D. Laude, D. Merino, B. Serrurrier, and J.L. Elghozi. 1986. Peripheral and central short-term effects of fusaric acid, a DBH inhibitor, on tryptophan and serotonin metabolism in the rat. Journal of Neural Transmission 65: 219-232.

Chavez-Sanchez, Ma.C., C.A. Martinez, and I. Osorio Moreno. 1994. Pathological effects of feeding young *Oreochromis niloticus* diets supplemented with different levels of aflatoxin B_1. Aquaculture 127:49-60.

Cole, R.J. 1986. Etiology of turkey "X" disease in retrospect: A case for the involvement of cyclopiazonic acid. Mycotoxin Research 2:3-7.

Colvin, B.M. and L.R. Harrison. 1992. Fumonisin-induced pulmonary edema and hydrothorax in swine. Mycopathologia 117:79-82.

Conway, H.F., R.A. Anderson, and E.B. Bagley. 1978. Detoxification of aflatoxin-contaminated corn by roasting. Cereal Chemistry 55:115-117.

Coulombe, R.A.1993. Biological action of mycotoxins. Journal of Dairy Science 76:880-891.

Coulombe, R.A., Jr., G.S. Bailey, and J.E. Nixon. 1984. Comparative activation of aflatoxin B_1 to mutagens by isolated hepatocytes from rainbow trout (*Salmo gairdneri*) and coho salmon (*Onchorynchus kisutch*). Carcinogenesis 5:29-33.

Doster, R.C., R.O. Sinnhuber, and J.H. Wales. 1972. Acute intraperitoneal toxicity of ochratoxins A and B in rainbow trout (*Salmo gairdneri*). Food and Cosmetic Toxicology 10:85-92.

Dwivedi, P. and R.B. Burns. 1984. Pathology of ochratoxin A in young broiler chicks. Research Veterinary Science 36:92-103.

Gelderblom, W.C.A., K. Jaskiewicz, W.F.O. Marasas, P.G. Thiel, R.M. Horak, R. Vleggar, and N.P.J. Kriek. 1988. Fumonisins- novel mycotoxins with cancer promoting activity produced by *Fusarium moniliforme*. Applied Environmental Microbiology 54:1806-1811.

Giambrone, J.J., D.L. Ewert, R.D. Wyatt, and C.S. Eidson. 1978. Effect of aflatoxin on the humoral and cell-mediated immune systems of the chicken. American Journal of Veterinary Research 39:305-308.

Goel, S., S.D. Lenz, S. Lumlertdacha, R.T. Lovell, R.A. Shelby, M. Li, R.T. Riley, and B.W. Kemppainen. 1994. Sphingolipid levels in catfish consuming *Fusarium moniliforme* corn culture material containing fumonisin. Aquatic Toxicology 30:285-294.

Halver, J.E. 1969. Aflatoxicosis and trout hepatoma. In: *Aflatoxin; Scientific background, Control and Implications* (L.A. Goldblatt, ed). Academic Press, New York, New York, pp. 265-306.

Harvey, R.B., T.S. Edrington, L.F. Kubena, M.H. Elissalde, and G.E. Rotttinghaus. 1995. Influence of aflatoxin and fumonisin B_1-containing culture material on growing barrows. American Journal of Veterinary Research 56:1668-1672.

Hendricks, J.D., R.O. Sinnhuber, J.H. Wales, M.E. Stack, and D.P.H. Hsieh. 1980. Hepatocarcinogenicity of sterigmatocystin and versicolorin A to rainbow trout (*Salmo gairdneri*) embryos. Journal of National Cancer Institute 64:1503-1509.

Huff, W.E., R.D. Wyatt, T.C. Tucker, and P.B. Hamilton. 1974. Ochratoxicosis in the broiler chicken. Poultry Science 53:1585-1591.

Hughes, D.M., M.J. Gahl, C.H. Graham, and S.L. Grieb. 1999. Overt signs of toxicity to dogs and cats of dietary deoxynivalenol. Journal of Animal Science 77:693-700.

Huwig, A., S. Freimand, O. Kappeli, and H. Dutler. 2001. Mycotoxin detoxification of animal feed by different adsorbents. Toxicology Letters 122:179-188.

Jantrarotai, W. and R.T. Lovell. 1990a. Subchronic toxicity of dietary aflatoxin B_1 to channel catfish. Journal of Aquatic Animal Health 2:248-254.

Jantrarotai, W. and R.T. Lovell. 1990b. Acute and subchronic toxicity of cyclopiazonic acid to channel catfish. Journal of Aquatic Animal Health 2:255-260.

Krogh, P., 1991. Porcine nephropathy associated with ochratoxin A. In: *Mycotoxins in Animal Foods* (J.E. Smith and R.S. Hendersone eds). CRC Press, Boca Raton, Florida, pp. 627-645.

Kuiper-Goodman, T. and P.M. Scott. 1989. Risk assessment of the mycotoxin ochratoxin A. Biomedical and Environmental Sciences 2:179-248.

Ledoux, D.R. and G.E. Rottinghaus. 1999. *In vitro* and *in vivo* testing of adsorbents for detoxifying mycotoxins in contaminated feedstuffs. In: *Biotechnology in the Feed Industry, Proceedings of Alltech's Fifteenth Annual Symposium* (T.P. Lyons and K.A. Jacques, eds). Nottingham University Press, Nottingham, U.K. pp. 369-379.

Lee, D.J., J.H. Wales, J.L. Ayres, and R.O. Sinnhuber. 1968. Synergism between cyclopropenoid fatty acids and chemical carcinogens in rainbow trout (*Salmo gairdneri*). Cancer Research 28:2312-2318.

Li, M.H., S.A. Raverty, and E.H. Robinson. 1994. Effects of dietary mycotoxins produced by the mold *Fusarium moniliforme* on channel catfish *Ictalurus punctatus*. Journal of the World Aquaculture Society 25:512-516.

Loveland, P.M., J.E. Nixon, and G.S. Bailey. 1984. Glucuronides in rainbow trout (*Salmo gairdneri*) injected with [^{3}H] aflatoxin B_1 and effects dietary beta-naphthoflavone. Comparative Biochemistry and Physiology C, Comparative Pharmacology 78:13-19.

Luk, K.C., B. Kobbe, and J.M. Townsend. 1977. Production of cyclopiazonic acid by *Aspergillus flavus* Link. Applied and Environmental Microbiology 33:211-212.

Lumlertdacha, S. and R.T. Lovell. 1995. Fumonisin-contaminated dietary corn reduced survival and antibody production by channel catfish challenged with *Edwardsiella ictaluri*. Journal of Aquatic Animal Health 7:1-8.

Lumlertdacha, S., R.T Lovell, R.A. Shelby, S.D. Lenz, and B.W. Kemppainen. 1995. Growth, hematology, and histopathology of channel catfish, *Ictalurus punctatus*, fed toxins from *Fusarium moniliforme*. Aquaculture 130:201-218.

Manning, B.B. 1998. Fumonisin B_1 from *Fusarium moniliforme* culture material varies in toxicity to channel catfish and rats with purity. Dissertation. Auburn University, Alabama.

Manning, B.B., M.H. Li, E.H. Robinson, P.S. Gaunt, A.C. Camus, and G.E. Rottinghaus. 2003a. Response of channel catfish *Ictalurus punctatus* to diets containing T-2 toxin. Journal of Aquatic Animal Health 15:230-239.

Manning, B.B., R.M. Ulloa, M.H. Li, E.H. Robinson, and G.E. Rottinghaus. 2003b. Ochratoxin A fed to channel catfish causes reduced growth and lesions of hepatopancreatic tissue. Aquaculture 219:739-750.

Marquardt, R.R. and A.A. Frohlich. 1992. A review of recent advances in understanding ochratoxicosis. Journal of Animal Science 70:3968-3988.

Meronek, R. and W. Xie. 1999. Mycotoxins in feed. Feedstuffs 71(31):123-130.

Nguyen, A.T., J.M. Grizzle, R.T. Lovell, B.B. Manning, and G.E. Rottinghaus. 2002. Growth and hepatic lesions of Nile tilapia (*Oreochromis niloticus*) fed diets containing aflatoxin B_1. Aquaculture 212:311-319.

Nguyen, A.T., B.B. Manning, R.T. Lovell, and G.E. Rottinghaus. 2003. Responses of Nile tilapia (*Oreochromis niloticus*) fed diets containing different concentrations of moniliformin or fumonisin B_1. Aquaculture 217:515-528.

Norred, W.P. 1993. Fumonisins: mycotoxins produced by *Fusarium moniliforme*. Journal of Toxicology and Environmental Health 38:309-328.

Norred, W.P., C.W. Bacon, J.K. Porter, and K.A. Voss. 1990. Inhibition of protein synthesis in rat primary hepatocytes by extracts of *Fusarium moniliforme* contaminated corn. Food and Chemical Toxicology 28:89-94.

Norred, W.P., E. Wang, H. Yoo, R.T. Riley, and A.H. Merrill, Jr. 1992. *In vitro* toxicology of fumonisins and the mechanistic implications. Mycopathologia 177:73-78.

Pier, A.C. 1991. The influence of mycotoxins on the immune system. In: *Mycotoxins in Animal Foods* (J.E. Smith and R.S. Hendersone, eds). CRC Press, Boca Raton, Florida, pp. 489-497.

Post, G. 1987. Neoplastic diseases of fishes. In: *Textbook of Fish Health*. T.F.H. Publications, Neptune City, New Jersey. pp. 244-246.

Poston, H.A., J.L. Coffin, and G.F. Combs, Jr. 1982. Biological effects of dietary T-2 toxin on rainbow trout, *Salmo gairdneri*. Aquatic Toxicology 2:79-88.

Riley. R.T., E. Wang, and A.H. Merrill. 1994. Liquid chromatographic determination of sphingosine: use of free sphinganine to sphingosine ratio as a biomarker for consumption of fumonisin. Journal of Association of Analytical Chemist International 77:533-540.

Ross, P.F., A.E. Wedet, D.L. Owens, L.G. Rice, H.A. Nelson, G.D. Osweiler, and T.M. Wilson.1993. Experimental equine leukoencephalopathy caused by corn naturally contaminated with fumonisin. Journal of Veterinary Diagnostic Investigation 5:69-74.

Sahoo, P.K. and S.C. Mukherjee. 2002. Influence of high dietary α-tocopherol intakes on specific immune response, nonspecific resistance factors and disease resistance of healthy and aflatoxin B_1-induced immunocompromised Indian major carp, *Labeo rohita* (Hamilton). Aquaculture Nutrition 8:159-167.

Schoenhard, G.L., J.D. Hendricks, J.E. Nixon, D.J. Lee, J.H. Wales, R.O. Sinnhuber, and N.E. Pawlowski. 1981. Aflatoxicol-induced hepatocellular carcinoma in rainbow trout (*Salmo gairdneri*) and the synergistic effects of cyclopropenoid fatty acids. Cancer Research 41:1011-1014.

Smith, J.E. 1997. Aflatoxins. In: *Handbook of Plant and Fungal Toxicants* (J.P. Felix D'Mello, ed).CRC Press, Boca Raton, Florida, pp. 269-285.

Smith, T.K. 1992. Recent advances in the understanding of *Fusarium* trichothecene mycotoxicoses. Journal of Animal Science 70:3989-3993.

Smith, T.K. and M.G. Sousadias. 1993. Fusaric acid content of swine feedstuffs. Journal of Agricultural and Food Chemistry 41:2296-2398.

Smith, T.K. and E.J. MacDonald. 1991. Effect of fusaric acid on brain regional neurochemistry and vomiting behavior in swine. Journal of Animal Science 69:2044-2049.

Smith, T.K. and I.R. Seddon. 1998. Synergism demonstrated between *Fusarium* mycotoxins. Feedstuffs 70(25):12-17.

Smith, T.K., E.G. McMillan, and J.B. Castillo. 1997. Effect of feeding blends of *Fusarium* mycotoxin-contaminated grains containing deoxynivalenol and fusaric acid on growth and feed consumption of immature swine. Journal of Animal Science 75:2184-2191.

Strength, D.R., D.V. Saradambal, Shoou-Liz Wang, H.H. Daron, and W.P. Schoor. 1982. Glucuronosyl- and sulfo-transferases in fish exposed to environmental carcinogens. Federation Proceedings 41 (4):1147.

Taylor, D.R. 1999. Mycotoxins binders: what are they and what makes them work? Feedstuffs 71 (3):41-45.

Voss, K.A., J.K. Porter, C.W. Bacon, F.I. Meredith, and W.P. Norred. 1999. Fusaric acid and modification of the subchronic toxicity to rats of fumonisins in *F. moniliforme* culture material. Food and Chemical Toxicology 37:853-861.

Woodward, B., L.G. Young, and A.K. Lun. 1983. Vomitoxin in diets for rainbow trout (*Salmo gairdneri*). Aquaculture 35:93-10.

Wyatt, R.D., 1991. Poultry. In: *Mycotoxins in Animal Foods* (J.E. Smith and R.S. Henderson, eds). CRC Press, Boca Raton, Florida, pp. 553-606.

Wyatt, R.D., P.B. Hamilton, and H.R. Burmeister. 1973. The effects of T-2 toxin in broiler chickens. Poultry Science 52:1853-1859.

Yildirim, M., B. Manning, R.T. Lovell, J.M. Grizzle, and G.E. Rottinghaus. 2000. Toxicity of moniliformin and fumonisin B_1 fed singly and in combination in diets for channel catfish. Journal of World Aquaculture Society 31:599-608.

7

PRINCIPLES AND APPLICATIONS OF MYCOTOXIN ANALYSIS

Keith A. Scudamore
KAS Mycotoxins, Taplow, Maidenhead, Berkshire, UK

Introduction

Analysis forms an integral part of almost all studies involving mycotoxins. There are many reasons for mycotoxin analysis that range from the need to enforce legislative limits, to on-line monitoring for quality control of materials in the food and feed industries and surveillance to monitor occurrence to estimate consumer exposure. In addition, research studies require, for example, measurements to determine the fate of the toxins during food processing, means for their prevention and control, research into the relationship between fungal species and the toxin produced, toxicological studies and veterinary investigations.

Analysis is often regarded as something best left to the specialist chemist to provide a service whose results are accepted without question by the customer. Analysts are often regarded as cautious about their results and liable to wrap these up in all sorts of provisos and warnings. This chapter will show why this caution is justified and will illustrate that the determination of mycotoxins, in common with many other types of natural contaminants, is a complex and difficult field. It is thus essential that the analyst has full confidence in his methods and takes all necessary precautions to ensure the reliability of his results and the suitability of the method for the task in hand. At the same time the customer should demand evidence that appropriate quality checks are incorporated as routine. This is incumbent on all parties because important decisions that sometimes affect national and international trade relations and involve large amounts of money may need to be made on the basis of a few analytical results.

This chapter does not merely provide a list of selected methods and references but aims to discuss many of the factors that influence how a method is selected, developed and operated effectively, together with the controls necessary to ensure that it continues to work reliably and accurately. It is aimed to help those new to the field to understand and avoid some of the pitfalls in analytical development and method application and to assist in general improvement in the standard of mycotoxin analysis. In addition, it is hoped that managers and food quality technicians may gain an insight into the complexities of mycotoxin analysis and assist in communication with the analytical fraternity.

Aims and preparation for mycotoxin analysis

OBJECTIVE OF THE ANALYSIS

In this section it is assumed that the need for the determination of mycotoxins has arisen and that the organisation requiring this has no previous experience of mycotoxin analysis. This undertaking can be an expensive commitment and it is important that the correct approach is decided at an early stage. The full quantitative analysis of mycotoxins is a specialised and skilled activity and is most successful if carried out on a regular basis. However, developments over the past few years have enabled some methods to be carried out by technicians in quality control laboratories unequipped with sophisticated instrumentation and in the field situation.

Methods may be classified as fully quantitative, semi-quantitative, simply a yes/no answer at a set specified concentration or qualitative. In general the sophistication needed, the expertise required and the equipment and labour costs decrease from the first category to the last.

The requirements for a method will depend on the reason for analysis. The most demanding requirements that supply the most precise analytical values will be needed where methods are required for enforcement or trading purposes when contracts are legally binding. Consignments may be accepted or rejected on the basis of these results and large sums of money may be involved. Further, trade between countries may be affected and there have been a number of instances in international trade in recent years where, for example, the EC has imposed temporary bans on imports of commodities such as nuts and dried fruit from trading partners because of persistent, unacceptably high contamination with mycotoxins. As a consequence, the results in such circumstances will need to stand up to legal scrutiny and must be demonstrated to conform to the highest analytical standards. Similar standards are demanded where information is supplied to international bodies charged with assessing consumer exposure as part of the legislative process of setting maximum permissible limits for mycotoxins. Information about consumer exposure to a range of mycotoxins has been collected within the European Community through a series of **S**cientific **Coop**eration ('SCOOP') initiatives (e.g. EC, 2002a, 2002b, 2002c, 2003). Only 'sound' data are accepted for inclusion in these assessments.

In comparison, monitoring of food commodities to achieve a reduction in consumer exposure to mycotoxins, may only require methods that operate reliably at a preset level so that a yes/no, acceptance/rejection, quality control system can be operated. When a result is in doubt, or close to the rejection value, a fully quantitative analysis can be carried out. This approach is particularly cost effective when most consignments are expected to contain no, or trace levels of, contamination and the re-analysis frequency is low. Rapid, labour saving methods may then be suitable, relying on fully quantitative analysis for any consignments that appear to be contaminated at levels close to the agreed limit. However, it is essential that these rapid methods be fully validated, particularly to ensure that the occurrence of false negatives or positives does not occur. A further advantage is that an almost instant result can be obtained in contrast to a turnaround time that can be several days for fully quantitative instrumental methods.

Requirements within the research laboratory will vary enormously depending on the study in hand and decisions concerning the sophistication of the analysis can be left to the discretion of the scientists involved. It is clear that precise methods will be required for studies such as determining the stability of standards, reaction rate experiments or fate during processing, while those used in studying relations between fungal species and the mycotoxins they produce, metabolic studies and genetic variation of fungi or toxicological studies may need less rigorous standards or, at the extreme, may merely need to confirm the identity of the toxin rather than its amount.

IN HOUSE OR CONTRACT ANALYSIS?

In-house quantitative analysis requires suitable laboratory facilities and expertise together with the safe working protocols necessary for handling toxins, particularly when preparing standards. This has become easier with the commercial availability of calibrated dilute standard solutions, which preclude the need to handle and store highly toxic materials. Mycotoxin determination needs technical knowledge and skills to be acquired and maintained to guarantee long-term reliable results even in large analytical laboratories with long experience of other trace analysis. However, it normally will be cost-effective to set up in-house facilities for this work in laboratories carrying out research in this area or when frequent analyses and rapid turnaround times are required. In addition, quality control of the analyses can be conducted internally and methods adapted as necessary, e.g. to accommodate new commodities or additional toxins. The choice will ultimately depend on the national or local circumstances e.g. in developing economies, appropriate resources and expertise may be concentrated in central laboratory facilities.

SELECTING AN ANALYTICAL CONTRACT LABORATORY

In many cases it will be clear to the food quality control manager, local food standards officer or government food authority that contracting out the sample analysis will be the only or most suitable and cost-effective arrangement. It is thus important to select a reliable and proven contractor who can provide the service and turnaround time when it is required at all times, including during public holidays.

Because the customer is placing the analyses with a specialist contractor he may not always have the expertise to assess the competence of the service offered. There are several safeguards that can be taken to assist here. Firstly an accredited laboratory should be used whenever possible. This does not guarantee that results will always be correct but does at least ensure that approved working procedures are in place. In the past, each country or professional body has tended to manage its own quality assurance scheme, including laboratory accreditation. A major effort is currently being made to harmonise such schemes. The EN ISO CEI 17025 guide (requirements for the competence of testing and calibration laboratories) is now one of the main harmonised tools for accreditation bodies, and most countries have accreditation programmes for mycotoxins in food and feed.

Secondly, enquire whether the analyst takes part in a quality check programme such as the Food Analysis Performance Assessment Scheme (FAPAS®), Bureau interprofessionnel d´études analytiques (BIPEA), American Oil Chemists Society (AOCS) or Comparative Laboratory Tests (LVU), and if so, request a copy of recent results. Reluctance to provide such information should raise concern about the overall competence of the laboratory.

Thirdly, agree on the performance characteristics that are required for the method(s). Quality control and proficiency testing are discussed in more detail later.

Techniques used in mycotoxin analysis

FULLY QUANTITATIVE INSTRUMENTAL METHODS

HPLC or GLC are the favoured choices for the analyst when sensitive, reliable results are required with minimal variability. These methods must include optimised extraction and cleanup stages and must be suitably validated. Sometimes these procedures are linked with mass spectrometry to provide improved confirmation and sensitivity. The use of HPLC/MS is steadily increasing and methods published recently include those for deoxynivalenol and zearalenone (Pallaroni *et al.*, 2002), for cyclopiazonic acid in milk (Losito *et al.*, 2002) and for type A trichothecenes in maize and oats (Fuchs *et al.*, 2003).

It is not uncommon in trade for food commodities to be analysed more than once at different times during processing or marketing. If these results differ significantly it can lead to disagreement and has the potential for legal action, so it is essential that methods are consistently sound and reliable. The inherent variability in analysis needs to be understood and must be taken into account. A summary of the main techniques used for analysis of mycotoxins and comments on their usefulness is given in Table 1.

Most stages in these methods including cleanup, sample injection and data processing can be automated, saving in labour costs and time and improving reproducibility. Many recent methods include automated steps while a robotic method for determination of fumonisins in corn and corn products that allows operation of the system for 40 samples before system refuel has recently been published (Dilkin *et al.*, 2001). However, despite significant savings in time and costs that can come from automation, a major effort is still usually required to obtain the representative sample required for analysis.

OTHER QUANTITATIVE METHODS

Many of the pioneering studies on mycotoxins relied on TLC methods well

Table 1. Review of some method types available for mycotoxin analysis.

Method	*Advantages*	*Disadvantages*
TLC	Simple, cheap, rapid Detects most mycotoxins Sensitive for aflatoxins and ochratoxin A number of samples can be run together	Identity of spots may need confirmation Insensitive for some toxins Separation may not be satisfactory and require 2- dimensional approach Poor precision
HPTLC	Quantitative when used with densitometry Number of samples can be run together Improved separation compared with TLC	Insensitive for some toxins
ELISA	Visual assessments possible Plate reader assists quantifying results and improves detection limits	Matrix interference may affect result Limited to a few solvents Antibodies may cross react
HPLC	Sensitive, selective Easy to automate	Compounds must have UV absorption or fluoresce or require derivatisation
HPLC/MS	Provides high level of confirmation Multi-detection? Very sensitive	Expensive Specialist expertise required
GC/MS	Provides high level of confirmation Very sensitive	Expensive Specialist expertise required Compounds must be volatile
Capillary electro-phoresis	Very low solvent use Alternative separation technique Rapid	Not yet widely evaluated

before the general availability of HPLC and immunological methods. Otherwise well-equipped laboratories may not always be set up with HPLC or GLC particularly in developing parts of the world. Here the need to determine mycotoxin levels in key food commodities to protect vital exports and to monitor human and animal diets is equally important and excellent results for most of the key mycotoxins can be obtained at relatively low cost by using TLC or ELISA-based methods. Many such methods have been published and a significant number of these have been fully validated and collaboratively tested. The availability of a densitometer for assessment of TLC spots or a plate reader for optical density measurement in ELISA methods assists in accurately quantifying results at relatively low additional initial cost.

Chu (1984) described the development of immunoassays for mycotoxins while the current status of commercially available immunoassay kits has been assessed (Wilson *et al.*, 1998). Organisations such as the US Food and Drug Administration and the US Department of Agriculture have set some of the criteria to evaluate commercial test procedures (Trucksess and Koeltzow, 1995) while the AOAC International/IUPAC has sponsored several international collaborative studies to evaluate immunoassay kits. A wide range of these tests have been evaluated by the USDA Federal Grain Inspection Service (FGIS) and are available on their web site (*www.usda.gov/gipsa/programsfgis/services.htm*) and are also reviewed on the European Mycotoxin Awareness Network (EMAN) web site (*www.lfra.co.uk/eman2/index.asp*).

RAPID METHODS

Simpler, less time consuming and cheaper methods that include some of the commercial immunoassay kits may be suitable for routine monitoring, 'due diligence' or quality control purposes especially when the frequency of contamination is low or concentrations normally found are well below the maximum acceptable values. The sample can be re-checked by HPLC or GLC methods if a concentration appears close to the acceptable limit. This approach can be very useful to monitor a detection level set to accept or reject a consignment or to make rapid commercial decisions. There have been major advances in this area recently with rapid methods based on ELISA and immunoaffinity columns. The development of molecular imprint materials such as those for deoxynivalenol and zearalenone (Weiss *et al.*, 2003) and of lateral flow devices appear promising for the future (Danks *et al.*, 2003).

Sibanda *et al.* (2002) have developed a solid phase cleanup system and portable rapid flow-through enzyme immunoassay for the detection of ochratoxin A in coffee and this and similar field kits have been collaboratively studied (Desaeger *et al.*, 2002). A rapid simple method for detecting deoxynivalenol in wheat by fluorescence polarization has been published (Maragos and Plattner, 2002) and for aflatoxins in grain (Nasir and Jolley, 2002). For ultimate simplicity, transmittance or reflectance spectroscopy can detect aflatoxin in those single corn kernels that make the main contribution to the contamination, enabling them to be sorted and removed (Pearson *et al.*, 2001; Radova *et al.*, 2001).

The analytical procedure

Analysis of naturally occurring organic compounds in food materials involves a number of key steps. Some or all of the following stages are necessary: sampling, method preparation, extraction, cleanup, concentration, separation, detection, quantitation, confirmation and reporting. The design of each step will depend on the commodity, the chemical properties of the contaminant and the method used.

SAMPLING

The ultimate success of any analytical procedure depends on the ability to obtain a sample that is truly representative of the whole consignment under test. Subsequent analysis is likely to be of limited value unless the variability inherent in taking the sample can be minimised and restricted to an acceptable level. Mycotoxins in particular can be distributed in a highly heterogeneous manner within a food material because of the way fungi colonise and grow so that proper sampling plans are particularly important. Statistical evaluation of the distribution of some mycotoxins in a limited range of commodities has been carried out and sampling plans devised. However, sampling is a complex issue in itself and will not be covered here in detail. Suffice it to say that it is of paramount importance and is discussed more fully for animal feeds in another chapter. Design of sampling plans for the determination of mycotoxins in foods and feeds have been considered in detail and reviewed (e.g. Coker, 1998; Wilson *et al.*, 1998; Romer Laboratories, 2001).

Directives on sampling and analysis increasingly accompany regulations setting maximum permissible limits for mycotoxins. The European Community lays down sampling methods and methods of analysis for aflatoxins and ochratoxin A, respectively, in foodstuffs (European Commission, 1998; 2002a). These plans are for official control/enforcement purposes. They are statistically based to ensure that the results obtained when sampling according to these Directives give results that will fall within an accepted and agreed range of the 'true' mean. Methods of analysis to be used must meet agreed specific criteria but any method meeting these criteria may be used. Sampling plans have been developed to cover anything from bulk consignments of several thousand tonnes down to retail packages where these may be 0.1 kg to several kg in size. These regulations acknowledge that alternative, 'less onerous', sampling plans could be used by industry for routine quality control and to demonstrate 'due diligence'. However, in such cases, if a dispute should arise, it would be for a court of law to decide whether this approach is acceptable. Methods for analysis and sampling have also been further evaluated (FAO, 1997).

LABORATORY SAFETY

Handling mycotoxins

Mycotoxins are extremely toxic chemicals, while crystalline standards maybe electrostatic. Their handling should be avoided as far as possible and dilute standard solutions should be purchased in preference from reliable suppliers. Preparation of in-house solutions from concentrated standard materials should only be attempted in specially designated toxin laboratory facilities where staff are fully protected from contact with the concentrates and are fully trained in the precautions necessary. Mycotoxins in fungal cultures may be present in high concentration and in addition, the fungal spores may present a further hazard.

Risk assessment should be carried out before any work is started. Continuous exposure to low concentrations of mycotoxins presents a potential chronic hazard to analysts who may be working with these chemicals for many years because they may be carcinogenic, genotoxic or immuno-suppressant. Working protocols should aim for zero operator exposure.

Decontamination and waste disposal

Used solvents, samples and other waste should be decontaminated and disposed of by specialist contractors. Sodium hypochlorite of 10-20% concentration is often used for decontamination. Most methods use large volumes of solvents and these should be separated into chlorinated and nonchlorinated units although chlorinated solvents should only be used if alternatives are unsuitable.

Cleaning and decontamination

All glassware and equipment should be scrupulously clean and decontaminated to avoid cross contamination in common with other trace analyses. Glassware is commonly soaked in sodium hypochlorite and may also be rinsed with dilute hydrochloric acid to acidify surfaces, as some mycotoxins are degraded by alkali. Some mycotoxins are also sensitive to light so may be degraded during extraction or subsequent processing. Amber glassware or the use of aluminium foil is thus often used to protect solutions from light.

SAMPLE PREPARATION

The sample received from which the analytical test sample must then be prepared may be anything from 0.1 to 30 kg in weight, either as a single bulk sample or as a number of packs that should share a common batch number or clear identification. The analyst would not normally be responsible for any deficiencies in the sampling procedure.

Liquid samples by their nature are usually more homogeneous, however when these are cloudy or consist of slurries containing particulate matter, mycotoxins can be unequally distributed between liquid and solid phases. In solid food commodities such as cereal grains, nuts and dried fruit or processed foods such as bread, breakfast cereals, polenta and snack foods, mycotoxins may be distributed in a very heterogeneous manner. Thus ideally, the whole should be ground or comminuted and then thoroughly mixed to reduce sample error as much as possible while a portion can be retained if required for reference purposes.

There are two widely used approaches for extraction, either shaking with a solvent for up to one hour or high speed blending with a solvent for a few minutes. Blending is a rapid process and may allow grinding and extraction to be carried out as a single step for suitable food samples. However, care must be taken to ensure that the solid particles are continually wetted by the solvent. Shaking for up to an hour, although slower, may have an advantage when a large number of samples are being analysed as this can be done automatically with a number of samples being handled simultaneously. There are conflicting views on which method is most efficient. In an example from the specimen FAPAS® proficiency test report described later, the use of blending and shaking was split almost equally between participants.

EXTRACTION

The purpose of extraction is to remove as much of the mycotoxin from the food matrix as possible into a solvent suitable for subsequent cleanup and determination. The efficiency of the extraction process influences the overall recovery figure obtained for the method and is usually, although not always, less than 100%. However, the important factor is not the recovery figure itself but its consistency. It is important to choose the best solvent or solvent mixture for the product being analysed to ensure that the maximum amount of toxin is extracted, or the best compromise solvent when a range of products or mycotoxins are being sought. The solvent must have as low a toxicity as possible and be volatile and stable because the solution will be concentrated later using heat and reduced pressure.

Recent literature reports show that acetonitrile and methanol are by far the most common solvents used (Table 2) while acetone is sometimes used. Mixtures with water are often more effective than a single organic solvent. The presence of water appears to swell the food matrix and helps to disrupt the product cells assisting the release of the mycotoxins. Ethyl acetate is commonly used for patulin. The use of halogenated solvents such as chloroform and dichloromethane has diminished because of safety and

Table 2. Selected examples of extraction and cleanup procedures reported in the literature, 2000-2003.

Mycotoxin	*Commodity*	*Extraction solvent*	*Cleanup method*	*Reference*
Aflatoxins	corn, cotton-seed, nuts	methanol/water, 80/20, v/v	basic aluminium oxide	Sobolev and Dorner, 2002
Aflatoxins	spices	acetonitrile/water, 90/10, v/v	Multi-Sep #228 column	Akiyama *et al.*, 2001
Fumonisins	corn and corn flakes	acetonitrile/methanol/water, 25/25/50, v/v	immunoaffinity	Visconti *et al.*, 2001b
Fumonisins	corn	methanol/water, 80/20, v/v	immunoaffinity	Preis and Vargas, 2000
Ochratoxin A	coffee beans	methanol/3% sodium bicarbonate	Phenyl silane column + immunoaffinity	Martins *et al.*, 2003
Trichothe-cenes	wheat	acetonitrile/water, 84/16, v/v	Mycosep #227 column	Fuchs *et al.*, 2003
Trichothe-cenes	maize,oats	acetonitrile/water, 84/16, v/v	2 successive Mycosep columns	Schothorst and Jekel, 2001
Zearalenone	maize	acetonitrile/water, 75/25, v/v	immunoaffinity	Kryska *et al.*, 2003b
Zearalenone	wheat	acetonitrile/water, 84/16, v/v	GPC bio beads	Radova *et al.*, 2001

environmental concerns and it is questionable whether this should continue, as alternative solvents are mostly acceptable. Ionisable compounds such as moniliformin, tenuazonic acid or cyclopiazonic acid need buffering to control the pH.

Initial extraction of liquid samples such as fruit juices and milk is often carried out by liquid-liquid partitioning, the solvents being selected so that the mycotoxin is separated from the principle interfering substances. However, this uses large volumes of solvent and is labour intensive. As a result, this is now often replaced by solid phase or immunoaffinity column extraction/cleanup.

EXTRACTION EFFICIENCY

The main factors that affect the extraction efficiency are the chemical properties of the mycotoxin, the food commodity, the ratio of commodity weight to solvent volume and the solvent composition, particularly the percentage of water included. Naturally acquired residues have often been reported to be more difficult to extract than spiked samples so that, for example in a study of deoxynivalenol in wheat, it was found that the naturally contaminated sample required considerably longer to approach the same values (Trenholm *et al.*, 1985).

The ratio of solvent volume to product weight will affect the final value obtained. This will also depend on the solvent used and has been investigated in detail for aflatoxin B_1 in a range of products (Stroka *et al.*, 1999). The value for aflatoxin determined using acetonitrile and water almost doubled in changing from 10 g to 40 g of sample per 100 ml of solvent while with methanol/water the value decreased somewhat. A concentration effect occurred when water was absorbed preferentially by the product, as happened with the acetonitrile/water mixtures. Sometimes apparent recoveries of more than 100% are reported and this has occurred with dried fruit and figs that contain large amounts of water soluble sugars.

Cole and Dorner (1994) found that methanol and water gave more efficient extraction of aflatoxins from peanuts than acetonitrile and water. Pierard *et al.* (2004) investigated the effect of different proportions of water in acetonitrile for the extraction of ochratoxin A and a range of *Fusarium* mycotoxins from cereals and found the recovery varied both with the mycotoxin and the percentage of water. Different extraction solvents were investigated for removing fumonisins from corn and it was found that the temperature of extraction markedly affected the result so that for some products this increased 3-fold going from 23°C to 80°C although there were some problems when working at higher temperatures (Lawrence *et al.*, 2000).

Supercritical fluid extraction has been investigated by a number of authors but has yet to be used widely (Taylor *et al.*, 1993 [aflatoxin B_1 in corn]; Taylor *et al.*, 1997 [aflatoxin M_1 in beef liver]; Josephs *et al.*, 1998 [trichothecenes in wheat]).

CLEANUP

General

Cleanup involves the preparation of the analytical solution in a pure enough state to proceed to the final separation, detection and quantitation steps. This can be regarded as an additional separation stage prior to TLC, HPLC or GLC and may not always be necessary when initial extracts are very clean. However, the more effective the extraction procedure is for removing the mycotoxin from the food matrix, the more unwanted co-extractives are likely to migrate into the solution, to interfere subsequently in detection and quantitation. This will depend on the chemical properties of the mycotoxin and the solvent chosen. Some examples of cleanup methods reported recently are given in Table 2.

Cleanup procedures and strategies include classical liquid-liquid partitioning, solid phase extraction cleanup columns packed with a variety of materials, ion exchange and immunoaffinity columns. Many of these techniques are long established and packing materials are commercially available. Stoloff (1982) listed silica gel, acidic alumina, and florosil together with suitable cleanup solvents that are appropriate for use in the determination of aflatoxins; and these together with a variety of other materials have been used for most mycotoxins. More recently, multifunctional columns have been developed, e.g. containing both lipophilic and charged active sites, for determination of aflatoxins in spices (Akiyama *et al.*, 2001).

When isolation of very pure solutions is required, methods may be combined so that TLC or HPLC may be used for an initial cleanup and the spot of interest or the HPLC solvent eluting at the correct retention can then be examined further using a second separation technique.

Immunoaffinity columns

Immunoaffinity columns have become invaluable for cleanup. These are small flow-through tubes packed with an inert material to which antibodies raised against a specific mycotoxin are bound. When a solution containing the target mycotoxin is passed through the column the analyte binds to active sites that recognise the compound, allowing other co-extractives to pass through and out of the column. The mycotoxin is then eluted from the column with a small volume of a suitable solvent. Extremely pure solutions are usually obtained.

The success of the immunoaffinity column depends on the properties of the antibody used. These are raised by injecting a non-lethal dose of the mycotoxin into an animal, often rabbit, mouse or rat. The mixture of antibodies produced is purified, processed and tested and used to prepare either a polyclonal or a monoclonal product. Polyclonal antibodies are less selective and, when stocks are used up, will need to be replaced by freshly raised antibodies that are likely to differ in properties from the original. Monoclonal antibodies in contrast are more selective and are prepared from cells that are immortalised so that the stock is in theory unlimited. While immunoaffinity columns can result in extremely pure solutions of mycotoxins, extracts containing high concentrations of lipids or other constituents such as carbohydrates and peptides may still need an initial cleanup with traditional solid phase columns.

Although excellent purification can be achieved, the analyst needs to be aware of several potential pitfalls. There may be some cross reactivity with closely related compounds, particularly when using polyclonal antibodies. This can be both an advantage when the related compounds are of interest or a disadvantage for quantifying results in methods where no separation of analytes subsequently takes place. For example a rapid test procedure for aflatoxin B_1 that relies on optical density measurement would need to use an antibody that is specific to aflatoxin B_1, with low cross reactivity to aflatoxins B_2, G_1 and G_2 but for use in an HPLC method that is designed to determine all four metabolites, the antibody would need a high affinity for each aflatoxin.

Because a column contains a finite number of active sites it can become overloaded, especially with highly concentrated solutions, so that not all the toxin is retained by the column and low recovery occurs. Columns may also fail to perform effectively in solutions containing high concentrations of other co-extractives and it is considered that this occurs, at least in part, because these physically prevent the mycotoxin from linking to the antibody.

Other cleanup techniques

Other techniques described include a microwave-assisted extraction method for zearalenone in wheat and corn using a 50:50 mixture of methanol and acetonitrile at 80°C for 5 minutes (Pallaroni *et al.*, 2002). This enabled extraction and cleanup in a single step, as was also achieved using supercritical fluid extraction (Taylor *et al.*, 1993., 1997; Josephs *et al.*, 1998). Gel permeation has also been used for cleanup (Scudamore and Hetmanski, 1992; Dunne *et al.*, 1993).

CONCENTRATION

During analysis, solutions are likely to require concentration particularly prior to separation and determination. Rotary evaporation or other solvent concentrators are used to remove large volumes of solvent, often under nitrogen and reduced pressure and sometimes protected from direct sunlight to avoid the risk of breakdown of the mycotoxins present. Care is then needed to ensure that the residue or small volume of solvent remaining is completely re-dissolved in the analytical solvent and that glass surfaces are rinsed to avoid loss due to insolubility.

Immunoaffinity columns are also used to concentrate mycotoxins while removing interfering substances. Large volumes of liquids such as milk or beer can be passed through immunoaffinity columns if suitable precautions are taken. The residue bound to the column is subsequently washed out with a small volume of solvent, commonly methanol or acetonitrile. Such methods have been used (for example Hansen, 1990 [aflatoxin M_1 in milk]; Scott and Lawrence, 1995 [fumonisins in beer]; Scott and Lawrence, 1997 [aflatoxins in beer]; Bresch *et al.*, 2000 [ochratoxin A in beer]).

SEPARATION

TLC, HPTLC, HPLC, and GLC are the techniques most commonly used for separating mycotoxins prior to detection and quantification. Other approaches reported include capillary electrophoresis, for ochratoxin A (Corneli and Maragos, 1998) and over pressure layer chromatography for separation of fumonisins 1-4 (Katay *et al.*, 2001). The reader will need to consult specialist publications for full descriptions and technical details of these techniques.

The two key parameters in separation are the solvent or gas that carries the contaminant and the separating material over or through which the carrier passes. The degree of separation depends on the interaction of the analyte with the separating material and the efficiency of the column or the TLC plate. With HPLC, packing material size is most commonly 5 μ or 3 μ, the smaller particle materials being more efficient but generating more backpressure within the column. Higher efficiency enables shorter columns to be used and this has the advantage that the volume of solvent needed is much decreased.

In contrast to TLC where plates are used once, the life of GLC and HPLC columns depends on the operating conditions. Injection of relatively 'dirty' samples will drastically shorten column life and may lead to broader peaks if residues build up in the injector or column. GLC can be operated either at constant temperature or with temperature programming while HPLC can be run with constant solvent composition (isocratic operation) or with solvent programming.

Operation under constant conditions is more convenient for simple separations and is used most frequently, requiring less time between injections. However, for complex separations, temperature programming (e.g. for the separation of trichothecenes by GLC), or solvent programming for multi-toxin separations by HPLC may be necessary. Other constituents present on injection may need to be eluted from the column after each run, which can be done by increasing temperature or changing solvent composition. Because re-equilibrium at the starting conditions must be carried out after each run, the number of samples that can be processed is reduced, although the procedure can be automated for continuous operation.

Analytical solutions may still contain compounds not removed during cleanup that may or may not be related to the target analyte. These may have retention times similar to the target analytes that can lead to false positives or mis-identification. An example of one such compound that has caused problems in the determination of patulin is 5-hydroxy-2-furaldehyde (HMF) formed during the production and storage of apple juice although HPLC conditions can be easily modified to avoid this (Wilson *et al.*, 1998). Spices have caused problems during analysis for aflatoxins because of the large number of volatile compounds that are naturally present. Often the identity of the interfering compound is unknown.

The chance of interfering compounds occurring in TLC may similarly prove troublesome depending on the effectiveness of the cleanup procedure used and is particularly likely when using 1-dimensional TLC. If this is a problem, the analyst may need to use 2-dimensional TLC, HPTLC and/or suitable confirmatory tests to avoid mis-identification and improve quantitation. Two-dimensional TLC is a procedure in which a plate is developed with one solvent mixture in one direction, then rotated through 90° and run with a second solvent system.

DETECTION, QUANTITATION AND CONFIRMATION

The detection, quantitation and confirmation of mycotoxins in the clean analytical solution are separate activities but may in practice sometimes be carried out in a single step.

TLC

Spots on TLC plates should closely coincide in R_f values with those from standards applied in neighbouring channels and should have a characteristic colour or fluorescence. A TLC spot could be visible under daylight or may fluoresce either under short wave or long wave UV light. Sometimes a plate will need treating with a reagent, possibly with heating, to make the spot visible or to induce a response. Quantitation is usually achieved by comparing the intensity of the spot with standards run alongside on the TLC plate but is made more precise if a densitometer is used. A response at the expected position on a plate may not always be due to the species being sought as other compounds may migrate to the same place and have the same colouration, especially when there are many co-extractives present on the plate.

Increased confidence can be obtained by using a reagent that causes a change in colour or by using a second solvent system, type of plate or by using 2-dimensional TLC. If a standard is over-spotted at the sample origin before developing the plate, only a single spot should occur. Any distortion of the spot shape suggests that the sample may not contain the suspect mycotoxin. HPTLC provides a more efficient separation but confirmation should still be carried out.

HPLC and GLC

HPLC or GLC of clean sample extracts are capable of high resolution and the exact coincidence in retention time with a standard may often be sufficient confirmation, especially when the mycotoxin is expected. Immunoaffinity columns used during cleanup are highly selective, further reducing, but not entirely eliminating, the chance of interfering compounds. A HPLC or GLC peak at the mycotoxin retention time can also be due to another compound. Adding a standard to the sample solution injected may detect any closely eluting compound by distortion of the peak shape.

It is thus crucial that safeguards are in place to confirm the identity of the peak and obtain reliable quantitation. The presence of chromatographic interference can mask the presence of the mycotoxin. Even with optimised extraction solvents, cleanup procedures and separation, it is still possible for compounds to co-elute.

Detectors

The principle detectors used for HPLC are UV, UV diode array, fluorescence or mass spectrometric, and for GLC are electron capture and mass spectrometric. Electrochemical detection for HPLC and flame photometric detection for GLC are used less frequently. Specialised publications should be consulted for description of these detectors. It is sometimes necessary or an advantage to use derivatisation, either post- or pre-column, to assist sensitive detection of the mycotoxin. Early HPLC methods for determination of aflatoxins used UV detection but this proved insensitive for trace analysis so that a variety of post-column techniques such as bromination, iodination and the 'Kobra' cell (Kok *et al.*, 1984) have become well-established procedures in order to induce increased fluorescence of the eluting peaks.

One advantage of post-column techniques is that the reagent is fed in after the separation has occurred. It is important to keep dead space in the connective tubing to a minimum to avoid peak broadening. A recent method reports the use of cyclodextrins added to the LC eluant solvent to enhance the fluorescence of aflatoxins B_1 and G_1 by up to 15-fold (Chiavaro *et al.*, 2001). Other reagents that have been used include ammonia for ochratoxin A (Burdaspal, 1997) and aluminium chloride for zearalenone (Hetmanski and Scudamore, 1991). Because mycotoxins are usually not very volatile, HPLC is more frequently used than GLC although this is used for separation of trichothecene mixtures after derivatisation with any of a number of different reagents.

Detectors such as electron capture for GLC or fluorescence and UV for HPLC are inherently selective and further adjustment of detector parameters can assist in confirmation. The HPLC diode array detector can be used to obtain the spectrum of each compound or can be used to check the response ratio at different wavelengths. The use of mass spectrometry probably provides the ultimate but expensive detection and confirmatory system and can be used with single or multiple ion monitoring. The use of small mass, commonly found ions should be avoided.

Particular care is required when using ELISA methods where optical density is used to measure concentration. Co-extractives from the sample can give a background reading so that standard calibration curves should be carefully prepared in extracts from a similar sample material.

Reporting

Information in an analytical report must meet the customer's need. It should be complete enough to support the data provided but should not include superfluous information that makes the report cumbersome and difficult to interpret. A common malpractice is to include too many significant figures in the report, often because results are transferred directly from a computer printout (a result such as 20.32 μg/kg can rarely if ever be justified or substantiated in trace analysis but alas is seen too frequently!). Any supporting data can be attached as an annex if required for reference, or archived in the event of a future query. The recovery value for the mycotoxin obtained from a spiked sample should be included, clearly indicating whether the values given have been corrected.

Important decisions must often be made on the basis of a chemical analysis. In many sectors of analytical chemistry it is now a formal and often legislative requirement for laboratories to introduce quality assurance measures to ensure that they are capable of and are providing data of the required quality. Chemists are fully aware that any analytical result is subject to some degree of uncertainty and an indication of the degree of uncertainty should be included, if appropriate. This is discussed later.

Method performance, validation and quality assurance

METHOD DEVELOPMENT

The development of a successful analytical method should offer a distinct advantage over any existing methodology. This can range from better sensitivity and selectivity, application of a current method to an expanded range of foodstuffs, easier operation and saving of materials, reduced time and labour costs, automation, use of less hazardous reagents, more reliable operation, improved performance characteristics and better confirmation of identity. As more mycotoxins are regulated, the advantages of multi-mycotoxin analytical schemes are increasingly worth considering. Any method developed should be fully validated and collaboratively tested if it is to be accepted and widely adopted.

It is a common misunderstanding by non-analysts that a method developed for one food matrix can be used 'as is' for a whole range of other foods or feeds. The method should only be used for those materials for which it was developed unless its performance with these additional products is checked. For example, a method developed for the determination of aflatoxins in peanuts may not work as well for peanut butter. Similarly, the extraction efficiency and cleanup of mycotoxins from whole wheat grains or other cereals may be considerably different than for white flour, wholemeal flour, bran, bread dough, bread, biscuits, cake, whole wheat breakfast cereals or extruded cereal products. An excellent example of this is provided (de Girolamo *et al.*, 2001) who compared results using three methods for the determination of fumonisins in corn flour and in processed corn flakes. Using the 'AOAC' method (Sydenham *et al.*, 1996), the 'Borax' method (Scott and Lawrence, 1994) and the 'SMT' method (Solfrizzo *et al.*, 2000) acceptable, similar results were obtained from all methods for the corn flour with the 'AOAC' value being the highest. However, when applied to processed corn flakes, only about 10% and 67% of the values were obtained from the 'AOAC' and 'Borax' methods, respectively, when compared with the 'SMT' method.

MULTI-MYCOTOXIN METHODOLOGY

The ultimate goal in mycotoxin analysis could be to analyse all the common mycotoxins to the sensitivities required in a method with a single extraction, cleanup and detection procedure. The diverse chemical structures and properties of mycotoxins make this an extremely challenging task, yet to be achieved.

Multi-mycotoxin determination of specific mycotoxin groups of related compounds, such as trichothecenes, is commonly carried out using GC/MS while the determination of the four naturally occurring aflatoxins using HPLC has been long established. However, combining these two groups into a single method is extremely difficult. Even within the trichothecene group the individual metabolites exhibit a wide range of polarity from the non-polar T2-toxin to the highly hydroxylated nivalenol. Problems are compounded when other unrelated mycotoxins are included so that the method may be unsuccessful even at the initial extraction stage where at the best, a compromise solvent mixture must be found. If this hurdle is overcome, the cleanup necessary is likely to be complex and the end detection of all toxins in one system is in itself a major problem. Here, the diode array UV detector is able to detect most mycotoxins although the detection limits required would not be achieved for all. HPLC/MS may ultimately be the technique that comes closest to achieving this aim, but would not be a cheap or simple method available to all.

There have been a number of successful attempts to integrate several mycotoxin groups into a single method. Most include complex cleanup, separation and detection schemes that cater to all the mycotoxins required so that the method soon loses its intrinsic advantage over simpler 'single toxin' methodology. A multi-method using TLC has been widely used for screening animal feeds but involves the use of a dual split cleanup stage and different TLC systems for end detection (Patterson and Roberts, 1979). TLC has also been used for determination of aflatoxins, ochratoxin A, citrinin, zearalenone and deoxynivalenol in beer (Odhav and Naicker, 2002) and Gel Permeation Chromatography for aflatoxins, ochratoxin and zearalenone (Dunne *et al.*, 1993). Gel Permeation Chromatography was also studied as a cleanup and separation technique for a range of mycotoxins (Scudamore and Hetmanski, 1992) and this method was later used for determination of mycotoxins in animal feedstuffs (Scudamore *et al.*, 1997a). However, the authors concluded that while this methodology had proved useful for surveying a wide range of animal feed samples, the method was unsatisfactory for some toxins and cumbersome to operate. Deoxynivalenol (DON), nivalenol, 3- and 15-acetyl-DON, zearalenone, zearalenol and ochratoxin A have been determined using acetonitrile and water for extraction with UV diode array and fluorescence for detection (Pierard *et al.*, 2004). Sensitive detection and acceptable recoveries were obtained but a complex scheme for cleanup was required, again emphasising the inherent difficulties involved.

Perhaps the realistic approach with current technology is to limit multi-methods to groups of mycotoxins that commonly occur together as illustrated (Pierard *et al.*, 2004). Immunoaffinity columns for aflatoxins and ochratoxin A were successfully linked in tandem in a single extraction, cleanup and detection system for determination of these toxins in dry pet foods (Scudamore *et al.*, 1997b). The authors suggested that a commercial product containing a mixture of such antibodies would be a valuable tool. Expanding this idea to include, for example, zearalenone would be worth investigating.

VALIDATION STUDIES

Once a new method is working satisfactorily it needs to be validated internally and then by collaborative testing. Requirements for conducting inter-laboratory validation studies are published by the AOAC International. Properly designed studies not only establish how well the method performs but also show any problems and provide the opportunity for rectifying or minimising these. A recent report provides an overview of the validation of analytical methods to determine aflatoxin, ochratoxin and patulin (Stroka and Anklam, 1998). Most of the methods cited later have been collaboratively tested and a good understanding of how these studies are organised can be obtained through the references.

INTER-LABORATORY TESTING OF METHODS

Collaborative testing helps to determine the current status of the methodology for a mycotoxin and indicates those methods that have significant weaknesses or need particular care in their operation. In one example selected from many, an inter-laboratory study was carried out among 28 laboratories from 12 different countries for the determination of the *Fusarium* mycotoxins zearalenone in maize and deoxynivalenol in maize and wheat, employing each laboratory's in-house method. This study produced coefficients of variation (CV) between laboratory mean results (outliers rejected) ranging from 28 to 41% for zearalenone and from 32 to 38% for deoxynivalenol (Josephs *et al.*, 2001). The high CVs for zearalenone were traced back partly to problems occurring in determination of the concentration of the participants' own calibrant solutions. Experience suggests that precision in collaborative testing will usually be improved when a specific method is designated for all laboratories together with precise operating instructions.

The use of GC/MS for the determination of trichothecenes in cereals is an example where problems with variable results and low recoveries of some components have often been met. Many of the factors involved have been studied (Eskola, 2002) and the role of these factors in the quality of data assessed (Eskola *et al.*, 2000).

QUALITY ASSURANCE

Analytical laboratories are expected to implement appropriate measures for quality control to ensure the production of reliable and consistent data. Food control laboratories within the EU are required to do so by law. The CODEX Committee on Methods of Analysis and Sampling has developed objective criteria for assessing the competence of testing laboratories involved in the official import and export control of foods. These were recommended (FAO, 1997) and mirror closely the EU recommendations for laboratory quality standards and methods of analysis. These have been discussed more comprehensively (Wood, 1998) and the internal quality control (IQC) of analytical data has also been discussed (Royal Society of Chemistry, 1995). A more detailed discussion of the principles involved is provided by these and other publications (such as Gilbert and Anklam, 2002).

The sophistication of the quality assurance instigated needs to be appropriate for the aims of the analysis. Thus when monitoring product quality or for 'due diligence' purposes only, sufficient safeguards are required to ensure that the procedure is working satisfactorily; and the inclusion of suitable standards and spiked samples may suffice. Confirmation of the identity of a metabolite in studies of fungal cultures will be of much greater importance than its actual numerical value.

Quality control is the term to describe the practical steps undertaken to ensure that the analytical data are adequately free from error. The basic approach to IQC involves the analysis of control materials alongside the test materials of interest. A normal approach in mycotoxin analysis is to include 'spiked' or naturally contaminated samples of accurately determined content. Spiking is usually carried out by adding a known amount of a mycotoxin solution to material of similar composition to the test samples and allowing the solvent to evaporate. Ideally, spiked samples should be left overnight so that the mycotoxin is completely absorbed before analysis. Studies have shown that mycotoxins in spiked samples may be recovered more readily than those from naturally contaminated products. Therefore the use of naturally contaminated 'in-house' standards is preferable, although such material is often only obtainable in limited quantities. Logistical problems arise when many types of sample are analysed together because the number of spikes or controls theoretically required could form a significant proportion of the total test samples. The analyst must use his experience to decide what is appropriate for each situation. However, one spiked check sample should be included in every analytical run to indicate that the method is operating correctly.

METHOD PERFORMANCE PARAMETERS

It is now widely accepted that any method meeting agreed performance parameters is acceptable for quantitative analysis, which enables the analyst to select the most appropriate method for his laboratory. The main performance characteristics of a method are the **recovery**, the **precision** as shown by the standard deviation of the method, the **limit of quantitation** and the **limit of detection**. The method must also be specific, accurate and practical under normal laboratory conditions.

The range of recovery usually accepted for mycotoxin analysis is 70 to 110% and this value should always be stated. Whether or not to correct results for recovery has been discussed widely. It might be expected that a corrected result would be closer to the true value and this is supported by a statistical analysis of results for aflatoxins obtained in a series of collaborative trials when the relative between-laboratory standard deviations were less after results were corrected for recovery (Vonholst *et al.*, 2002).

The relative standard deviation is established by analysing a number of replicate samples of the particular matrix either within a laboratory (RSD_r -repeatability) or among laboratories (RSD_R -reproducibility). Inter-laboratory comparisons will determine the repeatability and reproducibility of the method and will also show analytical bias by any participant.

The limit of quantitation and the limit of detection must have a meaningful basis. For example, the limit of detection could be set at three times the electronic background signal of the measuring device and the limit of quantitation at three times the limit of detection.

ANALYTICAL VARIABILITY

All methods are subject to variability due to the errors occurring at each stage in a complex series of operations that comprise the analytical method. A value 'a', representing the true value of the concentration measured, is qualified by a value '±b', the measurement uncertainty, that is the range within which the true value is estimated to fall with a given probability. The value of 'a' depends on the accuracy of the method used while 'b' is dependent on the inherent precision of the analytical method used and the number of replicates. The effect of variability must be considered when applying an enforcement limit so that the value 'b' should be subtracted from the analytical result obtained before deciding whether action is needed. If the value obtained still exceeds the limit, it indicates beyond reasonable doubt that the limit has been exceeded. The reader may like to consult a paper in which the concept of uncertainty has been discussed (Horwitz, 2003).

To assess whether the precision characteristics of a method are acceptable the values obtained can be compared with those calculated on an empirical basis (Horwitz, 1982). This worker devised a predictive equation that gives the likely variability over a large range of analyte concentrations, by examination of results from a very large number of inter-laboratory studies. The values for the RSD_R for concentrations 1 g/kg to 1 μg/kg are given in Table 3 together with an indication of values that fall within the acceptable range. This shows that variability increases as concentrations decrease and that these values are somewhat greater than for repeatability RSD_r (not given). Careful analysis should produce results well within these ranges.

Table 3. Relative standard deviation for reproducibility (RSD_R) as predicted by the Horwitz equation.

Concentration		*RSD_R*	*Range of acceptable concentrations, μg/kg*	
10^{-3}	1 g/kg	5.6	944-1056	($x10^3$)
10^{-4}	100 mg/kg	8	92-108	($x10^3$)
10^{-5}	10 mg/kg	11	8.9-11	($x10^3$)
10^{-6}	1 mg/kg	16	840-1160	
10^{-7}	100 μg/kg	22	78-122	
10^{-8}	10 μg/kg	32	6.8-13.2	
10^{-9}	1 μg/kg	44	0.56-1.44	

Horwitz, 1982

PROFICIENCY TESTING AND INHERENT VARIABILITY IN ANALYTICAL RESULTS

Results obtained should be normally distributed when a sample is analysed several times or by different operators. That is to say the majority of results will be centred on a mean value, but inevitably some results will lie at the extremes of the distribution. Testing schemes were discussed briefly earlier and proficiency testing within the European Union National Reference laboratories e.g. for milk has been discussed (Dragacci *et al.*, 2001).

One proficiency scheme is FAPAS® that operates on a concept that the statistics of a normal distribution imply that about 95% of data points can be considered acceptable and fall within a 'z-score' of –2 and +2 as defined within the proficiency scheme. To allow computation of these values, a consensus mean value is calculated from the results of all participants, after correction for recovery. At the end of the test, the analyst is informed of his/her performance score and is able to seek advice if problems have been encountered. However, because of the inherent variability the participant should not consider a single result but the record over successive tests.

Table 4 presents a section from the results (for participating laboratories numbers 46-63, results given only for aflatoxin B_1) from a FAPAS® exercise for aflatoxins B_1, B_2, G_1 and G_2 in maize for which 73 laboratories requested samples (Report No. 0453, May-June 2003). In this scheme there is no restriction on which laboratories may take part or the method that is used. There is no guarantee that all participants regularly carry out mycotoxin analysis although it is likely that special care will be taken by participants. With these reservations, the results obtained by most laboratories fall within the acceptable range (e.g. those values in Table 4 that lie within the range 3.8 to 9.4 μg/kg for the assigned consensus mean value of 6.56 μg/kg). These results are representative of the rest of the data set not shown and are typical of results from other FAPAS® mycotoxin rounds.

Table 4. Extract of results for aflatoxin B_1 taken from all laboratories (001-073) participating in a proficiency scheme exercise for aflatoxins in maize[1].

Lab. No.	*Aflatoxin B_1, assigned value is 6.56 μg/kg* Result (μg/kg)	Recovery (%)	z-score
046	7.2	91.0	0.4
047	**10.7**	61.2	**2.9***
048	6.68	76	0.1
049	4.14	Not stated	-1.7
050	5.5	98	-0.7
051	6	99	-0.4
052	6	99	-0.4
053	7.5	90	0.6
054	6.28	86.5	-0.2
055	**9.59**	70	**2.1***
056	5.57	100.9	-0.7
057	7.74	44.5	0.8
058	5.79	77.1	-0.5
059	5.6	Not stated	-0.7
060	4.78	89.4	-1.2
061	6.0	84.0	-0.4
062	9.4	79.0	2.0
063	6.68	90	0.1

*z-score considered unacceptable
[1] (Series 4, Round 53, 2003: data published with permission of the FAPAS® proficiency testing scheme directorate, FAPAS®, CSL, York, UK)

The customer commissioning contract analyses must be aware that a spread of values is to be expected even from competent analysts working in accredited laboratories. Therefore caution is needed before action is taken when working to an agreed limit (here for illustration purpose set at 6.56). In this example action would not be legally defendable, using the Z score of ± 2 criterion, unless a value of at least 9.4 had been obtained.

In circumstances where resources and equipment is limited, less stringent criteria may be acceptable. This is illustrated by a collaborative study of a simplified method for determination of aflatoxin M_1 in milk and milk powder using immunoaffinity column cleanup and TLC in which 14 laboratories from developing countries took place. Recoveries obtained were in the range 32-120%; repeatability was 27-48% and reproducibility 35-54%. Problems were encountered with matrix sample reconstitution and extract evaporation. Despite this, the method was considered suitable for publication as standard ISO/DIS 14674-IDF 190.

Overall variability in mycotoxin determination is a combination of the sampling and the analytical method and these factors have been studied for a range of mycotoxins and commodities (Whitaker *et al.*, 1991; 2000, and in a series of other papers by these authors).

STANDARDS

Reliable standards are crucial to ensure good quality control. They are used at different stages in the analysis so that calibrated solutions are used for spiking to establish recovery while inclusion of in-house standard material checks that the method is giving the expected quantitative result. Calibrated solutions are also used for quantification and for checking retention times. Certified reference standards are used for occasional checks on the overall performance of the method and for the calibration of other in-house standards.

Availability of standards

Standards of most of the common mycotoxins are available commercially but may be impure or contain related compounds, especially when originally prepared from fungal cultures. A figure for the purity is not always given. Accurate concentrations must thus be established in the laboratory by measurement of the UV absorbance of the solvents containing the mycotoxins using published literature values (Egan, 1982). Standard materials are relatively expensive but only small quantities are used. Standards may not be available for some less studied mycotoxins, so for research purposes toxins may need to be isolated from fungal culture and purified in-house. Calibrated dilute solutions of mycotoxins are offered by some commercial companies and provide an alternative to handling hazardous solid material.

In-house reference material

In-house reference materials are invaluable for routine quality control of analytical methods and should be prepared as far as possible from naturally contaminated samples. If these are unobtainable they can be prepared by spiking the appropriate food material with a solution of the mycotoxin. Mycotoxins in commodities and in suitable solvents are stable if stored frozen at –20°C; therefore a batch of material can be prepared and used for a number of years. For example, the stability of fumonisins in methanol or acetonitrile/water at temperatures from –18 to +40°C was studied and it was found that, while considerable breakdown occurred in methanol at the higher temperatures, solutions in acetonitrile and water were completely stable over a 6 month period between –18 and +25°C (Visconti *et al.*, 1993).

To prepare an in-house reference material, several kilograms of the target sample are ground or chopped finely and then blended thoroughly. The homogeneity and concentration is best checked by drawing 10 samples of about 50 g at random and analysing, ideally establishing the concentration by reference to a certified material or solution. If the standard deviation for the sample set is unacceptable, this process is repeated. The remaining material is separated for convenience into 25-50 g lots, sealed in aluminium foil sachets under nitrogen and stored at –20°C until required. A sachet taken at random should be checked occasionally against a certified reference material as an on-going quality control check. The acquisition of naturally contaminated material such as aflatoxins in groundnuts or fumonisins in maize should cause no problem because of their widespread occurrence but it may be more difficult to find sufficient quantities of mycotoxin-contaminated samples containing some of the other mycotoxins.

Certified reference materials and standards

Certified reference materials (CRM) provide primary references for quality control in the chemical laboratory. The care and effort needed to produce these materials means that they are expensive and unavailable in large quantities and should only be used as primary references.

The success and preparation of one of the first certified reference materials for deoxynivalenol in wheat and maize flour was assessed (Gilbert *et al.*, 1992). The stages used by the BCR (Bureau Communautaire de Reference) in developing and testing such material are described in detail (Josephs *et al.*, 2004) and require typically the participation of 10-15 expert laboratories. A feasibility study may be required once the clear need for a CRM is identified and, if positive, the candidate material will be produced. This will then be tested by the chosen laboratories for homogeneity, short and long term stability, value assignment and, after peer review, will be packaged and dispatched.

Many problems must be solved in the preparation of certified mycotoxin standards (Krska *et al.* (2003a). These workers showed that an important problem in developing a certified standard for zearalenone proved to be the purity of the crystalline material, as a result of this being originally isolated from fungal cultures (Krska *et al.*, 2003b).

Examples of CRMs produced by BCR are aflatoxins in peanut meal, aflatoxin M_1 in milk powder, deoxynivalenol and ochratoxin A in wheat flour, deoxynivalenol and zearalenone in maize flour and a solution of zearalenone in acetonitrile. A source of certified reference materials is the IRMM (Institute for Reference Materials and Measurements) of the European Communities Joint Research Council.

Recent and future developments

Introduction of demanding statutory limits for mycotoxins, particularly relating to food for babies and children, is likely to continue and the need remains for ever more sensitive and reliable methods. The need to enforce these regulations requires the development of easier and cheaper methods that can provide the appropriate tests at acceptable cost. The validation and development of these methods is an on-going requirement.

In the future, attention may turn to further mycotoxins and improved methods for these will then be required. A few examples of toxins of possible future concern, with methods cited in the literature, are beauvericin, enniatins and fusaproliferin (Monti *et al.*, 2000), tenuazonic acid and cyclopiazonic acid (Damotta and Soares, 2000), mycophenolic acid, griseofulvin, roquefortine C, chaetoglobosin B, verruculogen and penitrem A, (Runderget and Wilkins, 2002) and xanthomegnin, viomellein and vioxanthin (Scudamore *et al.*, 1986). The xerophilic mould *Wallemia sebi* appears to occur commonly on a number of food commodities including cereals (personal research) and has been shown to produce some highly toxic metabolites, 'the walleminols' (Wood *et al.*, 1990).

The development of methods that can incorporate groups of mycotoxins rather than all embracing multi-methods is likely to gather impetus. Universal multi-mycotoxin analysis appears unlikely in the near future. Most steps in the analytical procedure are readily automated and the use of automation and robotic systems is the logical solution for high throughput laboratories.

Efficient cleanup of mycotoxin extracts is essential to use the full potential of instrumental detection systems. Further development of single or mixed antibody immunoaffinity columns is expected and alternative approaches such as binding antibodies to magnetic beads could provide an alternative means of removing interfering substances. Molecular imprint technology is expected to develop further so that templates manufactured to recognise specific chemical structures could provide an alternative technique (Jodlbauer *et al.*, 2002) probably with reusable plates.

The availability of 'dip stick' tests would be ideal for liquid samples such as mycotoxins in beer, patulin in apple juice, aflatoxin M_1 in milk or ochratoxin A in coffee. The development of lateral flow tests for mycotoxins, based on simple devices already widely used in medical diagnostics, is in its infancy but the simplicity of the method is attractive.

Capillary electrophoresis with laser-induced fluorescence has been examined for the determination of ochratoxin A in corn, sorghum and roasted coffee with similar sensitivity to HPLC methods (Corneli and Maragos, 1998). The authors cite the advantages of excellent separation of ochratoxin A from interferences, use of very small sample volumes, the absence of organic solvent during the determinative step and the use of less expensive capillaries. This appears to offer an alternative technique to current methods but remains to be evaluated and developed.

Further reading and directory of selected methods

The following review articles and selected methods provide further comprehensive information on the art of mycotoxin analysis.

REVIEW ARTICLES

Betina, V. 1985. Thin-layer chromatography of mycotoxins. Journal of Chromatography 334:211-276.

Krska, R., S. Baumgartner, and R.D. Josephs, 2001. The state of the art in the analysis of type A and type B trichothecene mycotoxins in cereals. Fresenius Journal of Analytical Chemistry 371:285-299.

Krska, R. and R.D. Josephs. 2001. The state of the art in the analysis of estrogenic mycotoxins in cereals. Fresenius Journal of Analytical Chemistry 369:469-476

Langseth, W. and T. Rundberget. 1998. Instrumental methods for determination of nonmacrocyclic trichothecenes in cereals, foodstuffs and cultures. Journal of Chromatography A, 815:103-122.

Lin, L., J. Zhang, P. Wang, Y. Wang, and J. Chen 1998. Thin layer chromatography of mycotoxins and comparison with other chromatographic methods (Review). Journal of Chromatography A 815:3-20.

Scott, P.M. 2001. Analysis of agricultural commodities and foods for *Alternaria* mycotoxins. Journal of the Association of Official Analytical Chemists International 84:1809-1817.

Shephard, G.S. 1998. Chromatographic determination of the fumonisin mycotoxins (Review). Journal of Chromatography A 815:31-40.

Trucksess, M.W. 2003. Mycotoxins. Committee on natural toxins and food allergens. General Referee Reports. Journal of the Association of Official Analytical Chemists International 86:129-138 (Report published yearly).

Valenta, H. 1998. Chromatographic methods for the determination of ochratoxin A in animal tissues and fluids (Review). Journal of Chromatography A, 815:75-92.

Wilson, D.M., E.W. Sydenham, G.A. Lombaert, M.W. Trusksess, D. Abramson, and G.A. Bennett 1998. Mycotoxin analytical techniques. In: *Mycotoxins in Agriculture and Food Safety*. (K.K. Sinha, K.K. and D. Bhatnagar, eds). Marcel Dekker Inc, New York, pp. 135-182.

SPECIFIC METHODS

The inclusion or omission of any method by the author in the following list neither endorses those included nor dismisses those omitted. Indeed examination of successive reports from the FAPAS® assessment scheme shows that participants use a wide range of methods, many of which are 'in-house' modifications of earlier published methods. The primary requirement is for the analyst to obtain consistent results with his chosen method and for it to meet agreed performance characteristics. In selecting methods here, the emphasis has been put on recently published work, particularly where the methods have been the subject of collaborative testing, that enables the reader to judge their performance, both in strengths and weaknesses. These publications provide references to earlier or related work from which the new method has often been derived. Method development is a continuous process so that these methods will undoubtedly be developed further and modified to suit an ever-increasing range of foods or feeds. Advances in mycotoxin methodology are reviewed yearly by the General Referee for mycotoxins of AOAC International and provide an excellent overview of recent developments (e.g. Trucksess, 2003). The Report of the CAST Task Report on mycotoxins (CAST, 2003) cites that the following organisations have programs that evaluate and approve analytical methods for mycotoxins: the AOAC International, the International Union for Pure and Applied Chemistry (IUPAC), the American Association of Cereal Chemists (AACC), and the American Oil Chemists Society (AOCS). In addition, the 17th edition of the Methods of Analysis of the AOAC International (Horwitz, 2000) provides detailed descriptions of well-established methods for mycotoxin analysis.

The proscribing of a validated official method in a regulatory procedure is a useful guide for analysts and as a reference method but has the disadvantage of being difficult and slow to update as methods improve and develop. The approach now commonly adopted and accepted by customers is to agree on the performance characteristics required of a method and then to allow free choice of any method that consistently achieves the required performance values. This allows good new methods to be adopted and used quickly.

Aflatoxins

Dragacci, S., F. Grosso, and J. Gilbert. 2001. Immunoaffinity column cleanup with liquid chromatography for determination of aflatoxin M_1 in liquid milk: Collaborative study. *Journal of the* Association of Official Analytical Chemists International 84:437-443.

Grosso, F., J.M. Fremy, S. Bevis, and S. Dragacci. 2004. Joint IDF-IUPAC-IAEO (FAO) interlaboratory validation for determining M_1 in milk using immunoaffinity cleanup before thin layer chromatography. Food Additives and Contaminants 21:348-357.

Sharman, M. and J. Gilbert. 1991. Automated aflatoxin analysis of foods and animal feeds using immunoaffinity column clean-up and high-performance liquid chromatographic determination. Journal of Chromatography 543:220-225.

Stroka, J., E. Anklam, U. Jorissen, and J. Gilbert. 2000. Immunoaffinity column cleanup with liquid chromatography using post-column bromination for determination of aflatoxins in peanut butter, pistachio paste, fig paste, and paprika powder: collaborative study. Journal of the Association of Official Analytical Chemists International 83:320-40.

Stroka, J., E. Anklam, U. Jorissen and J. Gilbert. 2001. Determination of aflatoxin B_1 in baby food (infant formula) by immunoaffinity column cleanup liquid chromatography with post column bromination: collaborative study. Journal of the Association of Official Analytical Chemists International 84: 1116-23.

Trucksess, M.W., M.E. Stack, S. Nesheim, S.W. Page, R.H. Albert, and T.R. Romer. 1994. Multifunctional column coupled with liquid chromatography for determination of aflatoxins B_1, B_2, G_1 and G_2 in corn, almonds, Brazil nuts, peanuts, and pistachio nuts: Collaborative study. Journal of the Association of Official Analytical Chemists 77:1512-1521.

Cyclopiazonic acid

Zambonin, G.C., L. Monaci, and A. Aresta. 2001. Determination of cyclopiazonic acid in cheese samples using solid-phase microextraction and high performance liquid chromatography. Food Chemistry 75:249-254.

Deoxynivalenol and other trichothecenes

Josephs, R.D., R. Schuhmacher, and R. Krska. 2001. International interlaboratory study for the determination of the *Fusarium* mycotoxins zearalenone and deoxynivalenol in agricultural commodities. Food Additives and Contaminants 18:417-430.

Mateo, J.J., A. Llorens, R. Mateo, and M. Jimenez. 2001. Critical study of and improvements in chromatographic methods for the analysis of type B trichothecenes. Journal of Chromatography A 918:99-112.

Pettersson, H. and W. Langseth. 2002. Intercomparison of trichothecene analysis and feasibility to produce certified calibrants. Method studies. BCR information, EU Report EUR 20285/1 EN 82 pages.

Fumonisins

Bird, C.B., B. Malone, L.G. Rice, P.F. Ross, R. Eppley, and M.M. Abouzied. 2002. Determination of total fumonisins in corn by competitive direct enzyme-linked immunosorbent assay. Journal of the Association of Official Analytical Chemists International 85:404-410.

Scott, P.M. and G.A. Lawrence. 1994. Stability and problems in recovery of fumonisins added to corn-based foods. Journal of the Association of Official Analytical Chemists International 77:541-545.

Solfrizzo, M., A. De Girolamo, A. Visconti, E. Sizoo, and H.P. Van Egmond. 2000. Determination of fumonisin B_1 and B_2 in corn-based foodstuffs by HPLC with immunoaffinity cleanup: in house and interlaboratory validation, *X International IUPAC Symposium on Mycotoxins and Phycotoxins*, Guarujà (Brazil). pp. 87-96.

Sydenham, E.W., G.S. Shephard, P.G. Thiel, S. Stockenstrom, W. Snijman, and D.J. Van Schalkwik. 1996. Liquid chromatographic determination of fumonisins B_1, B_2, and B_3 in corn: AOAC –IUPAC collaborative study. Journal of the Association of Official Analytical Chemists International 79:688-696.

Visconti, A., M. Solfrizzo and A. De Girolamo. 2001b. Determination of fumonisins B_1 and B_2 in corn and corn flakes by liquid chromatography with immunoaffinity colunn clean-up: Collaborative study. Journal of the Association of Official Analytical Chemists International 84:1828-1837.

Ochratoxin A

Burdaspal, P., T.M. Legarda, and J. Gilbert. 2001. Determination of ochratoxin A in baby food by immunoaffinity column cleanup with liquid chromatography: Interlaboratory study. Journal of the Association of Official Analytical Chemists International 84:1445-1452.

Entwisle, A.C., A.C. Williams, P.J. Mann, J. Russell, P.T. Slack, and J. Gilbert. 2001. Combined phenyl silane and immunoaffinity column cleanup with liquid chromatography for determination of ochratoxin A in roasted coffee. Collaborative study. Journal of the Association of Official Analytical Chemists 84:444-450.

Entwisle, A.C., A.C. Williams, P.J. Mann, P.T. Slack, and J. Gilbert. 2000. Liquid chromatographic method with immuno-affinity column cleanup for determination of ochratoxin A in barley: collaborative study. Journal of the Association of Official Analytical Chemists 83:1377-1383.

Larsson, K. and T. Möller. 1996. Liquid chromatographic determination of ochratoxin A in barley, wheat bran and rye by the AOAC/IUPAC/NMKL method: NMKL collaborative study. Journal of the Association of Official Analytical Chemists 79:1102-1105.

Leitner, A., P. Zollner, A. Paolillo, J. Stroka, A. Papadopoulou-Bouraoui, S. Jaborek, E. Anklem, and W. Lindner. 2002. Comparison of methods for the determination of ochratoxin A in wine. Analytica Chimica Acta 453:33-41.

Scudamore, K.A. and S.J. Macdonald. 1998. A collaborative study of an HPLC method for the determination of ochratoxin A in wheat using immunoaffinity column clean-up. Food Additives and Contaminants 15:401-410.

Visconti, A., M. Pascale, and G. Centonze. 2001a. Determination of ochratoxin A in wine and beer by immunoaffinity column clean up and liquid chromatographic analysis with fluorometric detection: Collaborative study. Journal of the Association of Official Analytical Chemists International 84:1818-1827.

Patulin

MacDonald, S., M. Long, and J. Gilbert. 2000. Liquid chromatographic method for the determination of patulin in clear and cloudy apple juices and apple puree: collaborative study. Journal of the Association of Official Analytical Chemists, 86:1387-1394.

Sewram, V., J.J. Nair, T.W. Nieuwoudt, N.L. Leggott, and G.S. Shephard. 2002. Determination of patulin in apple juice by high performance liquid chromatography-atmospheric pressure chemical ionization mass spectrometry. Journal of Chromatography A 897:365-374.

Sterigmatocystin

Scudamore, K.A., M.T. Hetmanski, P.T. Clarke, K.A. Barnes, and J.R. Startin. 1996. Analytical methods for the determination of sterigmatocystin in cheese, bread and corn products using HPLC with atmospheric pressure ionisation mass spectrometric detection. Food Additives and Contaminants 13:343-358.

Zearalenone

Kruger, S.C., B. Kohn, C.S. Ramsey, and R. Prioli. 1999. Rapid immunoaffinity-based method for determination of zearalenone in corn by fluorometry and liquid chraomatography. Journal of the Association of Official Analytical Chemists International 82:1364-1368.

Ostry, V. and J. Skarkova. 2003. A HPTLC method for the determination of the mycotoxin zearalenone in cereal products. Mycotoxin Research 19:64-68.

Visconti, A. and M. Pascale. 1998. Determination of zearalenone in corn by means of immunoaffinity clean up and high performance liquid chromatography with fluorescence detection. Journal of Chromatography A 815:133-140.

General references

Akiyama, H., Y. Goda, T. Tanaka and M. Toyoda. 2001. Determination of aflatoxins B_1, B_2, G_1 and G_2 in spices using a multifunctional column cleanup. J. Chrom. A 932:153-157.

Analytical Methods Committee of the Royal Society of Chemistry. 1995. Internal Quality Control of Analytical Data. Analyst 120:29-34.

Bresch, H., M. Urbanek and K. Hell. 2000. Ochratoxin A in coffee, tea and beer. Archiv fur Lebemsmittelhygiene 51:81-128.

Burdaspal, P. 1997. Determination of Ochratoxin A in Baby Food, CEN/TC 275/WG5, N 219. Food Add. Contam. 14:237-248.

CAST. 2003. Mycotoxins: Risks in plant, animal and human systems. Task Force Report, Council for Agricultural Science and Technology, Ames, Iowa, USA, No. 139, January 2003.

Chiavaro, E., C. Dallasta, G. Galaverna, A. Biancardi, E. Gamberelli, A. Dossena and R. Marchelli. 2001. New reverse phase liquid chromatographic method to detect aflatoxins in food and feed with cyclodextrins as fluoscence enhancers added to the effluent. J. Chromat. A 937:31-40.

Chu, F.S. 1984. Immunoassays for analysis of mycotoxins. J. Food Prot. 47:562-9.

Coker, D. 1998. Design of sampling plans for the determination of mycotoxins in foods and feeds. In: *Mycotoxins in Agriculture and Food Safety* (K.K. Sinha and D. Bhatnagar, eds). New York: Marcel Dekker Inc., pp. 109-133.

Cole, R.J. and J.W. Dorner. 1994. Extractions of aflatoxins from naturally contaminated peanuts with different solvents and solvent/peanut ratios. J. Assoc. Offic. Anal. Chem. Intl. 77:1509-1511.

Corneli, S. and C.M. Maragos. 1998. Capillary electrophoresis with laser induced fluorescence: method for the determination of ochratoxin A. J. Agric. Food Chem. 46:3162-3165.

Damotta, S. and L.M.V. Soares. 2000. Simultaneous determination of tenuazonic acid and cyclopiazonic acid in tomato products. Food Chem. 71:111-116.

Danks, C., S. Ostoja-Starzewska, J. Flint and J.N. Banks. 2003. The development of a lateral flow device for the discrimination of OTA producing and non-producing fungi. In: *Mycotoxins in Food Production Systems*. Aspects Appl. Bio. 68:21-28.

De Girolamo, A., M. Solfrizzo and A. Visconti. 2001. Effect of processing on fumonisin concentration in corn flakes. J. Food Prot. 64:701-705.

Desaeger, S., L. Sibanda and C. VanPeteghem. 2002. A collaborative study to validate novel field immunoassay kits for rapid mycotoxin detection. Intl. J. Food Microb. 75:135-142.

Dilkin, P., C.A. Mallmann, C.A.A. Dealmeida and B. Correa. 2001. Robotic automated clean up for detection of fumonisins B_1 and B_2 in corn and corn-based feed by high performance liquid chromatography. J. Chrom. 625:151-157.

Dragacci, S., F. Grosso, N. Pfauwathel-Marchmond, J.M. Fremy, A. Venant and B. Lombard. 2001. Proficiency testing for the evaluation of the ability of the European Union National Reference laboratories to determine aflatoxin M_1 in milk at levels corresponding to the new European Union legislation. Food Add. Contam. 18:405-415.

Dunne, C., M. Meaney, M. Smith and L. Tuinstra. 1993. Multimycotoxin detection and cleanup method for aflatoxins, ochratoxin and zearalenone in animal feed ingredients using high performance liquid chromatography and gel permeation chromatography. J. Chrom. 629:229-235.

Egan, H. 1982. Preparation and testing of mycotoxin standard reference materials, Chapter 4 (adapted in major part from official methods of analysis of the AOAC, Chaper 26, 1980). In: *Environmental Carcinogens Selected Methods of Analysis, Volume 5, Some Mycotoxins* (L. Stoloff, M. Casregnaro, P. Scott, I.K. O'Neill and H. Bartsch, eds). Lyon, IARC Scientific Publications, pp. 63-84.

Eskola, M. 2002. Study on trichothecenes, zearalenone and ochratoxin A in Finnish cereals: Occurrence and analytical techniques. Academic thesis, ISSN 1458-6878, EELA publications, Helsinki.

Eskola, M., G. Boonzaaijer, W.A. van Osenbruggen, A. Rizo and G. Tijmensen. 2000. A study of the suitability of gas chromatographic-electron capture detection for the analysis of deoxynivenol in cereals. Mycotoxin Res. 16:103-120.

European Commission. 1998. Commission Directive (98/53/EC) of 16 July. Off. J. Europ. Comm. L201:93-101.

European Commission. 2002a. Commission Directive (2002/27/EC) of 13 March 2002 amending Commission Directive (98/53/EC). Off. J. Europ. Comm. L75:44-54.

European Commission. 2002b. Assessment of dietary intake of ochratoxin A. EUR Report 17523, Reports on tasks for Scientific Cooperation Office of Official Publications of the EC, L-2985, Luxembourg.

European Commission. 2002c. Assessment of dietary intake of patulin by the population of EU Member States. Report on tasks for Scientific Cooperation. Report of experts participating in task 3.2.8, March 2002.

European Commission. 2003. Collection of occurrence data of *Fusarium* toxins in food and assessment of dietary intake by the population of EU Member States. Report on tasks for Scientific Cooperation. Report of experts participating in task 3.2.10, April 2002.

FAO. 1997. Report of the Twenty-first Session of the Codex Committee on Methods of Analysis and Sampling. ALINORM 97/23A, FAO, Rome.

Fuchs, E., B. Rabus, J. Handl and E-M. Binder. 2003. Type A-trichothecenes-Quantitative ananlysis using LC-MS and occurrence in Austrian maize and oats. Mycotoxin Res. 19:56-59.

Gilbert, J. and E. Anklam. 2002. Validation of analytical methods for determining mycotoxins in foodstuffs. Trends Anal. Chem. 21:468-486.

Gilbert, J., A. Boenke and P.J. Wagstaffe. 1992. Deoxynivalenol in wheat and maize flour reference materials. 1. An intercomparison of methods. Food Add. Contam. 9:71-81.

Hansen, T.J. 1990. Affinity column cleanup and direct fluorescence measurement of aflatoxin M_1 in raw milk. J. Food Prot. 53:75-77.

Hetmanski, M.T. and K.A. Scudamore. 1991. A novel method for detection of zearalenone in cereal extracts using HPLC with post-column derivatisation. J. Chrom. 588:47-52.

Horwitz, W. 1982. Evaluation of analytical methods for regulation of food and drugs. Anal. Chem. 54:67A-78A.

Horwitz, W. 2003. The certainty of uncertainty. J. Assoc. Offic. Anal. Chem. Intl. 86:109-111.

Horwitz, W. (ed). 2000. Official Methods of Analysis of AOAC International. 17th Ed. Assoc. Offic. Anal. Chem.

Jodlbauer, J., M.N. Maier and W. Lindner. 2002. Towards ochratoxin A selective molecularly imprinted polymers for solid-phase extraction. J. Chrom. A 945:45-63.

Josephs, R.D., R. Koeber, A. Bernreuther, T.P.F. Linsinger and H. Schimmel. 2004. Deveopment of certified reference materials for mycotoxins. In: *Meeting the Mycotoxin Menace*. (D. Barug, H.P. van Egmond, R. Lopez-Garcia, W.A. van Osenbruggen and A. Visconti, eds). Wageningen Academic Publishers, Wageningen, The Netherlands, pp. 237-254.

Josephs, R.D., R. Krska, M. Grasserbauer and J.A.C. Broekaert. 1998. Determination of trichothecene mycotoxins in wheat by use of supercritical fluid extraction and high-performance liquid chromatography with diode array detection or gas chromatography with electron capture detection. J. Chrom. A 957:297-304.

Josephs, R.D., R. Schuhmacher and R. Krska. 2001. International interlaboratory study for the determination of the *Fusarium* mycotoxins zearalenone and deoxynivalenol in agricultural commodities. Food Add. Contam. 18:417-430.

Katay, G., A. Szecsi and E. Tyihak. 2001. Separation of fumonisins by OPLC. J. Planar Chromatography-Modern TLC 14:53-56.

Kok, W., U.A. Brinkman and R.F. Frei. 1984. On-line electrochemical reagent production for detection in liquid chromatography and continuous flow systems. Anal. Chim. Acta 162:1984.

Krska, R., R.D. Josephs, S. MacDonald and H. Pettersson. 2003b. Development of certified calibrants for the analysis of zearalenone and B-trichothecenes. Mycotoxin Res. 19:60-63.

Krska, R., H. Pettersson, R.D. Josephs, M. Lemmens, S. MacDonald and E. Welzig. 2003a. Zearalenone in maize: stability testing and matrix characterisation of a certified reference material. Food Add. Contam. 20:1141-1152.

Lawrence, J.F., B. Niedzwiadek and P.M. Scott. 2000. Effect of temperature and solvent composition on extraction of fumonisins B_1 and B_2 from corn products. J. Assoc. Offic. Anal. Chem. Intl. 83:604-611.

Losito, I., L. Monaci, A. Aresta and C.G. Zambonin. 2002. LC-ion trap electrospray MS-MS for the determination of cyclopiazonic acid in milk samples. Analyst 127:499-502.

Maragos, C.M. and R.D. Plattner. 2002. Rapid fluorescence polarization immunoassay for the mycotoxin deoxynivalenol in wheat. J. Agric. Food Chem. 50:1827-1832.

Martins, M.J., H.M. Martins and A. Gimeno. 2003. Incidence of mycoflora and of ochratoxin A in green coffee beans (*Coffea arabica*). Food Add. Contam. 20:1127-1131.

Monti, S.M., V. Fogliano, A. Logrieco, R.

Ferracane and A. Ritieni. 2000. Simultaneous determination of beauvericin, enniatins and fusproliferin by high performance liquid chromatograhy. J. Agric. Food Chem. 48:3317-3320.

Nasir, M.S. and M.E. Jolley. 2002. Development of a fluorescence polarization assay for the determination of aflatoxins in grain. J. Agric. Food Chem. 50:3116-3121.

Odhav, B. and V. Naicker. 2002. Mycotoxins in South African traditionally brewed beers. Food Add. Contam. 19:55-61.

Pallaroni, L., E. Bjorklund and C. Vonholst. 2002. Optimisation of atmospheric pressure chemical ionization parameters for the simultaneous determination of deoxynivalenol and zearalenone using HPLC/MS. J. Liq. Chrom. Rel. Tech. 25:913-926.

Patterson, D.S.P. and B.A. Roberts. 1979. Mycotoxins in animal feedstuffs: sensitive thin layer chromatographic detection of aflatoxin, ochratoxin A, sterigmatocystin, zearalenone and T-2 toxin. J. Assoc. Offic. Anal. Chem. 62:1268-1273.

Pearson, T.C., D.T. Wicklow, E.B. Maghirang, F. Xie and F.E. Dowell. 2001. Detecting aflatoxin in single corn kernels by transmittance and reflectance spectrometry. Transactions of the ASAE 44:1247-1254.

Pierard, J-Y., C. Depasse, A. Delafortrie and J-C. Motte. 2004. Multi-mycotoxin determination methodology. In: *Mycotoxins in Agriculture and Food Safety* (K.K. Sinha and D. Bhatnagar, eds). Marcel Dekker Inc., New York, pp. 255-268.

Preis, R.A. and E.A. Vargas. 2000. A method for determining, fumonisin B_1 in corn using immunoaffinity column cleanup and thin layer chromatography/densitrometry. Food Add. Contam. 17:463-468.

Radova, Z., J. Hajslova, J. Kralova, L. Papouskova and S. Sykorova, S. 2001. Analysis of zearalenone in wheat using high performance liquid chromatography with fluorescence detection and/or enzyme-linked immunoassay. Cereal Res. Commun. 29:435-442.

Romer Labs. 2001. Sampling and aample preparation for mycotoxin analysis. Romer Labs, Inc., Union, Missouri.

Runderget, T. and A. Wilkins. 2002. Determination of *Penicillium* mycotoxins in foods and feeds using liquid chromatography - mass spectrometry. J. Chrom. A:189-197.

Schothorst, R.C. and A.A. Jekel. 2001. Determination of trichothecenes in wheat by capillary gas chromatography with flame ionisation detection. Food Chem. 73:111-117.

Scott, P.M. and G.A. Lawrence. 1994. Stability and problems in recovery of fumonisins added to corn-based foods. J. Assoc. Offic. Anal. Chem. Intl. 77:541-545.

Scott, P.M. and G.A. Lawrence. 1995. Analysis of beer for fumonisins. J. Food Prot. 58:1379-1382.

Scott, P.M. and G.A. Lawrence. 1997. Determination of aflatoxins in beer. J. Assoc. Offic. Anal. Chem. Intl 80:1229-1234.

Scudamore, K.A. and Hetmanski, M.T. (1992). Analysis of mycotoxins in cereals and animal feed using a multi-toxin gel permeation chromatography clean-up method. Mycotoxin Res. 8:37-47.

Scudamore, K.A., J.H. Clarke and P.M. Atkin. 1986. Natural occurrence of fungal naphthoquinones in cereals and animal feedstuffs. In: *Proceedings of the Biodeterioration Society Meeting on Spoilage and Mycotoxins of Cereals and Animal Feedstuffs*. Intl. Biodeterioration 22(Suppl):71-81.

Scudamore, K.A., M.T. Hetmanski, H.K. Chan and S. Collins. 1997a. Occurrence of mycotoxins in raw ingredients used for animal feedingstuffs in the United Kingdom in 1992. Food Add. Contam. 14:157-173.

Scudamore, K.A., M.T. Hetmanski, S. Nawaz, J. Naylor and S. Rainbird. 1997b. Determination of mycotoxins in pet foods sold for domestic pets and wild birds using linked-column immunoassay clean-up and HPLC. Food Add. Contam. 14:175-186.

Sibanda, L., S. Desaeger, I. Barna-Vetro and C. Van-Peteghem. 2002. Development of a solid phase cleanup and portable rapid flow-through enzyme immunoassay for the detection of ochratoxin A in roasted coffee. J. Agric. Food Chem. 50:6964-6967.

Sobolev, V.S. and J.W. Dorner. 2002. Cleanup procedure for determination of aflatoxins in major agricultural commodities by liquid chromatography. J. Assoc. Offic. Anal. Chem. Intl. 85:642-645.

Stoloff, L. 1982. Analytical methods for aflatoxins- an overview. In: *Environmental Carcinogens Selected Methods of Analysis, Volume 5, Some Mycotoxins* (H. Egan, ed). International Agency for Research on Cancer, Lyon, pp. 33-62.

Stroka, J. and E. Anklam. 1998. EC EUR 18640 EN. Validation of analytical methods to determine the content of aflatoxin, ochratoxin and patulin in foodstuffs of vegetable origin. Luxembourg, Office for Official Publications of the European Communities, BCR information series ISBN 92-828-5085-4.

Stroka, J., M. Petz, U. Joerissen and E. Anklam. 1999. Investigations of various extractants for the analysis of aflatoxin B_1 in different food and feed matrices. Food Add. Contam. 16:331-338.

Taylor, S.L., J.W. King, J.L. Richard and J.I. Greer. 1993. Analytical scale supercritical fluid extraction of aflatoxin B_1 from field-inoculated corn. J. Agric. Food Chem. 41:910-913.

Taylor, S.L., J.W. King, J.I. Greer and J.L. Richard. 1997. Supercritical fluid extraction of aflatoxin M_1 from beef liver. J. Food Prot. 60:698-700.

Trenholm, H.L., R.M. Warner and D.B. Prelusky. 1985. Assessment of extraction procedures in the analysis of naturally contaminated grain products, deoxynivalenol. J. Assoc. Offic. Anal. Chem. 68:645-649.

Trucksess, M.W. 2003. Mycotoxins. Committee on natural toxins and food allergens. General Referee Reports. J. Assoc. Offic. Anal. Chem. Int 86:129-138 (Report published yearly).

Trucksess, M.W. and D.E. Koeltzow. 1995. Evaluation and application of immunochemical methods for mycotoxins in food. In: *Immunoanalysis of Agrochemicals Emerging Technologies* (J.O. Nelson, A.E. Karu and R.B. Wong, eds). American Chemical Society, Washington International, Gaithersburg, Maryland, p. 326.

Visconti, A., M.B. Doko, C. Bottalico, B. Schurer and A. Boenke. 1993. The stability of fumonisins (fumonisin B_1 and fumonisin B_2) in solution. In: *United Kingdom Workshop on Occurrence and Significance of Mycotoxins* (K.A. Scudamore, ed). Central Science Laboratory, MAFF: Slough, UK, 21-23 April 1993, pp. 200-202.

Vonholst, C., J. Stroka and E. Anklam. 2002. Correction of analytical results for recovery - a comparison of the method performance characteristics from recent collaborative trials studies for aflatoxin quantification using conventional and robust statistics. Food Add. Contam. 19:701-708.

Weiss, R., M. Freudenschuss, R. Krska and B. Mizaikoff. 2003. Improving methods of analysis of mycotoxins: molecularly imprinted polymers for deoxynivalenol and zearalenone. Food Add. Contam. 20:386-395.

Whitaker, T.B., J.W. Dickens and F.G. Giesbrecht. 1991. Testing animal feedstuffs for mycotoxins: sampling, subsampling and analysis. In: *Mycotoxins and Animal Foods* (J.E. Smith and R.S. Henderson, eds). CRC Press Inc., Boca Raton, pp. 153-164.

Whitaker, T.B., W.M. Hagler, F.G. Giesbrecht and A.S. Johansson. 2000. Sampling, sample preparation and analytical variability associated with testing wheat for deoxynivalenol. J. Assoc. Offic. Anal. Chem. Intl. 83:1285-1292.

Wilson, D.M., E.W. Sydenham, G.A. Lombaert, M.W. Trusksess, D. Abramson and G.A. Bennett. 1998. Mycotoxin analytical techniques. In: *Mycotoxins in Agriculture and Food Safety* (K.K. Sinha and D. Bhatnagar. eds). Marcel Dekker Inc, New York, pp. 135-182.

Wood, G., P.J. Mann, D.F. Lewis, W.J. Reid and M. Moss. 1990. Studies on a toxic metabolite from the mould Wallemia. Food Add. Contam. 7:69-77.

Wood, R. 1998. Quality assurance. In: *Natural toxicants in Food* (D.H. Watson, ed). Sheffield Academic Press, Sheffield, UK, pp. 236-264.

8

MYCOTOXINS IN THE HUMAN FOOD CHAIN

Fabio Galvano[1], Alberto Ritieni[2], Gianfranco Piva[3] and Amedeo Pietri[3]
[1]*Department of Agro-forestry and Environmental Science and Technology, Mediterranean University of Reggio Calabria, Italy;* [2]*Department of Food Science, Federico II University, Napoli, Italy;* [3]*Institute of Food Science and Nutrition, Catholic University of the Sacred Heart, Piacenza, Italy*

Introduction

The impact of mycotoxins in the human food chain is an important issue worldwide. In general, consumers perceive less risk from mycotoxins than from other food-related threats such as pesticides, additives, heavy metals and microbial agents. This is due to the fact that, at least in developed countries, mycotoxins very rarely cause acute intoxication outbreaks or health emergencies, the impact of which is often amplified by the media. Nevertheless, the real danger of mycotoxins is potentially very high, which is why they have been called 'hidden killers'. Mycotoxins are highly undesirable substances that should not be present in food and for which a zero tolerance would be ideal. However, even good agricultural, storage and processing practices cannot completely prevent contamination; and it is impossible to achieve a truly mycotoxin-free food chain. Consequently, small quantities of mycotoxins believed not to be dangerous are legally tolerated. Despite incomplete toxicological, epidemiological and exposure data needed to fully assess human exposure and health risks, and establish causal relationships between mycotoxins in foods and human disease, responsible risk management makes it necessary to take action (Kuiper-Goodman, 1995). For these reasons and in order to achieve realistic goals for mycotoxin management, public authorities of many countries have fixed legal limits for mycotoxin presence in foods. Worldwide legal limits have been recently reviewed by van der Westhuizen *et al.* (2002), who recorded that 77 countries have specific regulations for mycotoxins, 13 countries have no specific regulations, and no data are available for about 50 countries, many of them in Africa. However, the matter of the legal limits is questionable. In the recent past the regulatory limits were highly variable depending on the degree of development and economic involvement of the countries. In recent years serious efforts at harmonization have been made worldwide, but further work is still needed, as regulatory limits are often a practical compromise between the need for safe commodities and the economic consequences of the regulatory level chosen.

Suspected human mycotoxicosis

Mycotoxins can enter the human food chain directly by cereals, seeds, spices, fruits, beverages and other plant materials, and indirectly by food products obtained from animals given contaminated feeds through residues in milk, meat, eggs and their derivates. Consumers in developed countries are surely less exposed to mycotoxins than those in developing countries. This is due to several factors, e.g. abundance of food resources, modern food handling and preservation technology, and effective regulation and control of food quality.

However, commodities imported from developing countries can increase the overall risk of exposure if rigorous control policies are not enforced.

Diseases caused by mycotoxins are termed **mycotoxicoses** and can be acute, chronic or sub-chronic. Acute mycotoxicosis is rare, and largely limited to developing countries. In contrast, long-term exposure to low levels of mycotoxins is a devious and widespread problem that scientists are called to face worldwide. Indeed, chronic mycotoxicoses are difficult to recognise, as the epidemiological evidence is not easily demonstrated, involves several epidemiological factors, and there are uncertainties in assessing extent of exposure. Although sufficient evidence of mycotoxins as etiological agents of human diseases has arisen from both laboratory and epidemiological studies, caution is needed when attributing human disease to mycotoxins, as the literature is inconclusive in establishing direct and unequivocal cause-effect relationships. There is even more uncertainty in our understanding of the immunosuppressive properties of mycotoxins. As laboratory and animal studies have demonstrated that mycotoxins are immunosuppressive, they are strongly suspected to interfere with immune processes in humans, thus increasing overall susceptibility to infective agents (Pfohl-Leszkowicz, 1999). However, it is difficult to discern their role, because the causal involvement of the mycotoxin is likely overshadowed by the infectious disease and thus is not considered in the overall syndrome (CAST, 2003). This may explain why data are scarce in the literature on the natural incidence of mycotoxins causing immunosuppression in humans (CAST, 2003). Table 1 provides a list of human diseases suspected to be related to mycotoxin-contaminated foodstuffs.

The first historically recorded case of human mycotoxicosis was **ergotism**, also known as **St. Anthony's Fire,** caused by the ingestion of grains contaminated by sclerotia of *Claviceps purpurea*. After periodic outbreaks the disease became endemic in central Europe during the Middle Ages. Nowadays the risk of ergotism appears to be limited to animals as the last literature reports on human incidence date to the 1970s.

Two recent epidemiological suspicions of

Table 1. Human diseases suspected to be related to mycotoxin-contaminated foods.

Disease	*Food*	*Etiological agent*	*Toxin*
Alimentary toxic aleukia	Cereal grains	*Fusarium* spp.	Trichothecenes
Balkan nephropathy and chronic interstitial nephropathy	Cereal grains	*Penicillium*	Ochratoxin A
Cardiac beriberi	Rice	*Aspergillus* and *Penicillium* spp.	Citreoviridin
Ergotism	Rye, cereal grains	*Claviceps purpurea*	Ergot alkaloids
Oesophageal tumors	Corn	*Fusarium verticillioides*	Fumonisin B_1
Hepatocarcinoma (acute aflatoxicosis)	Cereal grains, peanuts	*Aspergillus flavus A. parasiticus*	Aflatoxin B_1
Human neural tube defects	Corn	*Fusarium verticillioides*	Fumonisin B_1
Kashin Beck disease or 'Urov disease'	Cereal grains	*Fusarium spp.*	Trichothecenes
Kwashiorkor	Cereal grains	*Aspergillus flavus* and *A. parasiticus*	Aflatoxin B_1
Onyalai	Millet	*Phoma sorghina*	Undefined
Reye's syndrome	Cereal grains	*Aspergillus*	Aflatoxin B_1
Testicular cancer	Various	*Penicillium*	Ochratoxin A

mycotoxin involvement in human disease have been reported. One is the involvement of ochratoxin A (OTA) in the etiology of testicular cancer hypothesised by Schwartz (2002). The second is the more recent case of fumonisins and human neural tube defects reported by Marasas *et al.* (2004), who proposed that fumonisins affect folate utilization, thus becoming risk factors for neural tube defects and other birth defects arising from neural crest cells.

Sufficient epidemiological evidence exists for involvement of three mycotoxins in the etiology of certain human diseases with high regional incidence. These are OTA and human nephropathies, aflatoxin B_1 (AFB_1) and hepatocarcinoma, and fumonisins and oesophageal cancer, which are summarised in the following sections.

OCHRATOXIN A AND NEPHROPATHY

Ochratoxin A is suspected in the etiology of human nephropathies and tumors of urinary organs. The persistence of OTA in the human body is prolonged as it has a half-life in blood of 35 days after a single oral dosage, due to unfavorable elimination toxicokinetics (Studer-Rohr *et al.*, 2000). The long half-life of OTA, together with frequent exposure by ingestion of OTA-contaminated foods, result in a high frequency of OTA-positive blood samples collected around the world (Speijers and van Egmond, 1993; WHO, 2001). OTA is neither stored nor deposited in tissues, but numerous laboratory and animal studies have clearly demonstrated that it is distributed via the blood mainly to the kidneys (Hult and Fuchs, 1986).

In several mammalian species, even at low doses, OTA is nephrotoxic. While it is also likely to be nephrotoxic to humans, a cause-effect relationship has never been conclusively proven (WHO, 2001). The usual approach is to analyse OTA in foods and in biological fluids (mainly plasma) of healthy populations and/or restricted groups of patients suffering from nephropathies. Many studies reported possible involvement of OTA in the etiology of a fatal human disease referred to as **Balkan endemic nephropathy** (BEN), which is found in certain regions of Bosnia and Herzegovina, Bulgaria, Croatia, Serbia, and Romania (Petkova-Bocharova *et al.*, 2002) (Figure 1). BEN is characterised by strict geographical limits, familial occurrence of bilateral lesions of the kidney, a high mortality rate from uraemia, and the fact that it affects the agricultural population (Ceovic *et al.*, 1992). Due to the insidious nature of BEN, it has not been possible to establish the onset of the disease. The earliest pathological changes in the kidney are an increase in acellular material throughout the cortex interstitium and degenerative and regenerative changes in the proximal tubule epithelium (Vukelic *et al.*, 1992). In the advanced stage of the disease kidney weight is significantly reduced and a diffuse cortical fibrosis is the most prominent feature. The first clinical signs are mild proteinuria and scarce cellular elements in urine (Radonic and Radoševic, 1992). Blood pressure is usually normal until the terminal uraemic phase; and the first sign of renal lesions is impaired urine concentration, thereafter blood flow and glomerular filtration are impaired. First in Bulgaria, and subsequently in other countries where BEN is present, an unusually high incidence of urinary tract tumors was noticed (Ceovic *et al.*, 1992; Pfohl Leszkowicz *et al.*, 2002). The mechanism involved in the toxicity of OTA indicates three major effects: a) inhibition of mitochondrial respiration correlated with a depletion of ATP, b) inhibition of tRNA-synthetase accompanied by reduced protein synthesis, and c) enhanced lipid peroxidation (Roschenthaler *et al.*, 1984; Marquardt and Frohlich, 1992; Fink-Gremmels *et al.*, 1995). The mechanism of OTA-induced tumor formation is unknown and conflicting results have been obtained regarding the potential of OTA to bind to DNA. Using a ^{32}P-postlabeling method, putative DNA

Figure 1. Primary geographic regions of Balkan endemic nephropathy incidence.

adducts have been detected in kidneys of experimental animals and humans suffering from BEN (Petkova-Bocharova *et al.*, 2003; Arlt *et al.*, 2002), although the chemical structure of adducts is not known. Other authors found that formation of DNA adducts through covalent binding of OTA with DNA does not occur; and concluded it was unlikely to represent a mechanism of OTA-induced renal tumor formation (Mally *et al.*, 2004).

Several epidemiological studies in North African countries including Egypt (Wafa *et al.*,1998), Algeria (Khalef *et al.*,1993), Morocco (Filali *et al.*, 2002) and particularly Tunisia (Eko-Ebongue *et al.*, 1994; Maarou *et al.*, 1995a,b) have been conducted in the attempt to define the role of OTA in a human nephropathy similar to BEN and tentatively called **chronic interstitial nephropathy** (CIN). Two contradictory studies from Tunisian researchers (Abid *et al.*, 2003; Grosso *et al.*, 2003) have recently been published, which demonstrate the uncertainty that exists on the subject. Data collected by Abid *et al.* (2003) over the period 1991-2000 on a large number of volunteers supported involvement of OTA in the Tunisian CIN. To these authors, additional evidence arises from the significant detection of OTA in plasma of people living in conditions similar to those existing in the Balkans and consuming OTA-contaminated foods. In contrast, Grosso *et al.* (2003) observed minimal levels of OTA in sera from volunteers hospitalised for nephropathic damage. Very recently, Hassen *et al.* (2004) found that high blood OTA and ß(2)-microglobulinuria levels seem to be strongly associated with CIN, thus supporting OTA involvement in the outcome of CIN and underlining the importance of ß(2)-microglobulinuria in the characterization of this disease. According to Abouzied *et al.* (2002), OTA alone would not be able to induce BEN, but it would act synergistically with other environmental toxicants. A new approach, based on the duplicate-diet method, was reported in a very recent study of Vrabcheva *et al.* (2004). In an attempt to confirm exposure of a population to OTA in the Vratza district of Bulgaria, the authors monitored the frequency of OTA-contaminated food consumption by selected families in the BEN-afflicted area and determined the intake level during a single month. Results confirmed the extremely high exposure of the population to OTA and the potential involvement of this toxin in the disease.

It can be concluded that despite intensive research at both the epidemiological and experimental/mechanistic level, it is still open to question whether OTA plays a causative or only a subordinate role in the induction of human nephropathy. Problems related to i) detection limits and overall analytical reliability of OTA measurement in biological fluids, and ii) qualitative (e.g. type and severity of nephropathy) and quantitative sampling of the population relative to foods, make study results very difficult to compare. Seasonal variations of OTA in foods and consequently in sera should also be considered, as demonstrated by Peraica *et al.* (2001) in a survey involving a Croatian population. Finally, but equally importantly, it would be necessary to examine a possible synergistic effect of other mycotoxins (e.g. citrinin) simultaneously present in foods, as evidenced, for example, by a study of Vrabcheva *et al.* (2000), who detected levels of citrinin up to 200-fold higher than OTA in cereals from Bulgarian villages with a history of Balkan endemic nephropathy.

AFLATOXIN B_1 AND HEPATOCELLULAR CARCINOMA

Hepatocellular carcinoma (HCC) is a widespread malignancy with the 4th highest mortality rate worldwide. It is estimated to cause about 500,000 deaths annually. Geographical areas of particular concern include Asia, southern China and sub-Saharan Africa; but increasing incidence of HCC has also recently been observed in the US (el Serag and Mason, 1999).

Early epidemiological studies conducted in areas with high HCC incidence basically investigated the possible correlation between dietary exposure to AFB_1 and the occurrence of HCC. The involvement of other factors, such as hepatitis B virus (HBV) infection, were verified only in subsequent studies. Recently, the possibility of a synergistic carcinogenic interaction between HBV chronic infection and dietary exposure to AFB_1 arose from the observation of their co-existence in countries with high incidence of HCC and was confirmed by further experimental and epidemiological studies (Kew, 2003).

Recent studies highlighted that AFB_1 induces epigenetic changes involved in the molecular pathogenesis of HCC, including somatic mutations in the p53 tumor suppressor gene (TP53). Liver P450 enzyme system metabolises AFB_1 leading to the formation of the electrophilic intermediate AFB_1-8,9-epoxide, which can easily bind to guanine in DNA, resulting in typical mutations such as G:C to T:A transversions at the third base in codon 249 of TP53. The detection of this mutation in plasma and tissues was used as a reliable biomarker that facilitated researchers in understanding the role of AFB_1 (Olivier *et al.*, 2004). It was found that this mutation is common in HCC in areas of high exposure to AFB_1, whereas it is absent from HCC in regions with negligible levels of AFB_1 exposure, thus providing evidence of a carcinogenic role for the toxin (IARC, 2002). These data led to the conclusion that chronic hepatitis B or C infections, exposure to dietary AFB_1 and alcoholic cirrhosis are, in order of importance, the major risk factors in the multifactorial etiology of HCC. However, the carcinogenic potency of AFB_1 is considered much lower in populations where chronic hepatitis infections are rare.

FUMONISINS AND OESOPHAGEAL CANCER

Fumonisins occur naturally in corn worldwide. In areas where corn is a dietary staple, fumonisin contamination has been linked with oesophageal cancer (OC). The countries where this relationship has been more deeply investigated are South Africa (especially the Transkei region) (Chelule *et al.*, 2001; Marasas, 2001) and China (Qiu *et al.*, 2001; Abnet *et al.*, 2001; Chu and Li, 1994; Wang *et al.*, 2000; Yoshizawa *et al.*, 1994). However, the possible role of fumonisin-contaminated corn has been also investigated in other areas with high OC incidence such as northeastern Italy (Franceschi *et al.*, 1995; Visconti and Doko, 1994), southern Brazil (van der Westhuizen *et al.*, 2003) and Iran (Shephard *et al.*, 2002a). In Italy, Visconti and Doko (1994) found relatively high fumonisin levels in some corn-based foodstuffs and suggested that fumonisin in polenta, a popular food in the northeastern part of the country, may contribute to the rising OC rates in this region. In Brazil, the fumonisin levels in corn from the state of Santa Catarina, an area with high OC incidence, were similar to those recorded in other high OC regions of the world (van der Westhuizen *et al.*, 2003). Shephard *et al.* (2002a) reported that in Iran the fumonisin levels in maize from Mazandaran, an area of high OC incidence, were significantly higher than those found in maize from Isfahan, an area of low OC. Analogous results were noted in China by Wang *et al.* (2000) in the OC high-risk area of the Hebei Province, where the frequency of fumonisin contamination was about twice that in the low-risk areas, and the average fumonisin content of samples from OC high-risk areas was approximately three times that in low-risk areas. High levels of fumonisin contamination in corn were also reported by Chu and Li (1994) and Yoshizawa *et al.* (1994) in the provinces of Cixian and Linxian, two OC high-risk areas of China.

The typical approach in the studies noted above was to verify the relationship between fumonisin-contaminated corn and the incidence of OC, often by comparing data from high and low OC risk areas. A more recent approach examined the relationship between the altered ratio of sphingolipids (sphinganine:sphingosine, SA:SO ratio) and OC incidence. Fumonisin B_1 (FB_1), which structurally resembles sphingoid bases, is an inhibitor of ceramide synthetase, a key enzyme involved in *de novo* sphingolipid synthesis and in the reacylation of free sphingoid bases derived from sphingolipid turnover. This inhibitory effect leads to accumulation of free sphinganine and sphingosine and alteration of the sphinganine: sphingosine ratio. This ratio has been demonstrated to be an early biomarker of fumonisin exposure that is relatively easy to analyse in the urine. Basing on this finding, Qiu *et al.* (2001) found that the urinary SA:SO ratios of males in the OC-endemic area of Henan province in China were significantly higher than those of the population in Suixian county, a low OC incidence area, whereas no difference was observed in females. Conversely, Abnet *et al.* (2001) found no significant association between serum sphingolipid levels and risk of OC in Linxian (China). Obviously, differences related to biological fluids and analytical methods must be considered when comparing experimental results.

In 1993, the International Agency for Research on Cancer (IARC) concluded there is inadequate evidence for carcinogenicity in humans from oral exposure to fumonisin B_1 (FB_1) and a role for fumonisin in any other human disease has likewise not been proven (IARC, 1993). However, the conspicuous amount of data produced in the last decade demonstrates that naturally occurring levels of fumonisin present a potential threat to human health and also suggest that fumonisin can cause OC. As underlined in the conclusions of the recent task force report of the US Council for Agricultural Science and Technology (CAST, 2003) additional research on the relationships among fumonisins, disruption of sphingolipid metabolism, and apoptosis are necessary in order to understand the carcinogenicity of fumonisin and to determine the relevance of these mycotoxins to human health.

Occurrence of mycotoxins in foods

MILK AND MILK PRODUCTS

The occurrence of mycotoxins in human breast milk, commercially available milk, and milk products is one of the most serious problems of food hygiene, as milk is a key source of nutrients for humans. This is especially significant for infants and children, who are potentially more sensitive and have less variety in their diets. Major concerns about contamination of cow's milk by mycotoxins are limited to AFB_1. Negligible, if any, amounts of other mycotoxins such as deoxynivalenol, fumonisins and zearalenone are detectable in cow's milk apart from experimentally-induced intoxication using massive toxin dosages.

The rumen has a fundamental role in defending cattle and sheep from mycotoxins, and can be considered the first barrier against transfer of mycotoxins from contaminated feeds to milk. The rumen microflora acts as a biological filter, transforming or degrading (as in the case of OTA) mycotoxins into less toxic substances.

Mammals that ingest AFB_1-contaminated diets eliminate via milk amounts of the main hepatic 4-hydroxylated metabolite known as 'milk toxin' or aflatoxin M_1 (AFM_1). AFM_1 residues in milk are a variable percentage (0.3-6%) of AFB_1 ingested. AFM_1 is usually considered to be a detoxification product of AFB_1 as its toxicity is about one order of magnitude less than that of AFB_1; however, the International Agency for Research on Cancer (IARC, 1993) has classified AFM_1 as a possible human carcinogen.

OTA transfer to milk has been demonstrated in rats and humans. OTA can also be transferred to cow's milk, but only at very low levels except when massive doses are ingested. Indeed, OTA is largely transformed by the rumen microflora into ochratoxin-α. However, Skaug (1999) found that OTA levels in cow's milk in Norway were sufficient to result in OTA intake above the suggested total daily intake of 5 ng/kg BW/day, e.g. in small children who consume large quantities of milk.

Cow's milk and milk products

According to Stoloff (1980), milk demonstrates the greatest potential for introducing aflatoxin residues from edible animal tissues into the human diet. Data in the literature on AFM_1 contamination of milk and milk products are copious; and surely much more extensive than for any other mycotoxin or food (for extensive reviews see Fremy and Dragacci, 1999; Galvano *et al.*, 1996; van Egmond, 1989). In the last decade, the incidence of AFM_1 contamination seems to have been reduced both by increasing the accuracy of detection procedures and by establishment of increasingly strict regulatory limits (mainly in the EU) for AFB_1 in animal feeds and AFM_1 in milk. The incidence of AFM_1 contamination is often higher in commercial milk than in raw milk, because of the dilution of uncontaminated bulk milk by a few contaminated samples. For the same reason, high AFM_1 levels in commercial milk seldom occur. Nevertheless, the occurrence of AFM_1 in cow's milk and milk products is not uncommon, even if contamination levels do not seem to be a serious health hazard according to current scientific information. The above considerations have been confirmed by recent surveys conducted in many countries (Carvajal *et al.*, 2003a,b; Garrido *et al.*, 2003; Roussi *et al.*, 2002; Oruc and Sonal, 2001; Galvano *et al.*, 1998, 2001; Pietri *et al.*, 2003). However, as AFM_1 may or may not be present in dairy products in a particular year depending on the weather for that period, wide and frequent monitoring programs using accurate and reliable analytical techniques remain the primary strategy to protect consumers.

When dairy products are manufactured from milk contaminated with AFM_1, the toxin is transmitted to the resulting products. AFM_1 is stable in raw milk and processed milk products and is generally unaffected by pasteurization or processing into cheese, yogurt, cream and butter (Wood, 1991; Galvano *et al.*, 1996). Association of AFM_1 with casein causes the cheese to contain a higher concentration of the toxin than the whey. This concentration is 2.5- to 3.3-fold higher in many soft cheeses and 3.9- to 5.8-fold higher in hard cheeses than in milk from which these cheeses were made (Yousef and Marth, 1989). In addition, cheese can also be directly contaminated by toxins produced by undesired and/or intentionally added mycoflora during ripening. For example, the production of some French cheeses (e.g. Camembert, Brie and Blue) is achieved by the addition of selected fungi, such as *Penicillium camemberti* and *roqueforti*, which are able to produce numerous mycotoxins including aflatoxins, cyclopiazonic acid, and citrinin. The toxicological relevance of directly contaminated cheeses remains unclear.

Milk powder for infant formulae is another route of exposure to AFM_1. Galvano *et al.* (1996) reviewed literature from 1980 to 1995. Although few studies were available, the authors concluded that the incidence and contamination level of AFM_1 in dried milk and infant formulae should not have been a health hazard; and that since infants are more vulnerable and sensitive than adults, monitoring of infant foods should be conducted more frequently and extensively. Since 1995 only six surveys have been conducted, two in Italy (Galvano *et al.*, 1998; 2001) and one in Brazil, (Oliveira *et al.*, 1997), Korea (Kim *et al.*, 2000), Turkey (Aksit *et al.*, 1997) and India (Rastogi *et al.*, 2004). Consistent with data on liquid milk, only low levels of contamination were found, apart from a recent report from India noting that almost all samples tested exceeded the OC limit (50 ng/l). However, Aksit *et al.* (1997) correctly concluded that, although aflatoxin concentrations in infant formula were found to be within acceptable limits for most countries, its presence must be carefully evaluated, because future influences of very small amounts of aflatoxin on the growing organism have not been fully elucidated.

Human breast milk

Humans who ingest aflatoxin-contaminated foods eliminate a variable amount of the toxin and its metabolites in body fluids or accumulate them in tissues. The occurrence of aflatoxin in human tissues or fluids is more of a problem in tropical and subtropical countries than in cold or temperate climates. There are reports in the literature of AFM_1 detection in breast milk, cord blood and maternal blood in African countries (Sudan, Ghana, Kenya, Nigeria, Sudan, Gambia), in the Guangxi region of China, (see review by Galvano *et al.*, 1996) as well as in the United Arab Emirates (Abdulrazzaq *et al.*, 2003) and Australia and Thailand (el Nezami *et al.*, 1995). Indications of a serious prenatal health hazard due to aflatoxin exposure has been reported, as aflatoxin can cross the placental membrane and may be concentrated by the developing foetoplacental unit. The interaction between dietary aflatoxin intake and exposure of mother, foetus, and newborn infant is very complex, depending on the physiological status of the mother and on food composition. However the presence of AFB_1 and its metabolites in human blood and breast milk presents serious a health hazard.

While AFB_1 appears to be a concern mainly in tropical or subtropical countries, OTA residues in human milk are more typical of countries in the temperate and cold areas of the Northern hemisphere, such as Italy (Micco *et al.*, 1991; 1995; Miraglia *et al.*, 1995), Switzerland (Zimmerli and Dick, 1995), Sweden (Breitholtz-Emanuelsson *et al.*, 1993), and Germany (Gareis *et al.*, 1988). However, Jonsyn *et al.* (1995) reported that due to high OTA levels in human milk, infants in Sierra Leone are exposed to OTA and aflatoxin at levels that in some cases far exceed those permissible in animal feed in developed countries. In Norway, Skaug *et al.* (1998; 2001) examined the relationship between OTA contamination of human milk and dietary intake, finding that the risk of OTA contamination was related to dietary intake of breakfast cereals, processed meat products, cheese, liver paste, cakes, cookies and juice. Considering data on both breast and dried milk sources, it must be noted that infants often receive both milk sources concurrently in early life; therefore an uncontrolled additional consumption of mycotoxins (e.g. OTA plus AFM_1) is possible.

MEAT AND MEAT PRODUCTS

Meat and meat products from animals given mycotoxin-contaminated feeds are potential routes of exposure for humans. However, meat from ruminant animals can be almost excluded from consideration due to the degrading/converting action of rumen microflora, which drastically reduces transfer to tissues. A minor and indirect route of mycotoxin exposure via meat is in processed and/or cured meats that develop surface moulds and/or contain contaminated spices.

Transfer of mycotoxins into edible tissues of poultry is not considered to be of toxicological signficance. Among meat from non-ruminant animals pork is the most susceptible to mycotoxin contamination, specifically to OTA. Indeed, OTA residues can easily appear in pork tissues due to its high incidence in pig feeds and the unfavorable elimination toxicokinetics that lead to a relatively long half-life in edible animal tissues. OTA contamination of pork is a potential concern particularly in northern European countries and in other parts of the world where climatic conditions lead to OTA contamination of feeds. In Denmark, post-mortem inspection programs were indicated to be good precautionary practices (Mousing *et al.*, 1997); and it was decided that presence of 15 ppb OTA in pig liver or kidney would result in confiscation of the organ, and that levels exceeding 25 ppb would result in the confiscation of the entire carcass. Some recent data are available in the literature. Surveys performed on pig meat in Denmark (Jorgensen and Petersen, 2002; Jorgensen, 1998) and Romania (Curtui *et al.*, 2001), on sausages in Germany (Frank, 1991), on ham in Italy (Chiavaro *et al.*, 2002), on pig liver in France (Dragacci *et al.*, 1999), and on liver patés in Spain (Jimenez *et al.*, 2001), reported detection of acceptably low levels of OTA. An evaluation of the sources of OTA intake in France revealed that meat would represent only ~3% of total OTA intake (Verger *et al.*, 1999). It can be concluded that pork meat is a route for human exposure to OTA and continuous surveillance is needed, but its contribution is surely much less important than those of other foods.

EGGS

Very little information is available on the transfer of mycotoxins into eggs apart from

AFB_1. The occurrence of AFB_1 residues in eggs could be a particularly important potential human health hazard. AFB_1 biotransformation in the liver of hens generates a variety of hydroxylated toxic metabolites that can also be transferred to eggs (Micco *et al.*, 1987). Moreover, these AFB_1 metabolites can occur as glucuronic acid or sulphate conjugates (Raj and Lotlikar, 1984). Therefore both the free and conjugated forms of the parent AFB_1 and of its toxic metabolites should be taken into account in quantification of total aflatoxins in products of animal origin. By providing hens a diet containing 100 or 400 μg AFB_1/kg feed, Jacobson and Wiseman (1974) found that the transmission ratio (AFB_1 in feed:AFB_1 residue) for eggs was between 250:1 and 430:1. These results are in contrast with the data presented by Lotzsch and Leistner (1976), who found a transmission ratio in the range 55000-125000:1. Other studies indicated wide variation in transmission ratios, such as 5000:1, Oliveira *et al.* (2000); 2200:1, Park and Pohland (1986); 66200:1, Wolzak *et al.* (1985), and 40000:1, Trucksess *et al.* (1983).

Pietri *et al.* (2001) evaluated the presence of AFB_1 and its metabolites in free and conjugated forms in eggs from hens fed a contaminated diet and calculated a new transmission ratio. Naturally contaminated maize was added to a clean layer feed to obtain final AFB_1 concentrations of about 20 and 100 μg/kg. The hens were fed the two contaminated diets *ad libitum* for three weeks. The average AFB_1 level in eggs produced in the last week of the experimental period was about 0.002 μg/kg; and on this basis a transmission ratio from feed to eggs of about 48000:1 was calculated. These results fall in the higher range of values reported in studies carried out with laying hens.

In order to prevent high concentrations of AFB_1 in eggs, in 1974 the Commission of the European Communities set a limit for AFB_1 of 20 μg/kg for layer feed. The last study indicated that if the official limit is respected, no trace of AFB_1 or metabolites can be found in eggs. AFB_1 residues at the low ppt range are only found when the corresponding level in the feed is about 100 μg/kg. In examining these data and previously published information, it can be said that the contribution of eggs to overall AFB_1 intake is negligible.

CEREALS

Cereal grains and related by-products have a unique importance as they are consumed by millions of people and are considered, from a nutritional point of view, the primary source of carbohydrates for humans and farm livestock. Thus, microbiological, chemical and mycotoxicological safety of cereals are considered very important for both human and animal food.

Cereals are easily colonised by phytopathogenic moulds; as they provide excellent growth conditions for fungi in the field, after harvest and during storage. The primary fungi associated with cereal grains belong to the *Fusarium, Penicillium, Alternaria* and *Aspergillus* genera, which are involved in mycotoxin accumulation; while other genera such as *Acremonium, Phomopsis* and *Pithomyces* are considered ancillary.

Many excellent reviews of fungi occurrence and mycotoxin production are available (Abramson *et al.*, 1997; D'Mello *et al.*, 1997; Panigrahi, 1997; Smith, 1997). A great deal of attention has been given to aflatoxin-producing fungi such as *A. flavus* and *A. parasiticus*, because AFB_1 is considered by IARC to be the most carcinogenic compound produced by non-human activities. This special focus has resulted in regulation for aflatoxin content of several foods; and many countries have such legislation (Smith, 1997). The worldwide occurrence of aflatoxins in cereals is well documented, with the major contamination occurring in countries with high moisture and/or temperatures. Among cereals, corn is most frequently contaminated by aflatoxins, whereas sorghum, rice, barley and wheat are less susceptible. Except for seasonally elevated contamination of corn, aflatoxin contamination of cereals is rarely a concern *per se*, although its frequent co-occurrence with various combinations of other mycotoxins can cause serious problems. Nevertheless, very recently Park *et al.* (2004) reported that rice is the major contributor to the dietary intake of AFB_1 in Korea and calculated that the probable daily intake of AFB_1 for Koreans exceeds the estimated provisional maximum tolerable daily intakes.

Toxins produced by *Fusarium* fungi growing on cereals represent a serious concern. Temperate climates are associated with the development of *Fusarium* in cereals; and lower temperatures are required by these fungi to grow and produce mycotoxins in comparison with *Aspergillus* (Placinta *et al.*, 1999). The most important and largest family of fusariotoxins are trichothecenes, which are comprised of several compounds divided into four subgroups, with types A and B being the most significant (D'Mello *et al.*, 1997). The type A trichothecenes include T-2 toxin, HT-2 toxin, neosolaniol (NEO) and diacetoxyscirpenol (DAS), while the type B group includes deoxynivalenol (DON or vomitoxin) and its 3-acetyl and 15-acetyl derivatives (3-ADON and 15-ADON, respectively), nivalenol (NIV) and fusarenon-X. The chemical structures of trichothecenes are related to *Fusarium* producing species, and this permits chemotaxonomic identification. Type A trichothecenes predominate in cultures of *F. sporotrichioides* and also in *F. poae*, whereas type B trichothecenes occur mainly in *F. culmorum* and *F. graminearum* culture.

Cereal safety is closely related to trichothecene occurrence as summarised by Scott (1989) and Yoshizawa (1991). Many outbreaks of acute human disease involving nausea, vomiting, gastrointestinal upset, dizziness, diarrhoea and headache have been reported in Asia, which have been attributed to consumption of cereals contaminated with *Fusarium* species and, more recently, to the presence of DON at concentrations of 3000-93000 μg/kg in grains for human consumption. Occasionally, other trichothecenes were present as well, but less frequently and at much lower concentrations (WHO, 2002). Some selected survey data for DON, NIV and zearalenone (ZEN) are reported in Table 2. As shown in Table 2, trichothecene occurrence in cereals, especially in corn, is a worldwide problem, both for highly industrialised countries such as Canada, Japan, US, Germany, Norway etc. and for developing countries such as South Africa, India, Vietnam, etc.

Zearalenone is an important oestrogenic toxin that often co-contaminates cereals along with trichothecenes and moniliformin. Moniliformin is also synthesised by *F. oxysporum*, which in addition is a recognised source of the mycotoxins wortmannin and fusaric acid (D'Mello *et al.*, 1997). The diversity of *Fusarium* mycotoxins is further illustrated by the production of fusarochromanones (TDP-1, TDP-2 and TDP-6) by *F. equiseti*, which also synthesises several trichothecenes as well as zearalenone (Flannigan, 1991; D'Mello *et al.*, 1997). Extensive data on global contamination of cereal grains indicate that these *Fusarium* mycotoxins (Muller and Schwadorf, 1993; Chulze *et al.*, 1996; Viquez *et al.*, 1996) do not pose a secondary risk to consumer health.

Since 1988 fumonisins, a novel mycotoxin family, have been under the scrutiny of mycotoxicologists and chemists (Gelderblom *et al.*, 1988). This class of compounds consists of structurally diverse polar substances, having amino and carboxylic groups (Figure 2). Fumonisin occurrence is continuously monitored and numerous papers report natural occurrence mainly in maize (Marasas, 1995). Table 3 shows survey data on the occurrence of fumonisin in many countries.

Fumonisin B_1 (FB_1) is the most common fumonisin. High risk comes from corn simultaneously contaminated with FB_1 and

Table 2. Selected data on worldwide trichothecene and zearalenone contamination of cereals (μg/kg).

	Cereal/feed	*DON*	*Nivalenol*	*Zearalenone*	*Reference*
Argentina	Wheat	100-9250			Pacin *et al.* (1997)
Brazil	Wheat	470-590	160-400	40-210	Furlong *et al.* (1995)
Bulgaria	Wheat	Up to 1800		Up to 120	Vrabcheva *et al.* (1996)
Canada	Wheat and barley	Up to 500		Up to 300	Stratton *et al.* (1993)
	Wheat (hard)	10-10500			Scott (1997)
	Wheat (soft, winter)	10-5670			
	Wheat (soft, spring)	10-1510			
	Corn	20-4090			
	Animal feeds	13-200	65-311		Abramson *et al.* (1997)
China	Corn	490-3100	600		Wang *et al.* (1995b)
China	Wheat	2850†	578	209†	Li *et al.* (2002)
Denmark	Wheat	20-527			Rasmussen *et al.* (2003)
Finland	Cereals	5-111	10-20		Eskola *et al.* (2001)
Finland	Grain	272†	150†		Jestoi *et al.* (2004)
Finland	Feed grains	7-300		22-95	Hietaniemi and Kumpulainen (1991)
	Oats	1300-2600			
Germany	Cereal foods	103†	109†		Schollenberger *et al.* (1999)
Germany	Wheat	4-20500	3-32	1-8040	Muller and Schwadorf (1993)
India	*Paspalum palidosum* straw			422	Phillips *et al.* (1996)
	Mixed			0.843	
Italy	Corn	20-9400		20-2500	Pietri *et al.* (2004)
Italy	Rice	207†			Cirillo *et al.* (2003b)
Japan	Wheat	30-1280	40-1220	2-25	Sugiura *et al.* (1993)
	Barley			10-658	
Japan	Wheat	29-11700	10-4400	53-510	Yoshizawa and Jin (1995)
	Barley	61000-71000	14000-26000	11000-15000	
Korea	Barley	5-361	40-2038		Ryu *et al.* (1996)
	Corn	145	168		
Netherlands	Wheat	20-231	7-203	2-174	Tanaka *et al.* (1990)
	Barley	4-152	30-145	4-9	
	Oats	56-147	17-39	16-29	
	Rye	8-384	10-0.034	11	
New Zealand	Corn	3400	4400	2700	Lauren *et al.* (1996)
Norway	Wheat	450-4300	Max 54		Langseth and Elen (1996)
	Barley	2200-13330	Max 770		
	Oats	7200-62050	Max 670		
Norway	Oat	104†	56†		Langseth and Rundberget (1999)
	Barley	155†	30†		
Philippines	Corn		18-102	59-505	Yamashita *et al.* (1995)
Poland	Wheat	2000-40000	10	10-2000	Perkowski *et al.* (1990)
Poland	Maize kernels	4000-320000			Grabarkiewicz-Szczesna *et al.* (1996)
	Maize cobs: axial stems	9000-927000			

Table 2. Continued.

	Cereal/feed	DON	Nivalenol	Zearalenone	Reference
South Africa	Corn	Up to 1830	Up to 370		Rheeder et al. (1995)
	Cereals (animal feed)			50-8000	Dutton and Kinsey (1996)
Thailand	Corn			923	Yamashita et al. (1995)
USA	Wheat	Up to 9300			Fernandez et al. (1994)
USA	Wheat (winter) 1991	<100-4900			Trucksess et al. (1995a)
	Wheat (spring) 1991	<100-900			
	Wheat 1993	<500-18000			
	Barley 1993	<500-26000			
USA	Wheat		nd-23		Manthey et al. (2004)
Vietnam	Corn powder	1530-6510	780-1950		Wang et al. (1995a)

†Mean value.

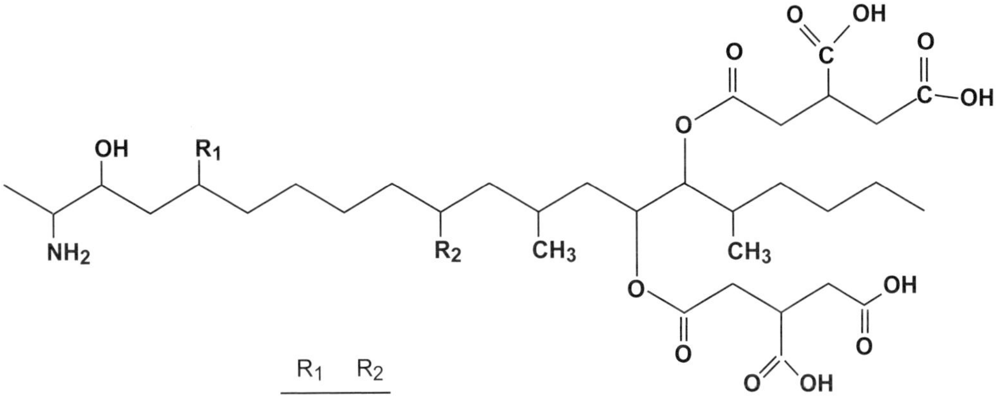

	R_1	R_2
Fumonisin B_1	OH	OH
Fumonisin B_2	OH	H
Fumonisin B_3	H	OH
Fumonisin B_4	H	H

Figure 2. Structures of the B series of fumonisins.

many combinations of other mycotoxins, as reported in China (along with AFB_1, Wang et al., 1995b), Vietnam (with DON, NIV and AFB_1, Wang et al., 1995a), Indonesia (with AFB_1, DON, NIV and ZEN, Ali et al., 1998), Korea (with DON, NIV, other trichothecenes and ZEN, Sohn et al., 1999; with aflatoxins and OTA, Park et al., 2002), Brazil (with aflatoxins and ZEN, Vargas et al., 2001; with aflatoxins, Ono et al., 2001), Ghana (with aflatoxins, Kpodo et al., 2000), USA (with moniliformin, Gutema et al., 2000), Thailand (with aflatoxins, Yoshizawa et al., 1996) and Southeast Asia (with aflatoxins, ZEN and nivalenol, Yamashita et al., 1995).

Recently, Cirillo et al. (2003a) reported frequencies and levels of contamination of DON and fumonisins B_1 and B_2 in cereal-based foods in Italy. Of 202 samples investigated, including raw materials and processed cereal foods (bread, pasta, breakfast cereals, biscuits, baby and infant foods), 84% were

Table 3. Selected data on worldwide fumonisin contamination of corn (μg/kg).

Country	*Fumonisin B_1*	*Fumonisin B_2*	*Fumonisin B_3*	*Reference*
Argentina	85-8791	nd-11 300	85-16760	Chulze *et al.* (1996)
Benin	nd-2630	nd-680		Doko *et al.* (1995)
Botswana	35-255	nd-75	nd-30	Doko *et al.* (1996)
Botswana	20-1270			Siame *et al.* (1998)
Brazil	0.096-22.6			Ono *et al.* (2001)
Brazil	200-6100			Vargas *et al.* (2001)
Brazil	0.09-17.69	0.05-5.24		Almeida *et al.* (2002)
Brazil	3.2	3.4	1.7	van der Westhuizen *et al.* (2003)
China	160-25970	160-6770	430-36870	Wang *et al.* (1995b)
China	≥0.5	≥0.5	≥0.5	Groves *et al.* (1999)
China	1.4-88.9			Wang *et al.* (2000)
Costa Rica	1700-4780			Viquez *et al.* (1996)
Denmark*	1-1000	4-250		Petersen and Thorup (2001)
Egypt	10-780			el Sayed *et al.* (2003)
Ghana	70-4222			Kpodo *et al.* (2000)
Guatemala	0.2-1.7	<0.1-1.6	<0.1-0.2	Trucksess *et al.* (2002)
Honduras	68-6555			Julian *et al.* (1995)
India	0.25-64.7			Bhat *et al.* (1997)
Indonesia	226-1780	231-556		Yamashita *et al.* (1995)
Iran	0.68-7.66			Shephard *et al.* (2002a)
Italy	10-2330	nd-520	10-2850	Doko *et al.* (1995)
Italy	54-51690			Pietri *et al.* (2004)
Korea	23.3	7.5	6.7	Sohn *et al.* (1999)
Korea	74†			Park *et al.* (2002)
Korea*	41.8-256			Kim *et al.* (2000)
Malawi	nd-115	nd-30	nd	Doko *et al.* (1996)
Mozambique	240-295	75-110	25-50	Doko *et al.* (1996)
Nepal	>1000			Desjardins *et al.* (2000)
Nigeria	70-1780	114†		Bankole and Mabekoje (2004)
Philippines	57-1820	58-1210		Yamashita *et al.* (1995)
Portugal	90-3370	nd-1080	90-4450	Doko *et al.* (1995)
South Africa	60-70	nd	nd	Doko *et al.* (1996)
South Africa	max 2000		nd	Dutton and Kinsey (1995)
South Africa	0.1-22-2			Chelule *et al.* (2001)
Taiwan	109-1148	222-255		Tseng and Liu (1999)
Taiwan	334-1614			Tseng and Liu (2001)
Taiwan*	73-2395	120-715		Tseng and Liu (1997)
Tanzania	nd-160	nd-60	nd	Doko *et al.* (1996)
Thailand	63-18800	50-1400		Yamashita *et al.* (1995)
The Netherlands	8-1430			De Nijs *et al.* (1998)
Uruguay	nd-3688			Pineiro *et al.* (1997)
USA	nd-350			Trucksess *et al.* (1995b)
USA	<75-5,916			Castelo *et al.* (1998)
USA	43-1622	nd		Gutema *et al.* (2000)
Vietnam	268-1516	155-401	524-2185	Wang *et al.* (1995a)
Zambia	20-1420	nd-290		Doko *et al.* (1995)
Zimbabwe	55-1910	nd-620	nd-205	Doko *et al.* (1996)

† Mean value.
*Various corn-based foods
nd - none detected

contaminated with DON at levels between 7 and 930 μg/kg (median 65 μg/kg); 26% contained FB_1 ranging from 10 to 2870 μg/kg (70 μg/kg); 35% contained FB_2 at 10-790 μg/kg (80 μg/kg). The highest levels of DON and FB_1 were detected in raw cereals and whole meal flours. The highest levels of FB_2 were detected in durum wheat pasta. A widespread DON contamination was found in baby and infant foods at levels varying from 7 to 166 μg/kg.

Pietri *et al.* (2004), surveying maize produced in Italy over the period 1995-1999, found that FB_1 was by far the most frequently identified mycotoxin. The average concentration was 3064 μg/kg, with 69.6% of samples exceeding 1000 μg/kg and 16.9% over 5000 μg/kg. Other studies have reported fumonisin occurrence in Italy, Portugal, and African countries, as observed by Doko *et al.* (1995). In Italy the co-contamination of FB_1 and cyclodepsipeptides or sesterterpenes such as beauvericin or fusaproliferin was also observed (Ritieni *et al.*, 1997).

OILSEEDS

Since the discovery of mycotoxins, their occurrence in oilseeds has been noted because these crops are considered an excellent substrate for synthesis and accumulation of mycotoxins produced by moulds. Aflatoxins produced by *A. flavus* and *A. parasiticus* are the main concerns for oilseeds (Hesseltine *et al.*, 1966). Several oilseeds are used as cocoa butter substitutes, which represent a serious risk, because many processed products based on cocoa are targeted for consumption by children and infants.

In 1981, 21 end-of-season commercial samples of peanut kernels coming from different countries were analysed in west Africa. The results were considered satisfactory, as they were very low (≤5 ppb in 86% of samples). The most contaminated products were groundnuts from Brazil; and *A. parasiticus* isolated from this plant was potentially able to produce >1100 μg/kg of aflatoxin B_1 (Kershaw, 1982). Other investigations of mycotoxin occurrence in groundnut/peanut crops were performed by Tang and co-workers in 1996 in China (Tang *et al.*, 1998), while Verma *et al.* (1996) focused their studies on India.

Recently, Begum and Samajpati (2000) investigated mycotoxin occurrence in rye, mustard, soybean and groundnuts. All samples were collected from local markets of Calcutta (India). A total of 195 potentially toxigenic fungi were isolated from eight different oilseed samples; and 14 of these were found to be able synthesise aflatoxin B_1 or sterigmatocystin (Begum and Samajpati, 2000).

Gamanya and Sibanda (2001), in a national survey carried out in 1995 and 1996, evaluated the distribution of *Fusarium verticillioides* and fumonisin B_1 levels in oilseeds from Zimbabwe. Results revealed that the incidence of *F. verticillioides* and other *Fusarium* species and levels of FB_1 generally decreased from regions with high rainfall and annual moderate temperatures to low rainfall regions. There was no *Fusarium* contamination and FB_1 was not detected in sunflower and soyabean.

WINE, GRAPE JUICE AND GRAPE PRODUCTS

Wine, grape juice and grape products are foods frequently contaminated by OTA. The occurrence of OTA in human serum was measured in a Swiss study to estimate the exposure of the population to the toxin (Zimmerli and Dick, 1995). The researchers found that Swiss males living south of the Alps (Ticino) had higher serum OTA concentrations than women and people living in the rest of the country. A possible explanation was that wine (whose consumption by males living south of the Alps is higher than north of the Alps) could be a source of OTA. The authors analysed wine samples and demonstrated for the first time the presence of OTA in wine. A subsequent survey of 133 wines obtained from retail outlets in Switzerland showed a higher OTA concentration in red wines than in white or rosé wines (Zimmerli and Dick, 1996). Regarding red wines, they deduced that wine OTA concentration depended on latitude of the region of origin: the further south the origin, the more frequent the occurrence and the greater the concentration. After the first detection of OTA in wine, several surveys were conducted, mainly in Europe. The data collected are summarised in Table 4, where the region of origin, the total number of samples, the median and/or the mean value (in parentheses) of white, rosè, red and dessert wines analysed, the maximum level detected and the authors of the papers are reported.

Otteneder and Majerus (2000) analysed 420 wines from across the world between 1996 and 1999.The authors divided the red wine samples originating from the OC into the Wine Growing Zones in OC regulation No. 822/87, which are defined as geographical regions with distinct climatic conditions for growth. Zone A covers areas of Germany, Luxembourg and the UK, whilst Zone B includes southern Germany and the north of France. Finally, CIIIa and CIIIb include the most southern growing zones in Europe, namely Sicily, Greece and the south of Spain. Contamination rates, along with mean and maximum values increased from the northern zones (A, B and CI) to the southern zones (CII and CIII). This trend was not very evident amongst the white wines, but was distinct within the red wines.

The survey conducted by Pietri *et al.* (2001) confirmed that geographic region of origin had a strong influence on OTA contamination, both for red and for dessert wines. In fact, wines produced in southern

Table 4. Results of selected surveys carried out from 1996 for ochratoxin A content in white, rosé, red and dessert wines (μg/kg)[1].

Regions monitored	*No. of samples*	*White*	*Rosé*	*Red*	*Dessert*	*Maximum*	*Authors*
Austria	117					20	Eder *et al.* (2002)
Europe	267	(20)	(31)	(54)	(1048)	603	Burdaspal and Legarda (1999)
France	138					1700	Rosari *et al.* (2000)
Greece	268	60(250)	80(175)	90(340)	330(580)	2690	Stefanaki *et al.* (2003)
Italy	31					3800	Tateo *et al.* (2000)
Italy	111			90(419)	8(736)	3856	Pietri *et al.* (2001)
Mediterranean basin	133	<3(11)	19(25)	13(39)	337(250)	388	Zimmerli and Dick (1996)
Morocco	30	55(73)	90(223)	785(912)		3240	Filali *et al.* (2001)
Northern Spain	40					316	Lopez de Cerain *et al.* (2002)
Portugal	64					80	Festas *et al.*, (2000)
South America	80					71	Rosa *et al.* (2004)
Southern Europe	31					3400	Markaki *et al.* (2001)
Southern Italy	56	80(290)	670(720)	760(1240)		7630	Visconti *et al.* 1999)
World	420	10(108)	10(119)	20(201)		7000	Otteneder and Majerus, (2000)

[1]Note: the values represent the median (mean) of OTA levels found in each survey.

Italy were markedly more contaminated. The overall median (mean) OTA concentration in the red wines produced in the four Italian areas (northwest, northeast, central and south) was 2 (11), 90 (81), 134 (295) and 1264 (1233) ng/l. The same trend was observed for the white dessert wines. OTA concentrations of over 1000 ng/l were found in 4 of 5 samples from southern Italy, while central and northern samples had very low contamination. It was clear that the difference in latitude (from 46° to 36° parallel) and the consequent climatic differences influenced mould contamination and OTA production. Similarly, Stefanaki *et al.* (2003) observed a trend toward increasing OTA contamination for red wines from northern to southern Greece. Some authors indicated that generally low cost red table wines had significantly higher OTA concentrations than good quality bottled wines.

Obviously, OTA can contaminate not only wine, but also other grape products. Apart from wine samples, Zimmerli and Dick (1996) analysed OTA content of eight red and three white commercial grape juices. The white grape juice samples contained <5 ng/l, but the red ones ranged from <3 to 311 ng/l, with a median (mean) of 235 (188) ng/l. Burdaspal and Legarda (1999) analysed 10 samples of grape juice and found OTA levels in the range of 15 to 102 ng/l. Twenty samples of grape juice were analysed by MAFF (1999). One sample contained no OTA and the others ranged from <20 to 2050 ng/l, with a mean of 480 ng/l. In the period 1995-98 many samples of grape juice were analysed in Germany (Wolff *et al.*, 2000). Seven out of 38 white grape juices tested contained less than 10 ng/l OTA, while in the rest of the samples OTA content ranged from 10 to 1300 ng/l. In red grape juices, 8 of the 73 samples analysed were free of OTA, while the rest ranged from 10 to 5300 ng/l. Rosa *et al.* (2004) analysed 48 red grape juice samples produced in Brazil. OTA was detected in 29.2% of the samples, with a maximum value of 100 ng/l.

Vinegar can also contain OTA. Majerus *et al.* (2000) found OTA in 19 of 38 wine vinegar samples (range 10-1900 ng/l, 90th percentile 220 ng/l). Balsamic vinegar samples (29 samples analysed) were the most contaminated (range 10-4350 ng/l, 90th percentile 3110 ng/l). Markaki *et al.* (2001) analysed 15 vinegar samples and found that the most contaminated were three samples of balsamic vinegar (range 102-252 ng/l), while the other 12 samples ranged between 8 and 46 ng/l.

Dried vine fruit (currants, raisins and sultanas) can be a further important dietary source of OTA. MacDonald *et al.* (1999) determined OTA in 60 samples of retail dried vine fruits purchased in the UK. OTA was found in excess of 0.2 μg/kg in 19 of 20 currant samples, 17 of 20 sultana and 17 of 20 raisin samples examined, with average (maximum) levels of 9.19 (53.6), 4.86 (18.1) and 2.79 (20.0) μg/kg, respectively. In 1999, a further survey was conducted in the UK (MAFF, 1999). OTA was determined in 301 dried vine fruit retail products and was found to be in excess of 0.2 μg/kg in 96 of 100 currant, 92 of 100 sultana and 98 of 101 raisin samples examined, with average (maximum) levels of 4.97 (40.8), 3.42 (25.1) and 2.87 (29.8) μg/kg, respectively.

Important surveys were also conducted in Greece and the US. OTA was present in 50 of 69, 76 of 114 and 90 of 133 samples of raisins analysed in 1997, 1998 and 1998-99, respectively (FAO, 2001). The average (maximum) levels were 0.42 (3.1), 0.82 (8.1) and 1.27 (29.0) μg/kg, respectively. Stefanaki *et al.* (2003) conducted a survey of the presence of OTA in 81 samples of Greek dried vine fruits collected between 1998 and 2000. OTA was found in excess of 0.5 μg/kg in 43 of 54 currant and 17 of 27 sultana samples examined, with average (maximum) levels of 2.8 (13.8) and 2.1 (13.2) μg/kg, respectively. It is evident that dried vine fruit can be highly contaminated. Dried vine fruit can be an important dietary source of OTA for people with high levels of consumption, in particular children. Due to this, the Commission of the

European Communities (2002) fixed a limit for OTA in dried vine fruit of 10 μg/kg.

BEER

Differently from wine, which is almost exclusively contaminated by OTA, beer can be contaminated by a number of mycotoxins. Indeed, barley as well as other cereals (e.g. corn, wheat and sorghum) used to produce brewed beer, are good substrates for the growth of many toxigenic fungi. Consequently, beer can be contaminated by all key cereal mycotoxins except perhaps citrinin, which does not seem to survive the mashing step. Scott (1996) reviewed data on beer contamination up to 1996, reporting: i) high incidence and concentrations of aflatoxins and zearalenone in local beers brewed in Africa; ii) no detection of aflatoxins and ZEN in European beers; iii) rare and/or trace detection of OTA; iv) high incidence of DON in Canadian and European beers, especially German beers; and v) limited occurrence of FB_1 and FB_2.

Since 1996, numerous surveys have been conducted on mycotoxin contamination of beers produced across the world from different grains and under different brewing conditions. Tangni *et al.* (2002) analysed the OTA content of 62 Belgian beers and 20 beers imported from Denmark, France, Germany, Ireland, Mexico, the Netherlands and Scotland. None of these beers exceeded the proposed EU limit of 200 ng/l. However, OTA was detected in 60 Belgian beers and in all imported beers. The average levels of contamination were 33 and 32 ng/l, respectively. The highest level found was 185 ng/l.

Schothorst and Jekel (2003), following the observation of high concentrations of DON in wheat and other cereals sold on the Dutch market in 1999 and 2000, determined the DON content of 51 beer samples. Low quantities of DON were found in only three samples, at levels ranging from 26 to 41 μg/l. Apart from DON, no trichothecenes were found in the beers at levels above the limit of quantification (25 μg/l). All positive samples were from so-called top fermentation beers.

In Italy, Visconti *et al.* (2000), analysed the OTA content of 61 samples of 10 domestic and 51 imported beers and found OTA levels ranging from <10 to 135 ng/l, with an incidence of contamination of 50% and no substantial difference between strong and pale beers. In Morocco, Filali *et al.* (2001) found that none of five beers sampled was contaminated by OTA.

A significant co-occurrence of OTA and AFB_1 has been observed by Nakajima *et al.* (1999) in 22 Japanese beer samples and in 94 beer samples imported from 31 producing countries. OTA was detected in 91.5% of imported beer samples (mean = 10.1 ng/l) and in 95.5% of Japanese beer samples (mean = 12.5 ng/l), whereas AFB_1 was detected in 11 of 94 of imported beer samples (0.5-83.1 ng/l) and in two Japanese beer samples (0.5 and 0.8 ng/l).

The occurrence of mycotoxins in traditionally brewed alcoholic beer regularly consumed by most ethnic black South Africans and obtained from maize and barley, was investigated by Odhav and Naicker (2002), who found that 2 of 6 commercial beer samples contained aflatoxins (200 and 400 μg/l) and 45% (13 of 29) of the home-brewed beers had ZEN (range 2.6-426 μg/l) and/or OTA (range 3-2340 μg/l).

Zollner *et al.* (2000) found a detectable level of ZEN in only 1 of 23 beer samples imported into Austria from different countries. Torres *et al.* (1998) reported that FB_1 occurred in 43.8% of 32 Spanish beers, including some non-alcoholic beers, in the range of 4.76-85.53 μg/l.

In the US, Hlywka and Bullerman (1999) analysed FB_1 and FB_2 in 25 domestic and 4 imported brands of beer, reporting that 86% were positive for FB_1 and 41% were positive for FB_2. The total fumonisin (FB_1 + FB_2) content of positive samples ranged from 0.3 to 12.7

μg/l. Scott and Lawrence (1997) analysed AFB_1 content of 24 beer samples, the majority from the US and Mexico; and found that four Mexican samples and one Brazilian sample contained AFB_1 at low levels (<10 ng/l). Molto *et al.* (2000) observed that the only mycotoxin detected in Argentinean beer was DON. It was present in 44% of the samples, 18% with more than 20 μg/l. Toxin levels ranged from 4 to 221 μg/l.

A total of 54 samples of Korean and imported beers were analysed for the natural occurrence of *Fusarium* mycotoxins by Shim *et al.* (1997). DON was detected in 14 samples (26%) and nivalenol was detected in 43 samples (80%). ZEN was not detected in any of the samples. The incidence of DON was 19% for Korean beers and 50% for imported beers, whereas the incidence of NIV was 85% for Korean beers and 58% for imported beers.

Interestingly, Henderson *et al.* (2000) demonstrated that hops (*Humulus lupulus*), which are often added as aromatics in the production of beer, contain several unique flavonoid compounds that have been shown to have cancer chemopreventive properties due, in part, to inhibition of cytochrome P450 enzymes. As it has been demonstrated that AFB_1 and perhaps OTA need to be activated by specific cytochrome P450 enzymes to exert genotoxicity, it should be experimentally verified whether presence in beer of hop flavonoids, AFB_1 and OTA could somehow reduce the toxicity of the above mycotoxins.

Some conclusions can be drawn from these data, though comparisons among experiments are difficult for many reasons including differing analytical procedures. With few exceptions, overall low levels of contamination have been reported. It is evident that there is great variability in extent of contamination, due to the numerous factors that play a role in determining the kind and the amount of mycotoxins occurring in beer. Climatic conditions and cereal source are the decisive factors that selectively favor growth of one mould rather than others, which explains the clear geographical incidence. The importance of other factors, such as storage and pretreatment of cereal source and brewery procedures, is still to be elucidated. The possible co-occurrence of mycotoxins in beer is an issue that should be considered. In any case, it seems unavoidable that, at least in some specific geographical areas, beer must be taken into account when estimating probable daily intake of mycotoxins other than OTA.

FRUITS AND JUICES

Fruits and vegetables are an important part of a balanced diet, but they may be colonised by moulds that produce mycotoxins (Drusch and Ragab, 2003). In fruit juice an important mycotoxin is byssochlamic acid, produced by *B. fulva* in canned fruits and harvested grapes (Delage *et al.*, 2003). Byssochlamic acid is also accumulated in prune, grape and pineapple juices, but is less important in apple, orange and tomato juices (Samson *et al.*, 1996). Another important mycotoxin is citrinin, produced by *Penicillium expansum* in apple juice and to a lesser extent in grape juices (Vinas *et al.*, 1993). Ochratoxin A, biosynthesised by *Aspergillus carbonarius*, is an important risk for apple and grape juice (Pitt, 2000). Patulin is considered the most important mycotoxin in fruit juices, and is produced by several species of fungi including *B. fulva*, *B. nivea* and *P. expansum*; the latter being the major patulin-producing microorganism (Frisvad and Thrane, 1996).

Patulin is partially destroyed by the fermentation process during cider production (Ough and Corison, 1980; Stinson *et al.*, 1978). Patulin has been widely studied in fruit juices (Mortimer *et al.*, 1985; Larsen *et al.*, 1998; de Sylos and Rodriguez-Amaya, 1999), particularly in apple juice (Scott *et al.*, 1972; Mortimer *et al.*, 1985; de Sylos and Rodriguez-Amaya, 1999; Beretta *et al.*, 2000). More recently, a survey of patulin and chaetoglobosins was perfomed by Ritieni (2003) and Andersen *et al.* (2004). Patulin occurrence in fruit juices is related to several factors such as temperature and water activity (Northolt *et al.*, 1996) or technological treatments during transformation, as described by Jackson *et al.* (2003). In addition, agronomic practises in fruit cultivation and juice production have a great impact on the occurrence of patulin in fruit juices as well as citrinin as described in Portuguese apples (Martins *et al.*, 2002). Many research groups have recommended that apple products for human consumption should not contain patulin in excess of 50 mg/L.

Bananas have been reported to be contaminated by *Fusarium* mycotoxins. Bananas can be invaded, following harvest, by various fungal species that produce 'crown rot' (Ogundero, 1987). *F. moniliforme* isolates have been found to produce potentially three 12,13-epoxytrichothecene mycotoxins (trichothecolone, DAS, and T-2 toxin), the palmitoyl esters of trichothecolone, scirpenetriol, and T-2 tetraol (T-2 TOL), and free and palmitic acid conjugated zearalenone (Chakrabarti and Ghosal, 1986).

Shenasi and coworkers have examined 25 varieties of dates (*Phoenix dactylifera*) at different maturation stages for aflatoxins and aflatoxigenic *Aspergillus* spp. In particular, the samples were examined both fresh and under simulated storage conditions of high humidity. Aflatoxins were detected in 12% of the samples and aflatoxigenic *Aspergillus* was detected in 40% of the varieties examined (Shenasi *et al.*, 2002).

COFFEE

Coffee can contain OTA produced by *A. ochraceus* or *A. carbonarius*. Studies in Brazil have shown that mould growth and OTA production occur only during drying of green coffee beans; and that if drying is rapid and effective OTA will not be produced. OTA has been found in green coffee for several years, but there has been and still is some uncertainty as to what extent OTA is degraded during the roasting process and further transmitted from roasted to brewed coffee. The reported reductions caused by roasting varied, but the major source of variation was the low OTA level (generally <10 μg/kg) in green coffee. Reports usually showed high rates of OTA reduction, ranging between 30 and 90%. Decaffeination seems to be an effective process, resulting in 92% reduction of OTA. During soluble coffee manufacture or brewing, nearly 80% of OTA initially present in roasted and ground coffee was found to be transferred into the cup (Van der Stegen *et al.*, 2001).

Many surveys of roasted coffee have demonstrated that the OTA incidence is generally low and that levels exceeding 5 μg/kg are very rare (Levi *et al.*, 1974; Galaz and Stalder, 1976; Cantafora *et al.*, 1983). Recent data obtained in Germany, the UK and Switzerland indicate that coffee on the European market contains a mean OTA content of around 0.8 μg/kg. On the basis of available data on levels of OTA contamination and on mean consumption of food categories, it was calculated that coffee contributed 2-3 ng/kg, while cereals and wine contributed about 25 and 10 ng/kg body weight (BW) per week, respectively, to average intake (WHO, 2001).

COCOA

Cocoa is a very important product in the economy of a number of countries such as Cote d'Ivoire, Ghana, Cameroon, Grenada and the Dominican Republic. In other more industrialised countries in South America and the Southeast Asia including Brazil, Ecuador, Colombia, Malaysia and Indonesia, cocoa is of regional importance, providing a major source of employment in agriculture in certain areas (Harrison, 2003). Cocoa beans are the source of cocoa powder, a common ingredient in several kinds of foods, cakes, biscuits, baby food, ice cream and sweets. In all these tropical regions, weather and agronomic conditions are favourable for mould growth, mainly *Penicillium* spp., and consequently OTA biosynthesis is not unexpected. Mould growth can occur on cocoa beans either during fermentation and drying or when moisture content of the beans is high during transport and storage. Because the regions where cocoa is grown are not near the processing plants, delivery times are long and therefore favourable for mould colonization and growth. An initial study of OTA in cocoa found it to be present at detectable levels in 67% of samples. There was no correlation between OTA and the visible mould on the beans; and beans of different origins were affected. OTA was found at lower levels in cocoa butter than in the nonfat fraction (powder or cake) (Beckett, 1994).

In 1998, a survey was conducted to assess the exposure of UK consumers to OTA in cocoa powder and chocolate (MAFF, 1999). The OTA contamination in chocolate bars and cocoa powder was generally low, although OTA was detected in all of the 20 samples of cocoa powder analysed and three out of four chocolate bars. Considering chocolate samples, 75% of samples were positive for OTA, but at low levels. It is concluded that there is little cause for concern about the level of OTA in cocoa and chocolate.

Human exposure and risk assessment

The danger posed by mycotoxins and their unavoidable presence in foods has led public authorities to adopt rigorous control policies. Though economic factors have an important influence on the final decision, scientific information is crucial in the establishment of meaningful regulatory levels for mycotoxins in foods. The overall scientific approach consists of combining exposure data for individuals or populations with toxicological data. In the estimation of the exposure levels the occurrence of a mycotoxin in a food and its dietary burden are taken into account. Toxicological data are obtained from laboratory studies on animals and safety/ uncertainty factors (multiples of 10) are used to extrapolate data on laboratory animals to humans and to provide an acceptable or tolerable daily intake (ADI/TDI).

On the basis of the available toxicological data and using the five regional diets Global Environmental Monitoring System/Food Contamination Monitoring and Assessment Programme (GEMS/Food) (WHO, 1998), the Joint Expert Committee on Food Additives (JEFCA) of the World Health Organization and of the Food and Agriculture Organization, has recently evaluated the hazard for the most significant mycotoxins (OTA, fumonisins, DON, T-2 toxin, patulin, and zearalenone) and established the **Provisional Maximum Tolerable Weekly Intake (PMTWI) and the Provisional Maximum Tolerable Daily Intake (PMTDI)**. These values are obtained by calculations subsequent to the quantification, by animal studies, of the **No Observed Adverse Effect Level (NOAEL)**. Then the NOAEL is divided for a safety/uncertainty factor of 100 (10 for extrapolation of animal data to humans and 10 for individual variation). As clearly indicated by the term *provisional*, the values cannot be considered definitive. In view of the further quantification of TDI, it will be necessary to establish the Risk Assessment, which is the product of hazard assessment and exposure assessment. The procedure includes the following steps: (1) hazard assessment based on animal toxicity and possibly epidemiological data, (2) exposure assessment based on data concerning actual concentrations in commodities, (3) extrapolation of hazard assessment from high-exposure animal data to low-exposure human data, and (4) comparison of the product of hazard assessment and exposure assessment with acceptable risk (Kuiper-Goodman, 1995). In the estimation of probable human exposure to such a mycotoxin, all potentially contaminated food sources must be considered. The estimated PMTWI or PMTDI and the exposure data allow establishment of a maximum permissible limit for a specific commodity or range of foodstuffs. However, due to worldwide variability in dietary patterns, it is difficult to harmonise limits. Indeed, a highly contaminated staple food consumed in one area requires a limit that would be too severe for the same food in another area where it is minimally consumed.

A different approach was applied for genotoxic mycotoxins such as AFB_1. As AFB_1 has carcinogenic properties, assuming that a no-effect concentration limit cannot be established for genotoxic compounds, any dose will have a proportional effect. Thus, JECFA did not establish a PMTWI or PMTDI, but recommended that its level in food should be **As Low As Reasonably Achievable (ALARA)**. As regards AFM_1, there is sufficient evidence that AFM_1 is a genotoxic carcinogen, although its carcinogenic potency is estimated to be approximately 10 times lower than aflatoxin B_1. Nevertheless, as milk and milk products are consumed in large quantities, particularly by infants and young children, the risks from aflatoxin exposure must be carefully considered. Thus, JEFCA assessed an estimation of dietary AFM_1 intake. Highest contamination levels were found in samples from the Far East, where milk consumption is generally low. The Committee concluded that even with using worst-case assumptions, the additional risks for liver cancer predicted with use of the proposed maximum levels of AFM_1 are very small, even for people who consume large quantities.

OCHRATOXIN A

Data on mycotoxin exposure and/or risk assessment are available in the literature for OTA and different tolerable daily intakes (TDI) have been suggested over the last decade. In Europe, the presence of OTA in food has been regulated by the Joint Expert Committee on Food Additives since 1995, and a provisional tolerable weekly intake of 0.1 mg/kg BW (around 14 ng/kg BW/day) has been established. In 1996 the JECFA, on the basis of the nephrotoxicity of OTA, proposed a PTDI of 14 ng/kg BW (WHO, 1996). In 1998 the European Commission's Scientific Committee on Food (SCFOO, 1998) recommended that it would be prudent to reduce exposure to OTA as much as possible, e.g. to below 5 ng/kg BW/day. In 1998, a survey was carried out to assess the exposure of UK consumers to OTA in retail products in the diet (MAFF, 1999). Reassuringly, the levels of OTA found were below the tolerable daily intake of 5 ng/kg BW/day recommended by the SCFOO. A conservative TDI (5 ng/kg BW) has also been proposed by the Working Group of the Nordic Council of Ministers (Olsen *et al.*, 1991), whereas the Canadian authority proposed a TDI in the range of 1.2-5.7 ng/kg BW (Kuiper-Goodman, 1996). These differences are also reflected in risk management measures established, resulting in variable legal limits applied to different commodities and to the same commodity in different countries (Walker, 2002). In any case, as pointed out by Skaug (1999), OTA risk assessments performed to date do not differentiate between risk to adults and children, which represent a particularly sensitive population that would need a specific TDI, especially in consideration of the unfavorable dose/BW ratio.

FUMONISINS

In 2003, JEFCA established a group TDI of 2 μg/kg BW for the total fumonisins (B_1, B_2, and B_3), alone or in combination. However, using GEMS/Food regional diet data, the fumonisin intake of people in the African region is 2.4 μg/kg BW/day based on mean levels of maize contamination. According to South African researchers at the PROMEC unit, the TDI established by JECFA is appropriate to protect consumers in developed countries, where maize consumption is low, whereas much lower tolerance levels of 200 μg/kg would be required to protect subsistence farming communities where maize is a dietary staple in developing countries, such as the Transkei region (Marasas, 1997; Shephard *et al.*, 2002b). Additional uncertainties in establishing risk assessments for fumonisin depend on the wide seasonal and geographical fluctuation in contamination levels. For example, when JECFA used estimates of contamination at the higher end of the scale (90th percentile), four out of five regions (Middle East, the Far East, Latin America and Africa) exceeded the PMTDI, with Africa reaching an alarming 7.3 mg/kg BW per day.

Experts from the Nordic countries (Denmark, Norway, Sweden, Finland and Iceland) have carried out a preliminary evaluation of fumonisins concluding that, at that time point, it was not possible to carry out a complete risk assessment. However, it was recommended that the human daily intake of fumonisins should be less than 1 μg/kg bw/day (Petersen and Thorup, 2001).

DON AND FUSARIOTOXINS

The joint FAO/WHO Expert Committee on Food Additives (JECFA, 2001) established for DON a PMTDI of 1 μg/kg BW and a group PMTDI of 0.06 μg/kg BW for T-2 toxin and HT-2 toxin. It must be pointed out that data were available only for the European region, so that more accurate information from other parts of the world and improved analytical methods are needed.

In a study conducted in the Netherlands, Pieters *et al.* (2002) derived a provisional TDI of 1.1 μg/kg BW and proposed a concentration limit of 129 μg DON/kg wheat based on this TDI and the high wheat consumption of children. The authors also highlighted that in the period September 1998-January 2000, due to high DON concentration in wheat, the dietary intake of DON exceeded the provisional TDI, especially in children, and that health effects could occur.

In France, Leblanc *et al.* (2002) showed that consumers of high amounts of organically produced cereals can exceed the PTDI for DON over their lifetimes. Very recently Schothorst and van Egmond (2004) presented the results of the SCOOP (Scientific Co-operation on Questions relating to Food) subcommittee on trichothecenes. Despite a significant lack of consumption data in some countries and very poor information on baby and children's food, the authors highlighted that: i) although among cereals, corn showed the highest level of contamination with trichothecenes, wheat and wheat-containing products (like bread and pasta) represent the major source of intake; ii) the overall mean intakes for DON are below the TDI, though for young children the mean intakes are sometimes (very) close to the TDI; iii) NIV intakes are far below the TDI; and iv) T-2 and HT-2 toxin intakes are in most cases above the TDI.

PATULIN

The joint FAO/WHO Expert Committee on Food Additives (JECFA, 2001) established for patulin a PMTDI of 0.4 μg/kg BW/day. In the recent report on patulin dietary intake assessment from the European expert task force (EU, 2002), it was concluded that exposure seems to be well below the value indicated by JECFA. However, as most countries did not provide information on all food products potentially contaminated by patulin, the total dietary intake should be generally considered underestimated. In addition, it was emphasised that in some countries people suffer from contamination despite intakes below the PMTDI, especially in worst-case situations and where specific groups of consumers, especially small children, are considered.

Conclusions

Overall, the worldwide scenario for potential mycotoxin contamination and consequent impact on human health is serious. Mycotoxins in the human food chain are still an unavoidable problem that affects almost all foods and people, albeit to varying degrees. Apart from peculiar regional circumstances, the overall situation is characterised by widespread contamination at low levels, leading to human health risks related to chronic exposure. Except for AFM_1, which is within tolerable limits of consumption, available data show that in a considerable part of the world PMDTIs are exceeded for fumonisin and DON. In addition, it is reasonable to hypothesise that further acquisition of data will show that dangerous levels of T-2/HT-2, OTA and patulin are being ingested. Additional concern arises from the fact that the single mycotoxin approach in the definition of PMDTI values ignores the potentially additive and/or synergic toxicity of mycotoxins commonly found simultaneously, especially in cereals. Very recently, Speijers and Speijers (2004) highlighted that mycotoxins of similar structure and of the same families, could likely have similar modes of action and/or toxicity profiles, thus exerting only additive effects. To the authors, in terms of risk assessment, these mycotoxins could be dealt with by establishing a group TDI or PTWI, whereas in terms of risk assessment those mycotoxins which interact in a synergistic manner would be of more concern.

Considering several factors such as: i) frequency and level of contamination; ii) relative importance of the food in a diet; and iii) available toxicological information, there are specific situations that should be noted. Among the mycotoxins considered, OTA should be monitored with the maximum attention. This is due to the fact that almost all foods and particularly many staple foods are contaminated by OTA, leading to potentially high total dietary intake. In addition, questions remain regarding its toxicological mechanism of action. As regards specific food categories, cereals should be first on a hypothetical 'warning list' due to their worldwide dietary importance, the natural co-occurrence of almost all possible combinations of the main mycotoxins, and the frequently high levels of contamination. The occurrence of fusariotoxins in cereals is another emerging issue. Finally, additional warnings regarding exposure of children merit the utmost consideration by researchers and public authorities. Some peculiar regional situations should also to be monitored with great attention. Thus, researchers are called to gather further toxicological and epidemiological information. The detection of biomarkers in biological fluids can be an useful tool for establishing effective mycotoxin exposure levels. Dietary strategies that can prevent or reduce the toxic effects of mycotoxins should also be further explored. Public authorities are charged to set worldwide harmonised regulatory limits apart from economic interests.

Acknowledgements

Authors express gratitude to Dr. Maja Perajca for critical evaluation of the data on the putative relationship between ochratoxin A and Balkan endemic nephropathy.

References

Abdulrazzaq, Y.M., N. Osman, Z.M. Yousif and S. Al-Falahi. 2003. Aflatoxin M_1 in breast-milk of UAE women. Ann. Trop. Paediatr. 23:173-179.

Abid, S., W. Hassen, A. Achour, H. Skhiri, K. Maaroufi, F. Ellouz, E. Creppy and H. Bacha. 2003. Ochratoxin A and human chronic nephropathy in Tunisia: is the situation endemic? Hum. Exp. Toxicol. 22:77-84.

Abnet, C.C., C.B. Borkowf, Y.L. Qiao, P.S. Albert, E. Wang, A.H. Merrill, Jr., S.D.

Mark, Z.W. Dong, P.R. Taylor and S.M. Dawsey. 2001. Sphingolipids as biomarkers of fumonisin exposure and risk of oesophageal squamous cell carcinoma in china. Cancer Causes Control. 12:821-828.

Abouzied M.M., A.D. Horvath, P.M. Podlesny, N.P. Regina, V.D. Metodiev, R.M. Kamenova-Tozeva, N.D. Niagolova, A.D. Stein, E.A. Petropoulos and V.S. Ganev. 2002. Ochratoxin A concentrations in food and feed from a region with Balkan Endemic Nephropathy. Food Addit. Contam. 19:755-764.

Abramson, D., J.T. Mills, R.R. Marquardt and A.A. Frohlich. 1997. Mycotoxins in fungal contaminated samples of animal feed from western Canada, 1982-1994. Can. J. Vet. Res. 61:49-52.

Aksit, S., S. Caglayan, I. Yaprak and S. Kansoy. 1997. Aflatoxin: is it a neglected threat for formula-fed infants? Acta Paediatr. Jpn. 39:34-36.

Ali, N., Sardjono, A. Yamashita and T. Yoshizawa. 1998. Natural co-occurrence of aflatoxins and *Fusarium* mycotoxins (fumonisins, deoxynivalenol, nivalenol and zearalenone) in corn from Indonesia. Food Addit. Contam. 15:377-384.

Almeida, A.P., H. Fonseca, A.L. Fancelli, G.M. Direito, E.M. Ortega and B. Correa. 2002. Mycoflora and fumonisin contamination in Brazilian corn from sowing to harvest. J. Agric. Food Chem. 50:3877-3882.

Andersen, B., J. Smedsgaard and J.C. Frisvad. 2004 *Penicillium expansum*: consistent production of patulin, chaetoglobosins, and other secondary metabolites in culture and their natural occurrence in fruit products. J. Agric. Food Chem. 52:2421-2428.

Arlt, V.M., D. Ferluga, M. Stiborova, A. Pfohl-Leszkowicz, M. Vukelic, S. Ceovic, H.H. Schmeiser and J-P. Cosyns. 2002. Is aristolochic acid a risk factor for Balkan endemic nephropathy-associated urothelial cancer? Int. J. Cancer 101:500-502.

Bankole, S.A. and O.O. Mabekoje. 2004. Occurrence of aflatoxins and fumonisins in preharvest maize from southwestern Nigeria. Food Addit. Contam. 21:251-255.

Beckett, S.T. 1994. *Industrial Chocolate Manufacture and Use. 2nd edition*, Blackie Academic & Professional/Chapman & Hall. London.

Begum, F. and N. Samajpati. 2000. Mycotoxin production on rice, pulses and oilseeds. Naturwissenschaften 87:275-277.

Beretta, B., A. Gaiaschi, C.L. Galli and P. Restani. 2000. Patulin in apple-based foods: Occurrence and safety evaluation. Food Addit. Contam. 17:399-406.

Bhat, R.V., P.H. Shetty, R.P. Amruth and R.V. Sudershan. 1997. A foodborne disease outbreak due to the consumption of moldy sorghum and maize containing fumonisin mycotoxins. J. Toxicol. Clin. Toxicol. 35:249-255.

Breitholtz-Emanuelsson, A., M. Olsen, A. Oskarsson, I. Palminger and K. Hult. 1993. Ochratoxin A in cow's milk and in human milk with corresponding human blood samples. J. AOAC Int. 76:842-846.

Burdaspal, P.A. and T.M. Legarda. 1999. Ochratoxina A en vinos, mostos y zumos de uva elaborados en España y en otros paises europeos. Alimentaria 6:107-113.

Cantafora, A., M. Grossi, M. Miraglia and L. Benelli 1983. Determination of ochratoxin A in coffee beans using reversed-phase high performance liquid chromatography. Riv. Soc. Ital. Sci. Aliment. 12:103-108.

Carvajal, M., A. Bolanos, F. Rojo and I. Mendez. 2003a. Aflatoxin M_1 in pasteurised and ultrapasteurised milk with different fat content in Mexico. J. Food Prot. 66:1885-1892.

Carvajal, M., F. Rojo, I. Mendez and A. Bolanos. 2003b. Aflatoxin B_1 and its interconverting metabolite aflatoxicol in milk: the situation in Mexico. Food Addit. Contam. 20:1077-1086.

CAST. 2003. Mycotoxins: Risks in Plant,

Animal, and Human Systems. Council for Agricultural Science and Technology. Task Force Report, n 139, Ames, IOWA, USA.

Castelo, M.M., S.S. Sumner and L.B. Bullerman. 1998. Occurrence of fumonisins in corn-based food products. J. Food Prot. 61:704-707.

Ceovic, S., A. Hrabar and M. Šaric. 1992. Epidemiology of Balkan endemic nephropathy. Food Chem. Toxicol. 30:183-188.

Chakrabarti, D.K. and S. Ghosal. 1986. Occurrence of free and conjugated 12,13-epoxytrichothecenes and zearalenone in banana fruits infected with *Fusarium moniliforme*. Appl. Environ. Microbiol. 51:217-219.

Chelule, P.K., N. Gqaleni, M.F. Dutton and A.A. Chuturgoon. 2001. Exposure of rural and urban populations in KwaZulu Natal, South Africa, to fumonisin B(1) in maize. Environ. Health Perspect. 109:253-256.

Chiavaro, E., A. Lepiani, F. Colla, P. Bettoni, E. Pari and E. Spotti. 2002. Ochratoxin A determination in ham by immunoaffinity clean-up and a quick fluorometric method. Food Addit. Contam. 19:575-581.

Chu, F.S. and G.Y. Li. 1994. Simultaneous occurrence of fumonisin B1 and other mycotoxins in moldy corn collected from the People's Republic of China in regions with high incidences of oesophageal cancer. Appl. Environ. Microbiol. 60:847-852.

Chulze, S.N., M.L. Ramirez, M.C. Farnochi, M. Pascale, A. Visconti and G. March. 1996. *Fusarium* and fumonisin occurrence in Argentinean corn at different ear maturity stages. J. Agric. Food Chem. 44:2797-2801.

Cirillo, T., A. Ritieni, F. Galvano and R. Amodio Cocchieri. 2003a. Natural co-occurrence of deoxynivalenol and fumonisins B_1 and B_2 in Italian marketed foodstuffs. Food Addit. Contam. 20:566-5671.

Cirillo, T., A. Ritieni, M. Visone and R.A. Cocchieri. 2003b. Evaluation of conventional and organic Italian foodstuffs for deoxynivalenol and fumonisins B(1) and B(2). J. Agric. Food Chem. 51:8128-8131.

Commission of the European Communities. 2002. OC Regulation 2002/472, 12.03.2002, Official Journal of the OC, L 75/18.

Curtui, V.G., M. Gareis, E. Usleber and E. Martlbauer. 2001. Survey of Romanian slaughtered pigs for the occurrence of mycotoxins ochratoxins A and B, and zearalenone. Food Addit Contam. 18:730-738.

de Nijs, M., E.A. Sizoo, A.E. Vermunt, S.H. Notermans and H.P. van Egmond. 1998. The occurrence of fumonisin B_1 in maize-containing foods in The Netherlands. Food Addit. Contam. 15:385-388.

de Sylos, C.M. and D.B. Rodriguez-Amaya. 1999. Incidence of patulin in fruits and fruit juices marketed in Campinas, Brazil. Food Addit. Contam. 16:71-74.

Delage, N., A. d'Harlingue, B. Colonna Ceccaldi and G. Bompeix. 2003. Occurrence of mycotoxins in fruit juices and wine. Food Contr. 14:225-227.

Desjardins, A.E., G. Manandhar, R.D. Plattner, C.M. Maragos, K. Shrestha and S.P. McCormick. 2000. Occurrence of *Fusarium* species and mycotoxins in Nepalese maize and wheat and the effect of traditional processing methods on mycotoxin levels. J. Agric. Food Chem. 48:1377-1383.

D'Mello, J.P.F., J.K. Porter, A.M.C. Macdonald and C.M. Placinta. 1997. *Fusarium* mycotoxins. In: *Handbook of Plant and Fungal Toxicants* (J.P.F. D'Mello, ed). CRC Press, Boca Raton, FL, pp. 287-301.

Doko, M.B., C. Canet, N. Brown, E.W. Sydenham, S. Mpuchane and B.A. Siame. 1996. Natural co-occurrence of fumonisins and zearalenone in cereals and cereal-based foods from eastern and southern Africa. J. Agric. Food Chem. 44:3240-3243.

Doko, M.B., S. Rapior, A. Visconti and J.E. Schjoth. 1995. Incidence and levels of

fumonisin contamination in maize genotypes grown in Europe and Africa. J. Agric. Food Chem. 43:429-434.
Dragacci, S., F. Grosso, R. Bire, J.M. Fremy and S.A. Coulon. 1999. French monitoring programme for determining ochratoxin A occurrence in pig kidneys. Nat. Toxins. 7:167-173.
Drusch, S. and W. Ragab. 2003. Mycotoxins in fruits, fruit juices, and dried fruits J. Food Prot. 66:1514-1527.
Dutton, M.F. and A. Kinsey. 1995. Occurrence of mycotoxins in cereals and animal feedstuffs in Natal, South Africa 1994. Mycopathologia 131:31-36.
Dutton, M.F. and A. Kinsey. 1996. A note on the occurrence of mycotoxins in cereals and animal feedstuffs in Kwazulu Natal, South Africa 1984-1993. S. Afr. J. Anim. Sci. 26:53-57.
Eder, R., E. Paar, W. Edinger and H. Lew. 2002. Untersuchungen uber den Gehalt an Ochratoxin A in Weinen, insbesondere Pradikatsweinen aus Osterreich. Mitteilungen Klosterneuburg, Rebe und Wein, Ostbau und Fruchteverwertung 52:25-131.
Eko-Ebongue, S., J.P. Antalick., M. Bonini, A.M. Betbeder, J.G. Faugere, K. Maarou, H. Bacha, M. Achour, Y. Grosse, A. Pfohl-Leszkowicz and E.E. Creppy. 1994. Determination des teneurs en ochratoxine A, vitamine E et vitamine A dans les sérums humains de France et de Tunisie: recherche d'une correlation. Annales de Falsification de l'Expertise Chimique 929:213-224.
el-Nezami, H.S., G. Nicoletti, G.E. Neal, D.C. Donohue and J.T. Ahokas. 1995. Aflatoxin M_1 in human breast milk samples from Victoria, Australia and Thailand. Food Chem. Toxicol. 33:173-179.
el-Sayed, A.M., E.A. Soher and A.F. Sahab. 2003. Occurrence of certain mycotoxins in corn and corn-based products and thermostability of fumonisin B_1 during processing. Nahrung. 47:222-225.
el-Serag H.B. and A.C. Mason. 1999. Rising incidence of hepatocellular carcinoma in the United States. N. Engl. J. Med. 340:745-750.
Eskola M., P. Parikka and A. Rizzo. 2001. Trichothecenes, ochratoxin A and zearalenone contamination and *Fusarium* infection in Finnish cereal samples in 1998. Food Addit. Contam. 18:707-718.
European Commission. 2002. Assessment of dietary intake of patulin by the population of EU Member States. Report of experts participating in Task 3.2.8. Directorate-General Health and Consumer Protection.
FAO. 2001. Food and Agriculture Organization. Food Nutr. Paper 74:81-415.
Fernandez, C., M.E. Stack and S.M. Musser. 1994. Determination of deoxynivalenol in 1991 U.S. winter and spring wheat by high performance thin layer chromatography. J. AOAC Int. 77:628-630.
Festas, I., P. Herbert, L. Santos, M. Cabral, P. Barros and A. Alves. 2000. Ochratoxin A in some Portuguese wines: method validation and screening in Port Wine and Winho Verde. Am. J. Enology Viticulture 51:50-154.
Filali, A., L. Ouammi, A.M. Betbeder, I. Baudrimont, R. Soulaymani, A. Benayada and E.E. Creppy. 2001. Ochratoxin A in beverages from Morocco: a preliminary survey. Food Addit. Contam. 18:565-568.
Filali, A., A.M. Betbeder, I. Baudrimont, A. Benayad, R. Soulaymani and E.E. Creppy. 2002. Ochratoxin A in human plasma in Morocco: a preliminary survey. Human Exper. Toxicol. 21:241-245.
Fink-Gremmels, J., A. Jahn and M.J. Blom. 1995. Toxicity and metabolism of ochratoxin A. Nat. Toxins 3:214-220.
Flannigan, B. 1991. Mycotoxins. In: *Toxic Substances in Crop Plants. The Royal Society of Chemistry* (J.P.F. D'Mello, C.M. Duffus and J.H. Duffus, eds). Cambridge, pp. 226-257.
Franceschi, S., E. Bidoli, A.E. Barón and C. La Vacchia. 1995. Maize and risk of cancer of the oral cavity, pharynx and

esophagus in northeastern Italy. J. Natl. Cancer Inst. 82:1407-1411.

Frank, H.K. 1991. Food contamination by ochratoxin A in Germany. IARC Sci. Publ. 115:77-81.

Fremy, J.M. and S. Dragacci. 1999. Mycotoxines et produits laitiers. In: *Les mycotoxines dans l'alimentation* (A. Pfohl-Leszkowicz, ed). Evaluation et gestion du risque. Editions TEC & DOC, Paris, pp. 353-369.

Frisvad, J.C. and U. Thrane. 1996. Mycotoxin production by foodborne fungi. In: *Introduction to Food-Borne Fungi* (R.A. Samson, E.S. Hoekstra, J.C. Frisvad and O. Filtenborg, eds). Centraalbureau voor Schimmelcultures: Baarn, The Netherlands, 5th Edition, pp. 251-260.

Furlong, E.B., L.M.V. Soares, C.C. Lasca and E.Y. Kohara. 1995. Mycotoxins and fungi in wheat harvested during 1990 in test plots in the state of São Paulo Brazil. Mycopathologia 131:185-190.

Gallaz, L. and R. Stalder. 1976. Ochratoxin A in Kaffee. Chem. Mikrobiol. Technol. Lebensm. 4:147-149.

Galvano, F., V. Galofaro, and G. Galvano. 1996. Occurrence and stability of aflatoxin M_1 in milk and milk products. A worldwide review. J. Food Prot. 59:1079-1090.

Galvano, F., V. Galofaro, A. De Angelis, M. Galvano, M. Bognanno and G. Galvano. 1998. Survey of the occurrence of aflatoxin M_1 in dairy products marketed in Italy. J. Food Prot. 61:738-741.

Galvano, F., V. Galofaro, M. Bognanno, A. De Angelis and G. Galvano. 2001. Survey of the occurrence of aflatoxin M_1 in dairy products marketed in Italy. Second year of observation. Food Addit. Contam. 18:644-646.

Gamanya, R. and L. Sibanda. 2001 Survey of *Fusarium moniliforme* (*F. verticillioides*) and production of fumonisin B_1 in cereal grains and oilseeds in Zimbabwe Int. J. Food Microbiol. 71:145-149.

Gareis, M., E. Martlbauer, J. Bauer and B. Gedek. 1988. Determination of ochratoxin A in human milk. Z. Lebensm. Unters. Forsch. 186:114-117.

Garrido, N.S., M.H. Iha, M.R. Santos Ortolani and R.M. Duarte Favaro. 2003. Occurrence of aflatoxins M(1) and M(2) in milk commercialised in Ribeirao Preto-SP., Brazil. Food Addit. Contam. 20:70-73.

Gelderblom, W.C.A., K. Jaskiewicz, W.F.O. Marasas, P.G. Thiel, R.M. Horak, R. Vleggaar and N.P.J. Kriek. 1988. Fumonisins - novel mycotoxins with cancer-promoting activity produced by *Fusarium moniliforme*. Appl. Environ. Microbiol. 54:1806–1811.

Grabarkiewicz-Szczesna, J., E. Foremska and Golinski. 1996. Distribution of trichothecene mycotoxins in maize ears infected with *F. graminearum* and *F. crookwellense*. Mycotoxin Res. 12:45-50.

Grosso, F., S. Said, I. Mabrouk, J.M. Fremy , M. Castegnaro,M. Jemmali and S. Dragacci. 2003. New data on the occurrence of ochratoxin A in human sera from patients affected or not by renal diseases in Tunisia. Food Chem Toxicol. 41:1133-1140.

Groves, F.D., L. Zhang, Y.S. Chang, P.F. Ross, H. Casper, W.P. Norred, W.C. You and J.F. Fraumeni, Jr. 1999. *Fusarium* mycotoxins in corn and corn products in a high-risk area for gastric cancer in Shandong Province, China. J. AOAC Int. 82:657-662.

Gutema, T., C. Munimbazi and L.B. Bullerman. 2000. Occurrence of fumonisins and moniliformin in corn and corn-based food products of U.S. origin. J. Food Prot. 63:1732-1737.

Harrison, T. 2003. The importance of quality from an international confectionery industry viewpoint. ICCO international seminar on quality and financial aspects of the cocoa campaign, Abidjan, 17-19 May 2000. Probert, T. Officials deride EU ochratoxin regulations. Public Ledger, (72, 336), 4, 1 September 2003.

Hassen, W., S. Abid, A. Achour, E. Creppy

and H. Bacha. 2004. Ochratoxin A and ß(2)-microglobulinuria in healthy individuals and in chronic interstitial nephropathy patients in the Centre of Tunisia: a hot spot of ochratoxin A exposure. Toxicology 199:185-193.

Henderson, M.C., C.L. Miranda, J.F. Stevens, M.L. Deinzer and D.R. Buhler. 2000. *In vitro* inhibition of human P450 enzymes by prenylated flavonoids from hops, *Humulus lupulus*. Xenobiotica 30:235-251.

Hesseltine, C.W., O.L. Shotwell, J.J. Ellis and R.D. Stubblefield. 1966. Aflatoxin formation by *Aspergillus flavus*. Bacteriol. Rev. 30:795-805.

Hietaniemi, V. and J. Kumpulainen. 1991. Contents of *Fusarium* toxins in Finnish and imported grains and feeds. Food Addit. Contam. 8:171-182.

Hlywka, J.J. and L.B. Bullerman. 1999. Occurrence of fumonisin B1 and B2 in beer. Food Addit. Contam. 16:319-324.

Hult, K. and R. Fuchs. 1986. Analysis and dynamics of ochratoxin A in biological systems. In: *Mycotoxins and Phycotoxins* (P.S. Steyn and R. Vleggaar, eds). Amsterdam: Elsevier Science Publishers BV, pp. 365-376.

IARC. 1993. *IARC Monograph on the evaluation of carcinogenic risk to humans*. Vol. 56: Toxins derived from *F. moniliforme*: Fumonisins B 1 and B 2 and fusarin C: In some naturally occurring substances: Food items and constituents, heterocyclic aromatic amines and mycotoxins. International Agency for Research on Cancer. IARC Press, Lyons, pp. 445-466.

IARC. 2002. *IARC monographs on the evaluation of carcinogenic risks to humans*. Vol 82: Some traditional herbal medicines, some mycotoxins, naphthalene and styrene. International Agency for Research on Cancer. IARC Press, Lyons.

Jackson, L.S., T. Beacham-Bowden, S.E. Keller, C. Adhikari, K.T. Taylor, S.J. Chirtel and R.I. Merker. 2003. Apple quality, storage and washing treatments affect patulin levels in apple cider. J. Food Prot. 66(4):618-624.

Jacobson, W.C. and H.G. Wiseman. 1974. The transmission of aflatoxin B_1 into eggs. Poult. Sci. 53:1743-1745.

Jestoi, M., A. Ritieni and A. Rizzo. 2004. Analysis of the *Fusarium* mycotoxins fusaproliferin and trichothecenes in grains using gas chromatography-mass spectrometry. J. Agric. Food Chem. 52:1464-1469.

Jimenez, AM., A. Lopez de Cerain, E. Gonzalez-Penas and J. Bello. 2001. Determination of ochratoxin A in pig liver-derived pates by high-performance liquid chromatography. Food Addit. Contam. 18:559-563.

Joint FAO/WHO Expert Committee on Food Additives (JECFA). 2001. Safety evaluation of certain mycotoxins in food. Prepared by the Fifty-sixth Meeting of JECFA. WHO food additives series 47/FAO food and nutrition 74-International Programme on Chemical Safety (IPCS). WHO, Geneva.

Jonsyn, F.E., S.M. Maxwell and R.G. Hendrickse. 1995. Ochratoxin A and aflatoxins in breast milk samples from Sierra Leone. Mycopathologia 131:121-126.

Jorgensen, K. 1998. Survey of pork, poultry, coffee, beer and pulses for ochratoxin A. Food Addit. Contam. 15:550-554.

Jorgensen, K. and A. Petersen. 2002. Content of ochratoxin A in paired kidney and meat samples from healthy Danish slaughter pigs. Food Addit. Contam. 19:562-567.

Julian, A.M., P.W. Wareing, S.I. Phillips, V.F.P. Medlock, M.V. MacDonald and L.E. del Rio. 1995. Fungal contamination and selected mycotoxins in pre- and post-harvest maize in Honduras. Mycopathologia 129:5-16.

Kershaw, S.J. 1982. Occurrence of aflatoxins in oilseeds providing cocoa-butter substitutes. Appl. Environ. Microbiol. 43:1210-1212.

Kew, M.C. 2003. Synergistic interaction between aflatoxin B_1 and hepatitis B virus in hepatocarcinogenesis. Liver Int.

23:405-409.

Khalef, A., C. Zidane, A. Charef, A. Gharbi, M. Tadjerouna, A.M. Betbeder and E.E. Creppy. 1993. Ochratoxicoses humaines en Algerie. In: *Human Ochratoxicosis and its Pathologies, INSERM Vol. 231* (E. Creppy, M. Castegnaro and G. Dirheimer, eds). John Libbey Eurotext Ltd, Paris, pp. 123-127.

Kim, E.K., D.H. Shon, D. Ryu, J.W. Park, H.J. Hwang and Y.B. Kim. 2000. Occurrence of aflatoxin M1 in Korean dairy products determined by ELISA and HPLC. Food Addit. Contam. 17:59-64.

Kpodo, K., U. Thrane, and B. Hald. 2000. Fusaria and fumonisins in maize from Ghana and their co-occurrence with aflatoxins. Int. J. Food Microbiol. 61:147-157.

Kuiper-Goodman, T. 1995. Mycotoxins: risk assessment and legislation. Toxicol. Lett. 82-83:853-859.

Kuiper-Goodman, T. 1996. Risk assessment of ochratoxin A: an update. Food Addit. Contam. 13(Suppl):53-57.

Langseth, W. and T. Rundberget. 1999. The occurrence of HT-2 toxin and other trichothecenes in Norwegian cereals. Mycopathologia 147:157-165.

Langseth, W. and O. Elen. 1996. Differences between barley, oats and wheat in the occurrence of deoxynivalenol and other trichothecenes in Norwegian grain. J. Phytopathol. 144:113-118.

Larsen, T.O., J.C. Frisvad, G. Ravn and T. Skaaning. 1998. Mycotoxin production by *Penicillium expansum* on black currant and cherry juice. Food Addit. Contam. 15:671–675.

Lauren, D.R., D.J. Jensen, W.A. Smith, B.W. Dow and S.T. Sayer. 1996. Mycotoxins in New Zealand maize: a study of some factors influencing contamination levels in New Zealand grain. J. Crop Hortic. Sci. 24:13-20.

Leblanc J.C., L. Malmauret, D. Delobel and P. Verger. 2002. Simulation of the exposure to deoxynivalenol of French consumers of organic and conventional foodstuffs. Regul. Toxicol. Pharmacol. 36:149-54.

Levi, C.P., H.L. Trenk and H.K. Mohr 1974. Study of the occurrence of ochratoxin A in green coffee beans. J. Assoc. Off. Anal. Chem. 57:866-870.

Li, F.Q., Y.W. Li, X.Y. Luo and T. Yoshizawa. 2002. *Fusarium* toxins in wheat from an area in Henan Province, PR China, with a previous human red mould intoxication episode. Food Addit Contam. 19:163-167.

Lopez De Cerain, A., E. Gonzáles-Penas, A.M. Jimenez and J. Bello. 2002. Contribution to the study of ochratoxin A in Spanish wines. Food Addit. Contam. 19:1058-1064.

Lötzsch, R. and L. Leistner. 1976. Aflatoxin-Rückstände in Hühnereiern und Eiprodukten. Fleischwirtschaft 12:1777-1785.

Maarou, K., A. Achour, A.M. Betbeder, M. Hamammi, F. Ellouz, E.E. Creppy and H. Bacha. 1995a. Foodstuffs and human blood contamination by the mycotoxin ochratoxin A: correlation with chronic interstitial nephropathy in Tunisia. Arch. Toxicol. 69:552-558.

Maarou, K., A. Achour, M. Hamammi, M. El May, A.M. Betbeder, F. Ellouz, E.E. Creppy and H. Bacha. 1995b. Ochratoxin A in human blood in relation to nephropathy in Tunisia. Human Exper. Toxicol. 14:609-615.

MacDonald, S., P. Wilson, K. Barnes, A. Damant, R. Massey, E. Mortby and M.J. Shepherd. 1999. Ochratoxin A in dried vine fruit: method development and survey. Food Addit. Contam. 16:253-260.

MAFF. 1999. 1998 survey of retail products for ochratoxin A. Food Surveillance Information Sheet No. 185. *http://archive.food.gov.uk/maff/archive/food/infsheet/1999/no185/185ochra.htm.*

Majerus, P., H. Bresch and H. Otteneder. 2000. Ochratoxin A in wines, fruit juices and seasonings. Arch. Lebensmittelhygiene 51:85-97.

Mally, A., H. Zepnik, P. Wanek, E. Eder, K.

Dingley, H. Ihmels, W. Volkel and W. Dekant. 2004. Ochratoxin A: lack of formation of covalent DNA adducts. Chem. Res. Toxicol. 17:234-242.

Manthey, F.A., C.E. Wolf-Hall, S. Yalla, C. Vijayakumar and D. Carlson. 2004. Microbial loads, mycotoxins, and quality of durum wheat from the 2001 harvest of the northern plains region of the United States. J. Food Prot. 67:772-780.

Marasas, W.F.O. 1995. Fumonisins: their implications for human and animal health. Nat. Toxins 3:193-198.

Marasas, W.F. 1997. Risk assessment of fumonisins produced by *Fusarium moniliforme* in corn. Cereal Res. Commun. 25:399-406.

Marasas, W.F. 2001. Discovery and occurrence of the fumonisins: a historical perspective. Environ Health Perspect. 109:239-243.

Marasas, W.F., R.T. Riley, K.A. Hendricks, V.L. Stevens, T.W. Sadler, J. Gelineau-Van Waes, S.A. Missmer, J. Cabrera, O. Torres, W.C. Gelderblom, J. Allegood, C. Martinez, J. Maddox, J.D. Miller, L. Starr, M.C. Sullards, A.V. Roman, K.A. Voss, E. Wang and A.H. Merrill, Jr. 2004. Fumonisins disrupt sphingolipid metabolism, folate transport, and neural tube development in embryo culture and *in vivo*: a potential risk factor for human neural tube defects among populations consuming fumonisin-contaminated maize. J. Nutr. 134:711-716.

Markaki, P., C. Delpont-Binet, F. Grosso and S. Dragacci. 2001. Determination of ochratoxin A in red wine and vinegar by immunoaffinity high-pressure liquid chromatography. J. Food Prot. 64:533-537.

Marquardt, R.R. and A.A. Frohlich. 1992. A review of recent advances in understanding ochratoxicosis. J. Anim. Sci. 70:3968-3988.

Martins, M.L., A. Gimeno, H.M. Martins and F. Bernardo. 2002. Co-occurrence of patulin and citrinin in Portuguese apples with rotten spots. Food Addit. Contam. 19(6):568-574.

Micco, C., MA. Ambruzzi, M. Miraglia, C. Brera, R. Onori and L. Benelli. 1991. Contamination of human milk with ochratoxin A. IARC Sci. Publ. 115:105-108.

Micco, C., M. Miraglia, C. Brera, S. Corneli and A. Ambruzzi. 1995. Evaluation of ochratoxin A level in human milk in Italy. Food Addit. Contam. 12:351-354.

Micco, C., C. Brera, M. Miraglia and R. Onori. 1987. HPLC determination of the total content of aflatoxins in naturally contaminated eggs in free and conjugated forms. Food Addit. Contam. 4:407-414.

Miraglia, M., A. de Dominicis, C. Brera, S. Corneli, E. Cava, E. Menghetti and E. Miraglia. 1995. Ochratoxin A levels in human milk and related food samples: an exposure assessment. Nat. Toxins 3:436-444.

Molto, G, M.M. Samar, S. Resnik, E.J. Martinez and A. Pacin. 2000. Occurrence of trichothecenes in Argentinean beer: a preliminary exposure assessment. Food Addit. Contam. 17:809-813.

Mortimer, D.N., I. Parker, M.J. Shepherd and J. Gilbert. 1985. A limited survey of retail apple and grape juices for the mycotoxin patulin. Food Addit. Contam. 2:165-170.

Mousing, J., J. Kyrval, T.K. Jensen, B. Aalbaek, J. Buttenschon, B. Svensmark and P. Willeberg. 1997. Meat safety consequences of implementing visual postmortem meat inspection procedures in Danish slaughter pigs. Vet. Rec. 140:472-477.

Muller, H.M. and K. Schwadorf. 1993. A survey of the natural occurrence of *Fusarium* toxins in wheat grown in a southwestern area of Germany. Mycopathologia 121:115-121.

Nakajima M., H. Tsubouchi and M. Miyabe. 1999. A survey of ochratoxin A and aflatoxins in domestic and imported beers in Japan by immunoaffinity and liquid chromatography. J. AOAC Int. 82:897-902.

Northolt, M.D., J.C. Frisvad and R.A. Samson. 1996. Occurrence of food-borne fungi and factors for growth. In: *Introduction to food-borne fungi* (5th ed, pp. 243-250). (R.A. Samson, E.S. Hoekstra, J.C. Frisvad and O. Filtenborg, eds.), Centraalbureau voor Schimmelcultures: Baarn, The Netherlands.

Odhav, B. and V. Naicker. 2002. Mycotoxins in South African traditionally brewed beers. Food Addit. Contam. 19:55-61.

Ogundero, V.W. 1987. Crown-rot fungi of Nigerian bananas cv. Robusta and the effects of benomyl on their exoenzymes. J. Basic Microbiol. 27:43-47.

Oliveira, C.A., P.M. Germano, C. Bird and C.A. Pinto. 1997. Immunochemical assessment of aflatoxin M1 in milk powder consumed by infants in Sao Paulo, Brazil. Food Addit. Contam. 14:7-10.

Oliveira, C.A.F., E. Kobashigawa, T.A. Reis, L. Mestieri, R. Albuquerque and B. Correa. 2000. Aflatoxin B_1 residues in eggs of laying hens fed a diet containing different levels of the mycotoxin. Food Addit. Contam. 17:459-462.

Olivier, M., S.P. Hussain, C. Caron de Fromentel, P. Hainaut and C.C. Harris. 2004. TP53 mutation spectra and load: a tool for generating hypotheses on the etiology of cancer. IARC Sci. Publ.157:247-270.

Olsen, M., I. Thorup, I. Knudsen, J. Larsen, B. Hald and J. Olsen. 1991. In: *Nordiske Seminar-og Arbejds-rapporter, Health Evaluation of Ochratoxin A in Food Products, Vol. 545*. Nordic Council of Ministers, Copenhagen.

Ono, E.Y., M.A. Ono, F.Y. Funo, A.E. Medinal, T.C. Oliveira, O. Kawamura, Y. Ueno and E.Y. Hirooka. 2001. Evaluation of fumonisin-aflatoxin co-occurrence in Brazilian corn hybrids by ELISA. Food Addit. Contam. 18:719-729.

Oruc, H.H. and S. Sonal. 2001. Determination of aflatoxin M_1 levels in cheese and milk consumed in Bursa, Turkey. Vet. Hum. Toxicol. 43:292-293.

Otteneder, H. and P. Majerus. 2000. Occurrence of ochratoxin A in wines: influence of the type of wine and its geographical origin. Food Addit. Contam. 17:793-798.

Ough, C.S. and C.A. Corison. 1980. Measurement of patulin in grapes and wines. J. Food Sci. 43:476-478.

Pacin, A.M., S.L. Resnik, M.S. Neira, G. Molto and E. Martinez. 1997. Natural occurrence of deoxynivalenol in wheat, wheat flour and bakery products in Argentina. Food Addit. Contam. 14:327-331.

Panigrahi, S. 1997. Alternaria toxins. In: *Handbook of Plant and Fungal Toxicants*. (J.P.F. D'Mello, ed), CRC Press, Boca Raton, FL, pp. 319-337.

Park, J.W., E.K. Kim and Y.B. Kim. 2004. Estimation of the daily exposure of Koreans to aflatoxin B1 through food consumption. Food Addit. Contam. 21:70-5.

Park, J.W., E.K. Kim, D.H. Shon and Y.B. Kim. 2002. Natural co-occurrence of aflatoxin B_1, fumonisin B_1 and ochratoxin A in barley and corn foods from Korea. Food Addit. Contam. 19:1073-1080.

Park, D.L. and A.E. Pohland. 1986. A rationale for the control of aflatoxin in animal feeds. In: *Mycotoxins and Phycotoxins* (P.S. Steyn and R. Vleggaar, eds). Amsterdam, Elsevier Applied Science, pp. 473-482.

Peraica, M., A.M. Domijan, M. Matasin, A. Lucic, B. Radic, F. Delas, M. Horvat, I. Bosanac, M. Balija and D. Grgicevic. 2001. Variations of ochratoxin A concentration in the blood of healthy populations in some Croatian cities. Arch. Toxicol. 75:410-414.

Perkowski, J., R.D. Plattner, P. Golinski and R.F. Vesonder. 1990. Natural occurrence of deoxynivalenol, 3-acetyldeoxynivalenol, 15-acetyl-deoxynivalenol, nivalenol, 4,7-dideoxynivalenol and zearalenone in Polish wheat. Mycotoxin Res. 6:7-12.

Petersen, A. and I. Thorup. 2001. Preliminary evaluation of fumonisins by the Nordic countries and occurrence of fumonisins (FB_1 and FB_2) in corn-based foods on the Danish market. Food Addit. Contam. 18:221-226.

Petkova-Bocharova, T., I.N. Chernozemsky and M. Castegnaro. 2002. Balkan endemic nephropathy and associated urinary tract tumours: a review on aetiological causes and the potential role of mycotoxins. Food Addit. Contam. 19:282-302.

Petkova-Bocharova, T., C. El Adlouni, V. Faucet, A. Pfhohl-Leszkowicz and P.G. Mantle. 2003. Analysis for DNA adducts, ochratoxin A content and enzyme expression in kidneys of pigs exposed to mild experimental chronic ochratoxicosis. Med. Biol. 10:111-115.

Pfohl-Leszkowicz, A. 1999. Effets biologiques et pathologies. In: *Les mycotoxines dans l'alimentation. Evaluation et gestion du risque* (A. Pfohl-Leszkowicz, ed). Editions TEC & DOC, Paris, pp. 69-118.

Pfohl-Leszkowicz, A., T. Petkova-Bocharova, I.N. Chernozemsky and M. Castegnaro. 2002. Balkan endemic nephropathy and associated urinary tract tumours: a review on aetiological causes and the potential role of mycotoxins. Food Add. Contam. 19:282-302.

Phillips, S.I., P.W. Wareing, A. Dutta, S. Panigrahi and V. Medlock. 1996. The mycoflora and incidence of aflatoxin, zearalenone and sterigmatocystin in dairy feed and forage samples from eastern India and Bangladesh. Mycopathologia 133:15-21.

Pieters, M.N., J. Freijer, B.J. Baars, D.C. Fiolet, J. van Klaveren, and W. Slob. 2002. Risk assessment of deoxynivalenol in food: concentration limits, exposure and effects. Adv. Exp. Med. Biol. 504:235-248.

Pietri, A., T. Bertuzzi, F. Gadolini and A. Gualla. 2001. Aflatoxin B_1 residues in eggs of laying hens fed a naturally contaminated diet. Proceedings of the A.S.P.A. XIV Congress, Firenze, June 12-15, pp. 448-450.

Pietri, A., T. Bertuzzi, M. Moschini and G. Piva. 2003. Aflatoxin M_1 occurrence in milk samples destined for Parmigiano Reggiano Cheese production. Ital. J. Food Sci. 15:301-306.

Pietri, A., T. Bertuzzi, L. Pallaroni and G. Piva. 2001. Occurrence of ochratoxin A in Italian wines. Food Addit. Contam. 18:647-654.

Pietri, A., T. Bertuzzi, L. Pallaroni and G. Piva. 2004. Occurrence of mycotoxins and ergosterol in maize harvested over 5 years in Northern Italy. Food Addit. Contam. 21:479-487.

Pineiro, M.S., G.E. Silva, P.M. Scott, G.A. Lawrence and M.E. Stack. 1997. Fumonisin level in Uruguayan corn products. J. AOAC Int. 80:825-828.

Pitt, J.I. 2000. Toxigenic fungi: Which are important? Med. Mycol. 38(Supp. 1):17-22.

Placinta, C.M., J.F.P. d'Mello and A.M.C. MacDonald. 1999. A review of worldwide contamination of cereal grains and animal feeds with *Fusarium* mycotoxins. Anim. Feed Sci. Tech. 78:21-37.

Qiu, M., X. Liu, Y. Wang and C. Zhang. 2001. Survey on the fumonisins intake and the urinary Sa/So ratio of people suffering from a high incidence of oesophageal cancer. Wei Sheng Yan Jiu. 30:365-367.

Radonic, M. and Z. Radoševic. 1992. Clinical features of Balkan endemic nephropathy. Food Chem. Toxicol. 30:189-192.

Raj, H.G. and P.D. Lotlikar. 1984. Urinary excretion of thiol conjugates of aflatoxin B_1 in rats and hamsters. Cancer Lett. 22:25-133.

Rasmussen, P.H., F. Ghorbani and T. Berg. 2003. Deoxynivalenol and other *Fusarium* toxins in wheat and rye flours on the Danish market. Food Addit. Contam. 20:396-404.

Rastogi, S., P.D. Dwivedi, S.K. Khanna and M. Das. 2004. Detection of aflatoxin M_1 contamination in milk and infant milk products from Indian markets by ELISA. Food Contr. 15:287-290.

Rheeder, J.P., E.W. Sydenham, W.F.O. Marasas, P.G. Thiel, G.S. Shephard, M. Schlechter, S. Stockenstrom, D.W. Cronje and J.H. Viljoen. 1995. Fungal infestation and mycotoxin contamination of South African commercial maize harvested in 1989 and 1990. S. Afr. J. Sci. 91:127-131.

Ritieni, A. 2003. Patulin in Italian commercial apple products. J. Agric. Food Chem. 51:6086-6090.

Ritieni, A., A. Moretti, A. Logrieco, A. Bottalico, G. Randazzo, S.M. Monti, R. Ferracane and V. Fogliano. 1997. Occurrence of fusaproliferin, fumonisin B_1, and beauvericin in maize from Italy. J. Agric. Food Chem. 45:4011-4016.

Rosa, C.A.R., C.E. Magnoli, M.E. Fraga, A.M. Dalcero and D.M.N. Santana. 2004. Occurrence of ochratoxin A in wine and grape juice marketed in Rio de Janeiro, Brazil. Food Addit. Contam. 21:358-364.

Rosari, R., C. Arzouyan and J. Estienne. 2000. Données recente sur la contamination des vins en France par l'ochratoxine A. Annales des Falsifications et de l'Expertise Chimique 93:401-408.

Roschenthaler, R., E.E. Creppy and G. Dirheimer. 1984. Ochratoxin A – on the mode of action of a ubiquitous mycotoxin. J. Toxicol. Toxin Rev. 1:53-86.

Roussi, V., A. Govaris, A. Varagouli and N.A. Botsoglou. 2002. Occurrence of aflatoxin M(1) in raw and market milk commercialised in Greece. Food Addit. Contam. 19:863-868.

Ryu, J.C., J.S. Yang, Y.S. Song, O.S. Kwon, J. Park and I.M. Chang. 1996. Survey of natural occurrence of trichothecene mycotoxins and zearalenone in Korean cereals harvested in 1992 using gas chromatography/mass spectrometry. Food Addit. Contam. 13:333-341.

Samson, R.A., E.S. Hoekstra, J.C. Frisvad and O. Filtenborg. 1996. *Introduction to food-borne fungi (5th ed).* The Netherlands: Centraalbureau voor Schimmelcultures, p. 322.

Schollenberger, M, S. Suchy, H.T. Jara, W. Drochner and H.M. Muller. 1999. A survey of *Fusarium* toxins in cereal-based foods marketed in an area of southwest Germany. Mycopathologia 147:49-57.

Schothorst, R.C. and A.A. Jekel. 2003. Determination of trichothecenes in beer by capillary gas chromatography with flame ionisation detection. Food Chem. 82:475-479.

Schothorst, R.C. and H.P. van Egmond. 2004. Collection of occurrence data of *Fusarium* toxins in food and assessment of dietary intake by the population of EU member states. Report from SCOOP task 3.2.10; Subtask: trichothecenes. Toxicol. Lett. 153:133-143.

Schwartz, G.G. 2002. Hypothesis: does ochratoxin A cause testicular cancer? Cancer Causes and Control. 13:91-100.

Scientific Committee on Food Opinion on Ochratoxin A, CS/CNTM/MYC/14 final, Annex II to Document XXIV/2210/98, European Commission, Bruxelles, 17 September 1998.

Scott, P.M. 1996. Mycotoxins transmitted into beer from contaminated grains during brewing. AOAC Int. 79:875-882.

Scott P.M. and G.A. Lawrence. 1997. Determination of aflatoxins in beer. J. AOAC Int. 80:1229-1234.

Scott, P.M. 1989. The natural occurrence of trichothecenes. In: *Trichothecene Mycotoxicosis: Pathophysiologic Effects, Vol. I* (V.R. Beasley, ed). CRC Press, Boca Raton, FL, pp. 1-26.

Scott, PM. 1997. Multi-year monitoring of Canadian grains and grain-based foods for trichothecenes and zearalenone. Food Addit. Contam. 14:333-339.

Scott, P.M., W.F. Miles, P. Toft and J.G. Dube. 1972. Occurrence of patulin in apple juices. J. Agric. Food Chem. 20:450-451.

Scott, P.M., J.M. Yeung, G.A. Lawrence and D.B. Prelusky. 1997. Evaluation of enzyme-linked immunosorbent assay for analysis of beer for fumonisins. Food Addit. Contam. 14:445-450.

Shenasi, M., K.E. Aidoo and A.A. Candlish. 2002. Microflora of date fruits and production of aflatoxins at various stages of maturation. Int. J. Food Microbiol.

15:113-119.

Shephard, G.S., W.F. Marasas, H. Yazdanpanah, H. Rahimian, N. Safavi, A. Zarghi, A. Shafaati and MA. Skaug. 1999. Analysis of Norwegian milk and infant formulas for ochratoxin A. Food Addit. Contam. 16:75-78.

Shephard, G.S., W.F. Marasas, H. Yazdanpanah, H. Rahimian, N. Safavi, A. Zarghi, A. Shafaati and H.R. Rasekh. 2002a. Fumonisin B(1) in maize harvested in Iran during 1999. Food Addit. Contam. 19:676-679.

Shephard, G.S., N.L. Leggott, S. Stockenström, N.I.M. Somdyala and W.F.O. Marasas. 2002b. Preparation of South African maize porridge: effect on fumonisin mycotoxin levels. S. Afr. J. Sci. 98:393-396.

Shim, W.B., J.C. Kim, J.A. Seo and Y.W. Lee. 1997. Natural occurrence of trichothecenes and zearalenone in Korean and imported beers. Food Addit. Contam. 14:1-5.

Siame, B.A., S.F. Mpuchane, B.A. Gashe, J. Allotey and G. Teffera. 1998. Occurrence of aflatoxins, fumonisin B_1, and zearalenone in foods and feeds in Botswana. J. Food Prot. 61:1670-1673.

Skaug, M.A. 1999. Analysis of Norwegian milk and infant formulas for ochratoxin A. Food Addit. Contam. 16:75-78.

Skaug, M.A., F.C. Stormer and O.D. Saugstad. 1998. Ochratoxin A: a naturally occurring mycotoxin found in human milk samples from Norway. Acta Paediatr. 87:1275-1278.

Skaug, M.A., I. Helland, K. Solvoll and O.D. Saugstad. 2001. Presence of ochratoxin A in human milk in relation to dietary intake. Food Addit. Contam. 18:321-327.

Smith, J.E. 1997. Aflatoxins. In: *Handbook of Plant and Fungal Toxicants* (J.P.F. D'Mello, ed). CRC Press, Boca Raton, FL, pp. 269-285.

Sohn, H.B., J.A. Seo and Y.W. Lee. 1999. Co-occurrence of *Fusarium* mycotoxins in mouldy and healthy corn from Korea. Food Addit. Contam. 16:153-158.

Speijers, G.J.A. and H.P. van Egmond. 1993. Worldwide ochratoxin A levels in food and feeds. Human ochratoxicosis and its pathologies. (E.E. Creppy, M. Castegnaro, G. Dirheimer, eds). John Libbey Eurotoext Lt., Montrouge, Vol 231, pp. 85-100.

Speijers, G.J. and M.H. Speijers. 2004. Combined toxic effects of mycotoxins. Toxicol. Lett. 153:91-98.

Srivastava, V.P., A. Bu-Abbas, Alaa-Basuny, W. Al-Johar, S. Al-Mufti and M.K. Siddiqui. 2001. Aflatoxin M_1 contamination in commercial samples of milk and dairy products in Kuwait. Food Addit. Contam. 18:993-997.

Stefanaki, I., E. Foufa, A. Tsatsou-Dritsa and P. Dais. 2003. Ochratoxin A concentrations in Greek domestic wines and dried vine fruits. Food Addit. Contam. 20:74-83.

Stinson, E.E., S.F. Osman, C.N. Huhtanen and D.D. Bills. 1978. Disappearance of patulin during alcoholic fermentation of apple juice. Appl. Environ. Microbiol. 36:620-622.

Stoloff, L. 1980. Aflatoxin M_1 in perspective. J. Food Prot. 43:226-230.

Stratton, G.W., A.R. Robinson, H.C. Smith, L. Kittilsen and M. Barbour. 1993. Levels of five mycotoxins in grains harvested in Atlantic Canada as measured by high performance liquid chromatography. Arch. Environ. Contam. Toxicol. 24:399-409.

Studer-Rohr, I., J. Schlatter and D.R. Dietrich. 2000. Kinetic parameters and intraindividual fluctuations of ochratoxin A plasma levels in humans. Arch. Toxicol. 74:499-510.

Sugiura,Y., K. Fukasaku, T. Tanaka, Y. Matsui and Y. Ueno. 1993. *Fusarium poae* and *Fusarium crookwellense*, fungi responsible for the natural occurrence of nivalenol in Hokkaido. Appl. Environ. Microbiol. 59:3334-3338.

Tanaka, T., S. Yamamoto, A. Hasegawa, N. Aoki, J.R. Besling, Y. Sugiura and Y. Ueno. 1990. A survey of the natural occurrence of *Fusarium* mycotoxins, deoxynivalenol, nivalenol and zearalenone, in cereals harvested in The Netherlands. Mycopathologia 110:19-22.

Tang, Z.X., G.X. Li, Y.M. Kang and Z.W. Zhong. 1998 Aflatoxin in groundnut crops in Fujian, China. J. Fujian Acad. Agric. Sci. 13:25-28.

Tangni, E.K., S. Ponchaut, M. Maudoux, R. Rozenberg, and Y. Larondelle. 2002. Ochratoxin A in domestic and imported beers in Belgium: occurrence and exposure assessment. Food Addit. Contam. 19:1169-1179.

Tateo, F., M. Bononi and E. Lubian. 2000. Survey on ochratoxin A in wines. Data concerning the market of table wines in brick. Bulletin de l'O.I.V. 73:773-783.

Torres, MR.,V. Sanchis and A.J. Ramos. 1998. Occurrence of fumonisins in Spanish beers analysed by an enzyme-linked immunosorbent assay method. Int. J. Food Microbiol. 39:139-143.

Trucksess, M.W., M.A. Dombrink-Kurtzman, V.H. Tournas and K.D. White. 2002. Occurrence of aflatoxins and fumonisins in Incaparina from Guatemala. Food Addit. Contam. 19:671-675.

Trucksess, M.W., M.E. Stack, S. Allen, and N. Barrion. 1995b. Immunoaffinity column coupled with liquid chromatography for determination of fumonisin B_1 in canned and frozen sweet corn. J. AOAC Int. 78:705-710.

Trucksess, M.W., L. Stoloff, K. Young, R.D. Wyatt, and B.L. Miller. 1983. Aflatoxicol and aflatoxins B_1 and M_1 in eggs and tissues of laying hens consuming aflatoxin-contaminated feed. Poultry Sci. 62:2176-2182.

Trucksess, M.W., F. Thomas, K. Young, M.E. Stack, W.J. Fulgueras and S.W. Page. 1995a. Survey of deoxynivalenol in U.S. 1993 wheat and barley crops by enzyme-linked immunosorbent assay. J. AOAC Int. 78:631-636.

Tseng, T.C. and C.Y. Liu. 1997. Occurrence of fumonisin B_1 and B_2 in corn-based foodstuffs in Taiwan market. Mycopathologia 137:57-61.

Tseng, T.C. and C.Y. Liu. 1999. Natural occurrence of fumonisins B(1) and B(2) in domestic maize of Taiwan J. Agric. Food Chem. 47:4799-4801.

Tseng, T.C. and C.Y. Liu. 2001. Occurrence of fumonisin B_1 in maize imported into Taiwan. Int. J. Food Microbiol. 65:23-26.

van der Stegen, G.H.D., P.J.M. Essens and J. van der Lijn. 2001. Effect of roasting conditions on reduction of ochratoxin A in coffee. J. Agric. Food Chem. 49:4713-4715.

van der Westhuizen, L., G.S. Shephard, V.M. Scussel, L.L. Costa, H.F. Vismer, J.P. Rheeder and W.F. Marasas. 2003. Fumonisin contamination and *Fusarium* incidence in corn from Santa Catarina, Brazil. J. Agric. Food Chem. 51:5574-5578.

van der Westhuizen, L., G.S Shephard, V.M. Scussel, L.L. Costa, H.F. Vismer, J.P. Rheeder and H.P. van Egmond. 2002. Worldwide regulations for mycotoxins. Adv. Exp. Med. Biol. 504:257-269.

van Egmond, H.P. 1989. Aflatoxin M_1: occurrence, toxicity, regulation. In: *Mycotoxins in dairy products* (H.P. Van Egmond, ed). Elsevier Applied Science, London and New York, pp. 11-55.

Vargas, E.A., R.A. Preis, L. Castro and C.M. Silva. 2001. Co-occurrence of aflatoxins B_1, B_2, G_1, G_2, zearalenone and fumonisin B_1 in Brazilian corn. Food Addit. Contam. 18:981-986.

Verger, P., J.L. Volatier and A. Dufour. 1999. Estimation des niveaux théoriques d'ingestion d'aflatoxines et d'ochratoxine. In: *Les mycotoxines dans l'alimentation: evaluation et gestion du risque* (A. Pfohl-Leszkowicz, ed). TEC & DOC Lavoisier, Paris, France, pp 371-384.

Verma, R.J., A.S. Kolhe and H.C. Dube. 1996. Aflatoxin contamination in different varieties of peanuts. Indian Phytopathol. 49:62-66.

Vinas, I., J. Dadon and V. Sanchis. 1993. Citrinin-producing capacity of *Penicillium expansum* strains from apple packing houses of Lerida (Spain). Int. J. Food Microbiol. 19:153-156.

Viquez, O.M., M.E. Castell-Perez and R.A. Shelby. 1996. Occurrence of fumonisin B_1 in maize grown in Costa Rica. J. Agric. Food Chem. 44:2789-2791.

Visconti, A., and M.B. Doko. 1994. Survey of fumonisin production by *Fusarium* isolated from cereals in Europe. J. AOAC Int. 77:546-550.

Visconti, A., M. Pascale and G. Centonze. 1999. Determination of ochratoxin A in wine by means of immunoaffinity column clean-up and HPLC. J. Chromatogr. A 864:99-101.

Visconti, A., M. Pascale and G. Centonze. 2000. Determination of ochratoxin A in domestic and imported beers in Italy by immunoaffinity clean-up and liquid chromatography. J. Chromatogr A. 888:321-326.

Vrabcheva, T., R. Gessler, E. Usleber and E. Martlbauer. 1996. First survey on the natural occurrence of *Fusarium* mycotoxins in Bulgarian wheat. Mycopathologia 136:47-52.

Vrabcheva, T., E. Usleber, R. Dietrich and E. Martlbauer. 2000. Co-occurrence of ochratoxin A and citrinin in cereals from Bulgarian villages with a history of Balkan endemic nephropathy. J. Agric. Food Chem. 48:2483-2488.

Vrabcheva, T., T. Petkova-Bocharova, F. Grosso, I. Nikolov, I.N. Chernozemsky, M. Castegnaro and S. Dragacci. 2004. Analysis of ochratoxin A in foods consumed by inhabitants from an area with Balkan endemic nephropathy: a 1 month follow-up study. J. Agric. Food Chem. 52:2404-2410.

Vukelic, M., B. Šoštaric and M. Belicza. 1992. Pathomorphology of Balkan endemic nephropathy. Food Chem. Toxicol. 30:193-200.

Wafa, E.W., R.S. Yahya, M.A. Sobh, I. Eraky, M. el-Baz, H.A. el-Gayar, A.M. Betbeder, and E.E. Creppy. 1998. Human ochratoxicosis and nephropathy in Egypt: a preliminary study. Hum. Exp. Toxicol. 17:124-129.

Walker, R. 2002. Risk assessment of ochratoxin: current views of the European Scientific Committee on Food, the JECFA and the Codex Committee on Food Additives and Contaminants. Adv. Exp. Med. Biol. 504:249-255.

Wang, D.S., Y.X. Liang, N.T. Chau, L.D Dien, T. Tanaka and Y. Ueno. 1995a. Natural co-occurrence of *Fusarium* toxins and aflatoxin B_1 in corn for feed in North Vietnam. Nat. Toxins 3:445-449.

Wang, D.S., Y.X. Liang, K. Iijima, Y. Sugiura, T. Tanaka, G. Chen, S.Z. Yu and Y. Ueno. 1995b. Co-contamination of mycotoxins in corn harvested in Haimen, a high risk area of primary liver cancer in China. Mycotoxins 41:67-70.

Wang, H., H. Wei, J. Ma and X. Luo. 2000. The fumonisin B_1 content in corn from North China, a high-risk area of oesophageal cancer. J. Environ. Pathol. Toxicol. Oncol. 19:139-141.

WHO, 1996. Ochratoxin A- toxicological evaluation of certain food additives and contaminants. WHO Food Additives series (Vol. 35), World Health Organization (WHO), Geneva, pp. 363-376.

WHO, 1998. Food safety issues, GEMS/FOOD Regional Diets.

WHO, 2001. Ochratoxin A. Safety evaluation of certain mycotoxins in food. International Program on Chemical Safety. WHO Food Additives Series (vol. 47), World Health Organization, Geneva, pp. 281-415.

WHO, 2002. Evaluation of certain mycotoxins in food. WHO Technical Report Series n. 906, World Health Organization, Geneva, pp. 8-42.

Wolff, J., H. Bresch, C. Cholmakov-Bodechtel, G. Engel, S. Erhardt, M. Gareis, P. Majerus, H. Rosner and R. Scheuer. 1997. Burden of ochratoxin A in food and in the consumer. Final report. Institute for the biochemistry of cereals and potatoes. Federal Institute for Research on Cereals, Potatoes and Fats, Detmold, Germany.

Wolff, J., H. Bresch, C. Cholmakov Bodechtel, G. Engel, S. Erhardt, M. Gareis, P. Majerus, H. Rosner and R. Scheuer. 2000. Burden of ochratoxin A in food and in the consumer. Final report. Institute for the Biochemistry of Cereals and Potatoes. Federal Institute for Research on Cereals, Potatoes and Fat.

Wolzak, A., A.M. Pearson, T.H. Coleman, J.J. Pestka, and J.I. Gray. 1985. Aflatoxin deposition and clearance in the eggs of laying hens. Food Chem. Toxicol. 23:1057-1061.

Wood, G.A. 1991. Aflatoxin M_1. In: *Mycotoxins and Phytoalexins*, (R.P. Sharma and D.K. Salunkhe, eds). Boca Raton, Florida: CRC Press, pp. 145-164.

Yamashita, A., T. Yoshizawa, Y. Aiura, P.C. Sanchez, E.I. Dizon, R.H. Arim and Sardjono. 1995. *Fusarium* mycotoxins (fumonisins, nivalenol and zearalenone) and aflatoxins in corn from southeast Asia. Biosci. Biotech. Biochem. 59:1804-1807.

Yoshizawa, T. 1991. Natural occurrence of mycotoxins in small grain cereals (wheat, barley, rye, oats, sorghum, millet, rice). In: *Mycotoxins and Animal Foods* (J.E. Smith and R.S. Henderson, eds). CRC Press, Boca Raton, FL, pp. 301-324.

Yoshizawa, T. and Y.Z. Jin. 1995. Natural occurrence of acetylated derivatives of deoxynivalenol and nivalenol in wheat and barley in Japan. Food Addit. Contam. 12:689-694.

Yoshizawa, T., A. Yamashita and Y. Luo. 1994. Fumonisin occurrence in corn from high- and low-risk areas for human oesophageal cancer in China. Appl. Environ. Microbiol. 60:1626-1629.

Yoshizawa, T., A. Yamashita and N. Chokethaworn. 1996. Occurrence of fumonisins and aflatoxins in corn from Thailand. Food Addit. Contam. 13:163-168.

Yousef, A.E. and E.H. Marth. 1989. Stability and degradation of aflatoxin M_1. In: *Mycotoxins in Dairy Products* (H.P. van Egmond, ed). Elsevier Applied Science, London and New York, pp. 11-55.

Zimmerli, B. and R. Dick. 1995. Determination of ochratoxin A at the ppt level in human blood, serum, milk and some foodstuffs by high-performance liquid chromatography with enhanced fluorescence detection and immunoaffinity column cleanup: methodology and Swiss data. J. Chromatogr. B Biomed. Appl. 666:85-99.

Zimmerli, B. and R. Dick. 1996. Ochratoxin A in table wine and grape-juice: occurrence and risk assessment. Food Addit. Contam. 13:655-668.

Zollner, P., D. Berner, J. Jodlbauer and W. Lindner. 2000. Determination of zearalenone and its metabolites alpha- and beta-zearalenol in beer samples by high-performance liquid chromatography-tandem mass spectrometry. J. Chromatogr. B. Biomed. Sci. Appl. 738:233-241.

9

MOULD GROWTH AND MYCOTOXIN PRODUCTION

Elizabeth Santin
Universidade Federal do Paraná, Curitiba, Paraná, Brazil

Introduction

The significant advance in intensive animal production systems is at once the cause and the consequence of the increase in grain production. Production of poultry and swine, in particular, requires a large amount of grain to fulfill their dietary needs. For this reason grain production increases every year; and following harvest, most of the grain is stored until utilized. Storage conditions are determined by the complex interaction among the grain, the macro- and microenvironments, and a variety of organisms including microorganisms, insects, mites, rodents, and birds. Grain provides an abundant source of nutrients, but the natural consequences of the ecosystem during storage often will be spoilage of the grain.

Mould growth in grain is a normal occurrence in both the field and in storage. Mould growth can spoil the nutritional aspects of the grain and also result in secondary metabolites that are highly toxic to animals, humans, and plants. These so-called mycotoxins have been extensively studied since 1961, when groups of highly toxic *Aspergillus flavus* metabolites, the aflatoxins, were isolated from groundnut meals that had been imported into the UK.

In general, pre-harvest control of mould growth is somewhat compromised by the inability of man to control the climate, since both insufficient and excessive rainfall during critical phases of crop development can lead to mould contamination, spoilage of grain, and mycotoxin production. However, the post-harvest handling of grain presents many more opportunities for controlling mould growth and its consequences. Careful drying of grains and good storage management should minimize post-harvest fungal growth and therefore, mycotoxin production.

This review introduces certain aspects of mould growth in grains, the production of mycotoxins by these moulds, and their relationship with animal production.

Grain mould pathogens

Moulds are fungi that grow by producing long filaments called **hyphae**. They are plants that contain no chlorophyl and can grow in the absence of natural light. Moulds grow from single cells to a body of branched hyphae. In general, hyphae are important to the survival and dispersal of fungi. A network of hyphae is referred to as **mycelium**. This hyphal network is responsible for 'cementing' kernels together, which results in columns of grain that cannot be separated. Grain mould fungi also produce **spores** (conidia) capable of aerial dispersal in the field as well as within a grain storage bin. It is usually masses of these spores that give the mould a characteristic colour. Spores are dispersed passively by wind and rain. Insects can serve as vectors of these fungi, usually by transporting the spores on their bodies. Insects also increase the surface area available for fungi growth by compromising the external protective covering of the kernels. Spores can lay dormant for months or years until the proper conditions for fungal development are available.

In the past, many researchers have divided fungal species into two groups: field fungi and storage fungi. **Field fungi** are those that invade the seeds while the crop is still in the field and require high moisture conditions (20-21%). These include species of *Fusarium, Alternaria, Clodosporium, Diplodia, Gibberella* and *Helminthosporium*. **Storage fungi** (also called storage moulds) are those that invade grains or seeds during storage, need less moisture than field fungi (13-18%) and usually do not present any serious problem before harvest. Storage fungi include species of *Aspergillus* and *Penicillium*. While the field/storage terminology is generally used to indicate the differences in temperature and moisture required by various fungi, in fact the proper conditions for growth of a specific organism can occur in either the field or storage bin. Ideal conditions for fungal growth depend on the species, but normally moulds need high temperature and moisture. Infection by fungi can occur at various stages of crop production: in the field, during transportation, or in storage.

It is important to note that fungal growth is not always visible to the naked eye. The first stage of fungal growth is microscopic. Most grain fungi are relatively non-specific; they colonize several species of plants and can utilize different types of organic matter for energy. A very large number of grain moulds can produce mycotoxins. The major genera of mycotoxin-producing fungi are *Aspergillus, Claviceps, Penicillium* and *Fusarium* (Marquardt, 1996). Mycotoxins are produced as secondary metabolites. The primary metabolites of fungi and other organisms are those compounds that are essential for growth. Secondary metabolites are formed in the final stages of the exponential growth phase (Jay, 2000). Under field conditions, stress and subsequently reduced vigor often predispose plants to infestation and colonization by toxigenic fungi. In stored grain, toxigenic fungal infection and mycotoxin production result from a complex interaction among moisture, temperature, substrate, oxygen (O_2) and carbon dioxide (CO_2) concentration, insect presence and fungal abundance.

Storage conditions predisposing grain to fungal damage

The interaction of temperature, O_2 and CO_2 concentrations and moisture content determine the extent of mould colonization (Tuite, 1994). In general, most fungi need at least 1-2% oxygen. However, *Fusarium verticillioides (moniliforme)* is an exception and is able to grow in an environment with 60% CO_2 and less than 0.5% O_2 (Tuite, 1994).

Usually, fungi grow at temperatures between 20 and 30°C. It is important to note that if the grain is at high temperature at harvest, it can maintain high temperature for several days or weeks after harvest unless the storage facility has cooling equipment.

Normally, in storage conditions fungi grow at 13-18% moisture. However, in the case of grains with high levels of oil (e.g. peanuts), fungal growth occurs at moisture contents as low as 7% (Muschen and Frank, 1994). Sometimes management conditions intended to reduce fungal growth actually increase the potential infection. A classic example of this is the practice of drying grain at high temperature, which causes a break in the protective pellicle of the grain, opening the door to mould invasion of the exposed endosperm. On the other hand, moulds utilize inter-granular water vapour as a moisture source inside the bin; the concentration of which is determined by the equilibrium between free water within the grain and water vapour immediately surrounding the grain. The inter-granular water concentration is described either in terms of the **equilibrium relative humidity** (ERM %) or **water activity** (A_w). Water activity describes the ratio of the vapour pressure of water in grain to that of pure water at the same temperature and pressure, while the ERM is equivalent to the water activity expressed as a percentage. At a given moisture content, different grains afford a variety of water activities, and consequently support different rates and types of mould growth. The passage of water from the grain into the vapour phase is enhanced by an increase in temperature. Consequently, for a given moisture content, the water activity, and thus the propensity for mould growth, will increase with temperature. In practical situations the storage bin is normally exposed to the sun, which increases the temperature inside the container as convection moves heat into the grain mass. This might carry water to cooler regions of the container. When water vapour moves from warmer to cooler regions, it condenses and can increase the moisture content of an area inside the container, creating a 'hot spot' where mould can more easily grow (Cruz, 1995). In this situation, even if the overall moisture content in the container is around 12% (e.g.), in some regions in the container the moisture content may be around 18%, providing the mould with good conditions for growth.

Finally, insects and mites (arthropods) make a significant contribution to mould growth through physical damage of grain, which predisposes mould invasion of the exposed endosperm. Also, the metabolic activity of insects and mites causes an increase in both moisture content and temperature of the infested grain, which is favourable for fungal development. Insects and mites can also carry mould spores; and their faecal material is an additional substrate for mould growth.

THE RESPIRATORY PROCESS IN STORED GRAIN

After harvest, the seed is still a live organism and continues the metabolic process of respiration. In the presence of O_2 (aerobic conditions), **respiration** occurs by oxidation of carbohydrate and fat to produce CO_2, water and energy in the form of heat (Puzzi, 1986). When O_2 access to the grain mass is interrupted, anaerobic metabolism occurs. The final products of anaerobic activity are CO_2 gas and organic compounds such as acetic acid. This process is called **fermentation** (Puzzi, 1986).

The intensity of the respiratory process depends on the temperature and humidity of the grain. An increase in temperature can increase respiration rate 2- to 3-fold, but stops when the high temperature destroys the enzymes needed for respiration (Puzzi, 1986). In a grain mass the increase in temperature (55-75°C) can be exacerbated by the development of thermophilic bacteria, which can lead to combustion of the grain mass. On the other hand, when stored grain is dry (e.g. 11-13% moisture for corn), respiration is reduced. Even at low temperatures, if the moisture level is high, fungal growth can occur. For this reason, refrigerating grains with high moisture content is not an efficient method of reducing fungal growth.

In wet grains, the most important factor in increasing respiration is the fungi themselves. These fungi can transform the kernel from a live organism to a simple substrate for the production of CO_2, water, and heat.

Respiration and heating are associated with deterioration of grain. Because of that, CO_2 production in a grain mass is an important indicator of the intensity of the respiratory process.

LOSSES DURING GRAIN STORAGE

In the past, losses in stored grain were attributed to insect activity. By 1950, storage fungi were considered the main cause of deterioration. The first phase of mould growth cannot be seen by the naked eye, but results in serious losses due to physical injury of the kernels, reduced nutritional quality and/or production of toxins by the fungi.

Fungi can affect the colour, smell and fat content of grain. The natural yellow colour of corn might change to white, pink, red or green, depending on the fungal species (Qasem and Christensen, 1958). Moulded grain also has a characteristic musty smell. Fungal attack is accelerated by the reduction in specific weight of the grain. Specific weight is calculated by determining the weight of a grain mass in a specific volume (kg/m^3).

Mould growth significantly reduces grain dry matter content; and grain infected by fungi could be half the weight of uncontaminated grains (Lazzari, 1993; Krabbe *et al.*, 1994). Krabbe *et al.* (1994) reported a loss of 4% of dry matter for grains stored at 18% moisture for 30 days, and losses of 8% over 60 days of storage.

Metabolic activity of fungi is associated with aerobic respiration, which utilises grain fat and carbohydrate (Dixon and Hamilton 1981b). As a result, the fat content of grains is markedly reduced; and diets containing these grains also have reduced fat content. Bartov and Paster (1986) reported a reduction in fat content of broiler feed after storage for 18 to 45 days. The losses in fat content were associated with an increase in CO_2 production. According to Coelho (1994a), the metabolisable energy of corn can be reduced by 5 to 25% as a result of fungal growth.

In a study in our laboratory, broilers fed diets formulated with mouldy feed showed an increase in feed intake, which was demonstrated only at 14 days and not during the first week of age (Table 1). It is possible that the reduction of 1.12% in fat content of feed formulated with mouldy grain resulted in an increase in feed intake by the birds to achieve the required energy intake. Maiorka (2002) observed that dietary energy influenced feed intake and feed conversion in broilers only after the second week of age because lipid digestion and absorption are immature in early hatch birds, resulting in poor utilization of lipids.

Table 1. Feed intake), weight gain and feed conversion (FCR) at 6 and 14 days of age in broilers fed diets formulated with mouldy or normal corn.

	6 days of age		*14 days of age*	
	Moulded	*Normal*	*Moulded*	*Normal*
Feed intake, g/day	99	103	522^a	447^b
Weight gain, g	81	85	316^a	365^b
FCR	1.222	1.211	1.651^a	1.224^b

^{ab}Means in a column with different letters differ ($P<0.05$)

Similarly, Kao and Robinson (1972) reported reductions in cysteine, lysine, and arginine content of mouldy wheat. These studies demonstrate that even in the absence of mycotoxins, losses due to fungal growth can significantly reduce the nutritional value of cereals and consequently reduce the performance of animals fed diets formulated with them.

Mycotoxin production

Mycotoxins are an important consequence of fungal growth. Actually, fungi produce mycotoxins under stressful conditions such as changes in temperature, moisture, or aeration and of course in the presence of aggressive agents. These secondary metabolites of fungi are classified based on the starting points of their production, as polyketides, terpenes, shikimic acid-derived substances and metabolites derived from amino acids. Their structural diversity is derived from a few reactions including condensation, oxidation, reduction, alkylation and halogenation and the multiple branching of the pathways (Steyn, 1998).

In animals or humans mycotoxins can lead to deterioration of the liver and kidneys. In addition, some mycotoxins act as neurotoxins, while others act by interfering with cellular protein synthesis, producing skin sensitivity and extreme immunodeficiency.

The toxigenic fungi involved in the human and animal food chains belong mainly to three genera: *Aspergillus*, *Fusarium* and *Penicillium* (Sweeney and Dobson, 1998). The important mycotoxins produced by *Aspergillus* species include aflatoxin B_1, B_2, G_1, and G_2, ochratoxin A, sterigmatocystin, and cyclopiazonic acid. Aflatoxins are produced mainly by *A. flavus*, *A. parasiticus* and *A. nominus*, and are considered very potent liver carcinogens in various animal species and humans. Lactating animals or humans that consume food containing aflatoxins B_1, B_2, G_1, or G_2, can secrete aflatoxins M_1 or M_2 in milk. Aflatoxins are produced at temperatures ranging from 12 to 40°C (Koehler *et al.*, 1985), at a pH from 3.5 to 8.0 (ICMSF, 1996) and with optimum water activity around 0.99. According to Gqaleni *et al.* (1997), *A. flavus* can co-produce aflatoxin and cyclopiazonic acid at 0.996 A_w. In addition, nutritional factors such as carbohydrate and nitrogen sources affect aflatoxin production, but the mechanisms by which these factors regulate aflatoxin biosynthesis are still unclear (Luchese and Harrigan, 1993). Cyclopiazonic acid is produced primarily by *A. flavus* and some *Penicillium* species and causes necrosis of the gastrointestinal tract.

Ochratoxin A is a potent nephrotoxin, teratogen, and carcinogen. It is produced by *A. ochraceus* and some *Penicillium* species. *Aspergillus* spp. can produce ochratoxins at 12-37°C at a minimum pH of 2.2 (Wheler *et al.*, 1991). According to Adebajo *et al.* (1994), *Aspergillus* spp. can produce ochratoxin at a water activity as low as 0.80. Ochratoxin also can be produced by *Penicillium* spp. at 4°C with a water activity as low as 0.86 (Sweeney and Dobson, 1998).

Another *Aspergillus* toxin is sterigmatocystin, which is a precursor of aflatoxin and is produced mainly by *A. versicolor*. This fungal species is less tolerant to low pH and grows at a water activity of 0.76 (ICMSF, 1996).

According to Pitt and Leistner (1991), *Penicillium* species can produce 27 different mycotoxins, with three being the most important: ochratoxin, patulin, and citrinin. Ochratoxin has been described before and can be produced by *P. verrucosum* in temperate climates while *Aspergillus* species are associated with ochratoxin production in warmer climates. Patulin produces adverse neurological and gastrointestinal effects and is produced by a fruit pathogen, *P. expansum*, that can grow at temperatures from −2 to 35°C. Patulin production needs minimum water activity of 0.95 (Damoglou and Campbell, 1986). Finally, citrinin is a nephrotoxin and is produced mainly by *P. citrinum*, *P. expansum*, and *P. verrucosum*. Citrinin is produced at temperatures ranging from 15 to 37°C, but the optimum temperature is 30°C (Pitt, 1997).

While *Aspergillus* and *Penicillium* species are very important mycotoxin producers, the *Fusarium* fungi are the most economically significant on a global basis. These fungi are the most prevalent in the crop prior to harvest and can produce a variety of different mycotoxins, the most important of which are trichothecenes, fumonisins, zearalenone, moniliformin and fusaric acid (D'Mello and Macdonald, 1997). *Fusarium* populations in agricultural soils can be very high and include saprophytes that break down plant residues in the soil as well as pathogens that can cause rots, wilts, and other plant diseases (CAST, 2003).

Mycotoxins from *Fusarium* species have traditionally been associated with cereals in temperate countries, since these fungi require somewhat lower temperatures for growth and mycotoxin production than aflatoxigenic *Aspergillus* species (D'Mello and Macdonald, 1997). *Fusarium* species can survive in corn residue, which is probably the most important source of inoculum for kernel infection. These fungi can be associated with every part of the corn plant; and 50 to 100% incidence of kernel infection is not uncommon, with the majority of kernels showing no visible damage (CAST, 2003). This fact may explain the presence of mycotoxins in normal kernels. In addition, spores produced pre-harvest enter storage bins with the crop and begin proliferating. Despite that, analyses of shelled kernels have found higher fumonisin levels in visibly mouldy corn than in good quality corn (Shephard *et al.*, 1996).

Fumonisin seems to be prevalent in grains grown in sub-tropical and tropical regions due to the relatively long and warm growing season (Henry and Wyatt, 1993). According to Placinta *et al.* (1999), fumonisin can be produced by *F. moniliforme* in association with moniliformin and fusarin C. The optimum growth temperature and pH for fumonisin production by a strain of *F. moniliforme* in a corn culture occurred in 7 weeks at 25°C over a pH range of 3 to 9.5 (Jay, 2000). Fumonisins originate by condensation of the amino acid alanine with an intermediate derived from acetate (CAST, 2003).

Trichothecenes include more than 100 different chemical structures that are divided into types A and B. Type A trichothecenes include T-2 toxin, HT-2 toxin, neosolaniol and diacetoxyscirpenol (DAS), while type B trichothecenes include deoxynivalenol (DON), nivalenol and fusarenon-X. The synthesis of the two types of trichothecenes appears to be characteristic for a particular *Fusarium* species. For example, production of type A trichothecenes predominates in *F. sporotrichioides* and possibly also *F. poae*, whereas synthesis of type B trichothecenes occurs principally in *F. culmorum* and *F. graminearum* (D'Mello and Macdonald, 1997). In a general way, trichothecene biosynthesis begins with the formation of the sesquiterpene trichodiene, which is then oxygenated, esterified, and cyclized to yield a large family of sesquiterpene epoxides (CAST, 2003).

Zearalenone is produced by a variety of *Fusarium* species and the structure of this toxin suggests biosynthesis by the acetate-polynalonate pathway, resulting in head-to-tail condensation of acetate units into a polyketide that requires relatively little modification to accomplish conversion to zearalenone (Shier, 1998). The maximum production of this toxin *in vitro* occurred after 40 days of incubation from an isolate of *F. graminearum*, at A_w of 0.97, and temperature of 28°C for two weeks followed by incubation at 12°C (Jímenez *et al.*, 1996).

Fusaric acid is probably the most widely distributed mycotoxin produced by strains in the genus *Fusarium*; and according to Bacon *et al.* (1996) may serve as a presumptive indicator of *Fusarium* contamination of food and grains. In addition, fusaric acid is known to have synergistic effects with other fusariotoxins. Finally, it is also important to note that the genetics of mycotoxin biosynthesis have been identified. Many genes related to production of mycotoxins including aflatoxin (Mahanti *et al.*, 1996), fumonisin (Desjardins *et al.*, 1996), and trichothecenes (McCormick *et al.*, 1996) are now characterized. These discoveries will be very important in better understanding the biosynthetic pathways of mycotoxins and to aid in finding controls for the production of mycotoxins.

Management of grain to avoid mould growth

PRE-HARVEST

Maintaining agronomic and management practices that promote the general health of crops can reduce, but not eliminate, pre-harvest mycotoxin contamination. For example, insect resistant germplasm such as corn transformed with the gene encoding *Bacillus thuringiensis* Berliner crystal protein (Bt maize) reduces levels of fumonisin. Irrigation of peanuts essentially prevents aflatoxin contamination, probably by preventing drought stress known to lead to fungal contamination in peanuts. However, optimization of management practices to control mycotoxins is not always possible for reasons of cost, geographical location or the nature of the production system (Cleveland *et al.*, 2003).

Hybrids that are less susceptible to grain mould should be planted where available in areas with grain mould and/or mycotoxin problems. Identifying the prevalent grain mould pathogens is necessary before the appropriate hybrids can be selected.

After planting, the overall strategy for grain mould management is to minimize plant stress during and after silking. Common stresses associated with grain mould include high moisture in mid-to-late season after a dry early season, moisture stress early in the season and during grain fill, high leaf disease pressure, and insect damage.

One strategy that has shown great promise for reducing mycotoxin contamination in crops involves application of non-mycotoxin-producing strains (atoxigenic strains) of fungi derived from the same species or taxonomic group that produces the harmful toxins. Current biocontrol strategies rely upon the ability of atoxigenic strains to competitively exclude mycotoxin producers in agricultural environments (Cleveland *et al.*, 2003).

POST-HARVEST

All storage facilities should be monitored regularly to detect grain mould development. Grain moulds rarely develop uniformly throughout a storage unit; and development of hot spots is common. Monitoring requires a systematic sampling plan to account for the unique design characteristics of each storage facility. It is important to look at all points of storage facilities to avoid fungal development, since it can occur anywhere including storage bins, the feed mill, mixed feed bins, and also in the feeders of animals. It is appropriate to observe the correct storage temperature and moisture conditions and clean bins and equipment frequently. The objective is to remove old grain and residues that harbour fungi.

To minimize the potential for mould growth and mycotoxin production, grain moisture content should be reduced to less than 15% within 48 hrs of harvest. This may be difficult to achieve in many storage facilities, especially in larger capacity elevators. Good air circulation throughout the storage bin is extremely important.

Harvest and post-harvest grain handling should be designed to minimize mechanical damage. Although most mould pathogens can directly penetrate plant tissues, mechanical damage provides additional entry sites, facilitating infection and spreading from kernel to kernel during shipping and storage.

Like mechanical damage, insect damage also provides additional entry sites for mould pathogens, facilitates infection and promotes distribution throughout the grain mass. Some insects also act as vectors for grain mould pathogens. An effective insect management plan during the growing season as well as during post-harvest shipping and storage of the grain, can reduce mould and mycotoxin contamination.

Use of fungistatic agents is a very common management practice and can be very efficient in reducing mould growth and its consequences of grain losses and mycotoxin production. Nevertheless, it should be emphasised that once the mould has already damaged grain and/or produced mycotoxins, the effectiveness of this practice is limited. However in some situations, a mould inhibitor can be useful for mould control from the time of application (Leeson and Summers, 1991).

Organic acids are important and efficient mould inhibitors. Organic acids are considered to be all acids that have carbon in their structure. The organic acid changes the pH of food, preventing development of fungi. Dixon and Hamilton (1981a) studied the effects of propionic, acetic, sorbic, and benzoic acids as mould inhibitors and reported that the efficiency of these organic acids depended on the moisture in a grain mass; since as moisture increases there is a need for higher levels of organic acid addition to avoid fungal development. However, a study by Krabbe *et al.* (1994) has shown that the use of organic acids reduced fungal growth even in grains stored at high moisture content (Table 2).

Table 2. Fat content, crude protein, and specific weight in maize stored for 60 days at different moisture contents with or without organic acid treatments.

Treatments	*Fat content (%)*	*Crude protein (%)*	*Specific weight (kg/m³)*
12% moisture	4.6[a]	8.7[a]	689[a]
12% moisture + organic acid	4.6[a]	8.6[a]	691[a]
15% moisture	4.0[b]	8.9[a]	622[b]
15% moisture + organic acid	4.5[a]	8.9[a]	670[a]

Adapted from Krabbe, 1994
[ab]Means in a column differ ($P<0.05$)

Overall, propionic acid proved to be the most effective organic acid for controlling mould growth (Krabbe, 1994). There are two forms of propionic acid: undissociated (CH_3CH_2COOH) and dissociated ($CH_3CH_2COO^-$) forms. Undissociated propionic acid is corrosive, caustic, volatile, and pungent, while the dissociated form is precisely the opposite. However, undissociated propionic acid is the only form that effectivly kills fungi, bacteria, and yeast. To do this, the undissociated radical attaches to the cell wall of the microorganism and draws out intercellular water. The cytoplasm has a pH of 7, and when it decomposes, the undissociated radical reverts to the dissociated form and a salt is produced that eventually becomes part of the pool of dissociated propionic acid in food.

Despite the fact that the undissociated acid is the active form, it is not necessary to have all undissociated propionic acid present at initial application, since dissociated propionic acid can be shifted immediately to the undissociated pool if undissociated acid is removed as a result of killing a microbe. If all propionic acid molecules are undissociated at application, most will evaporate and be wasted since often there are not many moulds that can be killed at any one time. If undissociated propionic acid is added to food, the major portion reacts with cations to form calcium propionate, magnesium propionate, or sodium propionate, but part of it will evaporate to the head space of the bin where it might react with water and cause significant corrosion (Coelho, 1994b). For these reasons, dissociated propionic acid is the appropriate form to add to food, since it is stable and readily available if and when mould begins to grow. This is important because if undissociated propionic acid is added to food while there are no moulds growing, very little or none will be available later when moulds start growing. If undissociated acid starts killing moulds and disappearing from the pool, the shift from dissociated to undissociated forms (to maintain equilibrium) will be determined by the pH of the food (Coelho, 1994b). In this situation, mould inhibitors should contain 10-15% undissociated propionic acid for an initial burst of activity to effectively stop mould growth as soon as they are applied to food, after which the acid should maintain the appropriate undissociated:dissociated ratio in food.

Conclusions

Fungal growth in cereals in the field as well as in stored grain causes nutritional and physical losses and can also produce mycotoxins that can be highly toxic in animals and humans. Therefore, the economic impact on animal performance due to fungal growth can be immense. Because of this, the control of fungal growth is vital in maintaining the nutritional values of cereals and in avoiding mycotoxin contamination of animal feeds.

References

Adebajo L.O, O.A. Bamgbelu, and R.A. Olowu. 1994. Mould contamination and the influence of water activity and temperature on mycotoxin production by two Aspergilli in melon seed. Natrung 38:209-217.

Bacon, C.W., J.K. Porter, W.P. Norred, and

J.F. Leslie. 1996. Production of fusaric acid by *Fusarium* species. Appl. Envirom. Microbiol. 62:4039-4043.

Bartov, I. and N. Paster. 1986. Effect of early stages of fungal development on the nutritional value of diets for broiler chicks. Brit. Poult. Sci. 27:415-420.

CAST (Council for Agricultural Science and Technology). 2003. Mycotoxins: Risks in plant, animal and human systems. Ames, Iowa, USA.

Cleveland, T.E., P.F. Dowd, A.E. Desjardins, B. Deepak and P.J. Cotty. 2003. United States Department of Agriculture - Agriculture research on pre-harvest prevention of mycotoxins and mycotoxigenic fungi in US crops. Pest. Man. Sci. 59:629-642.

Coelho, M.B. 1994a. Microbial deterioration of foodstuffs. In: *Moulds, Mycotoxins and Food Preservatives in the Food Industry*. Parsippany, New Jersey, BASF Corporation, p. 9-13.

Coelho, M.B. 1994b. Mode of action of mould inhibitors. In: *Moulds, Mycotoxins and Food Preservatives in the Food Industry*. Parsippany, New Jersey, BASF Corporation, p. 81-82.

Cruz, L.C.H. 1995. Características Gerais das Micotoxinas e Micotoxicoses. Reflexos na indústria avícola. Simpósio Internacional sobre Micotoxinas e Micotoxicoses em Aves. Anais, Curitiba, PR, p. 1-13.

D'mello, J.P.F and A.M.C. Macdonald. 1997. Mycotoxins. Anim. Feed Sci. Tech. 69:155-166.

Damoglou, A.P. and D.S. Campbell. 1986. The effect of pH on the production of patulin in apple juice. Lett. Appl. Microbiol. 9-11.

Desjardins, A.E., R.D. Plattner, and R.H. Proctor. 1996. Linkage among genes responsible for fumonisin biosynthesis in *Giberella fugikuroi* mating population. Appl. Environ. Microbiol. 62:2571-2576.

Dixon, R.C. and P.B. Hamilton. 1981a. Effect of food ingredients on the antifungal activity of propionic acid. Poult. Sci. 60:2407-2411.

Dixon, R.C. and P.B. Hamilton. 1981b. Evaluation of some organic acids as mould inhibitors by measuring CO_2 production from food and ingredients. Poult. Sci. 60:2182-88.

Gqaleni, N., J.E. Smith, J. Lavcey, and G. Gettinby. 1997. Effects of temperature, water activity and incubation time on production of aflatoxins and cyclopiazonic acid by an isolate of *Aspergillus flavus* in surface agar culture. Appl. Environ. Microbiol. 63:1048-1053.

Henry, M.H. and R.D. Wyatt. 1993. A review of fumonisin productin by *Fusarium moniliforme* and fumonisin toxicosis in animals. J. Appl. Poult. Res. 2:188-192.

ICMSF (International Commission on Microbiological Specification for Food). 1996. Toxigenic fungi: *Aspergillus*. In: *Microorganisms in Foods 5: Characteristics of Microbial Pathogens*. Academic Press, London, pp. 347-381.

Jay, J.M. 2000. *Modern Food Microbiology*. An Aspen Publication, Maryland.

Jímenez, M., M. Mañez and E. Henandéz. 1996. Influence of water activity and temperature on the production of zearalenone in corn by three *Fusarium* species. Food Microbiol. 29:417-421.

Kao, C. and J. Robinson. 1972. *Aspergillus flavus* deterioration of grain: Its effect on amino acids and vitamins in whole wheat. J. Food Sci. 37:261.

Koehler, P.E., L.R. Beuchat, and M.S. Chinnan. 1985. Influence of temperature and water activity on aflatoxin production by *Aspergillus flavus* in cowpea seeds and meal. J. Food Prot. 48:1040-1043.

Krabbe, E.L., A.M. Penz, Jr., F.A. Lazzari, and M.F. Reginatto. 1994. Efeito da Umidade e do Ácido Propiônico sobre as Características Bromatológicas e Microbiológicas de Grãos de Milho. Conferência APINCO de Ciência e Tecnologia Avícola. Anais, Santos, SP, p. 180.

Lazzari, F.A. 1993. *Umidade, fungos e*

micotoxinas na qualidade de sementes, grãos e rações. UFPR press.

Leeson, S. and J.D. Summers. 1991. In: *Commercial Poultry Nutrition*. University Books, Guelph, Ontario, Canada, p. 283.

Luchese, R.H. and W.F. Harrigan. 1993. Biosynthesis of aflatoxin - The role of nutritional factors. J. Appl. Bacteriol. 74:5-14.

Mahanti, N., D. Bhatnagar, J.W. Cary, J. Joubran, and J.E. Linz. 1996. Structure and function of fas-lA, a gene encoding a putative fatty acid synthetase directly involved in aflatoxin biosynthesis in *Aspergillus parasiticus*. Appl. Environ. Microbiol. 62:191-195.

Maiorka, A. 2002. Efeitos da idade da matriz, do jejum, da energia da ração e da glutamina sobre o desenvolvimento da mucosa intestinal e atividade enzimática do pâncreas de pintos de corte. Jaboticabal, FCAV/UNESP. 103f. Tese de doutorado - Produção Animal.

Marquardt, R.R. 1996. Effects of molds and their toxins on livestock performance: A western Canadian perspective. Anim. Feed Sci. Tech. 58:77-89.

Mccormick, S.P., T.M. Hohn, and A.E. Desjardins. 1996. Isolation and characterization of tri 3, a gene encoding 15-O-acetyl-transferase from *Fusarium sporotrichioides*. Appl. Environ. Microbiol. 62:353-359.

Muschen, H. and K. Frank. 1994. Mycotoxins in oilseeds and risks in animal production. In: *Moulds, Mycotoxins and Food Preservatives in the Food Industry*. Parsippany, New Jersey, BASF Corporation, p. 31-35.

Pitt, J.I. 1997. Toxigenic *Penicillium* species. In: *Food Microbiology, Fundamentals and Frontiers* (M.P. Doyle, L.R. Beuchat, and T.J. Montville, eds). ASM Press, Washington, DC, pp. 406-418.

Pitt, J.I. and L. Leistner. 1991. Toxigenic *Penicillium* species. In: *Mycotoxins and Animal Food* (J.E. Smith and R.S. Henderson, eds). CRC Press, Boca Raton, FL, USA, pp. 91-99.

Placinta, C.M., J.P.F. D'mello and A.M.C. Macdonald. 1999. A review of worldwide contamination of cereal grains and animal feed with *Fusarium* mycotoxins. Anim. Feed Sci. Tech. 78:21-37.

Puzzi, D. 1986. Abastecimento e armazenagem de grãos. Campinas, Instituto Campineiro de Ensino Agrícola., p. 603.

Qasem, S.A. and M. Christensen. 1958. Influence of moisture content, temperature and time on the deterioration of stored corn by fungi. Phytopath. 48:544-549.

Shephard, G.S., P.G. Thiel, S. Stockenstrom, and E.W. Sydenham. 1996. Worldwide survey of fumonisin contamination of corn and corn-based products. J. Assoc. Off. Anal. Chem. Intl. 79:671-687.

Shier, W.T. 1998. Estrogenic mycotoxins. Revue Med. Vet. 149:599-604.

Steyn, P.S. 1998. The biosynthesis of mycotoxins. Revue de medicine Veterinairé 149:469-478.

Sweeney, M.J. and A.D.W. Dobson. 1998. Mycotoxin production by *Aspergillus, Fusarium* and *Penicillium* species. Intern. J. Food Microbiol. 43:141-158.

Tuite, J. 1994. Epidemiology of moulds in grain. In: *Moulds, Mycotoxins and Food Preservatives in the Food Industry*. Parsippany, New Jersey, BASF Corporation, p. 5-8.

Wheler, K.A., B.F. Hurdman and J.I. Pitt. 1991. Influence of pH on the growth of some toxigenic species of *Aspergillus, Penicillium* and *Fusarium*. Int. J. Food Microbiol. 55:86-90.

10

CURRENT CONCEPTS IN MYCOTOXICOSES IN SWINE

Trevor K. Smith, Gabriel Diaz and H.V.L.N. Swamy
Department of Animal and Poultry Science, University of Guelph, Guelph, Ontario, Canada

Introduction

Mycotoxins are fungal metabolites that can reduce performance and alter metabolism of livestock and poultry (Wannemacher *et al.*, 1991). The pathological states arising from the consumption of feeds contaminated with mycotoxins are referred to as **mycotoxicoses**. Mycotoxins can be formed in the field pre-harvest and may continue to be formed under suboptimal storage conditions post-harvest. High moisture content often predisposes feedstuffs to fungal growth and mycotoxin formation. Temperature is another key factor. Some fungi, such as *Aspergillus flavus*, are usually found in tropical and semi-tropical climates. This mold produces the carcinogenic hepatotoxin aflatoxin. *Fusarium* fungi, however, are more common in temperate climates and *Fusarium* mycotoxins are likely the most common mycotoxins on a global basis (Wood, 1992).

Aflatoxicosis in swine

Aflatoxins are produced mainly by *Aspergillus flavus* and *A. parasiticus*. Aflatoxicosis has been investigated in many different animal species and in more depth than other mycotoxins (Smith and Ross, 1991). This is because aflatoxins are among the most acutely toxic of mycotoxins causing extensive liver pathology. There is also concern about residues of aflatoxin and metabolites in foods because of the well-documented carcinogenicity of these compounds.

EFFECTS OF DIETARY AFLATOXIN ON GROWING AND FINISHING PIGS

Chronic aflatoxin B_1 toxicity in growing and finishing pigs (40 to 140 kg live weight) was described by Bonomi *et al.* (1992). The feeding of 500, 650 and 800 μg/kg aflatoxin B_1 reduced weight gain, feed utilisation, lipid digestion and renal function. In an experiment of similar design, it was shown that chronic feeding of aflatoxin G_1 resulted in more severe liver pathology than was observed with aflatoxin B_1 (Bonomi *et al.*, 1993). The feeding of up to 400 μg/kg aflatoxin B_1, however, had little effect on performance and mycotoxin tissue residues in pigs grown from 65 to 95 kg live weight (Wu *et al.*, 1989). Studies of feeding aflatoxin contaminated corn to weanling and growing pigs described reduced growth rates and feed consumption as well as elevated serum γ-glutamyltransferase (GGT) activity at a dietary aflatoxin concentration of 922 μg/kg B_1 (Schell *et al.*, 1993). Feeding 800 μg/kg aflatoxin for four weeks to starter pigs resulted in changes in numerous serum parameters including albumin and total protein concentrations as well as activities of GGT and alkaline phosphatase indicating liver damage (Schell *et al.*, 1993). A similar study by Lindemann *et al.* (1997) described a 27.8% reduction in average daily weight gain when starter pigs were fed 500 μg/kg aflatoxin for 34 days.

Higher doses of aflatoxin (2.5 mg/kg) fed to growing boars for 32 days resulted in toxicity including reduced serum tocopherol and retinol concentrations (Harvey *et al.*, 1995). It was concluded that feeding aflatoxin-contaminated diets might exacerbate vitamin A and E deficiencies in pigs. Whole body autoradiography of ^{3}H-labeled aflatoxin B_1 in young pigs showed localisation in the nasal olfactory and respiratory mucosa as well as in the liver (Larson and Tjalve, 1996). A very significant aspect of aflatoxicosis in swine, however, may be reduced immune function. A study of weanling pigs fed up to 280 μg/kg aflatoxin indicated a linear reduction in skin thickness at 12 and 24 hrs following injection of phytohemagglutinin (Van Heugten *et al.*, 1994). The economic significance of such immunosuppression, however, will vary with the degree of disease challenge in different production units.

EFFECTS OF AFLATOXIN ON SWINE REPRODUCTION

The adverse effects of aflatoxin on hepatic metabolism, protein synthesis and immune status combine to reduce swine reproductive efficiency although low concentrations are tolerated. Sows were fed graded levels of aflatoxin B_1 up to 400 μg/kg from the first day of gestation to the end of lactation (Wu *et al.*, 1992). There were no significant effects on numbers of piglets per litter, piglets weaned at 28 days or survival rate. Residues of aflatoxins B_1 and M_1 were detected, however, in milk 14 days postpartum with the concentrations rising with dietary aflatoxin concentration. Increasing the dietary dose of aflatoxin B_1 up to 800 μg/kg, however, did result in some adverse effects (Bonomi *et al.*, 1995a). Similar effects were noted with the feeding of aflatoxin G_1 (Bonomi *et al.*, 1995b). The effects of aflatoxin B_1 and aflatoxin G_1 on swine reproduction, moreover, appear to be additive (Bonomi *et al.*, 1996).

Ochratoxicosis in swine

Ochratoxin A is produced by a number of species of both *Aspergillus* and *Penicillium*. Acute ochratoxicosis is characterised by nephropathy, enteritis and immuno-suppression (Terao and Ohtsubo, 1991). Interest in this mycotoxin has focused on the carcinogenic nature of the compound and, like aflatoxin, residues in foods of animal origin are a public health concern.

EFFECTS OF DIETARY OCHRATOXIN ON GROWING AND FINISHING PIGS

The effects of feeding up to 2.5 mg/kg ochratoxin A to barrows of 15 kg initial weight was monitored for 21 days (Lippold *et al.*, 1992). Weight gain, feed intake and feed efficiency were reduced particularly at the highest dietary concentration of ochratoxin A. Impaired renal function as indicated by hyperproteinemia and azotemia, however, was noted at even 0.5 mg/kg ochratoxin A. A detailed description of nephropathy caused by ochratoxin A in growing pigs in Hungary has been provided by Glavits (1993). The feeding of as little as 800 μg/kg ochratoxin A has been show to result in renal lesions over a one year period (Stoev *et al.*, 2002). There have been attempts to correlate pig blood concentrations of ochratoxin A with contamination of the feed. In Sweden, 14% of pigs had 2 ng ochratoxin A/mL blood with a maximum concentration of 215 ng/mL (Hult *et al.*, 1992). In western Canada, 36% of serum samples collected from 1600 pigs contained detectable levels of ochratoxin A (Ominski *et al.*, 1996). The Danish swine industry uses renal ochratoxin A residues as an effective quality control system to minimise potentially harmful residues in pork products (Jorgensen and Peteresen, 2002). Interest has also been centered, however, on the immunosuppressive effects of ochratoxin A in swine. It has been shown that feeding 2.5 mg/kg ochratoxin A to growing gilts for 35 days suppresses cell-mediated immune responses (Harvey et al., 1992).

Studies of the effect of dietary ochratoxin A on reproduction in swine have focused on boar semen quality. The feeding of five and ten times the human tolerable daily intake resulted in a reduction in initial motility and impaired longevity although it was not clear if ochratoxin A had a direct effect on the germinal epithelium or if it only delayed sperm cell maturation (Solti et al., 1999). A subsequent study came to similar conclusions (Biro et al., 2003).

Fusarium mycotoxicoses in swine

FUMONISIN

Effects on swine production

The fumonisins are a small family of relatively recently discovered *Fusarium* mycotoxins that are produced mainly by *F. moniliforme*. The chemical structure of the fumonisins is such that they inhibit lipid synthesis in biological membranes. This can cause a lethal condition, **equine leucoencephalomalacia**, which is characterised by massive atrophy of the brain and sudden death. Swine, however, are much less sensitive to acute fumonisin toxicosis, which is characterised by **pulmonary oedema** (Haschek et al., 2001). Likely of greater commercial significance, however, is the effect of fumonisins on swine immunity.

Feeding a diet containing 330 mg/kg fumonisin B_1 to weaned piglets resulted in hydrothorax and pulmonary oedema and mortality in 5-6 days (Fazekas et al., 1998). Subsequently, feeding a diet containing 40 mg/kg fumonisin B_1 resulted in no significant effect on weight gain or feed consumption and no mortality (Zomborszky et al., 2000). In a more chronic study, up to 10 mg/kg fumonisin B_1 was fed to starter pigs for eight weeks (Rotter et al., 1996). Average daily gain was reduced by 11% when the maximum dose of fumonisin was fed. The free sphinganine:free sphingosine ratio increased in liver, pancreas and adrenal glands. An 8-week feeding trial with diets up to 10 mg/kg fumonisin B_1 did not result in performance impairment or clinical signs (Zomborszky-Kovacs et al., 2002). The effects of feeding up to 1 mg fumonisin B_1/kg to growing and finishing pigs to market weight did not significantly affect carcass quality (Rotter et al., 1997).

Daily intravenous injections of fumonisin B_1 to pigs indicated that cardiovascular function is altered and that fumonisin-induced pulmonary oedema is caused by left-sided heart failure and not by altered endothelial permeability (Smith et al., 2000). This was further characterised by Constable et al. (2003) as being due to decreased cardiac output and characteristic impedance. The specific nature of fumonisin toxicity in swine has made possible the use of a biomarker for measuring exposure of pigs to fumonisin. Tissue and serum sphinganine:sphingosine ratios can be used as an early index of consumption of fumonisin-contaminated feeds (Riley et al., 1993). This is a great advantage as it overcomes problems related to feed analysis such as sampling error. Pharmacokinetic studies have indicated that systemic bioavailability of fumonisin B_1 in swine is limited to 3-6% following intragastric administration (Prelusky et al., 1994). In examining tissue residues of fumonisin B_1, it was concluded that the compound accumulated only in liver and kidney and residues were only barely detectable nine days following withdrawal of contaminated feed (Prelusky et al., 1996). It was not possible to demonstrate humoral or cellular specific and nonspecific immune response when weaned pigs were fed diets containing up to 100 mg/kg fumonisin B_1 for eight days or lower concentrations for longer periods (up to 10 mg/kg for up to 4 months) (Tornyos *et al.*, 2003). This is in contrast to reports with other animal species, which indicated a degree of immunosuppression. Recent reports, moreover, have indicated that oral doses of fumonisin B_1 when given to weanling pigs predispose the pig to infectious disease including intestinal colonisation by pathogenic *E. coli* associated with extraintestinal infection (Oswald *et al.*, 2003).

Effects of fumonisin on reproduction

Lactating sows were fed fumonisin B_1 at concentrations of up to 175 mg/kg to determine a non-lethal dietary concentration (Becker *et al.*, 1995). There were no detectable amounts of fumonisin B_1 in milk and no evidence of toxicosis in piglets or immunosuppression. Administration of fumonisin B_1 to gestating sows, however, resulted in considerable damage to foetuses *in utero* (Zomborszky-Kovacs *et al.*, 2000).

DEOXYNIVALENOL

Effects on swine production

Deoxynivalenol is one of the most common *Fusarium* trichothecene mycotoxins. This large family of compounds is generally associated with feed refusal, vomiting, and lesions of the gastrointestinal tract in swine. Intravenous infusion of deoxynivalenol to pigs resulted in peak concentrations in cerebral spinal fluid 30–60 minutes following infusion (Prelusky *et al.*, 1990). This was much longer than that observed for sheep (5-10 minutes). This was thought to be due to the very rapid and extensive tissue distribution of deoxynivalenol in swine. A subsequent study monitored tissue distribution of deoxynivalenol following intravenous dosing. It was observed that there was no extensive uptake or retention by any tissue, suggesting that accumulation of residues in swine would not happen upon chronic consumption of deoxynivalenol (Prelusky and Trenholm, 1991). This was confirmed by subsequent long term (up to seven weeks) swine feeding trials (Prelusky and Trenholm, 1992).

The effects of feeding deoxynivalenol to growing swine was determined by feeding up to 4 mg/kg deoxynivalenol in the form of contaminated oats for eight weeks. The highest dose resulted in decreased weight gain, feed intake and feed efficiency but this was not seen with dietary dosages up to 1 mg/kg deoxynivalenol (Bergsjo *et al.*, 1992). Feeding diets with up to 3.5 mg/kg deoxynivalenol to growing pigs increased liver weights, decreased serum protein and albumin concentrations (Bergsjo *et al.*, 1993). The subacute toxic effects of dietary deoxynivalenol (up to 3 mg/kg) were examined over a 32 day feeding period in grower pigs. The feeding of naturally contaminated grains resulted in more severe feed refusal than was seen when an equivalent amount of purified deoxynivalenol was fed (Prelusky *et al.*, 1994). Growing pigs (initial weight 18 kg) were fed deoxynivalenol-contaminated diets (4 mg/kg) for 42 days with performance and blood parameters monitored (Rotter *et al.*, 1995). Although reduced serum protein concentrations were seen after feeding, the pigs recovered to control levels. It was concluded that although consumption of deoxynivalenol reduced hepatic protein synthesis, animals were able to adapt to this challenge. Prelusky (1997) made a comparison of oral administration of deoxynivalenol with intraperitoneal (IP) infusion in growing pigs. Weight gains were less affected by the IP administration of the toxin, inferring that toxic effects of deoxynivalenol are greatest with oral administration. Barley naturally contaminated with deoxynivalenol was fed to growing and finishing pigs to market weight (House *et al.*, 2002). A dietary concentration of 2 mg/kg deoxynivalenol did not reduce growth rates or alter carcass composition of barrows. Deoxynivalenol was quite evenly distributed throughout the barley based on abrasive dehulling (House *et al.*, 2003).

The effect of acute intravenous administration of deoxynivalenol on brain regional neurochemistry was tested in pigs (Prelusky et al., 1992). The changes seen, however, were not indicative of known neurochemical changes associated with chemical-induced anorexia. Prelusky (1993), however, chronically administered low-level concentrations of deoxynivalenol to pigs and this resulted in changes in cerebral spinal fluid indicative of elevated brain serotonin turnover, which has been linked to reduced feed intake. No changes were seen in pig blood concentrations of serotonin or metabolites following deoxynivalenol administration thereby indicating that no peripheral effects of deoxynivalenol could account for increased serotonergic activity (Prelusky, 1994). It was shown using an *in vitro* membrane receptor-binding assay that deoxynivalenol possesses only weak affinity for the serotonin receptor subtypes. Unless deoxynivalenol concentrations are relatively high, therefore, pharmacological effects may be mediated by mechanisms other than interaction with serotonergic receptors (Prelusky, 1996). Certain specific serotonin-receptor antagonists were found to efficaciously prevent deoxynivalenol induced vomiting in pigs (Prelusky and Trenholm, 1993).

The effects of feeding graded levels of oats naturally contaminated with deoxynivalenol on immune responses were determined in growing pigs. Up to 4.7 mg/kg deoxynivalenol was fed for nine weeks. A dose-dependent reduction in secondary antibody response to tetanus toxoid was observed, but few other changes were noted (Overnes et al., 1997). Low doses of deoxynivalenol have been shown to be immunostimulatory in swine (Zielonka et al., 2003). Increased serum IgA concentrations were seen in response to feeding up to 1.2 mg/kg deoxynivalenol to piglets for eight weeks (Drochner et al., 2004).

ZEARALENONE

Effects on swine production

Zearalenone is a *Fusarium* fungal metabolite with oestrogenic properties. Reproduction in swine and other species can be adversely affected by zearalenone contamination of feedstuffs (Etienne and Dourmad, 1994). Clinical manifestations can include vulval swelling and reddening and rectal and vaginal prolapse (Rainey et al., 1991). The effect of feeding less than 1.1 mg zearalenone/kg feed to gilts, gestating sows and lactating sows on reproductive performance was determined by Kordic et al. (1990). Hyperoestrogenic syndrome in the form of vulvovaginitis was observed in 0.24% of gilts and this syndrome receded with time. Similar results were recorded in a study of the effect of very low levels of zearalenone on reproductive efficiency of gilts. It was concluded that 0.5 mg zearalenone/kg of feed would have no serious effects on reproductive efficiency of young gilts (Friend et al., 1990). In another study, prepubertal gilts were fed 10 mg zearalenone/kg of feed for two weeks (Green et al., 1990). Although prepubertal consumption of zearalenone suppressed mean serum concentrations of leuteinizing hormone, zearalenone did not delay attainment of puberty or adversely affect subsequent reproduction. Rainey et al. (1990) administered lower doses of zearalenone (up to 2 mg/kg) and noted that the hypothalamo-hypophysial axis was affected by zearalenone consumption but later reproductive performance was not affected. Feeding up to 22 mg/kg zearalenone to breeding gilts had an obvious harmful effect on reproductive performance including decreased number of corpora lutea, decreased weight of ovaries, decreased number of live embryos, increased number of piglets born dead and occurrence of abortions (Kordic et al., 1992).

Zearalenone is metabolised to α and ß-zearalenols in the pig but it was not possible to correlate blood concentrations of zearalenone and α-zearalenol with the occurrence of the external symptoms of hyperoestrogenism (Obremski *et al.* (2003). Zearalenone and metabolites were analysed in urine and tissues of pigs fed zearalenone-contaminated oats (Zollner *et al.* (2002). In urine, 60% of zearalenone was in the form of α and ß-zearalenol. Glucuronide conjugates of zearalenone and the zearalenols were also found in urine and liver. There is also extensive biliary secretion and enterohepatic cycling of zearalenone and metabolites in pigs (Biehl *et al.*, 1993). Identification of zearalenone toxicosis in swine is complicated by the presence of zearelenone glycosides, which release free zearalenone into the digestive tract upon digestion (Gareis *et al.*, 1990).

Toxicological synergism arising from combinations of mycotoxins

Increased international trading of feedstuffs has contributed to the severity of mycotoxicoses as this increases the chance that a given compound swine feed will contain components of widely varying geographical origins. Such blends of ingredients increase the chance of the feed containing mixtures of different mycotoxins. This can result in toxicological synergies that increase the severity of mycotoxicoses (Speijers and Speijers, 2004). Most recent research in the effects of mycotoxins on swine production have focused on the toxicological interactions between mycotoxins.

Symptoms typical of swine mycotoxicoses are often seen despite analysis of the feed that indicates only very low concentrations of mycotoxins (Trenholm *et al.*, 1983). In this situation, it is not clear if a mycotoxin problem really exists, or if poor performance is due to management or nutritional factors. It is now known that unexpected toxicity may be due to toxicological interactions between different mycotoxins that exaggerate the toxicity. The likelihood of this occurring is greatest for the *Fusarium* mycotoxins. It has been shown that feeding naturally-contaminated feedstuffs produces greater toxicity than an equivalent amount of purified mycotoxins (Trenholm *et al.*, 1994). It has been shown that fusaric acid, the most common of the *Fusarium* mycotoxins (Bacon *et al.*, 1996), can increase the toxicity of deoxynivalenol in starter pigs (Smith *et al.*, 1997). Fusaric acid concentration, however, is seldom determined in swine feeds due to its low toxicity when consumed in the absence of other toxins (Smith and MacDonald, 1991; Smith and Sousadias, 1993).

INTERACTIONS BETWEEN AFLATOXIN AND *FUSARIUM* MYCOTOXINS

The effects of aflatoxin and T-2 toxin singly and in combinations were evaluated in growing swine (Harvey *et al.*, 1990). It was concluded that when fed in combination, each mycotoxin appeared to have a sparing action on certain effects of the other and that the responses were either additive or less than additive. Feeding combinations of aflatoxin B_1 and fumonisin B_1 derived from culture material to growing pigs was investigated by Harvey *et al.* (1995). The mycotoxins given singly or in combination adversely affected clinical performance, biochemical, hematologic and immunologic values. In general, responses were affected more by the combination diet than by either mycotoxin fed alone, and the toxic responses could be described as additive or more than additive, particularly for induction of liver disease. More recent studies have also failed to describe major synergistic effects of aflatoxin and fumonisin in growing pigs (Dilkin *et al.*, 2003). The effects of aflatoxin B_1 and fumonisin B_1 on primary swine alveolar macrophages were investigated by Liu *et al.* (2002). Their results suggested that both are immunotoxic to swine alveolar macrophages but that they exert toxic effects through different biochemical mechanisms.

INTERACTIONS BETWEEN OCHRATOXIN AND OTHER MYCOTOXINS

Several studies have examined the potential interactions between ochratoxin A and *Fusarium* mycotoxins in swine. The effects of ochratoxin and T-2 toxin, singly and in combination, were evaluated in growing barrows (Harvey *et al.*, 1994). Ochratoxin A and T-2 toxin singly and in combination affected performance, serum biochemistry, haematology, immunology and organ weights. These effects, however, were described as additive and not synergistic. Combinations of 0.1 mg/kg ochratoxin A and 1.0 mg/kg deoxynivalenol were fed to growing pigs for 90 days (Lusky *et al.*, 1998). It was concluded that the toxins fed alone or in combination were toxic, indicating that toxicological synergy was absent. Weaner pigs were also fed combinations of ochratoxin A, deoxynivalenol, T-2 toxin and fumonisin in quantities that could be expected in feeds of central European origin (Mueller *et al.*, 1999). Little effect was seen beyond the effect of ochratoxin alone. It was concluded that synergistic effects on immuno-suppression should not be expected. The effects of ochratoxin A, deoxynivalenol and zearalenone were tested for 90 days in growing pigs (Lusky *et al.*, 2001). No detectable residues of deoxynivalenol or zearalenone were found in tissues although there was an effect of simultaneous administration of deoxynivalenol and zearalenone on secretion of ochratoxin A.

The potential synergistic effects of ochratoxin A and penicillic acid were investigated in young pigs by feeding diets contaminated with *Aspergillus ochraceus* (Stoev *et al.*, 2001). The microscopic lesions seen were somewhat different from classic Danish porcine nephropathy; and it was concluded that there may be a synergistic effect between ochratoxin A and penicillic acid. The combination of ochratoxin A and citrinin has also been tested in swine but no obvious toxicological synergy was detected (Sandor *et al.*, 1991).

INTERACTIONS BETWEEN FUMONISIN AND OTHER *FUSARIUM* MYCOTOXINS

The effects of dietary fumonisin B_1 and deoxynivalenol, singly and in combination, were determined in growing barrows (Harvey *et al.*, 1996). Performance, serum biochemistry, immunological responses and histopathological lesions were evaluated. It was concluded that for many variables, the response was additive but for some parameters the response was more than additive. It was suggested that caution should be exercised when feeding combinations of fumonisin and deoxynivalenol. It was not possible to demonstrate such a synergy when fumonisin was fed to growing barrows in combination with moniliformin (Harvey *et al.*, 2002).

INTERACTIONS BETWEEN DEOXYNIVALENOL AND OTHER *FUSARIUM* MYCOTOXINS

The toxicological interaction between deoxynivalenol and T-2 toxin in growing pigs was evaluated by Friend *et al.* (1992). In this study deoxynivalenol was fed at up to 2.5 mg/kg and T-2 toxin at up to 3.2 mg/kg but there was little evidence of toxicological synergy. A sophisticated attempt to demonstrate toxicological synergy between deoxynivalenol and different *Fusarium* mycotoxins was undertaken in growing swine by Rotter *et al.* (1992). Deoxynivalenol was fed singly and in combination with sambucinol, 15-acetyldeoxynivalenol, 3-

acetyldeoxynivalenol and culmorin. No significant interactions were seen.

NATURALLY-CONTAMINATED FEEDSTUFFS CONTAINING COMBINATIONS OF MYCOTOXINS

Corn naturally-contaminated with nivalenol (11.5 mg/kg) and zearalenone (3 m/kg) was fed to grower and pregnant pigs in five experiments (Williams and Blaney, 1994). Reduced feed intake, average daily gains and feed efficiencies were observed. The feeding of contaminated corn, however, had no adverse effects on conception rates or on fetal numbers and weights at slaughter. Diets containing blends of corn and wheat naturally contaminated with deoxynivalenol, 15-acetyldeoxynivalenol, zearalenone and fusaric acid were fed to starter pigs (Swamy *et al.*, 2002). Marked reductions in growth and feed consumption were observed coupled with changes in brain neurochemistry. It was possible to explain the different sensitivities of starter pigs and broiler chickens due to the different effects of these combinations of mycotoxins on brain neurochemistry (Swamy *et al.*, 2004). Pair feeding studies indicated that most of the reduced growth rate of pigs fed contaminated grains was due to reduced feed intake (Swamy *et al.*, 2003).

Conclusions

Research in mycotoxicoses in swine has progressed from the initial descriptive studies of pathologies arising from acute and chronic administration of individual mycotoxins to focus on the toxicological synergies arising from combinations of mycotoxins. This is most readily achieved by feeding naturally-contaminated feedstuffs, which most closely mimics conditions faced by the swine production industry worldwide. Advances in analytical chemistry are needed to more adequately characterise the mycotoxin challenge posed by naturally contaminated feeds. An area yet to be fully explored is the significance of secondary mycotoxic diseases arising from mycotoxin-induced immuno-suppression. Quantification of economic losses arising from such diseases will be important in understanding the significance of feed-borne mycotoxins in swine production.

References

Bacon, C.W., J.K. Porter, W.P. Norred, and J.F. Leslie. 1996. Production of fusaric acid by *Fusarium* species. Appl. Environ. Microbiol. 62:4039-4043.

Becker, B.A., L. Pace, G.E. Rottinghaus, R. Shelby, M. Misfeldt, and P.F. Ross. 1995. Effects of feeding fumonisin B_1 in lactating sows and their suckling pigs. Am. J. Vet. Res. 56:253-1258.

Bergsjo, B., W. Langseth, I. Nafstad, J. Jansen, and H.J.S. Larsen. 1993. The effects of naturally deoxynivalenol-contaminated oats on the clinical condition, blood parameters, performance and carcass composition of growing pigs. Vet. Res. Commun. 17:283-294.

Bergsjo, B., T. Matre, and I. Nafstad. 1992. Effects of diets with graded levels of deoxynivalenol on performance in growing pigs. J. Vet. Med. A 39:752-758.

Biehl, M.L., D.B. Prelussky, G.D. Koritz, K.E. Hartin, W.B. Buck, and H.L. Trenholm. 1993. Biliary excretion and enterohepatic cycling of zearalenone in immature pigs. Toxicol. Appl. Pharmacol. 121:152-159.

Biro, K., I. Barna-Betro, T. Pecsi, E. Szabo, G. Winkler, J. Fink-Gremmels, and L. Solti. 2003. Evaluation of spermatological parameters in ochratoxin A challenged boars. Theriogenology 60:199-207.

Bonomi, A., A. Quarantelli, I. Mazzali, L. Lucchelli, E. Cabassi, A. Corradi, A. Ubaldi, A. Fusari and R. Chizzolini. 1992. The effects of aflatoxin B_1 contaminated rations on the productive efficiency and on the meat yield and quality in fattening pigs. Rivista della Societa Italiana di Scienza dell'Alimentazione 21:501-534.

Bonomi, A., A. Quarantelli, I. Mazzali, E. Cabassi, A. Corradi, A. Ubaldi, A. Fusari, and R. Chizzolini. 1993. The effects of aflatoxin G_1 contaminated rations on the productive efficiency and on the meat yield and quality of fattening pigs. Rivista di Scienza dell'Alimentazione 22:351-377.

Bonomi, A., A. Quarantelli, E.M. Zambin, E. Cabassi, A. Corradi, R. Di Lecce, A. Ubaldi, and A. Fusari. 1995a. Effects of aflatoxin B_1 contaminated rations on productive and reproductive efficiency in swine. Rivista di Scienza dell'Alimentazione 24:361-384.

Bonomi, A., A. Quarantelli, E.M. Zambini, E. Cabassi, A. Corradi, R. Di Lecce, A. Ubaldi, and A. Fusari. 1995b. Effects of aflatoxin G_1 contaminated rations on productive and reproductive efficiency in swine. Rivista di Scienza dell 'Alimentazione 24:385-405.

Bonomi, A., A. Quarantelli, A., E.M. Zambini, E. Cabassi, A. Corradi, R. Di Lecce, A. Ubaldi, and A. Fusari. 1996. Effects of aflatoxin B_1 and G_1 contaminated rations on productive and reproductive efficiency in swine. Rivista di Scienza dell'Alimentazione 25:169-189.

Constable, P.D., G.W. Smith, G.E. Rottinghaus, M.E. Tumbeson, and W.M. Haschek. 2003. Fumonisin-induced blockade of ceramide synthase in sphingolipid biosynthetic pathway alters aortic input impedance spectrum of pigs. Am. J. Physiol. 284:H2034-H2044.

Dilkin, P., P. Zorzete, C.A. Mallmann, J.D.F. Gomes, E.E. Utiyama, L.L. Oetting, and B. Correa. 2003. Toxicological effects of chronic low doses of aflatoxin B_1 and fumonisin B_1 – containing *Fusarium moniliforme* culture material in weaned piglets. Food Chem. Tox. 41:1345-1353.

Drochner, W., M. Schollenberger, H.-P. Piepho, S. Gotz, U. Lauber, M. Tafaj, F. Klobasa, U. Weiler, R. Claus, and M. Steffl. 2004. Serum IgA-promoting effects induced by feed loads containing isolated deoxynivalenol (DON) in growing piglets. J. Tox. Env. Health A 67:1051-1067.

Etienne, M. and J.-Y. Dourmad. 1994. Effects of zearalenone or glucosinolates in the diet on reproduction in sows: A review. Livest. Prod. Sci. 40:99-113.

Fazekas, B., E. Bajmocy, R. Flavits, A. Fenyvesi, and J. Tanyi. 1998. Fumonisin B_1 contamination of maize and experimental acute fumonisin toxicosis in pigs. J. Vet. Med. B 45:171-181.

Friend, D.W., B.K. Thompson, H.L. Trenholm, H.J. Boermans, K.E. Hartin, and P.L. Panich. 1992. Toxicity of T-2 toxin and its interaction with deoxynivalenol when fed to young pigs. Can. J. Anim. Sci. 72:703-711.

Friend, D.W., H.L. Trenholm, B.K. Thompson, K.E. Hartin, P.S. Fiser, E.K. Asem and B.K. Tsang. 1990. The reproductive efficiency of gilts fed very low levels of zearalenone. Can. J. Anim. Sci. 70:635-646.

Gareis, M., J. Bauer, J. Thiem, G. Plank, S. Grabley, and B. Gedek. 1990. Cleavage of zearalenone glycoside a masked mycotoxin during digestion in swine. J. Vet. Med. B 37:236-240.

Glavits, R. 1993. Nephropathy caused by ochratoxin A in a Hungarian swine stock. Magyar Allatorvosok Lapja 48:343-349.

Green, M.L., M.A. Diekman, J.R. Malayer, A.B. Scheidt, and G.G. Long. 1990. Effect of prepubertal consumption of zearalenone on puberty and subsequent reproduction of gilts. J. Anim. Sci. 68:171-178.

Harvey, R.B., T.S. Edrington, L.F. Kubena, M.H. Elissalde, H.H. Casper, G.E. Rottinghaus, and J.R. Turk. 1996. Effects of dietary fumonisin B_1 – containing culture material, deoxynivalenol-contaminated wheat, or their combination on growing barrows. Am. J. Vet. Res. 57:1790-1794.

Harvey, R.B., T.S. Edrington, L.F. Kubena, M.H. Elissalde, and G.E. Rottinghaus. 1995. Influence of aflatoxin and fumonisin B-1- containing culture material on growing barrows. Am. J. Vet. Res. 56:1668-1672.

Harvey, R.B., T.S. Edrington, L.F. Kubena, G.E. Rottinghaus, J.R. Turk, K.J. Genovese, R.L. Ziprin, and D.J. Nisbet. 2002. Toxicity of fumonisin from *Fusarium verticilliodes* culture material and moniliformin from *Fusarium fujikuroi* culture material when fed singly and in combination to growing barrows. J. Food Protect. 65:373-377.

Harvey, R.B., M.H. Elissalde, L.F. Kubena, E.A. Weaver, D.E. Corrier, and B. Clement. 1992. Immunotoxicity of ochratoxin A to growing gilts. Am. J. Vet. Res. 53:1966-1970.

Harvey, R.B., L.F. Kubena, and M.H. Elissalde. 1995. Effects of aflatoxin on tocopherol and retinol concentrations in growing barrows. Agri-Practice 16:12-14.

Harvey, R.B., L.F. Kubena, M.H. Elissalde, G.E. Rottinghaus, and D.E. Corrier. 1994. Administration of ochratoxin A and T-2 toxin to growing swine. Am. J. Vet. Res. 55:1757-1761.

Harvey, R.B., L.F. Kubena, W.E. Huff, D.E. Corrier, G.E. Rottinghaus, and T.D. Phillips. 1990. Effects of treatment of growing swine with aflatoxin and T-2 toxin. Am. J. Vet. Res. 51:1688-1693.

Haschek, W.M., L.A. Gumprecht, G. Smith, M.E. Tumbleson, and P.D. Constable. 2001. Fumonisin toxicosis in swine: An overview of porcine pulmonary edema and current perspectives. Environ. Health Perspect. 109:251-257.

House, J.D., D. Abramson, G.H. Crow, and C.M. Nyachoti. 2002. Feed intake, growth and carcass parameters of swine consuming diets containing low levels of deoxynivalenol from naturally contaminated barley. Can. J. Anim. Sci. 82:559-565.

House, J.D., C.M. Nyachoti, and D. Abramson. 2003. Deoxynivalenol removal from barley intended as swine feed through the use of an abrasive perling procedure. J. Ag. Food Chem. 51:5172-5175.

Hult, K., E. Hokby, G. Sellyey, L. Rutqvist, and S. Gatenbeck. 1992. Ochratoxin A occurrence in slaughter-pigs in Sweden and its use as a tool for feed screening programs. J. Environ. Pathol. Toxicol. Oncol. 11:103-104.

Jorgensen, K. and A. Petersen. 2002. Content of ochratoxin A in paired kidney and meat samples from healthy Danish slaughter pigs. Food Addit. Contam. 19:562-567.

Kordic, B., S. Pribicevic, M. Muntanola-Cvetkovic, P. Nikolic, and B. Nikolic. 1992. Experimental study of the effects of known quantities of zearalenone on swine reproduction. J. Environ. Pathol. Toxicol. Oncol. 11:117-119.

Kordic, B., M. Simic, and D. Lazarevic. 1990. Effect of low doses of the mycotoxin zearalenone on reproductive characteristics of pigs. Veterinarski Glasnik 44:151-155.

Larsson, P. and H. Tjalve. 1996. Bioactivation of aflatoxin B_1 in the nasal and tracheal mucosa in swine. J. Anim. Sci. 74:1672-1680.

Lindemann, M.D., D.J. Blodgett, A.F. Harper, E.T. Kornegay, and J.A. Doerr. 1997. Appraisal of the value of selected clays and minerals in diets with and without aflatoxin-contaminated maize fed to young pigs. J. Anim. Feed Sci. 6:507-519.

Lippold, C.C., S.C. Stothers, A.A. Frohlich, R.J. Boila, and R.R. Marquardt. 1992. Effects of periodic feeding of diets containing ochratoxin A on the performance and clinical chemistry of pigs from 15 to 50 kg body weight. Can. J. Anim. Sci. 72:135-146.

Liu, B.-H., F.-Y. Yu, M.-H. Chan, and Y.-L. Yang. 2002. The effects of mycotoxins, fumonisin B_1 and aflatoxin B_1, on primary swine alveolar macrophages. Toxicol. Appl. Pharmacol. 180:197-204.

Lusky, K., R. Goebel, D. Tesch, K.-D.

Koberschuetz, K. Lusky, and W. Haider. 2001. The effects of mycotoxins ochratoxin A, zearalenone and deoxynivalenol on the health and residue values of pigs. Tieraerztliche Umschau 56:15-20.

Lusky, K., R. Goebel, D. Tesch, G. Tenner, W. Haider, M. Krueger, and A. Lippert. 1998. Studies on the effects of ochratoxin A and deoxynivalenol toxicity on the health of pigs and tissue residue concentrations. Tieraerztliche Umschau 53:623-630.

Mueller, G., P. Kielstein, H. Rosner, A. Berndt, M. Heller, and H. Koehler. 1999. Studies on the influence of combined administration of ochratoxin A, fumonisin B_1, deoxynivalenol and T-2 toxin on immune and defence reactions in weaner pigs. Mycoses 42:485-493.

Obremski, K., M. Gajecki, W. Zwierzchowski, T. Bakula, J. Apoznanski, and J. Wojciechowski. 2003. The level of zearalenone and α-zearalenol in the blood of gilts with clinical symptoms of toxicosis, fed diets with a low zearalenone content. J. Anim. Feed Sci. 12:529-538.

Ominshi, K.H., A.A. Frohlich, R.R. Marquardt, G.H. Crow, and D. Abramson. 1996. The incidence and distribution of ochratoxin A in western Canadian swine. Food Addit. Contam. 13:185-198.

Oswald, I.P., C. Desautels, J. Laffieet, S. Fournout, S.Y. Peres, M. Odin, P. Le Bars, J. Le Bars, and J.M. Fairbrother. 2003. Mycotoxin fumonisin B_1 increases intestinal colonization by pathogenic *Escherichia coli* in pigs. Appl. Environ. Microbiol. 69:5870-5874.

Overnes, G., T. Matre, T. Sivertsen, H.J.S. Larsen, W. Langseth, L.J. Reitan, and J.H. Jansen. 1997. Effects of diets with graded levels of naturally deoxynivalenol-contaminated oats on immune reponse in growing pigs. J. Vet. Med. A 44:539-550.

Prelusky, D.B. 1993. The effect of low-level deoxynivalenol on neurotransmitter levels measured in pig cerebral spinal fluid. J. Environ. Sci. Health B 28:731-761.

Prelusky, D.B. 1994. The effect of deoxynivalenol on serotoninergic neurotransmitter levels in pig blood. J. Environ. Sci. Health B 29:1203-1218.

Prelusky, D.B. and H.L. Trenholm. 1991. Tissue distribution of deoxynivalenol in swine dosed intravenously. J. Ag. Food Chem. 39:748-751.

Prelusky, D.B. and H.L. Trenholm. 1992. Nonaccumulation of residues in swine tissue following extended consumption of deoxynivalenol-contaminated diets. J. Food Sci. 57:801-802.

Prelusky, D.B. 1996. A study on the effect of deoxynivalenol on serotonin receptor binding in pig brain membranes. J. Environ. Sci. Health B 31:1103-1117.

Prelusky, D.B. 1997. Effect of intraperitoneal infusion of deoxynivalenol on feed consumption and weight gain in the pig. Nat. Toxins 5:121-125.

Prelusky, D.B., R.G. Gerdes, K.L. Underhill, B.A. Rotter, P.Y. Jui, and H.L. Trenholm. 1994. Effects of low-level dietary deoxynivalenol on haematological and clinical parameters of the pig. Nat. Toxins 2:97-104.

Prelusky, D.B., K.E. Hartin, and H.L. Trenholm. 1990. Distribution of deoxynivalenol in cerebral spinal fluid following administration to swine and sheep. J. Environ. Sci. Health B 25:395-414.

Prelusky, D.B., J.D. Miller, and H.L. Trenholm. 1996. Disposition of ^{14}C-derived residues in tissues of pigs fed radiolabelled fumonisin B_1. Food Add. Contam. 13:155-162.

Prelusky, D.B. and H.L. Trenholm. 1993. The efficacy of various classes of anti-emetics in preventing deoxynivalenol-induced vomiting in swine. Nat. Toxins 1:296-302.

Prelusky, D.B., H.L. Trenholm, and M.E. Savard. 1994. Pharmacokinetic fate of ^{14}C-labelled fumonisin B_1 in swine. Nat. Toxins 2:73-80.

Prelusky, D.B., J.M. Yeung, B.K. Thompson, and H.L. Trenholm. 1992. Effect of deoxynivalenol on neurotransmitters in discrete regions of swine brain. Arch. Environ. Contam. Toxicol. 22:36-40.

Rainey, M.R., R.C. Tubbs, L.W. Bennett, and N.M. Cox. 1990. Prepubertal exposure to dietary zearalenone alters hypothalamo-hypophysial function but does not impair postpubertal reproductive function of gilts. J. Anim. Sci. 68:2015-2022.

Rainey, M.R., R.C. Tubbs, D.K. Hardin, and N.M. Cox. 1991. Clinical manifestations of prepubertal exposure to zearalenone in gilts. Agri-Practice 12:35-36.

Riley, R.T., N.-H. An, J.L. Showker, H.-S. Yoo, W.P. Norred, W.J. Chamberlain, E. Wang, A.H. Merrill, and G. Motelin. 1993. Alteration of tissue and serum sphinganine to sphingosine ratio: An early biomarker of exposure to fumonisin-containing feeds in pigs. Toxicol. Appl. Pharmacol. 118:105-112.

Rotter, B.A., D.B. Prelusky, A. Fortin, J.D. Miller, and M.E. Savard. 1997. Impact of pure fumonisin B_1 on various metabolic parameters and carcass quality of growing-finishing swine. Can. J. Anim. Sci. 77:465-470.

Rotter, B.A., B.K. Thompson, and M. Lessard. 1995. Effects of deoxynivalenol-contaminated diet on performance and blood parameters in growing swine. Can. J. Anim. Sci. 75:297-302.

Rotter, B.A., B.K. Thompson, D.B. Prelusky, H.L. Trenholm, B. Stewart, J.D. Miller, and M.E. Savard. 1996. Response of growing swine to dietary exposure of pure fumonisin B_1 during an eight-week period: growth and clinical parameters. Nat. Toxins 4:42-50.

Rotter, R.G., B.K. Thompson, H.L. Trenholm, D.B. Prelusky, K.E. Hartin, and J.D. Miller. 1992. A preliminary examination of potential interactions between deoxynivalenol (DON) and other selected *Fusarium* metabolites in growing pigs. Can. J. Anim. Sci. 72:107-116.

Sandor, G., A. Busch, H. Watzke, J. Reek, and A. Vanyi. 1991. Subacute toxicity testing of ochratoxin A and citrinin in swine. Act. Vet. Hung. 39:149-160.

Schell, T.C., M.D. Lindemann, E.T. Kornegay, and D.J. Blodgett. 1993. Effects of feeding aflatoxin-contaminated diets with and without clay to weanling and growing pigs on performance, liver function, and mineral metabolism. J. Anim. Sci. 71:1209-1218.

Schell, T.C., M.D. Lindemann, E.T. Kornegay, D.J. Blodgett, and J.A. Doerr. 1993. Effectiveness of different types of clay for reducing the detrimental effects of aflatoxin-contaminated diets on performance and serum profiles of weanling pigs. J. Anim. Sci. 71:1226-1231.

Smith, G.W., P.D. Constable, R.M. Eppley, M.E. Tumbleson, L.A. Gumprecht, and W.M. Haschek-Hock. 2000. Purified fumonisin B_1 decreases cardiovascular function but does not alter pulmonary capillary permeability in swine. Toxicol. Sci. 56:240-249.

Smith, J.E. and K. Ross. 1991. The toxigenic aspergilli. In: *Mycotoxins and Animal Foods* (J.E. Smith and R.S. Henderson, eds). CRC Press, Boca Raton, FL, pp. 101-139.

Smith, T.K. and E.J. MacDonald. 1991. Effect of fusaric acid on brain regional neurochemistry and vomiting behavior in swine. J. Anim. Sci. 69:2044-2049.

Smith, T.K. and M.G. Sousadias. 1993. Fusaric acid content of swine feedstuffs. J. Ag. Food. Chem. 41:2296-2298.

Smith, T.K., E.G. McMillan, and J.B. Castillo. 1997. Effect of feeding blends of *Fusarium* mycotoxin-contaminated grains containing deoxynivalenol and fusaric acid on growth and feed consumption of immature swine. J. Anim. Sci. 75:2184-2191.

Solti, L., T. Pecsi, I. Barna-Vetro, F. Szasz, K. Biro, and E. Szabo. 1999. Analysis of serum and seminal plasma after feeding ochratoxin A with breeding boars. Anim. Reprod. Sci. 56:123-132.

Speijers, G.J.A. and M.H.M. Speijers. 2004. Combined toxic effects of mycotoxins. Toxicol. Lett.153:91-98.

Stoev, S.D., M. Paskalev, S. Macdonald, and P.G. Mantle. 2002. Experimental one year ochratoxin A toxicosis in pigs. Exp. Toxicol. Pathol. 53:481-487.

Stoev, S.D., S. Vitanov, G. Anguelov, T. Petkova-Bocharova, and E.E. Creppy. 2001. Experimental mycotoxic nephropathy in pigs provoked by a diet containing ochratoxin A and penicillic acid. Vet. Res. Commun. 25:205-223.

Swamy, H.V.L.N., T.K. Smith, and E.J. MacDonald. 2004. Effects of feeding blends of grains naturally contaminated with *Fusarium* mycotoxins on brain regional neurochemistry of starter pigs and broiler chickens. J. Anim. Sci. 82:2131-2139.

Swamy, H.V.L.N., T.K. Smith, E.J. MacDonald, H.J. Boermans, and E.J. Squires. 2002. Effects of feeding a blend of grains naturally contaminated with *Fusarium* mycotoxins on swine performance, brain regional neurochemistry, and serum chemistry and the efficacy of a polymeric glucomannan mycotoxin adsorbent. J. Anim. Sci. 80:3257-3267.

Swamy, H.V.L.N., T.K. Smith, E.J. MacDonald, N.A. Karrow, B. Woodward, and H.J. Boermans. 2003. Effects of feeding a blend of grains naturally contaminated with *Fusarium* mycotoxins on growth and immunological measurements of starter pigs, and the efficacy of a polymeric glucomannan mycotoxin adsorbent. J. Anim. Sci. 81:2792-2803.

Terao, K. and K. Ohtsubo. 1991. Biological activities of mycotoxins: Field and experimental mycotoxicoses. In: *Mycotoxins and Animal Foods* (J.E. Smith and R.S. Henderson, eds). CRC Press, Boca Raton, FL, pp. 455-488.

Tornyos, G., M. Kovacs, M. Rusvai, P. Horn, J. Fodor, and F. Kovacs. 2003. Effect of dietary fumonisin B_1 on certain immune parameters of weaned pigs. Acta. Vet. Hung. 51:171-179.

Trenholm, H.L., W.P. Cochrane, H. Cohen, J.I. Elliott, E.R. Farnworth, D.W. Friend, R.M.G. Hamilton, J.R. Standish, and B.K. Thompson. 1983. Survey of vomitoxin contamination of 1980 Ontario winter wheat crop: Results of survey and feeding trials. JAOAC 66:92-97.

Trenholm, H.L., B.C. Foster, L.L. Charmley, B.K. Thompson, K.E. Hartin, R.W. Coppock, and M.A. Albassam. 1994. Effects of feeding diets containing *Fusarium* (naturally) contaminated wheat or pure deoxynivalenol (DON) in growing pigs. Can. J. Anim. Sci. 74:361-369.

Wannemacher, R.W., D.L. Bunner, and H.A. Neufeld. 1991. Toxicity of trichothecenes and other related mycotoxins in laboratory animals. In: *Mycotoxins and Animal Foods* (J.E. Smith and R.S. Henderson, eds). CRC Press, Boca Raton, FL, pp. 499 – 522.

Williams, K.C. and B.J. Blaney. 1994. Effect of the mycotoxins, nivalenol and zearalenone, in maize naturally infected with *Fusarium graminearum* on the performance of growing and pregnant pigs. Aust. J. Agric. Res. 45:1265-1279.

Wood, G.E. 1992. Mycotoxins in foods and feeds in the United States. J. Anim. Sci. 70:3941-3949.

Wu, J.-F., C.-S. Cheng, I.-T. Yu, C.-C. Yen, and C.-C. Kuo. 1992. Effects of low dietary levels of aflatoxin on the reproductive performance of sows. J. Ag. Assoc. China New Series 159:82-90.

Wu, J.-F., I.-T. Yu, C.-S. Cheng, C.-C. Yen, and C.-C. Kuo. 1989. Effects of low dietary levels of aflatoxin on performance and tissue residues of finishing pigs. J. Ag. Assoc. China 146:42-50.

Zielonka, L., M. Gajecki, K. Obremski, and W. Zwierzchowski. 2003. Influence of low doses of deoxynivalenol applied *per os* on chosen indexes of immune response in swine. Polish J. Vet. Sci. 6:74-77.

Zollner, P., J. Jodlbauer, M. Kleinova, H. Kahlbacher, T. Kuhn, W. Hochsteiner, and

W. Lindner. 2002. Concentration levels of zearalenone and its metabolites in urine, muscle tissue, and liver samples of pigs fed with mycotoxin-contaminated oats. J. Ag. Food Chem. 50:2494-2501.

Zomborszky, M.K., F. Vetesi, I. Repa, F. Kovacs, A. Bata, P. Horn, A. Toth, and R. Romvari. 2000. Experiment to determine limits of tolerance for fumonisin B_1 in weaned piglets. J. Vet. Med. B 47:277-286.

Zomborszky-Kovacs, M., F. Vetesi, P. Horn, I. Repa, and F. Kovacs. 2002. Effects of prolonged exposure to low-dose fumonisin B_1 in pigs. J. Vet. Med. B 49:197-201.

Zomborszky-Kovacs, M., F. Vetesi, F. Kovacs, A. Bata, A. Toth, and G. Tornyos. 2000. Preliminary communication: Examination of the harmful effect to fetuses of fumonisin B_1 in pregnant sows. Teratog. Carcinog. Mutagen. 20:293-299.

11

MYCOTOXINS IN FORAGES

J. Fink-Gremmels
Utrecht University, Faculty of Veterinary Medicine, Utrecht, The Netherlands

Introduction

Each year, many million tonnes of forage consumed by animals is contaminated with fungal species invading forage plants prior to harvest or during storage as hay, straw, or silage. The interactions between fungal species and plants are diverse, and comprise a true symbiosis, as observed for many endophytes, an aggressive invasion of living plants resulting in visible plant disease and deterioration as being characteristic for many *Fusarium* species as well as *Rhizoctonia* spp., and finally the utilisation of organic biomass in the decay of plant material.

Fungi produce a wide diversity of secondary metabolites to increase their competitiveness in nature, and many of these metabolites exert antifungal and antibacterial activity. These specific properties have long been recognized and initiated the synthesis of medicinal products, the most prominent one being penicillin. In the definition of the term **mycotoxins**, these effects are often neglected. As most mycotoxin problems in animal husbandry involve contaminated feed grains and oilseeds that are used worldwide in the production of formulated feeds, research focused also on investigation of toxins, which are regularly observed in feeds. Ruminant species are less sensitive to many mycotoxins, as the forestomach flora effectively degrades large parts of the ingested toxins. The mycotoxins that exert antimicrobial effects, however, are of great significance in ruminants as they impair ruminal function, resulting not only in malnutrition, but also facilitate the unexpected absorption of other mycotoxins, which under normal circumstances do not pass the rumen. As forage is still the most economic way to feed ruminant animals, the economic losses encountered with mycotoxin contamination can be significant.

Mycotoxins in growing forage grasses

Natural grassland covers large areas in moderate climates across the world. Whilst grass and hay have been monitored for many decades for environmental pollutants, either naturally occurring substances or industrial products, natural toxins such as secondary plant metabolites or fungal metabolites were considered to be only of local interest. The two major diseases related to either tall fescue (fescue toxicosis) or perennial ryegrass (staggers) and the considerable local economic losses related to these disease complexes in grazing animals, like ruminants and horses, stimulated research activities in this area (Table 1).

TOXINOGENIC ENDOPHYTES IN PASTURE GRASS

Various endophytic fungi may infect growing pasture grass. In particular, species of the

Table 1. Seed-borne endophytes in grasses affecting animal health[1].

Grass	Endophyte	Herbivore	Prevailing effect	Continent
Achnatherum (Stipa) robusta Sleepy grass	Neotyphodium	Horses, sheep	Narcotic	NA
Achnatherum inebrians (drunken horse grass)	Neotyphodium	Cattle	Narcotic	AS
Lolium temelentum (Darnel)	Neotyphodium? Clavibacter toxicosis?	Livestock	Toxicosis	EU/AS
Lolium perenne (perennial ryegrass)	Neotyphodium	Cattle, sheep	Staggers	NA, NZ, AU, EU
Poa huecu	Neotyphodium?	Livestock	Staggers	SA
Melica descumbens (drunk grass)	Neotyphodium	Livestock	Staggers/ narcotic	AF
Echinopogon ovatus	Neotyphodium- like	Cattle, sheep	Staggers	AU/NZ
Festuca argentina	Neotyphodium?	Livestock	Toxicosis	SA
Festuca arundinacea	Neotyphodium	Cattle, sheep	Toxicosis	NA
Festuca hieronymi	Neotyphodium	Livestock	Toxicosis	SA

NA: North America; AS: Asia; EU: Europa; SA: South America; AU: Australia; NZ: New Zealand
[1]Adapted from Faeth *et al.*, 2000

genus *Balansia* and *Epichloe* produce mycelia in the intracellular space of leaf mesophyll of grasses and the parenchyma of culms. Mycelia may also be incorporated into ovules, infecting the embryo and subsequently being transmitted with the seeds to the next generation of the host. Partly because of seed transmission, many endophytes have become abundant and widespread in cool-season grasses and have been transferred from one continent to another with grass seeds. In contrast, *Balansia* endophytes are not seed-transmitted and are less common.

Endophytes may be classified according to the typical life cycles (Schardl, 2001). Type I endophytes (*Epichloe baconii, E. clarkii and E. typhina*) are those species of the *Epichloe* that are not seed-transmitted. Transmission can only occur with the help of plant-eating flies in the genus *Botanophila* (formerly *Phorbia*) to complete the sexual cycle. The life cycles of the endophyte and the flies are synchronised (Bultman *et al.*, 1995). Type II endophytes (*E. amarillans, E. brachyelytris, E. bromicola, E. elymi, E. festucae, E. glyceriae, E. sylvatica* etc.) show both stroma (the structure on which the spermatia and perithecia of the endophyte develop) on some culms and seed transmission on others (Funk and White, 1997).

In contrast, type III endophytes, the *Neotyphodium* spp. (formerly denoted *Acremonium*) including *N. lolii, N. starri, N. uncinatum* and *N. tembladerae*, very rarely or never form stromata, thus relying entirely on asexual reproduction and seed transmission. Permanent genetic changes (evolution through hybridisation) appear to have modified the genome of many endophytes, making them entirely incapable of forming stromata (Schardl, 2001). This is apparently the case for several Eurasian type II endophytes, such as *Neotyphodium coenophialum, N. lolii* and the South American *N. tembladerae* (Tsai *et al.*, 1994; Kuldau *et al.*, 1999; Cabal *et al.*, 1999;

Schardl and Wilkinson, 2000; Sullivan and Faeth, 2004).

Endophytes are often considered to be mutualists, as they confer a variety of benefits to the host (Schardl, 2001). Endophyte infection primarily increases plant resistance to pathogenic microorganisms, helminths and insects, mainly by the production of alkaloids like peramine and the lolines. Moreover, they increase drought tolerance of plants (Lewis *et al.*, 1997; Ravel *et al.*, 1997; Malinowski and Belesky, 2000). However, endophytes are also able to produce a wide range of other alkaloids, such as the ergot and indole diterpene-type alkaloids, inducing staggers, gangrene of extremities, reduced conception rates and heat intolerance in livestock (Porter, 1995; Tor-Agbidye *et al.*, 2001; Botha *et al.*, 2004).

A protective endophyte infection is essential in certain regions, as for example in New Zealand, where ryegrass pastures would not resist the damage caused by Argentine stem weevil (*Listronotus bonariensis*) and African black beetle (*Heterinychus arator*) without being protected by symbiotic endophytes (Eerens *et al.*, 1998; Easton, 1999).

Surveys of the grasses in North and South America indicated that 20-40% of all *Festuca* and *Poa* grass types are endophyte-infected (White, 1987), whereas later reports, using more advanced methods of detection, describe *Neotyphodium* infectivity in 60-100% of all fescue grasses, for example in Arizona. An increase in the prevalence of *Neotyphodium* has been found in agronomic grasses as compared to the wild type flora, which also might explain the apparent increase in endophyte positive grasses (Clay, 1998; Saikkonen *et al.*, 1999).

In Europe, *Neotyphodium* infected grass, particularly *Lolium perenne*, has only recently been described to occur in natural grassland, and intentionally on turf grasses (Ribeiro *et al.*, 1996; Oldenburg, 1997; Oliveira *et al.*, 1997). Infectivity varies, depending on season and climate, and the prevalence of infected plants is higher in the regions with Mediterranean climate (Lewis *et al.*, 1997) than in the northern part of Europe, although reports from Finland also describe highly infected *Festuca rubra* and *F. ovina* populations in the sub-arctic areas in the Northern part of the country (Wäli *et al.*, 2000). The mean incidence of *Neotyphodium* infections in wild populations of *Lolium perenne* varies between 49 and 95% (Lewis, 2000). With the emergence of rapid and sensitive methods to detect fungal endophytes, an even higher frequency of infections is expected (Moon *et al.*, 1999; Hahn *et al.*, 2000).

TOXICITY OF ENDOPHYTE INFECTED GRASS TO LIVESTOCK

The major mycotoxins formed by *Neotyphodium* species are ergot alkaloids, with ergovaline and ergovalinine being the predominant toxins. In most cases other ergot alkaloids such as ergine, erginine and the clavine alkaloids chanoclavine-I, agroclavine, elymoclavine, and penniclavine are present in infected fescue as well, but these alkaloids are obviously less toxic to mammalian species (Porter, 1995; Lane *et al.*, 2000). In addition, *Neotyphodium* species produce in a distinct synthetic pathway indole terpenoides, of which lolitrem B is the most prominent toxin involved in livestock intoxications following ingestion of infected perennial ryegrass (*Lolium perenne*).

As a group, these alkaloid-producing endophytes are commonly denoted e-endophytes (E+). A second group, less well described, are the a-endophytes, which are primarily detected on *Lolium* grasses (*L. temelentum, L. rigidum* etc.) in Europe. These a-endophytes seem to belong to the genus *Acremonium* according to the morphological characteristics of the mycelium, resembling *A. strictum* or *A. sclerotigenum* (Guillaumin *et al.*, 2000). Their toxin pattern has not yet been elucidated.

Ergovaline and fescue toxicosis

Tall fescue (*Festuca arundinacea*) is the major forage grass grown on millions of acres of pasture and hay land in the central and southeastern United States. Hence numerous reports describe the performance problems of livestock grazing endophyte-infected fescue. Other *Festuca* varieties such as fine fescue, hard fescue as well as perennial ryegrass and bent grass may serve as hosts for endophytes as well. The major toxin of concern formed by the endophytes is the ergopeptide **ergovaline**.

In general, cattle consuming endophyte infected tall fescue show reduced weight gain, lower conception rates, reduced milk production and intolerance to high temperatures with hyperthermia during the summer season. A secondary increase in respiration, long, dull haircoats and decreased heart rate is observed (Schmidt and Osborne, 1993; Paterson *et al.*, 1995; Stuedemann *et al.*, 1998; Schultze *et al.*, 1999; Browning, 2000).

Ergovaline acts merely as dopamine-receptor agonist and is associated also with 'fescue foot', the latter referring to gangrenous lesions at the extremities caused by the strong vasoconstrictive effect of ergovaline. The increased body temperature is thought to be a result of the peripheral vasoconstriction as well, whereas the loss of appetite may be related to an aminoacidemia, as increased tryptophan levels have been found in peripheral blood, associated with an increase of serotonin levels in the central nervous system.

A decreased level of prolactin has been found in all affected cattle, which explains the reduced milk production (Mizinga *et al.*, 1990). Subsequently, retarded growth rates of calves, particularly in beef cow herds, are observed. Growth retardation may persist even after weaning, but is compensated if cattle enter feedlots, providing energy-rich concentrates (Oliver *et al.*, 2000; Nihsen *et al.*, 2004). However, the carcass quality of exposed animals remains altered, as ergovaline exposure results in fat necrosis and watery, soft muscle tissue.

In horses, agalactia and increased mortality in foals (due to dystonia) have been reported in mares consuming endophyte contaminated tall fescue. Agalactia can also cause aggressive behaviour of mares toward their foals. Moreover, lower IgG levels are present in the milk, and the foal obviously poorly absorbs the remaining IgG, which explains the decreased immunocompetence of affected animals. Ergovaline is also held responsible for the prolonged luteal function, prolonged gestation, retained placentas and decreased conception rates in mares.

The sensitivity to ergovaline varies among species. Maximum tolerance levels are 300-500 ppb ergovaline in feed materials for horses, 500-700 ppb in cattle, and 1000-1500 ppb in sheep (Craig and Blythe, 1995).

Therapeutic approaches are based on counteracting the dopaminergic activity of ergotamine. Hence, the dopamine-antagonist domperidone was applied to affected horses, and has been successfully used in mares to overcome acute intoxications. This therapy, however, is expensive and labour intensive, which impedes its use in livestock. Symptomatic therapy consists of the use of seaweed (*Ascophyllum nodosum*) as a known source of plant growth regulators and stimulant of antioxidant superoxide dismutase (SOD) in plants and animals (Allen *et al.*, 2000). Given as a feed additive, seaweed is used widely to reduce the adverse effects of endophyte infected tall fescue.

Lolitrem B and ryegrass staggers

Perennial ryegrass (*Lolium perenne*) is the most common pasture species in New Zealand and parts of Australia, and thus most reports of ryegrass staggers originate from these countries. However, perennial ryegrass has a global distribution, habituating a wide range of climatic zones (Table 2). All varieties of perennial ryegrass may become infected by the endophyte

Neotyphodium lolii. Subsequently, the toxins isolated from infected grass were denoted **lolitrems**, in recognition of the producing endophyte and the major clinical symptom (tremor). Lolitrem B is the most prominent representative of a group of indole terpenoids that are formed by endophytes, comprising lolitrem A and B, paxilline, and also a number of related compounds such as lolicine A and B, lolitriol, lolitrem N, as well as 31-epilolitrem N and 31-epilolitrem F and various aflatrems (Munday-Finch *et al.*, 1998). As the typical clinical symptoms associated with staggers could be reproduced largely by the application of lolitrem B, toxin measures are confined to this toxin as a marker for endophyte contaminated pastures although the other toxins may contribute to the staggers syndrome as well (Gallagher *et al.*, 1981; Munday *et al.*, 1985; Gallagher and Hawkes, 1985; Gallagher *et al.*, 1985; Gallagher and Hawkes, 1986). Next to lolitrem B, ergovaline may be found at lower levels and data from Oregon (US) suggest a strong positive correlation between the two toxins in perennial ryegrass (Hovermale and Craig, 2001).

Table 2. Occurrence of lolitrem B in European samples of perennial ryegrass[1].

Country	*Lolitrem B (mg/kg)*	*Reference*
Germany	0.05 – 4.2	Reinhloz *et al.*, 1998a,b
Germany	0.02 - >2	Reinholz *et al.*, 2000
Germany	0.2 – 1.4	Oldenburg, 1995
Netherlands	1.5 – 2.5	Van Essen *et al.*, 1995
Netherlands	5 – 6*	Sloet *et al.*, 1999
UK	0.7	Holmes *et al.*, 1999
France	0.1 – 1.5	Durix *et al.*, 1998
France	0.5	Bony *et al.*, 1998

*Samples contained hay prepared from E+ turf grasses
[1]Adapted from Bony and Delatour, 2000

Early clinical signs of the stagger syndrome are head tremors and muscle fasciculation of the neck and shoulders and later of the legs. Young cattle tend to lay down, have difficulty rising again and show abnormal postures in the attempt to rise. Horses show muscle fasciculation, tremor and ataxia, and might lose the ability to stand unattended. As the disease progresses in sheep, visible head nods develop and the animals sway while standing. If these animals are forced to move or encounter other stresses, the severity of symptoms suddenly increases, progressing into titanic convulsions. Deaths rarely occur, and are in most cases caused by accidents, such as drowning of sheep during an episode of convulsions.

The mechanism of action of lolitrems is still not fully elucidated. Interaction with GABA-ergic pathways and an increase in excitatory amino acids in synaptosomes and the central nervous system have been described in affected animals (Cheeke, 1995; Smith *et al.*, 1997) as well as in *in vitro* models (Wang *et al.*, 2003). Ryegrass staggers have been reported to occur in sheep, cattle, deer and horses, primarily in New Zealand and Australia, but also in North and South America and Europe (Cunningham and Hartley, 1959; Gallagher *et al.*, 1981; Munday *et al.*, 1985; Galey *et al.*, 1991; van Essen *et al.*, 1995). The prevalence and severity of the disease, however, varies considerably between continents.

Despite the impressive clinical symptoms, typical neurological lesions are absent and animals may recover completely when withdrawn from the infected grass. A proper clinical diagnosis is essential in these cases, as other neurological disorders with comparable symptoms (viral diseases, early signs of fumonisin intoxication etc.) have a very poor prognosis. Particularly in horses, the clinical diagnosis should thus be supported by toxin analysis in the grass or hay. In sheep and cattle, the maximal tolerable toxin concentration is close to 2 mg/kg dry matter, whereas horses exhibited clinical symptoms in some cases following ingestion of 0.8 mg/kg dry matter, and more regularly if concentrations exceeding 1.2 ppm are found in feed.

Paspalitrems and paspalum staggers

In addition to ryegrass staggers associated with the ingestion of lolitrem B in perennial ryegrass, other forms of stagger diseases have been reported, but seem to occur with a lower prevalence. Paspalum staggers (also known as Dallis grass staggers) shows almost identical clinical symptoms to ryegrass staggers. Clinical intoxications in cattle have been related to toxins from *Claviceps paspali*, forming the group of **paspalitrems**, comprising paspalinine, paspalitrem A and paspalitrem B (Steyn and Vleggaar, 1985). Again, the staggers syndrome is not accompanied by gross or histological lesions and is thus considered to be entirely reversible.

A staggers syndrome of local importance is related to the consumption of Bermuda grass (*Cynodon dactylon*). Again symptoms and clinical presentation are very comparable to those in other staggers. As a causative factor the infection of pasture with sooty mould (*Pullularia* spp.) and probably other fungi has been discussed, but the actual toxins have not yet been identified.

MEASURES TO PROTECT LIVESTOCK FROM TOXIC METABOLITES OF ENDOPHYTES

As many endophytes are seed-transmitted, preservation of grass seed forms the first managerial option to reduce the spread of infectivity. Endophyte viability in stored grass seed declines within a few months. Loss of viability can be promoted by increasing temperature (to >20°C) and humidity (to >70%), whereas in ryegrass seed stored below 5°C and 11% moisture, endophytes remain viable for up to 15 years (Rolston *et al.*, 1986; Fletcher and Easton, 2000). This method of seed inactivation is successful in new grass cultures, but insufficient to reduce endophyte burden in continuous pastures.

Based on the observation that not all isolates of *Neotyphodium*, particularly those found in Europe and in the Northern hemisphere, form ergovaline has stimulated investigation of *Neotyphodium* species that produce the plant-protective pyramine and loline alkaloids, but not the toxic ergot alkaloids. Although in North America the introduction of endophyte-free cultivars of *Festuca arundinacea* was less successful (Parish *et al.*, 2003), the mutant AR-1, developed in New Zealand, producing pyramines but not the toxic lolitrems or ergot alkaloids, has been successfully implemented in pasture areas (Fletcher and Easton, 1997). The absence of toxic alkaloids was confirmed in animal trials with grazing sheep (Fletcher, 1999) and dairy cattle (Bluett *et al.*, 2000).

Besides the attempt to replace toxic endophyte strains by selected non-toxic isolates, degradation of the toxin by ammoniation, comparable to the process applied for the degradation of aflatoxins in groundnuts, has been investigated (Simeone *et al.*, 1998). Toxin levels declined following ammonia treatment, but the procedure is less practical as compared to the treatment of groundnuts.

INFECTIONS OF GRASSES WITH EPIPHYTIC FUNGI

Fusarium infection of grassland is very common, but analytical data on toxin concentrations in grasses are scarce.

A number of case studies suggest particular skin irritations related to the exposure to toxin-containing spores and dust. For example in dairy cattle, painful swelling of teats and lower legs, occurring when animals graze heavily infected pastures, have been attributed to the presence of trichothecenes (such as T-2 toxin and HT-2) in mould spores. Comparable skin irritations have also been described in cattle exposed to straw heavily infected with *Fusarium sporotrichoides* when used as bedding (Wu *et al.*, 1997).

Contamination of grass with the *Fusarium*

toxin zearalenone has been regularly described in reports from New Zealand. Animal exposure is associated with the typical symptoms caused by zearalenone, comprising signs of hyperoestrogenicity and impairment of fertility due to the agonistic effect of zearalenone on oestrogen receptors. The finding that zearalenone can be found on natural pastures has obtained wide interest, as the growth promoter zeralenol, which is licensed in some parts of the world, but banned in Europe, is derived from zearalenone. Rumen metabolism can convert zearalenone into zeralenol, complicating the analytical control of the use of the growth promoter (Kleinova *et al.*, 2002).

Mycotoxins in legumes

Legumes comprise the second largest group of forage and toxic plant varieties; and mould contamination is well recognized, but often of economic relevance only in distinct geographic regions. The most prominent fungal species infecting legumes is *Rhizoctonia leguminicola*. If plants become infected with *R. leguminicola*, dark spots on the leaves and stems appear resulting from fungal growth (hence described as 'black-patch disease'). The major toxins reported to occur in legumes are slaframine and swainsonine.

SLAFRAMINE AND SLOBBER DISEASE

Slobber disease, associated with the exposure to the indolizidine alkaloid slaframine and probably swainsonine, is one of the most prominent examples of legume toxicity. Livestock grazing infected plants develop the slobber syndrome, characterized by profuse salivation, followed by anorexia, bloating, diarrhoea, frequent urination and excessive lacrimation (Hibbard *et al.*, 1995). Secondary effects include weight loss, decreased milk production and abortion. The disease is still common in central, southeastern and southwestern states of the US in years when climatic conditions are favourable for mould invasion. Slaframine toxicosis has been described in cattle, sheep and horses. It was first detected in toxic red clover hay, where it is stable for several months (Broquist *et al.*, 1984).

The exact mechanism of slaframine toxicity remains to be elucidated. Upon ingestion, slaframine may be activated by liver microsomes to a ketoimine. The active ketoimine shows structural similarities to acetylcholine, and thus might act as an agonist on muscarinergic receptors, which would explain some of the typical clinical symptoms. However, treatment with atropine, an acetylcholine receptor antagonist, resulted in the suppression of some of the symptoms like the excessive salivation, but not in all (Chapa *et al.*, 1995). This suggested that the clinical syndrome exhibited as slobbers resulted from the combined presence of slaframine and swainsonine.

SWAINSONINE AND LOCOISM

Locoism is a disease in grazing livestock related to the consumption of perennial locoweeds belonging to the genera *Astragalus* and *Oxytropis*. Locoweeds have been considered to be prominent plants in the western US rangelands. The Australian legume species, *Swainsona canescens*, known as poison peas, has been associated with intoxications in sheep, which are generally known as 'pea struck'. In these plants the alkaloid swainsonine is regularly detected, and was considered initially to be a toxin of *Rhizococtonia leguminicola* as well (Gardner *et al.*, 2001). More recent investigations, however, suggest that swainsonine is produced by endophytes present in these legumes, most likely representing *Embellisia* species (Braun *et al.*, 2003; McLain-Romero *et al.*, 2004). Swainsonine is also found in *Ipomea carnea*, belonging to the group of *Convolvulaceae* that occur in many tropical countries (Ikeda *et al.*, 2003; Haraguchi *et al.*, 2003), whereas *Ipomea asarifolia* seems to produce different tremorgenic toxins (Medeiros *et al.*, 2003). Moreover, swainsonine or related toxins have been suggested to be involved in the aetiology of intoxications caused by *Sida carpinifolia* in horses (Loretti *et al.*, 2003a).

In livestock, intoxications related to locoweed ingestion and swainsonine exposure are characterized by ataxia and depression, proprioceptive deficits, intention tremors, accelerated response to stress situations, and occasionally even death as the animals have difficulty eating and drinking (Stegelmeier *et al.*, 1999; Pfister *et al.*, 2003). In addition to the neurological signs, foetal skeletal malformation, placental fluid accumulation, altered development of oocysts and arrhythmias have been described in dams exposed to infected locoweed (Panter *et al.*, 1999). An increased prevalence of congestive heart failure in cattle grazing at high altitudes has been reported, known as 'high mountain disease' (Ralphs and James, 1999).

Feeding swainsonine as a crystalline compound to animals can reproduce most of the clinical symptoms. The histological lesions found in these experiments are characteristic, as particularly in the central nervous systems vacuolisation due to oligosaccharide accumulation is found (Stegelmeier *et al.*, 1999). The principal oligosaccharide that accumulates in the central nervous system is mannose, due to the known inhibition of lysosomal α-D-mannosidase. Moreover, swainsonine inhibits mannosidase II, leading to the formation of hybrid oligosaccharides instead of complex oligosaccharides followed by the production of abnormal glycoproteins. The lesions caused by swainsonine in livestock resemble those of two genetically determined diseases: bovine α-mannosidosis (Heikinheimo *et al.*, 2003) and human mannosidosis (Sun and Wolfe, 2001). Affected human patients are mentally retarded and show psychomotor deteriorations.

As mentioned above, swainsonine inhibits the lysosomal enzyme α-mannosidase and membrane bound enzyme mannosidase II, altering glycoprotein synthesis, processing and transport. This has a broad variety of biological effects including dysfunction of cellular recognition and the expression of adhesion molecules, and modulation of intracellular and membrane-associate receptors. Among others, swainsoine triggers the production (or release) of cytokines such as IL-1 and IL-2 that induce cell proliferation. These immunomodulatory functions are appreciated in cancer chemotherapy and radiotherapy, as swainsonine seems to have the potential to overcome for example the bone marrow suppressive effects of these therapies (Oredipe *et al.*, 2003; Schraen-Maschke and Zanetta, 2003).

Mycotoxins in preserved forages

Mycotoxins originating in the pre-harvest stage may be present in forages prior to preservation and often remain stable in the preserved material over a long period. In addition, preserved forages are contaminated with a typical storage flora, consisting of diverse species of the genera *Penicllium* and *Aspergillus*. The pattern of toxins formed often encompasses a complex mixture of secondary metabolites originating from different fungal species.

MOULDS AND MYCOTOXINS IN HAY AND STRAW

Virtually all toxins that have been found in growing grass can be detected in hay as well, although with increasing storage time, the toxin burden is often reduced due to fermentation activity. For example, the concentration of lolitrem B found in hay from perennial ryegrass declined over a period of three months to residual, non-toxic values. In contrast, slaframine in legumes has been found to persist for months (Broquist *et al.*, 1984; Panter *et al.*, 1999).

Typical storage moulds infecting hay are *Alternaria* species (Muller, 1991), and *Aspergillus fumigatus*. In particular *A. fumigatus* (and to a lesser extent *A. terreus* and *Alternaria* spp. as well as other mould species) has been found to be involved in the pathogenesis of **recurrent airway disease** (heaves or RAO, previously denoted also COPD) in horses and cattle (McGorum *et al.*, 1993; Schmallenbach *et al.*, 1998; Hare *et al.*, 1999). *Aspergillus fumigatus* (Wright *et al.*, 2004) and *A. terreus* (Hachem *et al.*, 2004) are facultative pathogenic moulds that may, following inhalation of spores, invade tissues of the upper airways and the lungs in animals and humans, causing fatal infection particularly in immuno-compromised patients (Pusterla *et al.*, 1996; Shen *et al.*, 2004).

In straw, various *Fusarium* toxins have been found, originating from pre-harvest contamination, but being stable under normal storage conditions. As straw is commonly used as bedding and to a lesser extent as roughage in animals diets, most reports refer to either skin lesions produced by dermatotoxin mycotoxins such as T-2 toxin and to a lesser extent other planar trichothecences and macrocyclic trichothecenes, particularly satratoxin (Harrach, Bata *et al.*, 1983; Wu *et al.*, 1997).

ENSILING AND SILAGE CONTAMINATION

Ensiling is one of the conventional practices of feed conservation. Making use of the natural microflora on plant materials, oxygen deprivation by pressure initiates a fermentation process resulting in the production of lactic acid, acetic acid and propionic acid, which preserve the product by reducing the pH value and exerting antibacterial and antifungal (propionic acid) properties.

Depending on the substrate ensiled, including grass, maize (green leaves-whole plant), grains (whole plant) or sugar beet (leaves or plant fibres after sugar extraction), the technique of silage preparation may vary. In recent years, the use of starter cultures and/or addition of organic acids and sugars to promote the fermentation process is increasing.

Shortcomings in the ensiling procedure result in increased malfermentation by *Clostridium* species, excreting proteolytic enzymes and producing butyric acid. Overgrowth with *Clostridia* can result in infectivity in animals. Moreover, an increased risk exists also for extensive yeast growth and mould contamination.

Shortcomings in the handling of silage resulting in oxygen exposure (early opening, incomplete cover, non-professional cutting) may affect the quality of the product even at a later stage. In addition, undesirable weather conditions (extreme day-night temperature differences) often promote spoilage flora in open silage stocks.

MOULDS AND MYCOTOXINS IN SILAGE

At the time of harvest grass, maize and cereals may be contaminated with *Fusarium* species. However, *Fusarium* species are sensitive to pH changes and require oxygen, and thus moulds disappear within a few days. This was confirmed by various investigations that failed to detect *Fusarium* in silage, with the exception of one report, describing a contamination of ensiled sugar beets with *Fusarium* (Bosch and Mirocha, 1992).

The most predominant mould species in corn silage is *Penicillium roqueforti*, which also contaminates grass and sugar beet products. Based on the time of sampling, contamination rates vary between 70% and 80% of all positive samples under investigation from Germany (Gedek *et al.*, 1981), France (Pelhate, 1975; Pelhate, 1977), Czechoslovakia (Vesely *et al.*, 1981) and Japan (Ohomo and Kitamoto, 1984). In contrast, in investigations conducted in North America, *Aspergillus fumigatus* was the prevailing mould species (Lynch *et al.*, 1972; Cole *et al.*, 1977; Smith *et al.*, 1997). In grass silage the presence of *Monascus* spp. *(Monascus ruber)* is frequently observed (Frevel *et al.*, 1985; Armbruster, 1994). There is some evidence that the *Monascus* contamination reduces the prevalence of *P. roqueforti* in grass silage.

Other mould species reported to occur on silage are *Mucor* and *Asidici* species, *Rhizopus* and yeasts, *Byssochlamus*, as well as *Scopulariopsis, Neurospora and Paecilomyces* spp. In a comparison of different culture media for the identification of any undesirable mould flora on silage from Norway, *Penicillium* species (*P. roqueforti*) were confirmed as the dominant species, followed by *Aspergillus (A. fumigatus* and incidentally *A. versicolor* and *A. ochraceus*), *Mucor, Geotriculum and Rhizopus*. However, with refined culture conditions using three different media, more than 60 fungal species could be isolated from big bale silage samples (Skaar and Stenwig, 1996), pointing towards the incompleteness of early studies devoted to the description of the mould flora on silage.

Although individual reports show considerable variability in the mould species reported and almost all report the high variability of the data when comparisons were made between years, regions, or procedures or culture methods in the ensiling process, the overall trend in most countries was for *P. roqueforti* to be the prevailing mould species. A comparable incidence is observed for *A. fumigatus* and *Monascus ruber*, the latter being found regularly in grass silage. These three species have been most extensively investigated.

Penicillium roqueforti **and its toxins**

Penicillium roqueforti (Thom) was initially identified according to morphological criteria as part of the genus *Penicillium*. As clear identification was difficult due to the variability in colony shape, size and colour of conidia, the toxin patterns produced by individual strains were used as secondary criteria. Based on the toxin pattern, *P. roqueforti* (chemotype I) was discriminated from *P. roqueforti var. carneum* (chemotpye II) (Frisvad *et al.*, 1989). Chemotype I strains were considered to produce as secondary (toxic) metabolites PR toxin, mycophenolic acid, roquefortine B, C and D, marcofortine A,B, C, fumigaclavine A and B, eremofortine A-E, festuclavine and agroclavine. In contrast, Chemotype II strains were considered to produce patulin, mycophenolic acid, roquefortine C, penicillic acid and botryodiploidine. Both strains have been described to occur on silage. More recently, genetic analysis revealed that morphologically and chemically defined *P. roqueforti* might represent three different mould species: *P. roqueforti, P. carneum* and *P. paneum* (Boysen *et al.*, 1996).

Regardless of the subclassification into genetically distinct species, all *Penicillium* species modify the sensory quality of the substrate, which affects palatability. Generally, reduced feed intake, particularly in dairy cows fed mouldy silage, is observed. The reduction in feed intake may result in considerable production losses and even adverse health effects particularly in young, high producing dairy cows, as these animals easily develop a negative energy balance with the onset of milk production (Jorritsma *et al.*, 2003).

The clinically observed adverse effects on milk production were substantiated by investigations of Hölterschinken *et al.* (1995), demonstrating that *P. roqueforti*-contaminated silage modified ruminal fermentation processes. In a long-term *in vitro* experiment with an artificial rumen system (Rusitec) *P. roqueforti* contaminated silage resulted in an increase in pH, reduction of methane production (-91%) and reduced short-chain fatty acid production (-77%).

P. roqueforti strains have the ability to produce a variety of toxins, as mentioned above, the most prominent being patulin, mycophenolic acid, roquefortine C, PR toxin and penicillic acid. These toxins are also formed in silage, albeit at different stages in the ageing of a fungal colony (Ohomo and Kitamoto, 1984; Schneweis *et al.*, 2000). Experiments in inoculated silage revealed that mycotoxin production followed a time-dependent sequence starting with penicillic acid measurable in appreciable concentrations by day 13 after inoculation, followed by patulin (22-27 days), mycophenolic acid (approximately 36 days) and PR toxin at 49 days (Müller and Amend, 1997). This production pattern not only hinders routine toxin monitoring of silage, but also impairs the evaluation of actual adverse health effects in animals.

P. roqueforti metabolites such as roquefortine C and patulin are known neurotoxins, and clinical signs in ruminants may include muscle weakness and lack of coordination (Kellerman *et al.*, 1984). However, these symptoms may be confined to high dose exposure as the oral bioavailability of the toxins, particularly in ruminant species, is very low. In contrast, cattle may be the most sensitive species regarding the antimicrobial effects of the other *P. roqueforti* toxins, such as penicillic acid, mycophenolic acid and PR toxin (Kopp-Holtwiesche and Rehm, 1990). The antimicrobial activity of these toxins was recognized immediately after they were identified, and has recently been confirmed. Exposure of cattle to these toxins, which also exert local inflammatory responses, will result in ruminal dysbacterioses and ruminitis, as observed regularly as a consequence of exposure to mouldy silage.

Aspergillus fumigatus and its toxins

A. fumigatus is a very common mould species, which can be found on decaying soils and plant materials, and is also a prominent toxinogenic mould species in silage (Cole *et al.*, 1977; dos Santos *et al.*, 2003). Contamination may originate from soil contamination of the plant material or technical shortcomings in the preparation of the silage. Exposure to *A. fumigatus* spores may result in fungal infection, particularly of organs exposed readily by external invasion such as the upper airways, the mammary gland and the uterus during parturition. Antigens against *A. fumigatus* have been found in serum of cattle with experimentally and spontaneous systemic mycotic infections (Jenssen and Olsen, 1994). Exposure to *A. fumigatus* spores correlates with airway hypersentivity of horses, and systemic infections in cattle with bronchitis, mastitis and incidental abortion (Sarfati *et al.*, 1996).

The known mycotoxin spectrum comprises a variety of tremorgenic mycotoxins (Tüller *et al.*, 1995). This group of indole derivatives including for example verrucologen, fumitremorgens and penitrem A, has been reported to induce tremor and other signs of neurotoxity in laboratory animal species and livestock, including sheep (Fink-Gremmels and Blom, 1994). Again, the low oral bioavailability of these mycotoxins might provide the explanation as to why these neurological symptoms are rarely seen under practical conditions. In addition, *A. fumigatus* produces a number of epipolythiodioxo-piperazines, particularly gliotoxin.

Upon invasion of the gastrointestinal tract (fungal overgrowth) the barrier function of the gut will be impaired, as released gliotoxin induces apoptosis in endocytes (Upperman *et al.*, 2003). Moreover, *A. fumigatus* can also produce toxins (gliotoxin) in other tissues, as demonstrated in cattle suffering from therapy-resistant mastitis (Bauer *et al.*, 1989).

Gliotoxin exerts a variety of biological effects, as it is capable of forming mixed disulfides with free thiol groups thereby impairing the function of diverse enzymes. Gliotoxin induces a specific calcium release from intact (liver) mitochondria (Salvi *et al.*, 2004) and induces apoptotic cell death in numerous cell types. It has been found to be a strong immunosuppressive agent, depending on the concentration of exposure (Watanabe *et al.*, 2003). However, as gliotoxin is actively concentrated in cells in a glutathione-dependent manner, not only the actual dose, but particularly long-term exposure (as with feeding contaminated silage for a longer period) will result in acceleration of its toxic effects (Bernardo *et al.*, 2003).

Aspergillus clavatus and its toxins

On occasion *A. clavatus* has been isolated from preserved feeds, including hay and silage. Like *A. fumigatus*, *A. clavatus* occurs in the soil, and soil contamination of feed seems to be the major source (Wilson *et al.*, 2002). Moreover, *A. clavatus* has been found in residues from breweries. By-products of malted cereals are commonly used in animal nutrition, and several disease outbreaks in cattle and sheep related to *A. clavatus* contaminated brewing residues have been reported (Loretti *et al.*, 2003b; Odhav and Naicker, 2002; Shlosberg *et al.*, 1991, Gilmour *et al.*, 1989). The major toxin produced is patulin, a cyclic γ-lactone.

Intoxications with *A. clavatus* result in neurotoxicity, with characteristic degenerative lesions of the peripheral nervous system, which might be related to an interaction with sodium or proton transport mechanisms as suggested by the proven capacity to inhibit plasma membrane Na^+/K^+ ATPase *in vivo* and *in vitro* (Riley *et al.*, 1992). Besides its neurotoxic effects, patulin has been shown to be toxic to viruses, bacteria, and protozoa (Detroy and Still, 1976). It may contribute to the adverse effects that mould contaminated silage exerts in ruminants (Escoula, 1992; Morgavi *et al.*, 2003) and impairs the barrier function of the intestinal tract due to its ability to bind glutathione (Mahfoud *et al.*, 2002). Moreover, pronounced immune suppressive effects have been reported *in vivo* and *in vitro* and its potential carcinogenic effect has been debated, as patulin is also found as a frequent contaminant of tomato and apple juices and related products (Beretta *et al.*, 2000).

Fusarium species and their toxins

Considering the high prevalence in corn (maize) plants, it can be predicted that silage will contain various fusariotoxins. *Fusarium* toxins originate mainly from the pre-harvest phase, but may be produced, albeit to a smaller extent, on the surface or in oxygen containing cavities after harvest as well (Pelhate, 1977; Lepom *et al.*, 1990a; Lepom *et al.*, 1990b). In particular, the toxins deoxynivalenol and zearalenone have been detected as fungal metabolites originating from pre- and post-harvest fungal growth (Bosch and Mirocha, 1992) as well as the group of fumonisins. These toxins will exert detrimental effects, particularly in pigs but also in cattle and horses (as described elsewhere in this volume). Deoxynivalenol affects the gastrointestinal tract, resulting in reduced nutrient absorption and growth retardation and induced pro-inflammatory and immunosuppressive effects (Rotter *et al.*, 1996). Zearalenone is known to bind to oestrogen receptors, thereby causing hyperoestrogenism and impaired fertility (Diekman and Green, 1992). The forestomach flora of ruminants will degrade the toxins, but this process is saturable depending on the competence of the ruminal flora. Fumonisins, however, are poorly degraded in the rumen, but their overall bioavailability is low, even in monogastric animal species. Fumonisins are potent inhibitors of ceramide synthetase and disrupt sphingolipid metabolism and thus many cell-cell interactions (Merrill *et al.*, 1996), including the gastrointestinal barrier (Bouhet *et al.*, 2004). Fumonisins have been related to a variety of diseases in horses (equine encephalomalacia) (Christley *et al.*, 1993), pigs (pulmonary oedema) (Gumprecht *et al.*, 2001) and cattle and sheep (reduced performance and nephrotoxicity) (Bucci *et al.*, 1998) as reviewed by Diaz and Boermans, 1994) and Marasas, 2001), and discussed elsewhere in this volume.

Summary and conclusions

Mycotoxins as secondary metabolites of moulds have been recognized worldwide as undesirable substances in foods and animal feed. The large diversity of mycotoxins, differing not only in chemical structures but consequently in biological properties, hinders a consistent presentation of all relevant groups. At present, most investigations have focused on monogastric animal species, including pigs and poultry, not only in consideration of their economic value, but also based on the paradigm that ruminants are less susceptible to the detrimental effects of mycotoxins, as they will degrade toxins with the help of the forestomach flora.

Although this statement is still valid, the various toxins present in forages exert clinical signs of intoxication, as well as an impairment of productivity. While the effects of individual toxins have been described in some detail, the challenge for the future is in understanding the effects of mixtures of toxins, as this occurs frequently in mixed diets offered to ruminants. In this approach, special attention must be given to non-typical effects, such as antibiotic activity and impairment of the intestinal barrier function. At present, the number of reports on intoxication or undesirable toxin concentrations in edible products of animal origin indicate the relevance of this approach to animal and human health.

References

Allen, V.G., K.R. Pond, M.F. Miller, J.L. Montgomery, J.P. Fontenot, K.E. Saker, J.H. Fike, C.P. Bagley, R.R. Evans, and R.L. Ivy. 2001. Influence of Tasco-Forage on immune response and carcass characteristics of steers grazing endophyte-infected tall fescue - A review, In: The Grassland Conference 2000; Proceedings of the 4th International Neotyphodium Grass Interactions Symposium. (V.H. Paul and P.D. Dapprich, eds) 27-29 September, 2000, Soest, Germany. p. 245-253.

Armbruster, G. 1994. Futtermittelhygienische Untersuchungen von Silagen: Nachweis und Vorkommen des Mykotoxins Roquefortine. Dissertation. München, Ludwig-Maximilians-Universität.

Bauer, J., M. Gareis, A. Bott and B. Gedek. 1989. Isolation of a mycotoxin (gliotoxin) from a bovine udder infected with *Aspergillus fumigatus*. J. Med. Vet. Mycology 27:45-50.

Beretta, B., A. Gaiaschi, C.L. Galli and P. Restani. 2000. Patulin in apple-based foods: occurrence and safety evaluation. Acta Vet. Hungarica 17:399-406.

Bernardo, P.H., N. Brasch, C.L. Chai and P. Waring. 2003. A novel redox mechanism for the glutathione-dependent reversible uptake of a fungal toxin in cells. J. Bio. Chem. 278:46549-46555.

Bluett, S. J., E. R. Thom, et al. 2000. In: The Grassland Conference 2000; Proceedings of the 4th International Neotyphodium Grass Interactions Symposium. (V.H. Paul and P.D. Dapprich, eds) 27-29 September, 2000, Soest, Germany. pp. 195-199.

Bony, S. and P. Delatour. 2000. In: The Grassland Conference 2000; Proceedings of the 4th International Neotyphodium Grass Interactions Symposium. (V.H. Paul and P.D. Dapprich, eds) 27-29 September, 2000, Soest, Germany. pp. 207-218.

Bosch, U. and C.J. Mirocha. 1992. Toxin production by *Fusarium* species from sugar beets and natural occurrence of zearalenone in beets and beet fibers. Appl. Environ. Microb. 58:3233-3239.

Botha, C.J., T.W. Naude, M.L. Moroe and G.E. Rottinghaus. 2004. Gangrenous ergotism in cattle grazing fescue (*Festuca elatior L.*) in South Africa. J. South African Vet. Assoc. 75:45-48.

Bouhet, S., E. Hourcade, N. Loiseau, A. Fikry, S. Martinex, M. Roselli, P. Galtier, E. Mengheri and I.P. Oswald. 2004. The mycotoxin fumonisin B1 alters the proliferation and the barrier function of porcine intestinal epithelial cells. Toxicol. Sci. 77:165-171.

Boysen, M., P. Skouboe, J. Frisvad and L. Rossen. 1996. Reclassification of the

Penicillium roqueforti group into three species on the basis of molecular genetic and biochemical profiles. Microb. 142(3):541-549.

Braun, K., J. Romero, C. Liddell and R. Creamer. 2003. Production of swainsonine by fungal endophytes of locoweed. Mycolog. Res. 107:980-988.

Broquist, H.P., P.S. Mason, W.M. Hagler and T.M. Harris. 1984. Identification of swainsonine as a probable contributory mycotoxin in moldy forage mycotoxicoses. Appl. Environ. Microbiol. 48:386-388.

Browning, R., Jr. 2000. Physiological responses of Brahman and Hereford steers to an acute ergotamine challenge. J. Anim. Sci. 78(1):124-130.

Bucci, T.J., P.C. Howard, W.H. Tolleson, J.B. Laborde and D.K. Hansen. 1998. Renal effects of fumonisin mycotoxins in animals. Toxic. Path. 26:160-164.

Bultman, T.L., J.F. White, Jr., T.I. Bowdish, A.M. Welch and J. Johnston. 1995. Mutualistic transfer of *Epichloe spermatia* by *Phorbia files*. Mycologia 89:182-189.

Cabal, D., M. Cafaro, B. Saidman, M. Lugo, P.V. Reddy and J.F. White, Jr. 1999. Evidence supporting the occurrence of a new species of endophyte in some South American grasses. Mycologia 91:315-325.

Chapa, A.M., J.M. Fernandez, D.L. Thompson, Jr., R.J. Tempelman, L.F. Berrio, W.J. Croom, Jr. and W.M. Hagler, Jr. 1995. Endocrine and metabolic response to muscarinic stimulation and inhibition in the ruminant: effects of slaframine. J. Anim. Sci. 73(12):3673-3680.

Cheeke, P.R. 1995. Endogenous toxins and mycotoxins in forage grasses and their effects on livestock. J. Anim. Sci. 73:909-918.

Christley, R.M., A.P. Begg, D.R. Hutchins and D.R. Hodgson. 1993. Leukoencephalomalacia in horses. Austr. Vet. J. 70(6):225-226.

Clay, K. 1998. Fungal endophyte infection and the population dynamics of grasses. In: *The Population Biology of Grasses* (G.P. Cheplick, ed). Cambridge, Cambridge University Press, UK, pp. 255-285.

Cole, R.J., J.W. Kirksey, J.W. Dorner, D.M. Wilson, J.C. Johnson, Jr., A.N. Johnson, D.M. Bedell, J.P. Springer, K.K. Chexal, J.C. Clardy and R.H. Cox. 1977. Mycotoxins produced by *Aspergillus fumigatus* species isolated from molded silage. J. Agric. Food Chem. 25(4):826-830.

Cunningham, I.J. and W.J. Hartley. 1959. Ryegrass staggers. New Zealand Vet. J. 7:53.

DelCurto, T. 2001. Endophytes in grass seed straw. *http://www.agric.gov.ab.ca/$department/deptdocs.nsf/all/for1314*

Detroy, R.W. and P.E. Still. 1976. Patulin inhibition of mycovirus replication in *Penicillium stoloniferum*. J. Gen. Microb. 92(1):167-174.

Diaz, G.J. and H.J. Boermans. 1994. Fumonisin toxicosis in domestic animals: a review. Vet. Human Toxicol. 36(6):548-555.

Diekman, M.A. and M.L. Green. 1992. Mycotoxins and reproduction in domestic livestock. J. Anim. Sci. 70:1615-1627.

dos Santos, V.M., J.W. Dorner and F. Carreira. 2003. Isolation and toxigenicity of *Aspergillus fumigatus* from moldy silage. Mycopathologia 156(2):133-138.

Easton, H.S. 1999. Endophyte in New Zealand ryegrass pastures, an overview. Grassland Research and Practice Series 7:1-9.

Eerens, J.P.J., M.H.P.W. Visker, R.J. Lucas, H.S. Easton and J.G.H. White. 1998. Influence of the ryegrass endophyte (*Neophitodium lolii*) in a cool-moist environment. New Zealand J. Agric. Res. 41:209-217.

Escoula, L. 1992. Patulin production by *Penicillium granulatum* and inhibition of ruminal flora. J. Environ. Pathol. Toxicol. Oncol. 11(2):45-48.

Faeth, S.H., T.J. Sullivan, et al. 2000. In: The Grassland Conference 2000; Proceedings of the 4th International Neotyphodium Grass Interactions Symposium. (V.H. Paul and P.D. Dapprich, eds) 27-29 September, 2000, Soest, Germany. pp. 65-69.

Fink-Gremmels, J. and M. Blom. 1994. Rye grass staggers linked to loliterm B: a case report. 6th International Congress of European Association for Veterinary Pharmacology and Toxicoloy, Edinburgh, UK, Blackwell Scientific Publications, pp. 129-130.

Fletcher, L.R. 1999. "Non-toxic" endophytes in ryegrass and their effect on livestock health and production. Grassland Research and Practice Series 7:133-139.

Fletcher, L.R. and H.S. Easton. 1997. The evaluation and use of endophytes for pasture improvement. In: *Neotyphodium/Grass Interaction* (C.W. Bacon and N.S. Hill, eds). New York & London, Plenum Press, pp. 209-227.

Fletcher, L.R. and H.S. Easton. 2000. In: The Grassland Conference 2000; Proceedings of the 4th International Neotyphodium Grass Interactions Symposium. (V.H. Paul and P.D. Dapprich, eds) 27-29 September, 2000, Soest, Germany. pp. 149-162.

Frevel, H-J., G. Engel and M. Teuber. 1985. Schimmelpilze in Silagen und Rohmilch. Milchwissenschaft 40:129-132.

Frisvad, J.C., O. Filtenborg and U. Thrane. 1989. Analysis and screening for mycotoxins and other secondary metabolites in fungal cultures by thin-layer chromatography and high-performance liquid chromatography. Arch. Environ. Contam. Toxicol. 18:331-335.

Funk, C.R. and J.F. White, Jr. 1997. The use of natural and transformed endophytes for turf improvement. In: *Neotyphodium/Grass Interaction* (C.W. Bacon and N.S. Hill, eds). New York & London, Plenum Press, pp. 229-239.

Galey, F.D., M.L. Tracy, A.L. Craigmill, B.C. Barr, G. Markegrad, R. Peterson and M. O'Connor. 1991. Staggers induced by consumption of perennial ryegrass in cattle and sheep from northern California. J. Amer. Vet. Med. Assoc. 199(4):466-470.

Gallagher, R.T. and A.D. Hawkes. 1985. High-performance liquid chromatography with stop-flow ultraviolet spectral characterization of lolitrem neurotoxins from perennial ryegrass. J. Chromat. 322:159-167.

Gallagher, R.T. and A.D. Hawkes. 1986. The potent tremorgenic neurotoxins lolitrem B and aflatrem: a comparison of the tremor response in mice. Experientia 42(7):823-825.

Gallagher, R.T., E.P. White and P.H. Mortimer. 1981. Ryegrass staggers: isolation of potent neurotoxins lolitrem A and lolitrem B from staggers-producing pastures. New Zealand Vet. J. 29(10):189-190.

Gallagher, R.T., A.D. Hawkes and J.M. Stewart. 1985. Rapid determination of the neurotoxin lolitrem B in perennial ryegrass by high-performance liquid chromatography with fluorescence detection. J. Chromat. 321(1):217-226.

Gardner, D.R., R.J. Molyneux and M.H. Ralphs. 2001. Analysis of swainsonine: extraction methods, detection, and measurement in populations of locoweeds (*Oxytropis spp.*). J. Agric. Food Chem. 49(10):4573-4580.

Gedek, B., J. Bauer and H. Schreiber. 1981. Zur Mykotoxinbildung Silage-verderbender Schimmelpilze. Wiener Tierärztliche Monatschrift 68(8/9):299-301.

Gilmour, J.S., D.M. Inglis, J. Robb and M. Maclen. 1989. A fodder mycotoxicosis of ruminants caused by contamination of a distillery by-product with *Aspergillus clavatus*. The Veterinary Record 124(6):133-135.

Guillaumin, J-J., V. Gendraud, et al. 2000. In: The Grassland Conference 2000; Proceedings of the 4th International Neotyphodium Grass Interactions Symposium. (V.H. Paul and P.D. Dapprich, eds) 27-29 September, 2000, Soest, Germany. pp. 387-392.

Gumprecht, L.A., G.W. Smith, P.C. Constable and W.M. Haschek. 2001. Species and organ specificity of fumonisin-induced endothelial alterations: potential role in porcine pulmonary edema. Toxicology 160(1-3):71-79.

Hachem, R.Y., D.P. Kontoyiannis, M.R. Boktour, C. Afif, C. Cooksley, G.P. Bodey, I. Chatzinikolaau and C. Perego. 2004. *Aspergillus terreus*: an emerging amphotericin B-resistant opportunistic mold in patients with hematologic malignancies. Cancer 101(7):1594-1600.

Hahn, H., W. Huth, et al. 2000.In: The Grassland Conference 2000; Proceedings of the 4th International Neotyphodium Grass Interactions Symposium. (V.H. Paul and P.D. Dapprich, eds) 27-29 September, 2000, Soest, Germany. pp. 139-143.

Haraguchi, M., S.L. Gorniak, K. Ikeda, Y. Minami, A. Kato, A.A. Watson, R.J. Nash, R.J. Molyneux and N. Asano. 2003. Alkaloidal components in the poisonous plant, *Ipomoea carnea* (*Convolvulaceae*). J. Agric. Food Chem. 51(17):4995-5000.

Hare, J.E., L. Viel, P.D. Conlon and J.S. Marshall. 1999. *In vitro* allergen-induced degranulation of pulmonary mast cells from horses with recurrent airway obstruction (heaves). Amer. J. Vet. Res. 60(7):841-847.

Harrach, B., A. Bata, E. Bajmocy and M. Benko. 1983. Isolation of satratoxins from the bedding straw of a sheep flock with fatal stachybotryotoxicosis. Appl. Environ. Microbiol. 45(5):1419-1422.

Heikinheimo, P., R. Helland, H.K. Leiros, I. Leiros, S. Karlsen, G. Evjen, R. Ravelli, G. Schoehn, R. Ruigrok, O.K. Tollersrud, S. McSweeney and E. Hough. 2003. The structure of bovine lysosomal alpha-mannosidase suggests a novel mechanism for low-pH activation. J. Molecular Bio. 327(3):631-644.

Hibbard, B., J.P. Peters, S.T. Chester, J.A. Robinson, S.F. Kotarski, W.J. Croom, Jr. and W.M. Hagler, Jr. 1995. The effect of slaframine on salivary output and subacute and acute acidosis in growing beef steers. J. Anim. Sci. 73(2):516-525.

Hölterschinken, M., A. Mauruschat, J. Bauer, K.H. Böhm and H. Scholz. 1995. Untersuchungen zum Einfluss von *Penicillium roqueforti* und dessen Mycotoxin Roquefortine C auf die Pansenfermentation *in vitro*. DVG-Kongress, Bad Nauhein, pp. 265-272

Hovermale, J.T. and A.M. Craig. 2001. Correlation of ergovaline and lolitrem B levels in endophyte-infected perennial ryegrass (*Lolium perenne*). J. Vet. Diagn. Invest. 13(4):323-327.

Ikeda, K., A. Kato, I. Adachi, M. Haraguchi and N. Asano. 2003. Alkaloids from the poisonous plant *Ipomoea carnea*: effects on intracellular lysosomal glycosidase activities in human lymphoblast cultures. J. Agric. Food Chem. 51(26):7642-7646.

Jenssen, H.E. and S.N. Olsen. 1994. Gastrointestinal aspergillosis and zygomycosis of cattle. Vet. Pathol. 31:2836.

Jorritsma, R., T. Wensing, T.A. Kruip, P.L. Vos and J.P. Noordhuizen. 2003. Metabolic changes in early lactation and impaired reproductive performance in dairy cows. Vet. Res. 34(1):11-26.

Kellerman, T.S., S.J. Newsholme, J.A.W. Coetzer and G.C.A. Westhuizen. 1984. A tremorgenic mycotoxicosis of cattle caused by maize sprouts infested with *Aspergillus clavatus*. The Onderstepoort J. Vet. Res. 51(4):271-274.

Kleinova, M., P. Zollner, H. Kahlbacher, W. Hochsteiner and W. Lindner. 2002. Metabolic profiles of the mycotoxin zearalenone and of the growth promoter zeranol in urine, liver and muscle of heifers. J. Agric. Food Chem. 50(17):4769-4776.

Kopp-Holtwiesche, B. and H.J. Rehm. 1990. Antimicrobial action of roquefortine. J. Environ. Pathol. Toxicol. Oncol. 10(1-2):41-44.

Kuldau, G.A., H-F. Tsai and C.L. Schardl. 1999. Genome sizes of *Epichloe* species and anamorphic hybrids. Mycologia 91(5):776-782.

Lane, G.A., M.J. Christensen and C.O. Milles. 2000. In: *Microbial Endophytes* (C.W. Bacon and J.F. White, eds). Marcel-Deker, New York.

Lepom, P., O. Knabe and H. Baathe. 1990a. Occurrence of *Fusarium* strains and their mycotoxins in corn silage. 7. Formation of deoxynivalenol (DON) in a silage corn plot artificially inoculated with *Fusarium culmorum* and the effect of silaging on the stability of the DON formed. Archiv fur Tierernahrung 40:1005-1012.

Lepom, P., O. Knabe and H. Baathe. 1990b. Occurrence of *Fusarium* strains and their mycotoxins on corn silage. 6. Formation of zearalenone and trichothecenes (type A) by indigenous *Fusarium* isolates. Archiv fur Tierernahrung 40:871-883.

Lewis, G.C. 2000. In: The Grassland Conference 2000; Proceedings of the 4th International Neotyphodium Grass Interactions Symposium. (V.H. Paul and P.D. Dapprich, eds) 27-29 September, 2000, Soest, Germany. pp. 123-130.

Lewis, G.C., C. Ravel, W. Naffa, C. Astier and G. Charmet. 1997. Occurence of *Acremonium* endophytes of wild populations of *Lolium spp.* in European countries and a relationship between level of infection and climate in France. Annals of Appl. Bio. 130(2):227-238.

Loretti, A.P., E.M. Colodel, E.J. Gimeno and D. Driemeier. 2003a. Lysosomal storage disease in *Sida carpinifolia* toxicosis: an induced mannosidosis in horses. Equine Vet. J. 35(5):434-438.

Loretti, A.P., E.M. Colodel, D. Driemeier, A.M. Correa, J.J. Bangel, Jr. and L. Ferreiro. 2003b. Neurological disorder in dairy cattle associated with consumption of beer residues contaminated with *Aspergillus clavatus*. J. Vet. Diagnostic Invest. 15(2):123-132.

Lynch, G.P., F.T. Covey, D.F. Smith and B.T. Weinland. 1972. Response of calves to a single dose of aflatoxin. J. Anim. Sci. 35(1):65-68.

Mahfoud, R., M. Maresca, N. Garmy and J. Fantini. 2002. The mycotoxin patulin alters the barrier function of the intestinal epithelium: mechanism of action of the toxin and protective effects of glutathione. Toxicol. Appl. Pharmacol. 181(3):209-218.

Malinowski, D.P. and D.P. Belesky. 2000. Adaptation of endophyte-infected cool-season grasses to environmental stresses: mechanisms of drought and mineral stress tolerance. Crop Sci. 40:923-940.

Marasas, W.F. 2001. Discovery and occurrence of the fumonisins: a historical perspective. Environ. Health Perspectives 109(Suppl. 2):239-243.

McGorum, B.C., P.M. Dixon and R.E. Halliwell. 1993. Responses of horses affected with chronic obstructive pulmonary disease to inhalation challenges with mould antigens. Equine Vet. J. 25(4):261-267.

McLain-Romero, J., R. Creamer, H. Zepeda, J. Strickland and G. Bell. 2004. The toxicosis of *Embellisia* fungi from locoweed (*Oxytropis lambertii*) is similar to locoweed toxicosis in rats. J. Anim. Sci. 82:2169-2174.

Medeiros, R.M., R.C. Barbosa, I.M. Tabosa, F. Riet-Correa, S.S de Barros, D.R. Gardner and R.J. Molyneux. 2003. Tremorgenic syndrome in goats caused by *Ipomoea asarifolia* in Northeastern Brazil. Toxicon 41:933-935.

Merrill, A.H., Jr., D.C. Liotta and R.T. Riley. 1996. Fumonisins: fungal toxins that shed light on sphingolipid function. Trends in Cell Biology 6:218-223.

Mizinga, K.M., F.N. Thompson, J.A. Stuedemann, T.E. Kiser, C.K. Smith and R.G. Powell. 1990. Effects of endophyte infected seed of endocrine function, body weight, and milk production in postpartum beef cows. J. Anim. Sci. 68(Suppl. 1):402.

Moon, C.D., B.A. Tapper and B. Scott. 1999. Identification of *Epichloe* endophytes in plants by a microsatellite-based PCR fingerprinting assay with automated analysis. Appl. Environ. Microbiol. 65(3):1268-1279.

Morgavi, D.P., H. Boudra, J.P. Jouany and D. Graviou. 2003. Prevention of patulin toxicity on rumen microbial fermentation by SH-containing reducing agents. J. Agric. Food Chem. 51(23):6906-6910.

Müller, H-M. and R. Amend. 1997. Formation and disappearance of mycophenolic acid, patulin, penicillic acid and PR toxin in maize silage inoculated with *Penicillium roqueforti*. Archiv. Anim. Nutr. 50:213-225.

Muller, M. 1991. Alternaria infestation of corn silage and hay. Zentralblatt fur Mikrobiologie 146:481-488.

Munday, B.L., I.M. Monkhouse and R.T. Gallagher. 1985. Intoxication of horses by lolitrem B in ryegrass seed cleanings. Austr. Vet. J. 62(6):207.

Munday-Finch, S.C., A.L. Wilkins and C.O. Miles. 1998. Isolation of lolicine A, lolicine B, lolitriol and lolitrem N from *Lolium perenne* infected with *Neotyphodium lolii* and evidence for the natural occurrence of 31-Epilolitrem N and 31-Epilolitrem F. J. Agric. Food Chem. 46(2):590-598.

Nihsen, M.E., E.L. Piper, C.P. West, R.J. Crawford, Jr., T.M. Denard, Z.B. Johnson, C.A. Roberts, D.A. Spiers and C.F. Rosenkrans, Jr. 2004. Growth rate and physiology of steers grazing tall fescue inoculated with novel endophytes. J. Anim. Sci. 82:878-883.

Odhav, B. and V. Naicker. 2002. Mycotoxins in South African traditionally brewed beers. Food Add. Contam. 19:55-61.

Ohomo, S. and H.K. Kitamoto. 1984. Detection of roquefortines in *Penicillium roqueforti* from moulded maize silage. J. Sci. Food Agric. 64.

Oldenburg, E. 1997. Endophytic fungi and alkaloid production in perennial ryegrass in Germany. Grass Forage Sci. 52:425-431.

Oliveira, J.A., G.E. Rottinghaus, J. Collar and P. Castro. 1997. Perennial ryegrass endophyte in Galicia, Northwest Spain. J. Agric. Sci. 129(2):173-177.

Oliver, J.W., A.E. Schultze, B.W. Rohrback, H.A. Fribourg, T. Ingle and J.C. Waller. 2000. Alterations in hemograms and serum biochemical analytes of steers after prolonged consumption of endophyte-infected tall fescue. J. Anim. Sci. 78(4):1029-1035.

Oredipe, O.A., P.M. Furbert-Harris, I. Laniyan, W.R. Green, W.M. Griffin and R. Sridhar. 2003. Co-administration of swainsonine and doxorubicin attenuates doxorubicin-induced lethality in mice. Cell. Mole. Bio. 49:1037-1048.

Panter, K.E., L.F. James, B.L. Stegelmeier, M.H. Ralphs and J.A. Pfister. 1999. Locoweeds: effects on reproduction in livestock. J. Natural Toxins 8:53-62.

Paterson, J., C. Forcherio, B. Larson, M. Samford and M. Kerley. 1995. The effects of fescue toxicosis on beef cattle productivity. J. Anim. Sci. 73(3):889-898.

Pelhate, J. 1975. Determination of the development of mycoflora of silaged maize forages. Review Mycol. 38:65-95.

Pelhate, J. 1977. Maize silage: incidence of moulds during conservation. Folia Vet. Latina 7:1-16.

Pfister, J.A., B.L. Stegelmeier, D.R. Gardner and J.F. James. 2003. Grazing of spotted locoweed (*Astragalus lentiginosus*) by cattle and horses in Arizona. J. Anim. Sci. 81:2285-2293.

Porter, J.K. 1995. Analysis of endophyte toxins: fescue and other grasses toxic to livestock. J. Anim. Sci. 73:871-880.

Pusterla, N., P. Ossent and U. Braun. 1996. A case of acute disseminated pneumomycosis in a cow. Schweizer Archiv fur Tierheilkunde 138:189-194.

Ralphs, M.H. and L.F. James. 1999. Locoweed grazing. J. Natural Toxins 8:47-51.

Ravel, C., C. Courty, A. Coudret and G. Charmet. 1997. Beneficial effects of *Neotyphodium lolii* on the growth and the water status in perennial ryegrass cultivates under nitrogen deficiency of drought stress. Agronomie 17:173-181.

Ribeiro, M.A.M.D., K.A. Gurney and L.P. Bush. 1996. Endophyte *Acremonium lolii* in ecotypes of perennial ryegrass (*Lolium perenne L.*) collected in cold Irish pastures. Irish J. Agric. Food Res. 35(2):151-157.

Riley, R.T., D.E. Goeger, H. Yoo and J.L. Showker. 1992. Comparison of three tetramic acids and their ability to alter membrane function in cultured skeletal muscle cells and sarcoplasmic reticulum vesicles. Toxicol. Appl. Pharmacol. 114(2):261-267.

Rolston, M.P., M.D. Hare, K.K. Moore and M.J. Christensen. 1986. Variability of *Lolium* endophyte fungus in seed stored at different moisture contents and temperatures. New Zealand J. Exper. Agric. 14(3):297-300.

Rotter, B.A., D.B. Prelusky and J.J. Pestka. 1996. Toxicology of deoxynivalenol (vomitoxin).

J. Toxicol. Environ. Health 48(1):1-34.

Saikkonen, K., M. Helander, S.H. Faeth, F. Schulthess and D. Wilson. 1999. Endophyte-grass-herbivore interactions: the case of *Neotypodium* endophytes in Arizona fescue populations. Oecologia 121:411-420.

Salvi, M., A. Bozac and A. Toninello. 2004. Gliotoxin induces Mg^{2+} efflux from intact brain mitochondria. Neurochem. Intl. 45:759-764.

Sarfati, J., H.E. Jensen and J.P. Latgé. 1996. Route of infections in bovine aspergillosis. J. Med. Vet. Mycol. 34:379-383.

Schardl, C.L. 2001. *Epichloe festucae* and related mutualistic symbionts of grasses. Fungal Genet. Bio. 33:69-82.

Schardl, C.L. and H.H. Wilkinson. 2000. In: *Microbial Endophytes* (C.W. Bacon and J.F. White, eds). Marcel-Deker, New York.

Schmallenbach, K.H., I. Rahman, H.H.L. Sasse, P.M. Dixon, R.E.W. Halliwell, B.C. McGorum, R. Crameri and H.R.P. Miller. 1998. Studies on pulmonary and systemic *Aspergillus fumigatus*-specific IgE and IgG antibodies in horses affected with chronic obstructive pulmonary disease (COPD). Vet. Immunol. Immunopathol. 66(3/4):245-256.

Schmidt, S.P. and T.G. Osborne. 1993. Effects of endophyte-infected tall fescue on animal performance. Agriculture, Ecosystems and Environment 44:233-262.

Schneweis, I., K. Meyer, S. Hormansdorfer and J. Bauer. 2000. Mycophenolic acid in silage. Appl. Environ. Microbiol. 66(8):3639-3641.

Schraen-Maschke, S. and J.P. Zanetta, 2003. Role of oligomannosidic N-glycans in the proliferation, adhesion and signalling of C6 glioblastoma cells. Biochimie 85:219-229.

Schultze, A.E., B.W. Rohrbach, H.A. Fribourg, J.C. Waller and J.W. Oliver. 1999. Alterations in bovine serum biochemistry profiles associated with prolonged consumption of endophyte-infected tall fescue. Vet. Human Toxicol. 41(3):133-139.

Shen, D.K., A.D. Noodeh, A. Kazemi, R. Grillot, G. Robson and J.F. Brugere. 2004. Characterisation and expression of phospholipases B from the opportunistic fungus *Aspergillus fumigatus*. FEMS Microbiol. Letters 239:87-93.

Shlosberg, A., I. Zadikov, S. Perl, B. Yakobson, Y. Varod, D. Elad, E. Rapoport and V. Handji. 1991. *Aspergillus clavatus* as the probable cause of a lethal mass neurotoxicosis in sheep. Mycopathologia 114:35-39.

Simeone, A., M.L. Westendorf, R.E. Tucker, L.P. Bush and G.E. Mitchell, Jr. 1998. Ammoniation to reduce the toxicity of endophyte-infected tall fescue seed fed to rats. Drug Chem. Toxicol. 21:373-385.

Skaar, I. and H. Stenwig. 1996. Malt-yeast extract-sucrose agar, a suitable medium for enumeration and isolation of fungi from silage. Appl. Environ. Microbiol. 62:3614-3619.

Smith, B.L., L.M. McLeay and P.P. Embling. 1997. Effect of the mycotoxins penitrem, paxilline and lolitrem B on the electromyographic activity of skeletal and gastrointestinal smooth muscle of sheep. Res. Vet. Sci. 62(2):111-116.

Stegelmeier, B.L., L.F. James, B.L. Stegelmeier, L.F. James, K.E. Panter, M.H. Ralphs, D.R. Gardner, R.J. Molyneux and J.A. Pfister. 1999. The pathogenesis and toxicokinetics of locoweed (*Astragalus* and *Oxytropis* spp.) poisoning in livestock. J. Nat. Toxins 8:35-45.

Steyn, P.S. and R. Vleggaar. 1985. Tremorgenic mycotoxins. Fortschritte der Chemie Organischer Naturstoffe 48:1-80.

Stuedemann, J.A., N.S. Hill, F.N. Thompson, R.A. Fayrer-Hosken, W.P. Hay, D.L. Dawe, D.H. Seman and S.A. Martin. 1998. Urinary and biliary excretion of ergot alkaloids from steers that grazed endophyte-infected tall fescue. J. Anim. Sci. 76(8):2146-2154.

Sullivan, T.J. and S.H. Faeth. 2004. Gene flow in the endophyte *Neotyphodium* and implications for coevolution with *Festuca arizonica*. Molecular Ecology 13:649-656.

Sun, H. and J.H. Wolfe. 2001. Recent progress in lysosomal alpha-mannosidase and its deficiency. Exper. Mole. Med. 33:1-7.

Tor-Agbidye, J., L.L. Blythe and A.M. Craig. 2001. Correlation of endophyte toxins (ergovaline and lolitrem B) with clinical disease: fescue foot and perennial ryegrass staggers. Vet. Human Toxicol. 43(3):140-146.

Tsai, H.F., J.S. Liu, C. Staben, M.J. Christensen, G.C.M. Latch, M.R. Siegel and C.L. Schardl. 1994. Evolutionary diversification of fungal endophytes of tall fescue grass by hybridization with *Epichloe* species. Proc. Natl. Acad. Sci. USA 91(7):2542-2546.

Tüller, G., H. Hertkorn, et al. 1995. *Aspergillus fumigatus* in maize silage: Bildung von Verruculogen and Fumitremorgen B. Mykotoxine Workshop, Kulmbach.

Upperman, J.S., D.A. Potoka, X.R. Zhang, K. Wong, R. Zamora and H.R. Ford. 2003. Mechanism of intestinal-derived fungal sepsis by gliotoxin, a fungal metabolite. J. Ped. Surgery 38:966-970.

van Essen, G.J., M. Blom and J. Fink-Gremmels Gehrmann. 1995. Ryegrass cramps in horses. Tijdschrift voor Diergeneeskunde 120(24):710-711.

Vesely, D., D. Vesela and A. Adamkova. 1981. Occurrence of PR-toxin-producing *Penicillium roqueforti* in corn silage. Veterinarni Medicina 26(2):109-115.

Wäli, P., K. Saikkonen, et al. 2000. In: The Grassland Conference 2000; Proceedings of the 4th International Neotyphodium Grass Interactions Symposium. (V.H. Paul and P.D. Dapprich, eds) 27-29 September, 2000, Soest, Germany. pp. 93-96.

Wang, L., A.L. Cross, K.L. Allen, B.L. Smith and L.M. McLeay. 2003. Tremorgenic mycotoxins increase gastric smooth muscle activity of sheep reticulum and rumen *in vitro*. Res. Vet. Sci. 74:93-100.

Watanabe, A., K. Kamei, T. Sekine, M. Waku, K. Nishimura, M. Miyaji and T. Kuriyama. 2003. Immunosuppressive substances in *Aspergillus fumigatus* culture filtrate. J. Infect. Chemother. 9(2):114-21.

White, J.F., Jr. 1987. Widespread distribution of endophytes in Poaceae. Plant Dis. 71:340-342.

Wilson, D.M., W. Mubatanhema and Z. Jurjevic. 2002. Biology and ecology of mycotoxigenic *Aspergillus* species as related to economic and health concerns. Adv. Exper. Med. Bio. 504:3-17.

Wright, M.S., H.K. Clausen and T.G. Abrahamsen. 2004. Liver cells respond to *Aspergillus fumigatus* with an increase in C3 secretion and C3 gene expression as well as an expression increase in TLR2 and TLR4. Immunology Letters 95:25-30.

Wu, W., M.E. Cook, F.S. Chu, T. Buttles, J. Hunger and P. Sutherland. 1997. Case study of bovine dermatitis caused by oat straw infected with *Fusarium sporotrichioides*. Vet. Rec. 140(15):399-400.

12

MYCOTOXIN INTERACTIONS

Roger D. Wyatt
Department of Poultry Science, University of Georgia, Athens, Georgia, USA

Introduction

Mycotoxicology can be defined as the study of highly toxic metabolic by-products, resulting from the growth of moulds on a feedstuff, such that when an organism is exposed to these metabolic by-products in relatively minute amounts, an alteration of the structure or disruption of the function of cells, tissues, organs, or organ systems results. When animals are exposed to a mycotoxin, the net result of exposure to the toxin is defined as a response. This mycotoxin-related response may be subclinical with no visible changes in the animal. On the other hand, this response may result in clinical changes of the animal with these changes being easily detectable and of such a unique nature that they can be used to accurately diagnose the toxicity. It can be inferred that a response, whether subclinical or clinical, is associated with substandard productivity or poor health of the affected animal.

The manifestation and magnitude of a mycotoxin-related response is dependent upon three situations. First, the severity of the response will be dictated by the dose of the mycotoxin. Since mycotoxins are usually introduced into an animal via the feed, the dose is directly related to 1) the concentration of the mycotoxin in the feed source, 2) the daily intake of feed, and 3) the body size of the target animal. When these variables are known, a daily dose or intake can be accurately determined and is usually expressed as units of toxin intake/unit of body weight/unit of time (i.e. mg toxin/kg of BW/ day).

Although the dose may be accurately determined, another important variable is the duration or time frame that the determined dose is exposed to the animal. For example, an animal receiving a particular dose for one day will obviously exhibit a less severe response compared to an animal receiving the exact same dose for 30 days. The longer the administration of a particular dose the more severe will be the response.

The third determinant of a response to a feed-borne mycotoxin is typically referred to as interactions. Interactions constitute a very broad set of circumstances that can enhance, diminish or have no impact on the response of an animal to a defined dose for a defined length of time. For example, a group of animals receiving a feed with a defined mycotoxin concentration for a defined length of time may respond quite differently than a group of the same animal species receiving the same dietary concentration of the mycotoxin for the same length of time, but managed in a different manner, exposed to a different set of environmental conditions, or influenced by some other factor.

Specific interactions known to affect response to mycotoxins include; age and gender of the target animal, nutritional status at the time of mycotoxin exposure, genetic make-up, environmental extremes, presence of multiple mycotoxins in the ration, and health status including the presence of infectious disease situations.

GENETIC INFLUENCES

Several examples of genetic-related interactions can be cited. Smith and Hamilton (1970) demonstrated that among various strains of commercial broiler chickens greater than a 2-fold difference existed between the most susceptible and the most resistant to acute aflatoxicosis. Marks and Wyatt (1979) demonstrated a genetic component related to the susceptibility of Japanese quail to acute aflatoxicosis. The genetic basis for resistance to aflatoxin in quail was found to be related to only a few genes since selection for resistance could be easily accomplished over only a few generations. Wyatt *et al.* (1987) also made rapid progress in selection for aflatoxin resistance in genetically diverse lines of chickens. The presence or absence of a gene responsible for uptake, metabolism, distribution or elimination of the toxin or toxic metabolites can have a profound impact upon the response of an animal to mycotoxin exposure. Across species, genetic differences may also account for the carcinogenic activity observed in one species but not in another.

ENVIRONMENTAL EXTREMES

An example of an interaction related to environmental temperature can be gleaned from an experiment conducted by Manning and Wyatt (1990). In this experiment, two groups of young broiler chickens were raised in cages with an appropriate heat source, however, all birds in one group were required to eat and drink in an environment with a temperature of 28-30°C, whereas the other group of birds were required to eat and drink in a much cooler environmental (10-12°C). After exposure and acclimation to this temperature regime for 20 days, birds from both groups were administered an oral dose of aflatoxin sufficient to cause high mortality. In birds exposed to the moderate temperature environment, the mortality from this oral dose was 90%. However, in those birds that had been exposed to the low environmental temperature, mortality resulting from administration of the same dose of aflatoxin was only 20%. The results of the experiment indicate a strong interaction between aflatoxicosis and previous exposure to low environmental temperatures in broiler chickens. It was hypothesised that exposure to the low environmental temperature resulted in an induction of hepatic microsomal enzymes, thereby facilitating the rapid metabolism, degradation, and elimination of the administered aflatoxin. Rapid degradation and elimination would, in turn, minimise the residence time that toxic metabolites would be in the animal and consequently minimise the response to aflatoxin. From a practical point of view, this may partially explain why commercial broilers tend to have less clinical aflatoxicosis during the colder seasons of the year compared to seasons characterised by higher environmental temperatures.

INTERACTION OF MYCOTOXICOSES WITH NUTRITION

The interaction of various mycotoxicoses, especially aflatoxin, with various nutrients has been long recognised and has been the topic of numerous experiments. Despite a long-standing recognition of this relationship and extensive experimentation, the interaction of mycotoxicoses with nutrition is complex and not fully understood. What is known is that impaired absorption, distribution, and utilisation of nutrients of practically all nutrient classes are common during mycotoxicoses. For example, Osborne and Hamilton (1981) demonstrated that aflatoxicosis results in low activities of pancreatic trypsin, lipase and amylase in broiler chicks. These low activities were apparently due to failure of enzyme synthesis rather than specific enzyme inhibition. Poor digestion of feedstuffs results in lower concentrations of monomeric units (amino acids, fatty acids, simple sugars, etc.). Since nutrients for absorption are not attained through digestive processes, there is a lack of sufficient nutrients to support optimal growth and reproduction. Donaldson *et al.* (1972) also noted abnormally low production of fatty acid synthetase during aflatoxicosis in broiler chickens. Low concentrations of B vitamins and amino acids were observed in plasma, bile and liver of broiler chicks experiencing aflatoxicosis (Voigt *et al.*, 1980). Low concentrations of circulating levels of lipid components have also been consistently observed during aflatoxicosis. In the review of Schaeffer and Hamilton (1991), interactions between fat soluble vitamins, water soluble vitamins, protein, fat, and fiber were discussed with reference to a wide range of mycotoxins (aflatoxin, trichothecenes, zearalenone, ochratoxin) in relation to diverse animal species such as chickens, ducks, turkeys, rats, mice, guinea pigs, rabbits, and swine. Needless to say, interactions with normal animal nutrition are of paramount concern to animal nutritionists and production specialists.

Inhibition of protein synthesis associated with various mycotoxicoses leads one to infer that amino acid nutrition in animals is critical during outbreaks of mycotoxicosis. Furthermore, it can be assumed that the most limiting amino acid would be the single most important. This was demonstrated to be the case by Veltmann *et al.* (1981). In this experiment, a deficiency in total sulfur amino acids, brought about by manipulation of methionine, resulted in a more severe expression of aflatoxicosis compared to the response of animals receiving the same level of aflatoxin where the ration contained 100% of the National Research Council (NRC) recommendation for methionine. Further-more, dietary supplementation of a ration with excess methionine (134% of NRC recommendation) prevented growth depression associated with aflatoxicosis in broiler chicks. It was concluded that dietary methionine in excess of NRC recommend-ations has a sparing effect upon aflatoxicosis. NRC recommendations for various nutrients assumes a 'normal' animal raised under 'normal' environmental conditions. During aflatoxicosis, the broiler chicken is 'abnormal' and hence the bird's requirement for nutrients for maximum performance may change. In the case of methionine, the chicken with aflatoxicosis appears to have a higher requirement in order to maintain maximal growth compared to the toxicosis-free animal.

Of interest is the fact that when a mycotoxin is found to be contaminating a feed, at least one component of the feed has supported mould growth. As the mould grows on either a feed ingredient or complete feed, the nutrients to support mould growth and metabolism are derived by the mould directly from the feedstuff. This suggests that mouldy feed may possess lower than normal levels of specific nutrients. This alone acts to the detriment of an animal, regardless of whether a mycotoxin is present. However, if specific nutrients are at lower concentrations than expected, and a mycotoxin is also present, this interaction between mycotoxicoses and nutrition has added importance. Table 1 depicts the destruction of amino acids in poultry feed as a result of increasing degrees of moulding. Note that in the case of lysine, no significant loss of lysine content in the feed occurred with a 'Low Moulding' treatment, however, in the case of the 'Medium' and 'High Moulding' treatment, a significant moulding-related reduction in lysine occurred. In the case of methionine, a significant moulding-related decrease in the amino acid occurred even with the lowest degree of moulding. It should be noted that in the 'Low Moulding' treatment, no visible mould growth was detectable and only sporadic clumping was noted. It can be seen that mould-related destruction of methionine is more sensitive than the destruction of lysine. This destruction of lysine and methionine, in the absence of mycotoxins, could have a negative impact upon the productivity of animals fed such mouldy feedstuffs, however, when this nutrient destruction is coupled with mycotoxin contamination, this interaction would certainly cause greater loss and productivity.

Table 1. Amino acid concentrations in moistened poultry feed[1] following moulding by the natural mycoflora.

Grouping[2]	*Lysine* (%)	*Methionine* (%)
No moulding	1.355[a]	0.537[a]
Low moulding	1.344[a]	0.435[b]
Medium moulding	1.329[a,b]	0.419[b]
High moulding	1.317[b]	0.397[c]

Kilburn and Wyatt (unpublished data)
[a,b,c]Values with different superscripts differ ($P<0.05$)
[1]Commercial poultry feed was moistened to 17.1% moisture by the addition of sterile water.
[2]The No Moulding feed was frozen until analyzed. The Low, Medium, and High groupings were based upon respirometric measurements where carbon dioxide production and oxygen consumption were measured as indicators of the degree of moulding.

Table 2. Fatty acid concentrations in moistened poultry feed[1] following moulding by the natural mycoflora.

Grouping[2]	*Oleic* (%)	*Linoleic* (%)	*Linolenic* (%)
No moulding	1.352[a]	3.047[a]	0.114[a]
Low moulding	0.807[b]	1.993[b]	0.063[b]
Medium moulding	0.682[c]	1.725[c]	0.053[c]
High moulding	0.593[c]	1.511[d]	0.044[d]

Kilburn and Wyatt, (unpublished data)
[a,b,c,d]Values with different superscripts differ ($P<0.05$)
[1]Commercial poultry feed was moistened to 17.1% moisture by the addition of sterile water.
[2]The No Moulding feed was frozen until analyzed. The Low, Medium, and High groupings were based upon respirometric measurements where carbon dioxide production and oxygen consumption were measured as indicators of the degree of moulding.

INTERACTION OF MYCOTOXICOSES WITH INFECTIOUS DISEASE

Of extreme interest to the animal industries is the potential for mycotoxins to interfere with the native resistance mechanisms or the development of immunity. The aflatoxins, ochratoxin A and toxic mycotoxins of the trichothecene grouping are typically associated with interference of resistance to infectious diseases. The primary reasons for this lay in the fact that these mycotoxins are known to inhibit protein synthesis, cause hypoplasia of lymphatic tissue (bursa of Fabricius, thymus, and gut-associated lymphoid tissue), and cause alterations of the blood system including the bone marrow.

Metabolically activated aflatoxin is known to bind covalently to DNA, RNA, and complex proteins. This binding results in failure of the RNA polymerases to recognise DNA as the substrate and inhibition of protein synthesis results. Toxic trichothecenes exert their effect on protein synthesis by binding to 80S ribosomes and inhibition of peptidyltransferase activity (Ueno, 1980). Ochratoxin A is known to competitively inhibit phenylalanyl-tRNA synthetase leading to problems with the synthesis of proteins.

In addition to the impact on protein synthesis, aflatoxin has been shown to interfere with heterophil and monocyte phagocytosis (Chang and Hamilton, 1979a,b) and cause immunosuppression (Thaxton, *et al.*, 1979). Also related to native infectious disease resistance is the failure of birds with aflatoxicosis to develop and maintain an elevation of body temperature following exposure to *Salmonella* and pathogenic *E. coli* (Wyatt, 2004, unpublished).

The trichothecene toxins have been reported to cause severe necrosis of lymphoid tissue responsible for B-cell and T-cell formation, necrosis of spleen and bone marrow, and necrosis of the thymic cortex (Schiefer and Hancock, 1984). Boonchuvit et *al.* (1975) observed an interaction when high levels of T-2 toxin were fed to broiler chicks infected with several serotypes of *Salmonella*. The role of ochratoxin A in disruptions of the immune system was the subject of a review by Dwivedi and Burns (1986). The primary action of ochratoxin in poultry is to alter cell-mediated immunity with much less of an impact upon humoral immunity. Chang *et al.* (1979) related the feeding of ochratoxin A to broiler chicks to a marked leucopenia. In a subsequent investigation Chang and Hamilton (1980) also reported altered phagocytic activity in heterophils obtained from chickens with ochratoxicosis. This altered phagocytic activity was characterised by the heterophils failing to exhibit random and chemical-directed locomotion.

The actual manifestation of immuno-suppression, in the case of cell-mediated immunity and humoral immunity, appear to depend upon the time of mycotoxin exposure and presentation of the affected animal with an antigen. For example, development of a mycotoxicosis prior to exposure to an antigen will result in more severe immunosuppression compared to exposure to an antigen and followed by the onset of the mycotoxicosis. For these reasons, prediction of immuno-suppression under field conditions is difficult. Nevertheless, numerous field cases have been reported where mycotoxin-related immunosuppression resulted in vaccine failures and poor immunity following natural exposure to infectious agents.

MYCOTOXIN-MYCOTOXIN INTERACTIONS

Mycotoxin-mycotoxin interaction is a term used to describe those situations where more than one mycotoxin is present simultaneously in feed and where the response of affected animals is influenced by the presence of both mycotoxins. The response of affected animals to exposure to more than one mycotoxin can be the same as the response predicted from the summation of the response to each mycotoxin individually (additive), less than the predicted response from each toxin individually (antagonistic), or more than the predicted summation of the responses from each individual mycotoxin (synergistic).

There is an increased awareness among animal producers regarding co-contamination of feedstuffs with mycotoxins and the responses that may occur in animals consuming such feedstuffs. Reasons for this increased awareness include: 1) an understanding of the diversity of mycotoxins produced by ubiquitous moulds, 2) increasing knowledge regarding co-production of mycotoxins by moulds, 3) routine analyses of feedstuffs for mycotoxin contamination, and 4) improvements in analytical testing of feedstuffs for mycotoxin contamination, specifically the use of screening tests which detect multiple mycotoxins or groups of mycotoxins.

The chance for multiple mycotoxin contamination of feedstuffs is relatively high and can occur for different reasons. Numerous species of *Aspergillus* (Smith and Ross, 1991), *Penicillium* (Pitt and Leistner, 1991), and *Fusarium* (Marasas, 1991) are known to simultaneously produce more than one mycotoxin. When conditions are favorable for the growth of a single mould species, multiple mycotoxins could be produced in the suspect feedstuff. This phenomenon is exemplified by the fact that aflatoxin and ochratoxin are often detected as co-contaminants in samples of poultry feed manufactured in the southeastern United States.

If a feedstuff is stored under specific environmental conditions that are conducive for growth and mycotoxin production by one mould, these same environmental conditions could also permit the growth and mycotoxin production by other moulds in the feedstuff. This, in turn, would lead to multiple mycotoxin contamination of the same commodity with the mycotoxins originating from different moulds.

Furthermore, some animal feeds are manufactured from ingredients originating from diverse geographical locations. An ingredient from one locale could be contaminated with one mycotoxin while the other ingredient may have undergone spoilage by a totally different mould and be contaminated by a different mycotoxin. When both ingredients are used in the manufacture of a single animal feed, multiple mycotoxin contamination will result.

As previously stated, when multiple mycotoxins contaminate a single feed, the possibility exists for one mycotoxin to either have no impact upon the expression of the other mycotoxin, to have an additive effect with the other mycotoxin, or for one mycotoxin to act in a synergistic fashion with the other such that the final response of the affected animal is unique and not typical of either mycotoxin. When a synergistic response is present, a rapid and accurate diagnosis of the problem is difficult.

Thorough testing of feedstuff suspected of being contaminated with mycotoxins remains one of the most useful tools to successfully manage mycotoxin-related problems in the animal industries. Unfortunately, when feedstuff samples are analysed for suspected mycotoxin contamination, detection of one mycotoxin may lead to cessation of additional testing. If this occurs, the presence of other contaminating mycotoxins will be unknown. Consequently, an accurate assessment of the total mycotoxin contamination of the feedstuff will remain obscure, sufficient information for an accurate diagnosis will be lacking, and rapid and accurate corrective actions cannot be taken.

Numerous examples of mycotoxin-mycotoxin interactions have been reported in the scientific literature. Co-contamination of feedstuffs with aflatoxin and ochratoxin is a relatively frequent occurrence. Additionally, since aflatoxin is a hepatotoxin and ochratoxin A is a nephrotoxin, simultaneous exposure to these two mycotoxins would suggest a severe response due to the fact that two organ systems would be simultaneously affected. Huff and Doerr (1981) fed aflatoxin and ochratoxin A singly and in combination to young broiler chicks. In this experiment dietary aflatoxin resulted in an increase in liver lipid, whereas this response was absent when ochratoxin A was fed. However, when these two mycotoxins were fed simultaneously no increase in liver lipid was noted. In a separate experiment, Huff *et al.* (1984) also observed a sparing effect of ochratoxin A on the increase in liver lipid caused by aflatoxin. In essence, the presence of ochratoxin A in the diet of these chicks prevented the expression of one of the most consistent responses of chickens to dietary aflatoxin, an increase in liver lipid. This unusual symptomatology, characterised by the absence of a fatty liver when aflatoxin is present in the diet, could prevent a rapid and accurate diagnosis of the toxicity and hence impede the implementation of appropriate corrective measures.

Ochratoxin and citrinin have been found to be produced simultaneously by *Penicillium viridicatum* (Scott *et al.*, 1972; Ciegler *et al.*, 1973). This finding is suggestive of a situation when possible interaction between these two mycotoxins could occur. Furthermore, these two mycotoxins are both considered primarily nephrotoxins. Manning *et al.* (1985) fed ochratoxin A and citrinin to young broiler chicks singly and in combination. They found no additive or synergistic interaction between ochratoxin A and citrinin in broiler chicks, but did observe an amelioration of the severe growth depression caused by ochratoxin A and the increased water consumption caused by citrinin when ochratoxin A and citrinin were fed in combination.

Kubena *et al.* (1993) noted a synergistic interaction between aflatoxin and diacetoxyscirpenol in broiler chicks. Depression of body weight gain, lowered mean corpuscular volume and mean corpuscular haemoglobin, and depression of serum triglycerides and serum calcium were those physiological measurements shown to respond to the combination of these two mycotoxins in a synergistic manner.

Kubena *et al.* (1997) observed no synergistic responses in commercial broiler chicks fed a diet containing T-2 toxin with fumonisin B_1 or deoxynivalenol with fumonisin B_1. These findings are noteworthy since these three mycotoxins are produced by various species of *Fusarium*. Hence, there is a high probability for co-contamination of a single feed with these mycotoxins.

Safe levels of mycotoxins

What are 'safe' levels of mycotoxins? Without hesitation, this is the most frequently asked question posed during discussions regarding mycotoxins. As thoroughly discussed by Hamilton (1984) and Schaeffer and Hamilton (1991), the reasons for asking this question are diverse. For example, governmental officials, charged with the responsibility of adopting regulations regarding the disposition of mycotoxin contaminated feedstuffs have a genuine interest in knowing the minimum effective dose (MED) or the maximum dose of a mycotoxin that will cause no response. Furthermore, regulatory agencies are also concerned with human toxicity. Hence, when this question is asked by governmental officials the term 'safe' can have a rather broad meaning. Litigants who have either supplied mycotoxin contaminated feedstuffs to an animal producer or an animal producer who has experienced production losses in his animals and has documented economic damages also desire to know the maximum dose of a mycotoxin that could account for the observed losses and often refer to this dose as a 'safe' level. Animal production professionals who have possession of mycotoxin contaminated feedstuffs and who are faced with the necessity of utilisation rather than destruction of the feedstuff also have a need to know the maximum dose of the mycotoxin that will cause no loss in animal production and have no impact upon animal health. This is a 'safe' level from their perspective. Specialists in the area of mycotoxicology are asked this question; and there is a reasonable expectation of a accurate, precise, and definitive answer. Unfortunately, such an answer must take into account all mycotoxins commonly encountered, all interacting factors, and all animals that are targets of such contamination, and cannot be given at the present time.

From a practical point of view with regard to animal producers, one can read and study research reports and listen to scientific presentations with the hope of ascertaining a safe level of a particular mycotoxin for a particular animal species. When taken at face value, the results of a single scientific experiment will typically demonstrate various responses or effects of the administered mycotoxin and a dose that, under the specific experimental conditions, results in no observed response or effect. However, caution should be exercised when interpreting the results of laboratory-conducted experiments with the expectation of ascertaining knowledge of a 'safe' level. Laboratory experimentation investigating the effect of mycotoxins on animals are primarily conducted to establish a cause-and-effect relationship, not to ascertain 'safe' levels of the mycotoxin. For example, dietary administration of aflatoxin has been shown to result in a significant decrease in body weight of young broiler chickens. An example of an experiment that demonstrated this effect is that of Smith and Hamilton (1970). Feeding dietary aflatoxin at 2500 ppb continuously from day of hatching to 3 weeks of age resulted in poor growth, with this dietary level resulting in a statistically significant reduction in body weight compared to lower dietary levels of aflatoxin. This experiment was successful in establishing a cause (i.e. aflatoxin) resulting in an economically important response in commercial broiler production (i.e. poor growth). After decades of experience with the poultry industry and being involved in countless field cases of aflatoxicosis, cause-and-effect relationships have been proven. However, this same experience with field outbreaks of aflatoxicosis has also demonstrated that growth depression caused by dietary aflatoxin in broiler chickens can occur with dietary concentrations of $\geq$10 times less aflatoxin and often by feeding these low levels of aflatoxin for a time period measured in days rather than weeks. There is a major difference between information obtained from a laboratory experiment and information from a field outbreak. Does this difference imply that the laboratory experiment was incorrect? The answer to this question is a resounding No! The goal of the laboratory experiment was to start with a relatively small but adequate number of chickens, administer aflatoxin to them via the diet at a known and constant dietary concentration, provide optimum animal care and husbandry practices to the test animals with few, if any, interacting factors, and finally measure a response (i.e. body weight). In the field situation almost all of the experimental conditions are absent or poorly controlled. For example, much larger numbers of animals are typically involved in the field outbreak. This difference alone permits detection of an adverse response at much lower levels of mycotoxin contamination compared to the laboratory experiment. The exposure of animals to a mycotoxin under field conditions is usually very different than that of a laboratory experiment. Feedstuff contamination under field conditions is non-uniform. Results of feed sampling during field outbreaks of mycotoxicoses may range from zero to very high. The actual dose received by the animals can only be estimated and can never be precisely determined. Due to management practices used within the animal industries, even the duration of mycotoxin exposure of animals under field conditions is rarely ascertained. Perhaps the most important dissimilarity between laboratory experiments and field outbreaks is the relative difference in the level of animal care and husbandry practices. Under laboratory conditions, a constant environmental temperature is both attainable and typically used. Under field conditions, this is often impossible to achieve. Animal density is usually much higher under field conditions compared to the laboratory environment. Higher animal density usually leads to some degree of stress in the animals and constitutes an interacting factor with mycotoxicosis leading to accentuated responses of affected animals. Air quality under laboratory conditions is characterised by low humidity, minimum dust, and freedom from airborne pathogens. Such conditions are rarely encountered under field conditions and can also interact with mycotoxicoses. Schaeffer and Hamilton (1991) employed mathematical modeling to predict a response and ultimately calculate a minimal effective dose (MED) for aflatoxin in commercial chickens. Their overall conclusion was that "...the MED for aflatoxin is <10 ppb under field conditions." This conclusion appears to be accurate and as additional information is forthcoming regarding other mycotoxins, similar determinations can be made.

References

Boonchuvit, B., P.B. Hamilton, and H.R. Burmeister. 1975. Interaction of T-2 toxin with *Salmonella* infections in chickens. Poultry Science 54:1693-1696.

Ceigler, A., D.I. Fennell, G.A. Sansing, R.W. Detroy, and G.A. Bennett. 1973. Mycotoxin producing strains of *Penicillium viridicatum*: classification into subgroups. Applied Microbiology. 26:271-278.

Chang, C.F. and P.B. Hamilton. 1979a. Impaired phagocytosis by heterophils from chickens during aflatoxicosis. Toxicology and Applied Pharmacology 48:459-462.

Chang, C.F. and P.B. Hamilton. 1979b. Impairment of phagocytosis in chicken monocytes during aflatoxicosis. Poultry Science 58:562-566.

Chang, C.F., W.E. Huff, and P.B. Hamilton. 1979. A leucocytopenia induced in chickens by ochratoxin A. Poultry Science 58:555-558.

Chang, C.F. and P.B. Hamilton. 1980. Impairment of phagocytosis by heterophils from chickens during ochratoxicosis. Applied and Environmental Microbiology 39:572-575.

Donaldson, W.E., H.T. Tung, and P.B. Hamilton 1972. Depression of fatty acid synthesis in chick liver *(Gallus domesticus)* by aflatoxin. Comparative Biochemical Physiology 41B:843-847.

Dwivedi, P. and R.B. Burns. 1986. The natural occurrence of ochratoxin A and its effect in poultry. A review. Part 1. Epidemiology and Toxicity. World's Poultry Science Journal 42:32-47.

Hamilton, P.B. 1984. Determining safe levels of mycotoxins. Journal of Food Protection 47:570-575.

Huff, W.E. and J.A. Doerr. 1981. Synergism between aflatoxin and ochratoxin in broiler chickens. Poultry Science 60:550-555.

Huff, W.D., J.A. Doerr, C.J. Wabeck, G.W. Chaloupka, , J.D. May, and J.W. Merkley. 1984. The individual and combined effect of aflatoxin and ochratoxin A on various processing parameters of broiler chickens. Poultry Science. 63:2153-2161.

Kubena, L.F., T.S. Edrington, R.B. Harvey, S.A. Buckley, T.D. Phillips, G.E. Rottinghaus, and H.H. Caspers. 1997. Individual and combined effects of fumonisin B_1 present in *Fusarium moniliforme* culture material and T-2 toxin or deoxynivalenol in broiler chicks. Poultry Science 76:1239-1247.

Kubena, L.F., R.B. Harvey, W.E. Huff, M.H. Elissalde, A.G. Yersin, T.D. Phillips, and G.E. Rottinghaus. 1993. Efficacy of a hydrated sodium aluminosilicate to reduce the toxicity of aflatoxin and diacetoxy-scirpenol. Poultry Science 72:51-59.

Manning, R.O., T.P. Brown, R.D. Wyatt, and O.J. Fletcher. 1985. The individual and combined effects of citrinin and ochratoxin A in broiler chicks. Avian Diseases 29:986-997.

Manning, R.O. and R.D. Wyatt. 1990. Effect of cold acclimation on the broiler chick's resistance to acute aflatoxicosis. Poultry Science 69:388-396.

Marks, H.L. and R.D. Wyatt. 1979. Genetic resistance to aflatoxin in Japanese quail. Science 206:1329-1330.

Marasas, W.F.O. 1991. Toxigenic Fusaria. In: *Mycotoxins and Animal Foods*. (J.E. Smith and R.S. Henderson, eds). CRC Press, Inc. Boca Raton, FL, USA. pp. 119-139.

Osborne, D.J. and P.B. Hamilton. 1981. Decreased pancreatic digestive enzymes during aflatoxicosis. Poultry Science 60:1818-1820.

Pitt, J.I. and L. Leistner. 1991. Toxigenic *Penicillium* species. In: *Mycotoxins and Animal Foods*, (J.E. Smith and R.S. Henderson, eds). CRC Press, Inc. Boca Raton, FL USA pp. 81-99.

Schaeffer, J.L. and P.B. Hamilton. 1991. Interactions of mycotoxins with feed ingredients. Do safe levels exist? In: *Mycotoxins and Animal Foods*, (J.E. Smith and R.S. Henderson, eds). CRC Press, Inc., Boca Raton, FL, USA. pp.827-843.

Schiefer, H.B. and D.S. Hancock. 1984. Systemic effects of topical application of T-2 toxin in mice. Toxicology and Applied Pharmacology 76:464-467.

Scott, P.M., W. Walbeck, B. Kennedy, and D. Anyeti. 1972. Mycotoxins (ochratoxin A, citrinin and sterigmatocystin) and toxigenic fungi in grains and other agricultural products. Agricultural and Food Chemistry 20:1103-1109.

Smith, J.F. and K. Ross. 1991. The toxigenic Aspergilli. In: *Mycotoxins and Animal Foods*, (J.E. Smith and R.S. Henderson, eds). CRC Press, Inc., Boca Raton, FL, USA. pp. 101-118.

Smith, J.W. and P.B. Hamilton. 1970. Aflatoxicosis in the broiler chicken. Poultry Science 49:207-215.

Thaxton, J.P., H.T. Tung, and P.B. Hamilton. 1974. Immunosuppression in chickens by aflatoxin. Poultry Science 53:721-723.

Ueno, Y. 1980. Trichothecene mycotoxins: mycology, chemistry and toxicology. Advances in Nutrition Research 3:301-305.

Voigt, M.N., R.D. Wyatt, J.C. Ayres, and P.E.

Koehler. 1980. Abnormal concentration of B vitamins and amino acids in plasma, bile and liver of chicks with aflatoxicosis. Applied and Environmental Microbiology 40:870-874.

Veltmann, J.R., R.D. Wyatt, and M.N. Voigt. 1981. The effect of varying the total sulfur amino acid content in the diets of chicks with aflatoxicosis. Poultry Science 60:1748.

Wyatt, R.D. H.L. Marks, and R.O. Manning. 1987. Selection for resistance to aflatoxin in chickens. Poultry Science 66:1901-1904.

13

MYCOTOXINS: METABOLISM, MECHANISMS AND BIOCHEMICAL MARKERS

Ronald T. Riley[1] and James Pestka[2]
[1] Toxicology and Mycotoxin Research Unit, R.B. Russell Research Center, USDA-ARS, Athens, Georgia, USA
[2] Department of Food Science and Human Nutrition, Michigan State University, East Lansing, Michigan, USA

Introduction

There are probably tens of thousands of fungal metabolites, however, the food-borne fungal metabolites that are suspected or known to cause disease in humans or animals number in the hundreds (Riley, 1998). For this review, mycotoxins are defined as those metabolites that when consumed or absorbed by animals cause sickness or behavioural changes. This definition distinguishes between 'fungal metabolites', 'toxic fungal metabolites', and 'mycotoxins'.

The number of known mycotoxins that pose a measurable health risk to farm animals is quite limited because, while many fungal metabolites have the potential for toxicity, only a relative few are documented and confirmed to cause toxic effects in field situations. Nonetheless, the knowledge derived from studies with laboratory animals and *in vitro* studies serves as a warning for the possible contribution of mycotoxins in altering immune function (Bondy and Pestka, 2000), contributing to unexplained animal diseases, and in performance problems in farm animals (Osweiler, 2000). Several excellent reviews document the toxicology of mycotoxins in farm animals with extensive descriptions of the clinical manifestations (National Academy of Sciences, 1979; Richard and Thurston, 1986; Raisbeck *et al.*, 1991; JECFA 56th, 2001; CAST, 2003; Haschek *et al.*, 2002; Cousin *et al.*, in press). This review describes the main clinical signs in farm animals exposed to levels of mycotoxins encountered in field outbreaks, metabolism, mechanisms of action and biochemical markers of exposure. The purpose of the review is to highlight those observations that could help to identify the responsible mycotoxin when it is known that the outbreak is a result of consumption of mouldy feed. Some clinical signs are unique and some are highly suggestive of a particular mycotoxin. While it would be most useful to have simple mechanism-based biochemical markers that provide definitive identification of a specific mycotoxicosis, this is only possible for a very few mycotoxins. Thus, from a forensic perspective, the investigator must use all the tools available to identify the cause of a feed-borne disease outbreak.

The mycotoxins that present the greatest risk to farm animals are those that occur in commodities that are consumed in large amounts, and include aflatoxin B_1, ochratoxin A, fumonisins B_1 and B_2, deoxynivalenol, T-2 toxin, zearalenone, ergot alkaloids, ergot-like alkaloids, and macrocyclic trichothecenes. This review will cover only those mycotoxins for which exposure is known to be high and those that are known to cause disease in animals or are suspected to be modifying factors in disease processes. Mycotoxins with biochemical mechanisms of action that strongly suggest that the action could modulate disease processes will also be reviewed. Because there are literally thousands of research articles on this subject, we will use pertinent review articles wherever they are available, to minimise the number of citations.

Aflatoxins and cyclopiazonic acid

Aflatoxins were first identified in the early 1960s as the cause of a mysterious outbreak in England, dubbed 'Turkey X Disease' (Blount, 1961). Other mycotoxins may also have been involved in this outbreak, most notably cyclopiazonic acid. The overt symptoms of aflatoxin poisoning are not definitive. Animals do not eat well and therefore have reduced weight gains and decreased feed efficiency. The effects on growth are dose dependent and at low levels may be barely discernable. Other performance effects include decreased reproductive performance, abortion and reduced egg or milk production. In turkeys, a sensitive species, reduced weight gain, impaired immune response, increased mortality, and acute mortality are seen at 125 ppb, 250 ppb, 250 ppb and 500 ppb, respectively (Norred, 1986). A similar relative dose response occurs in pigs but at higher levels of exposure, as they are more resistant to the effects of aflatoxins. In cattle and chickens much higher levels are required to attain a decrease in performance; and in chickens impaired immune response can occur at levels that have no effect on the growth rate. In all species, aflatoxins are hepatotoxic with fatty changes, hepatocyte degeneration, necrosis, and altered liver function. A common clinical sign is jaundice. Grossly, the liver appears pale and swollen or fatty with variable texture. In chickens **fatty liver syndrome** is believed to be caused by aflatoxin (Norred, 1986), although ochratoxin A can also cause fatty liver in poultry (Trenholm *et al.*, 1988). Liver damage ultimately can lead to coagulopathy as evidenced by haemorrhaging and anaemia. In poultry and pigs, coagulopathy contributes to the appearance of internal bruising during handling. In pigs this can occur at levels as low as 150 ppb (Edds, 1979).

Aflatoxin B_1 (AFB_1) absorption, distribution and elimination is rapid. Aflatoxin B_1 is well absorbed; it accumulates in liver and is extensively metabolised in the liver and other tissues. The binding of metabolites to macromolecules including proteins and nucleic acids also occurs soon after absorption. Unbound water-soluble metabolites are excreted in urine and other fluids and the parent compound is excreted in faeces; however, some metabolites bound to nucleic acids can persist for relatively long periods of time in tissues.

The metabolism of aflatoxin B_1 has been extensively reviewed (IARC, 1993; IARC, 2002). Briefly, in the liver and other tissues, aflatoxin B_1 is metabolised in microsomal systems to aflatoxin P_1, M_1, and Q_1 and to the presumed metabolite the aflatoxin B_1 8,9 epoxide. In liver, cytochrome P450 enzymes are responsible for activation of aflatoxin B_1, aflatoxin M_1 and aflatoxin P_1, all of which can form nucleic acid adducts or undergo conjugation to glutathione, conversion to dihydrodiols, or binding to serum proteins or other macromolecules. Aflatoxin M_1 is the main unconjugated metabolite in urine of cows, pigs and sheep. In rodents, and presumably farm animals, aflatoxin B_1 nucleic acid adducts are also found in urine with 80% of the depurinated adducts being excreted within 48 hrs of dosing. There is a close correlation between levels of adducts in urine and levels in liver (IARC, 1993), however, the correlation between dietary intake and adducts in urine and serum is more variable (IARC, 2002). The relative sensitivity of animals/individuals to the toxic effects of aflatoxin B_1 is closely linked to differences in metabolism between species/individuals (IARC, 2002). Risk factors contributing to an individual's sensitivity to aflatoxin-induced liver tumors include level of exposure to aflatoxin, expression of aflatoxin activation/ detoxification pathways, nutritional status, and most importantly, chronic infection with hepatitis B or C virus (Henry *et al.*, 1999). AFB_1 induction of liver tumors in laboratory animals is closely correlated with mutations to specific genes that are known to control tumorigenicity.

The mechanism of action of aflatoxins involves their metabolism to reactive intermediates which bind to macromolecules (nucleic acids and proteins) with consequent disruption of transcriptional and translational processes (Figure 1). To a large extent the response of an animal to aflatoxin is dependent on the rate of metabolism and the type of metabolites produced. For example, quail are sensitive to aflatoxin and have a high rate of epoxide formation and a low rate of glutathione conjugation. In resistant species, even if the rate of epoxidation is high, a high rate of glutathione conjugation is protective. Nonetheless, the resulting clinical signs are similar although the dose-dependence may be quite different. Aflatoxin adducts in urine and blood of farm animals could be very useful as a biomarker of exposure during suspected field outbreaks.

Cyclopiazonic acid (CPA) is produced by several fungi, however, in maize and peanuts, the primary producers are *Aspergillus* species, including *A. flavus*. Although CPA is not as toxic as aflatoxin, and apparently not carcinogenic, concern over its presence in foods and feeds stems primarily from its co-occurrence with AFB_1 in peanuts and maize and its mechanism of action, which is similar to that of the known tumour promoter thapsigargin (Riley *et al.*, 1995). CPA has been implicated as a causative

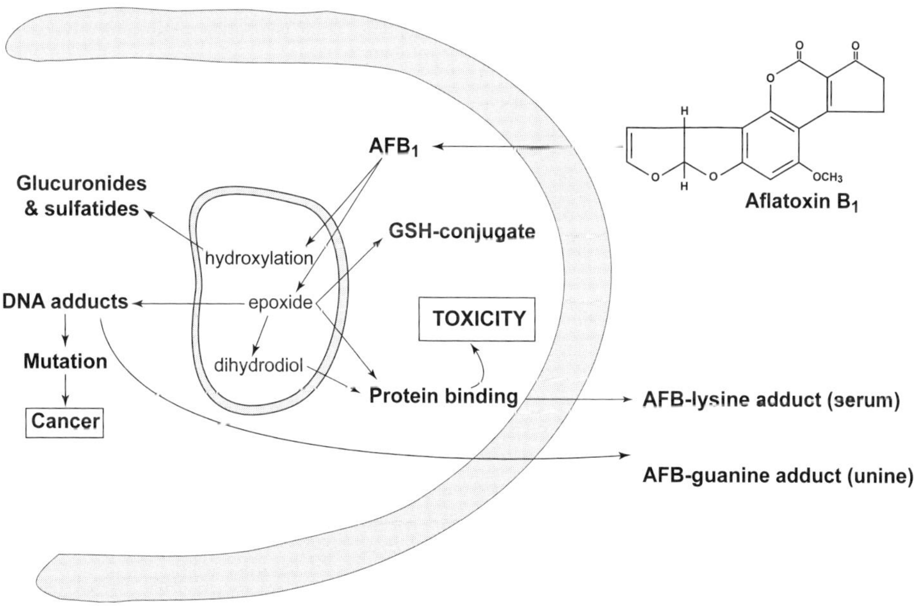

Figure 1. Mechanism of action of aflatoxin. The pathways of aflatoxin B_1 (AFB1) metabolism leading to cancer and cytotoxicity (after Eaton and Gallagher, 1994). Briefly, aflatoxin B_1 enters the cell and is either metabolised via monooxygenases in the endoplasmic reticulum to hydroxylated metabolites which are further metabolized to glucuronide and sulfate conjugates or is oxidised to the reactive epoxide which undergoes spontaneous hydrolysis to the AFB_1-8,9-dihydrodiol and can bind to proteins resulting in cytotoxicity. The epoxide can react with DNA or protein, or be detoxified by an inducible glutathione S-transferase to the glutathione (GSH)-conjugate. Both the DNA adduct and the protein adducts have proven useful as biomarkers in humans and laboratory animals.

agent in several cases of field intoxication, involving pigs, cattle, quail, and man (Bryden, 1991).

Acute doses of CPA administered to rats cause toxic lesions in liver, spleen, gastrointestinal tract and skeletal muscle. CPA is a potent inhibitor of Ca^{2+} uptake and Ca^{2+}-dependent ATP-ase activity in both sarcoplasmic and endoplasmic reticulum (Riley *et al.*, 1995) (Figure 2). Other than disrupted calcium metabolism, there is no useful biochemical marker for exposure to CPA, however, because it binds directly to the sarcoplasmic/endoplasmic reticulum calcium ATP-ase (SERCA), it is possible that the CPA-SERCA adduct could be detected in membrane vesicle preparations using electrophoresis.

Deoxynivalenol and the trichothecenes

Trichothecenes are a group of over 180 sesquiterpenoid mycotoxins produced by *Fusarium*, *Stachybotrys* and other fungi.

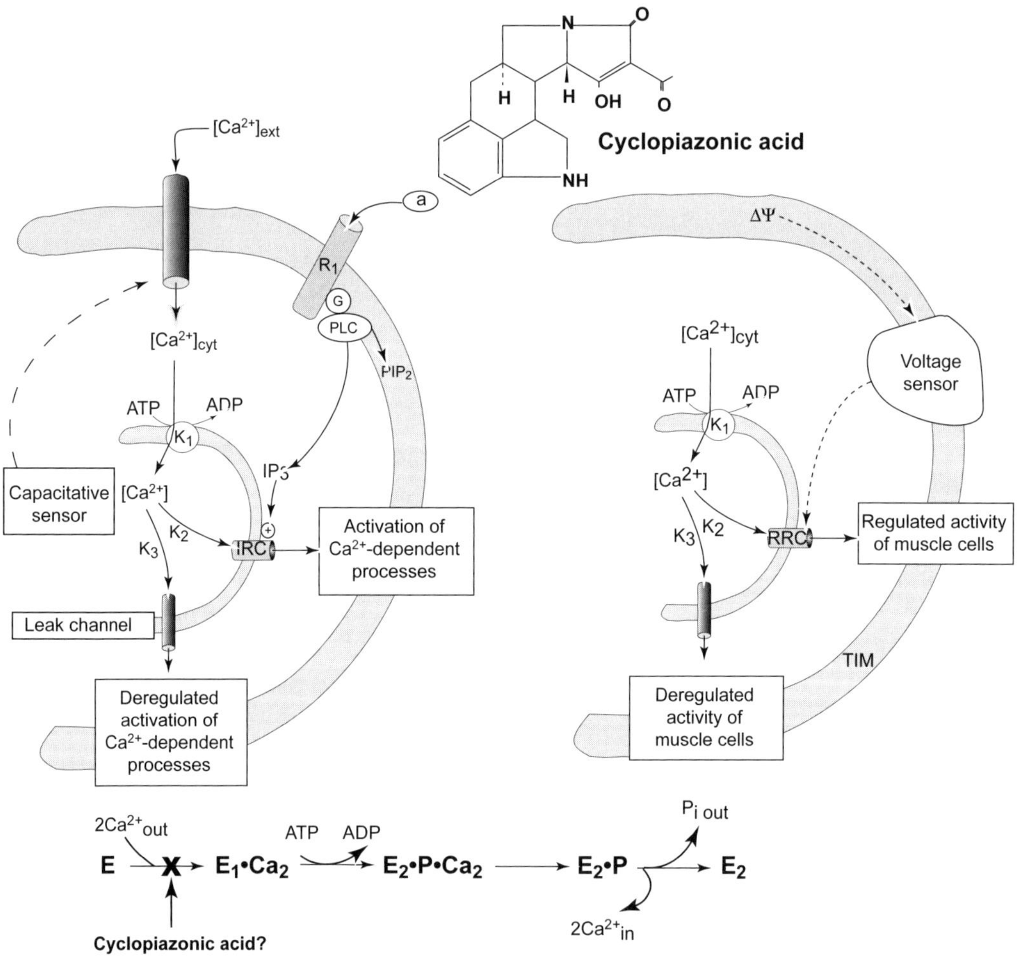

Figure 2. Biochemical mechanism of action of cyclopiazonic acid. The chemical structures of cyclopiazonic acid and a schematic showing inhibition of sarcoplasmic or endoplasmic reticulum calcium-dependent ATPases (Modified from Riley, 1998).

Deoxynivalenol is the most commonly encountered trichothecene worldwide. Even though deoxynivalenol is not considered to be acutely toxic to farm animals, it is considered to be a major cause of economic loss due to reduced performance. Food concentrations as low as 1 ppm have been associated with feed refusal in pigs, but, more typically, concentrations >2-5 ppm are required for decreased feed intake and reduced weight gain and >20 ppm for vomiting and feed refusal (Haschek *et al.*, 2002; Trenholm *et al.*, 1988). Feed refusal and emesis appear to be due to neurochemical imbalances in the brain. In pigs, feed refusal occurs even when deoxynivalenol is administered intraperitoneally, thus, feed refusal is not due to taste or learned responses (Prelusky, 1997). Clinical signs include gastrointestinal problems, soft stools, diarrhea, increased susceptibility to other diseases and decreased performance. In pigs, mild renal nephrosis, reduced thyroid size, gastric mucosal hyperplasia, increased albumin/α-globulin ratio, and sometimes mild changes in other haematological parameters have been reported (JECFA 56th, 2001). Numerous studies in laboratory animals demonstrate alterations in immune function induced by deoxynivalenol (Bondy and Pestka, 2000), but, there is little conclusive evidence that deoxynivalenol induces altered resistance to infectious diseases in the field or in farm animals experimentally (Osweiller, 2000). Nonetheless, the mechanism of action in laboratory animals suggests the potential for involvement in altered immune response in farm animals.

Laboratory animal studies have demonstrated that, depending on dose and exposure regimen, trichothecenes can be both immunosuppressive and immunostimulatory (Bondy and Pestka, 2000). The seemingly paradoxical effects of trichothecenes can be explained by their cellular and molecular modes of action (Figure 3) (reviewed in Pestka *et al.*, 2004). Exposure to low levels of trichothecenes appears to promote expression

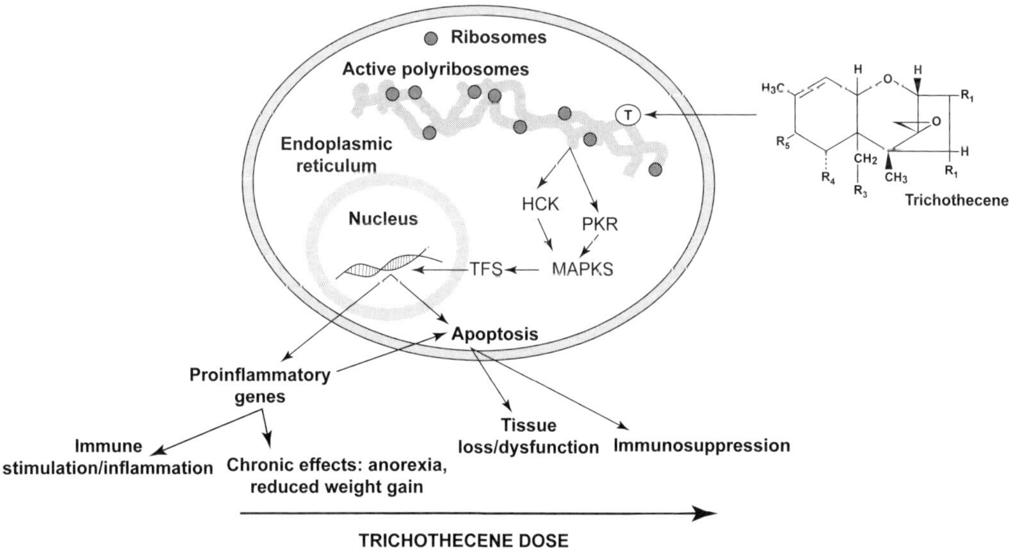

Figure 3. Molecular mode of action for deoxynivalenol and other trichothecenes. Deoxynivalenol enters cell via diffusion and binds to active ribosomes which transduce a signal to RNA–activated protein kinase (PKR) and haematopoietic cell kinase Hck. Subsequent phosphorylation of mitogen–actived protein kinases drives transcription factor activation and resultant chronic and immunotoxic effects (modified from Cousin *et al.*, in press and Pestka *et al.*, 2004).

of a diverse array of cytokines and proinflammatory genes *in vitro* and *in vivo* (reviewed in Pestka *et al.*, 2004). These genes have the potential to modulate many immune and inflammation-associated sequelae. For example, DON-induced aberrant IgA production may be mediated through superinduction of IL-6 gene expression in T- helper (TH) cells and macrophages (Pestka *et al.*, 2004). Up-regulation of proinflammatory genes appears to relate to the capacity to increase transcription and mRNA stability (reviewed in Pestka *et al.*, 2004). In contrast to these stimulatory effects, high doses of trichothecenes promote rapid onset of leukocyte apoptosis and this will undoubtedly be manifested as immunosuppression (reviewed in Pestka *et al.*, 2004). Trichothecene-induced apoptosis also occurs in tissues of the gastrointestinal tract including the gastric mucosa, gastric glandular epithelium, and intestinal crypt cell epithelium (Li *et al.*, 1997a, b; Li and Shimizu, 1997; Pestka *et al.*, 2004). Apoptotic loss in these tissues might result in the breakdown of non-specific mucosal defence mechanisms such as the epithelial barrier and mucus secretion and thus result in increased translocation of gut bacteria and endotoxin. The latter effects of trichothecenes appear to involve the activation of p53, a well-known apoptotic gene (Zhou *et al.*, 2003).

Trichothecenes bind to eukaryotic ribosomes (Bamburg, 1983) and, thus, inhibition of translation is an obvious outcome of this interaction. However, in addition, cell signalling appears to be rapidly affected via a 'ribotoxic stress response' that has been demonstrated for trichothecenes and other translation inhibitors such as anisomycin, shiga toxin, ricin, and α–sarcin (Iordanov *et al.*, 1997). In this model, alteration of 28s rRNA by these inhibitors was postulated to be an initiation signal for activation of SAPK/JNK, which is a mitogen-activated protein kinase (MAPK) (Figure 3). Similarly, Yang *et al.* (2000) observed a close correlation in protein synthesis inhibition capacity, MAPK activation and apoptosis in macrophages among a number of trichothecenes examined. Increased MAPK activity might also drive activation of transcription factors that promote expression of proinflammatory and stress-related genes as well as induce apoptosis. Zhou *et al.* (2003) have demonstrated the rapid sequential activation of MAPK transcription factor cytokine genes in mice given a single oral exposure to DON. Recently, it has been shown that trichothecenes induce the phosphorylation of double-stranded RNA-activated protein kinase (PKR) (reviewed in Pestka *et al.*, 2004) and a Src kinase (Hck), which are two kinases that are upstream of the MAPKs. Both PKR and Hck are likely to transduce the signal induced upon ribosomal binding to downstream MAPK-driven pathways (Figure 3). The transition between these two effects occurs with increasing concentration/dose of trichothecene with the net effects being immune stimulation or suppression, respectively.

Deoxynivalenol absorption, distribution and elimination are rapid following oral or parenteral dosing. There is no evidence for accumulation in tissues or transmission to eggs or milk. There is the potential for extensive metabolism in the gastrointestinal tract via de-epoxidation reactions. There is no evidence that deoxynivalenol is metabolised by microsomal enzymes; however, there is evidence for glucuronide conjugation in sheep and pigs (JECFA 56th, 2001; Eriksen *et al.*, 2003). In sheep, the glucuronide conjugate of deoxynivalenol in some studies can account for a large percentage of the total plasma DON. In pigs fed diets containing 3-acetyl deoxynivalenol, 42% and 33% of the deoxynivalenol in plasma and urine, respectively, was conjugated to the glucuronide (Eriksen *et al.*, 2003). In this same study, a significant portion of the deoxynivalenol in faeces was the de-epoxide. A urinary biomarker using ß-glucuronidase treatment has been developed to estimate the daily intake of deoxynivalenol in people (Meky *et al.*, 2003) and may also be useful in suspected field outbreaks in farm animals.

Ergot

Ergotism in farm animals can be divided, based on the major signs, into the gangrenous form and the convulsive form (Robbins *et al.*, 1985; Raisbeck *et al.*, 1991). Livestock consuming feed grains containing small seed grasses or foraging on pasture grasses are at particular risk. Cattle are at greatest risk, with sheep and horses being less frequently affected. The symptoms of ergotism include lameness, gangrene, agalactia, reduced weight gain, abortion, hypersensitivity, ataxia, convulsions and, in sheep, intestinal inflammation. In cattle grazing tall fescue, syndromes known as **fescue foot**, **summer slump** and **fat necrosis** have been described (Robbins *et al.*, 1985). Fescue foot is characterised by gangrene of the extremities (feet, ears, tail) and is caused by vasoconstriction resulting in reduced blood flow and ischaemia in the peripheral tissues. This is usually accompanied by increased core body temperature in cattle. Necrosis and loss of the tip of the tail is a common sign. Cold weather aggravates the problem of fescue foot, whereas, warm conditions are commonly associated with summer slump, which is a characterised by reduced milk production, heat intolerance, rough hair coat, increased respiration, decreased feed intake, reduced serum prolactin levels and reduced conception rates (Robbins *et al.*, 1985; Raisbeck *et al.*, 1991). A common sign in herds suffering from summer slump is that the cattle spend much of their time seeking shade or standing in ponds (Raisbeck *et al.*, 1991). Fat necrosis is characterised by hard fat masses in the adipose tissue of the abdominal cavity. Fat necrosis is believed to require long-term consumption of tall fescue (years), whereas, summer slump and fescue foot may occur within weeks of exposure (Raisbeck *et al.*, 1991). In all cases, it is the vasoconstrictive action of the ergot alkaloids that is the underlying cause of the three common fescue toxicities. In horses, reproductive problems resulting from consumption of ergot alkaloids in fescue include longer gestation, stillborn foals, agalactia, placental retention and reduced breeding efficiency (Cross, 2003). Laminitis can also be caused by consumption of toxic fescue; and like fescue foot in cattle, it is believed to be caused by ergot alkaloid-induced vasoconstriction in the extremities (Cross, 2003).

The absorption and metabolism of ergot alkaloids may be quite different in monogastrics and ruminants. The more water-soluble alkaloids appear to be better absorbed than the lipophilic ones and ruminal microbes appear to play an important role in the release and intestinal availability of soluble alkaloids in the rumen. The more soluble alkaloids are excreted in urine and the less soluble alkaloids can undergo enterohepatic recirculation and are excreted primarily in bile. Transmission is unlikely and accumulation in tissues, if any, is low. Once absorbed, ergot alkaloids such as ergotamine can be metabolised by cytochrome P450s. Ergopeptine alkaloids are metabolised in liver via cytochrome P 450s such as CYP3A and hydroxylated metabolites are excreted in bile (Haschek *et al.*, 2002; Moubarak and Rosenkrans, 2000). Ergolines (i.e. lysergic acid), which are more water soluble, are excreted in urine (Hill *et al.*, 2001).

Fumonisins

Equine leukoencephalomalacia (ELEM) syndrome is a fatal disease that apparently occurs only in horses and related species and is characterised by the presence of liquefactive necrotic lesions in the white matter of the cerebrum, although the gray matter may also be involved (WHO, 2000). Increased protein in the cerebrospinal fluid and other changes suggest that the cause of the brain lesion is vasogenic cerebral oedema (Haschek *et al.*, 2002). The first symptoms are lethargy, head pressing and decreased feed intake, followed by convulsions and death after several days.

Early clinical signs include mild proprioceptive dysfunction such as hindlimb ataxia, delayed forelimb placing reactions and tongue paresis (Foreman *et al.*, in press). Elevated serum enzyme levels indicative of liver damage are preceded by elevation in free sphingoid bases in serum. The elevation in free sphingoid bases occurs because fumonisins are inhibitors of the enzyme ceramide synthase. The elevation in serum and tissue free sphingoid bases is a biomarker for exposure to toxic levels of fumonisins and has been studied in horses, pigs, rabbits, poultry and other farm animals (WHO, 2000). Serum enzyme levels often return to near normal concentrations but usually increase markedly immediately prior to or at the onset of behavioural changes (JECFA 56th, 2001; WHO, 2000).

In addition to the brain lesions, histopathological abnormalities in liver and kidney have been reported in horses. Fatal liver disease in the absence of any brain lesions and ELEM concurrent with significant liver disease has been observed in horses and ponies. The length of exposure, level of contamination, individual animal differences, previous exposure, or pre-existing liver impairment may all contribute to the appearance of the clinical disease.

The minimum toxic dose in equids appears to be greater than 15 mg/kg diet and less than 22 mg/kg diet, however, analysis of feeds from confirmed cases of ELEM indicated that consumption of feed with a fumonisin B_1 concentration greater than 10 mg/kg diet is associated with increased risk of development of ELEM, whereas a concentration less than 6 mg/kg diet is not (JECFA 56th, 2001; WHO, 2000).

Clinical signs indicative of porcine pulmonary oedema syndrome typically occur soon (2–7 days) after pigs consume diets containing large amounts of fumonisins over a short period of time. Clinical signs usually include decreased feed consumption, dyspnoea, weakness, cyanosis and death. At necropsy, the animals exhibit varying degrees of interstitial and interlobular oedema, with pulmonary oedema and hydrothorax. Varying amounts of clear yellow fluid accumulate in the pleural cavity. Toxic hepatosis occurs concurrently with porcine pulmonary oedema and is also observed in animals that consume high levels of fumonisins but do not develop porcine pulmonary oedema. Two studies have reported nodular hyperplasia in the pig liver (JECFA 56th, 2001; WHO, 2000).

There is a strong correlation between fumonisin content of maize screenings obtained from different farms and outbreaks of porcine pulmonary oedema. The minimum toxic dose has not been clearly established. Concentrations of FB_1 as low as 17 mg/kg in culture material diets induced the disease in 5 days, however, a dose of 150 to 170 mg/kg diet for up to 210 days caused liver effects early on but no evidence of pulmonary oedema (WHO, 2000).

In pigs, tissues other than liver and lung have been reported to be targets for fumonisins, (e.g., pancreas, heart, kidney, pulmonary intravascular macrophages, and oesophagus). Altered growth and changes in selected haematological parameters in pigs were reported at dietary levels as low as 1 mg/kg (WHO, 2000).

Several reports have been published implicating *F. verticillioides* and by implication the presence of fumonisin contamination of feed in diseases of poultry. The clinical features of the disease often include diarrhoea, weight loss, increased liver weight and poor performance. Functional and morphological changes were observed in chickens exposed to fumonisin B_1. Several studies have confirmed that *F. verticillioides*, *F. proliferatum*, fumonisin B_1 and moniliformin are toxic to poultry (broiler chicks, turkeys, ducklings). The levels of fumonisins used in these studies were 75–644 mg/kg diet. Toxicity and altered haematological parameters have been reported in broiler chicks fed diets containing pure fumonisin B_1 (10 mg/kg) and fumonisin B_1 (30 mg/kg) from *F. verticillioides* culture material (JECFA 56th, 2001; WHO, 2000).

Other farm animals that have been studied using pure fumonisins, contaminated maize screenings or maize culture material of *F. verticillioides* include catfish, trout, cattle, mink, and rabbits. Rabbits are especially sensitive to renal toxicity. In all cases where toxicity was evident it involved liver and/or kidney or homologous organs and there was evidence of disruption of sphingolipid metabolism.

There is considerable evidence that the underlying mechanism by which fumonisins cause toxicity to animals is disruption of lipid metabolism (WHO, 2000; JECFA 56th, 2001). Most notably, fumonisins are specific inhibitors of ceramide synthase (sphinganine and sphingosine N-acyltransferase), a key enzyme in the pathway leading to formation of ceramide and more complex sphingolipids (Figure 4). ELEM, porcine pulmonary oedema, hepato- and nephrotoxicity that result from fumonisin exposure are closely correlated with the degree of disruption of sphingolipid metabolism (WHO, 2000; JECFA 56th, 2001). Equine leukoencephalomalacia is associated with sphingolipid-induced alterations in cardiovascular function; deregulation of cerebral arteries responsible for autoregulation of blood flow to the horse's brain (Haschek *et al.*, 2002). Porcine pulmonary oedema is hypothesised to be a result of acute left-sided heart failure as a consequence of sphingoid base-induced inhibition of L-type calcium channels (Haschek *et al.*, 2002). The elevation of free sphingoid bases in serum and urine has been proposed as a functional biomarker for exposure to fumonisins. This biomarker works well in farm animals. Recently it has been shown that elevation of sphinganine 1-phosphate in serum of pigs and horses may be a very sensitive biomarker for fumonisin exposure (Piva *et al.*, submitted; unpublished data). Fumonisins can be detected in human hair (Sewram *et al.*, 2001), and this may also be true in farm animals. If so, this could be a useful marker of exposure.

In most animals, fumonisin B_1 absorption, distribution and elimination is rapid. Fumonisins are poorly absorbed and while there is some evidence that fumonisins can be partially metabolised in the gut, there is currently no convincing evidence of metabolism *in vitro* or *in vivo* by animals even though they are clearly excreted in bile. However, there have been studies showing that fumonisins can alter P450 activity as a result of fumonisin-inhibition of the enzyme ceramide synthase. Both fumonisins of the B series and their hydrolysed counterparts can react with ceramide synthase; and hydrolysed fumonisin B_1 is a substrate for the enzyme (Merrill *et al.*, 2001) producing the compound known as PAP_1 (N-palmitoyl-AP_1).

Ochratoxin

The primary effect of ochratoxin A in all farm animals is **nephrotoxicity**. In pigs and poultry the proximal tubules are mainly affected and the kidney is pale and grossly enlarged. Fatty liver can occur in poultry. The most sensitive indicator of acute ochratoxicosis in chickens is the reduction in total serum proteins and albumin and in pigs, a decrease in phosphoenolpyruvate carboxykinase in kidney is a sensitive and specific indicator (Marquardt and Frohlich, 1992; Krogh, 1992). In pigs, large increases in proteins excreted in urine are indicative of glomerular proteinuria and were correlated with histological observations of renal damage.

Exposure to lower levels of ochratoxin in poultry and pigs can result in altered performance including reduced feed consumption and weight gain and at higher levels delayed response to immunisation, and increased susceptibility to infection (Stoev *et al.*, 2000 a,b). Other effects in poultry include decreased egg production, coagulopathy (increased susceptibility to bruising during processing), decreased force to break bones, decreased tensile strength of the large intestines, underpigmentation, and glycogen accumulation in liver.

The mechanism of action of ochratoxin is unclear, but its structural similarity to phenylalanine and the fact that it inhibits many enzymes and processes that are dependent on phenylalanine strongly suggest

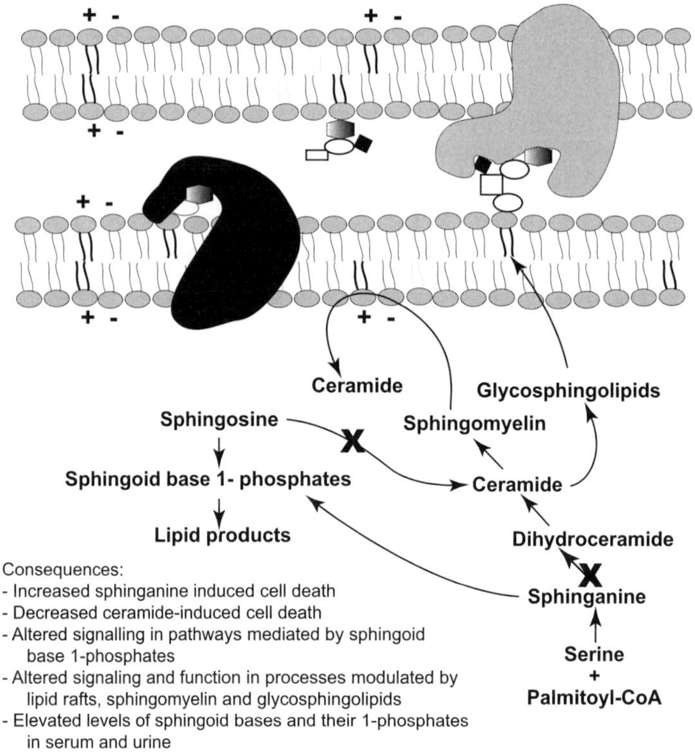

Figure 4. Biochemical mechanism of action of fumonisin. The structure of fumonisin B_1 and inhibition of sphinganine (sphingosine) N-acyltransferase (indicated by the large X) resulting in a blockage of *de novo* complex sphingolipid biosynthesis, accumulation of sphinganine (and sometimes sphingosine), and diversion of sphinganine to sphinganine 1-phosphate which is further broken down to lipid degradation products (fatty aldehydes and ethanolamine phosphate). Fumonisins also block the reacylation of dietary sphingosine and the sphingosine that is released by the turnover of more complex sphingolipids. As a consequence of disruption of sphingolipid metabolism, many bioactive lipid intermediates are created and some are lost. Some examples of the biological activity of these intermediates are increases in intracellular free sphingoid bases, which can inhibit protein kinase C and other kinases; increase in sphingosine 1-phosphate which triggers endoplasmic reticulum calcium release and serve as ligand for extracellular receptors in the vasculature (S1P receptors); increase in free sphinganine, which can affect cell proliferation, cell cycle progression, and initiate apoptosis or necrotic cell death; increase in kidney and serum sphinganine 1-phosphate which is also a ligand for S1P receptors; and depletion of complex sphingolipids (glycosphingolipids and sphingomyelin [indicated in the membrane by ceramide with a + or - on the head group) which are integral components of pathways that regulate cell growth, cell cycle progression, lipid raft function (for example inhibited folate uptake) and cell-cell interactions (for review see Merrill *et al.*, 2001).

that ochratoxin A acts by disrupting phenylalanine metabolism (CAST, 2003) (Figure 5). Several studies have shown that supplementation of feed with L-phenylalanine or protein protect against the toxic effects (including mortality) of ochratoxin A (JECFA 56th , 2001; Marquardt and Frohlich, 1992). Because ochratoxin binds tightly to albumin and serum proteins, the ochratoxin protein adduct in serum is a useful biomarker for exposure in pigs.

Comprehensive reviews of the pharmacokinetics and metabolism of ochratoxin are available (Marquardt and Frolich, 1992; Pfohl-Leszkowicz *et al.*, 1999). Ochratoxin A is rapidly and well absorbed.

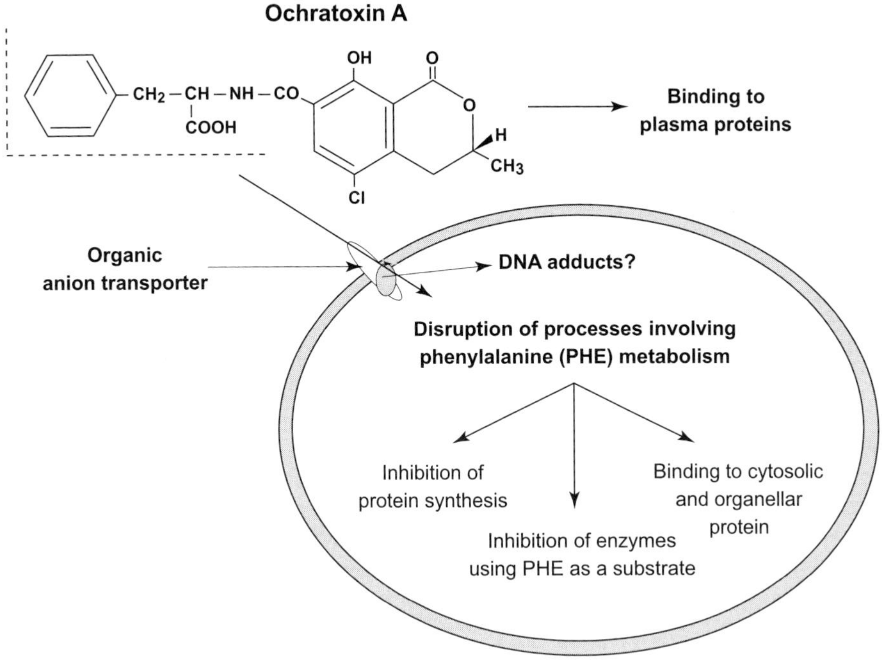

Figure 5. Hypothesized mechanism of action of ochratoxin A. Ochratoxin A is transported into cells via a multispecific organic anion transporter (Kobayashi *et al.*, 2002). The ochratoxin chemical structure contains an isocumarin moiety linked to phenylalanine (denoted by dashed line). Numerous studies have shown that ochratoxin A can alter processes that require phenylalanine and many of the biological effects of ochratoxin can be at least partially prevented by supplementation with phenylalanine or phenylalanine analogs (for example Zanic-Grubisic *et al.*, 2000; Schwerdt *et al.*, 1999; Baudrimont *et al.*, 2001). Ochratoxin albumen adducts are a useful marker for ochratoxin exposure.

The half-life in plasma is dependent on the extent of binding to plasma proteins. It is widely distributed and is accumulated in kidney and other tissues in pigs, and can occur in edible tissues. Ochratoxin A and its metabolites are reabsorbed by the kidney and excreted in urine and also undergo enterohepatic circulation and excretion in faeces. There is the potential for extensive metabolism to less toxic metabolites in the gastrointestinal tract by microbes. In addition to gastrointestinal degradation by microbes, ochratoxin A can be oxidised by cytochrome P450 enzymes to their less toxic hydroxyochratoxin A metabolites using microsomal preparations from rabbits and pigs (JECFA 56th, 2001).

Zearalenone

Zearalenone is an oestrogenic mycotoxin that often occurs with DON on scabby wheat and on maize. There is no doubt that zearalenone is a cause of farm animal performance problems. The primary effect of zearalenone is oestrogenic and prepubertal female pigs are clearly the most affected farm animal. The basis for the oestrogenic effect is well established and is due to a close structural similarity between zearalenone (and many of its metabolites) and oestradiol (Osweiller, 2000). α-Zearalenol is three times more potent in its oestrogenic activity compared to zearalenone and the relative binding affinity for oestrogen receptors was found to be greater in pigs than in other species; which may explain the interspecies differences in sensitivity to the oestrogenic effects (Fitzpatrick *et al.,* 1989). Clinical signs of oestrus can be induced in ovariectomised sows and doses as low as 1-5 ppm can induce vulvovaginitus, tenesmus, vaginal and rectal prolapse in young female pigs (Osweiller, 1986). Effects on prepubertal boars have also been reported and include reduced libido, plasma testosterone and other effects (Osweiller, 1986). Dietary levels of 3-10 ppm zearalenone can induce anoestrus in sows, reduced litter size, fetal resorption and implantation failure. Cattle are more resistant to the oestrogenic effects, however, conception rates can be reduced.

Zearalenone is rapidly absorbed and eliminated. It is metabolised in liver (and possibly the intestinal mucosa) and excreted in urine and feces as the glucuronide after considerable enterohepatic recirculation. It is metabolised by rumen microbes. Accumulation in tissues is minimal. The rapid hepatic conversion of zearalenone to the more easily excreted α- and ß-zearalenol derivatives in cattle, along with microbial metabolism in the rumen, could explain the resistance of cattle to the reproductive effects of zearalenone relative to pigs (Raisbeck *et al.,* 1991). The zearalenone metabolites are conjugated with glucuronic acid in pigs and recycled via the bile and there is some evidence that the intestinal mucosa is active in reducing zearalenone to α-zearalenol and conjugation with glucuronic acid (Biehl *et al.,* 1993). The glucuronide conjugates can also be detected in urine and liver (CAST, 2003;

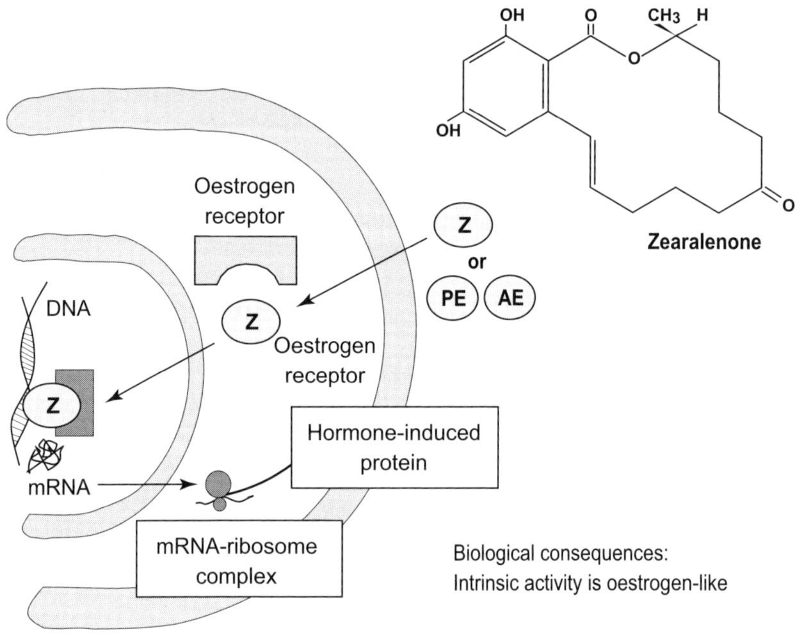

Figure 6. Mechanism of action of zearalenone. Interaction of zearalenone (Z) with the cytosolic oestrogen receptor illustrating hormonal mimicry (modified from Riley and Norred, 1996). Zearalenone (Z) like phytoestrogens (PE) and environmental oestrogens (AE), passively cross the cell membrane and bind to the cytosolic oestrogen receptor. The receptor-Z complex (or PE/AE complex) is rapidly transferred into the nucleus where it binds to specific nuclear receptors and generates oestrogenic responses via gene activation resulting in the production of mRNAs that code for proteins that are normally expressed by receptor-oestrogen complex binding.

Zollner *et al.*, 2002). The α-zearalenol metabolite is more frequently detected than the ß-zearalenol metabolite in liver and zeranol and α-zearalenol and trace amounts of zearalenol and zearalenone can be detected in muscle tissue of pigs fed zearalenone-contaminated oats (Zollner *et al.*, 2002). The glucoronide conjugates could have potential as a biochemical marker of exposure in suspected field outbreaks.

References

Bamburg, J.R. 1983. Biological and biochemical actions of trichothecene mycotoxins. Prog. Mol. Subcell. Biol. 8:41-110.

Baudrimont, I., B. Sostaric, C. Yenot, A.M. Betbeder, S. Dano-Djedje, A. Sanni, P.S. Steyn and E.E Creppy. 2001. Aspartame prevents the karyomegaly induced by ochratoxin A in rat kidney. Arch. Toxicol. 75:176-183.

Blount,W.P. 1961. Turkey 'X' Disease. Turkeys 9:52-77.

Bondy, G.S. and J.J. Pestka. 2000. Immunomodulation by fungal toxins. J. Toxicol. Environ. Health (Part B) 3:109-143.

Biehl, M.L., D.B. Prelusky, G.D. Koritz, K.E. Hartin, W.B. Buck and H.L. Trenholm. 1993. Biliary excretion and enterohepatic cycling of zearalenone in immature pigs. Toxicol. Appl. Pharmacol. 121:152-159.

Bryden,W.L. 1991. Occurrence and biological effects of cyclopiazonic acid. In: *Emerging Food Safety Problem Resulting from Microbial Contamination* (K. Mise and J.L. Richard, eds). Ministry of Health and Welfare, Tokyo, pp. 127-147.

CAST. 2003. Mycotoxins: Risks in Plant and Animal Systems. Task Force Report 139, Council for Agricultural Science and Technology, Ames, Iowa, p. 199.

Cross, D.L. 2003. Ergot alkaloid toxicity. In: *Clavicipitalean Fungi* (J.F. White, Jr., C.W. Bacon, N.L. Hywel-Jones and J.W. Spatafora, eds). Marcel Dekker, Inc. New York, pp. 475-494.

Cousin, M.A., R.T. Riley and J.J. Pestka. In press. Foodborne mycotoxins: chemistry, biology, ecology and toxicology. In: *Foodborne Pathogens: Microbiology and Molecular Biology* (P.M. Fratamico and A.K. Bhunia, eds). Horizon Scientific Press, Ltd., Norfolk, UK.

Eaton, D.L. and E.P. Gallagher. 1994. Mechanisms of aflatoxin carcinogenesis. Ann. Rev. Pharmacol. Toxicol. 34:135-172.

Edds, G.T. 1979. Biological effects of aflatoxins in swine. In: *Interactions of Mycotoxins in Animal Production*. National Academy of Sciences, Washington, pp. 67-76.

Eriksen, G.S., H. Pettersson and J.E. Lindberg. 2003. Absorption, metabolism and excretion of 3-acetyl DON in pigs. Arch. Anim. Nutr. (in press).

Fitzpatrick, D.W., C.A. Picken, L.C. Murphy and M.M. Buhr. 1989. Measurement of the relative binding affinity of zearalenone, alpha-zearalenol and beta-zearalenol for uterine and oviduct oestrogen receptors in swine, rats and chickens: An indicator of oestrogenic potencies. Comp. Biochem. Physiol. (C) 94:691-694.

Foreman, J.H., P.D. Constable, A.L. Waggoner, M. Levy, R.M. Eppley, G.W. Smith, M.E. Tumbleson and W.M. Haschek. In press. Neurological abnormalities and cerebrospinal fluid changes in horses administered fumonisin B_1 intravenously. J. Vet. Int. Med.

Haschek, W.M., K.A. Voss and V.R. Beasley. 2002. Selected mycotoxins affecting animal and human health. In: *Handbook of Toxicologic Pathology*. 2nd Ed, Vol. 1, Academic Press, pp. 645-699.

Henry, S., F.X. Bosch, T.C. Troxell and P.M. Bolger. 1999. Reducing liver cancer – global incidence of aflatoxin. Science 286:2453-2454.

Hill, N.S., F.N. Thompson, J.A. Stuedemann,

G.W. Rottinghaus, H.J. Ju, D.L. Dawe and E.E. Hiatt, III. 2001. Ergot alkaloid transport across ruminant gastric tissues. J. Anim. Sci. 79:542-549.

IARC. 1993. IARC Monographs on the Evaluation of Carcinogenic Risk to Humans, Vol. 56. Some Naturally Occurring Substances: Food Items and Constituents, Heterocyclic Aromatic Amines and Mycotoxins. Lyon, France.

IARC. 2002. IARC Monographs on the Evaluation of Carcinogenic Risk of Chemicals to Humans, Vol. 82. Some Traditional Herbal Medicines, Some Mycotoxins, Naphthalene and Styrene. Lyon, France.

Iordanov, M.S., D. Pribnow, J.L. Magun, T.H. Dinh, J.A. Pearson, S.L. Chen and B.E. Magun. 1997. Ribotoxic stress response: activation of the stress-activated protein kinase JNK1 by inhibitors of the peptidyl transferase reaction and by sequence-specific RNA damage to the α-sarcin/ricin loop in the 28S rRNA. Mol. Cell Biol. 17:3373-3381.

JECFA 56th. 2001. Joint FAO/WHO Expert Committee on Food Additives 56th Report. Safety evaluation of certain mycotoxins in food. Food and Agriculture Organization of the United Nations, paper 74. World Health Organization Food Additives Series 47, World Health Organization, Geneva, Switzerland.

Kobayashi, Y., N. Ohshiro, A. Shibusawa, T. Sasaki, S. Tokuyama, T. Sekine, H. Endou and T. Yamamoto. 2002. Isolation, characterisation and differential gene expression of multispecific anion transporter in mice. Mol. Pharmacol. 62:7-14.

Krogh, P. 1992. Role of ochratoxin A in disease causation. Fd. Chem. Toxicol. 30:213-224.

Li, J. and T. Shimizu. 1997. Course of apoptotic changes in the rat gastric mucosa caused by oral administration of fusarenon-X. J. Vet. Med. Sci. 59:191-199.

Li, G., J. Shinozuka, K. Uetsuka, H. Nakayama and K. Doi. 1997a. T-2 toxin-induced apoptosis in intestinal crypt epithelial cells of mice. Exp. Toxicol. Pathol. 49:447-450.

Li, J., T. Shimizu, N. Miyoshi and N. Yasuda. 1997b. Rapid apoptotic changes in the gastric glandular epithelium of rats administered intraperitoneally with fusarenon-X. J. Vet. Med. Sci. 59:17-22.

Marquardt, R.R. and A.A. Frohlich. 1992. A review of recent advances in understanding ochratoxicosis. J. Anim. Sci. 70:3968-88.

Meky, F.A., P.C. Turner, A.E. Ashcroft, J.D. Miller, Y-L. Qiao, M.J. Roth and C.P. Wild. 2003. Development of a urinary biomarker of human exposure to deoxynivalenol. Food Chem. Toxicol. 41:265-273.

Merrill, A.H., Jr., M.C. Sullards, E. Wang, K.A. Voss and R.T. Riley. 2001. Sphingolipid metabolism: roles in signal transduction and disruption by fumonisins. Environ. Health Persp. 109:283-289.

Moubarak, A.S. and C.F. Rosenkrans, Jr. 2000. Hepatic metabolism of ergot alkaloids in beef cattle by cytochrome P450. Biochem. Biophys. Res. Comm. 274:746-749.

National Academy of Sciences. 1979. *Interactions of mycotoxins in animal production*. National Academy of Sciences, Washington.

Norred, W.P. 1986. Occurrence and clinical manifestations of aflatoxicosis. In: *Diagnosis of Mycotoxicoses* (J.L. Richard and J.R. Thurston, eds). Martinus Nijhoff Publishers, Dordrecht, pp. 11-30.

Norred, W.P. and R.T. Riley. 2001. Toxicology – mode of action of mycotoxins. In: *Mycotoxins and Phycotoxins in Perspective at the Turn of the New Millennium* (W.J. de Koe, R.A. Samson, H.P. van Egmond, J. Gilbert and M. Sabino eds). W.J. de Koe, Hazekamp 2, Wageningen, The Netherlands. pp. 211-222.

Osweiler, G.D. 1986. Occurrence and clinical manifestations of trichothecene toxicoses and zearalenone toxicoses. In: *Diagnosis of Mycotoxicoses* (J.L. Richard and J.R. Thurston, eds). Martinus Nijhoff Publishers, Dordrecht, pp. 31-42.

Osweiler, G.D. 2000. Mycotoxins - contemporary issues of food animal health and productivity. Vet. Clin. North Am. Food Anim. Pract. 16:511-530.

Pfohl-Leszkowicz, A., P. Guerre and P. Galtier. 1999. Metabolisation des mycotoxins. In: *Les Mycotoxines dans l'alimentation: Evaluation et gestion di risqué*. Technique and Documentation, Paris, pp. 37-68.

Prelusky, D.B. 1997. Effect of intraperitoneal infusion of deoxynivalenol on feed consumption and weight gain in the pig. Natural Toxins 5:121-125.

Pestka, J.J., H-R. Zhou, Y. Moon and Y.J. Chung. 2004. Cellular and molecular mechanisms for immune modulation by deoxynivalenol and other trichothecenes: unravelling a paradox. Toxicol. Lett. 153:61-73.

Piva, A., G. Casadei, G. Pagliuca, E. Cabassi, F. Galvano, M. Solfrizzo, R.T. Riley and D.E. Diaz. Submitted. Inability of activated carbon to prevent the toxicity of culture material containing fumonisin B_1 when fed to weaned piglets. J. Agric. Fd. Chem.

Raisbeck, M.F., G.E. Rottinghaus and J.D. Kendall. 1991. Effects of naturally occurring mycotoxins on ruminants. In: *Mycotoxins and Animal Foods* (J.E. Smith and R.S. Henderson, eds). CRC Press, Boca Raton, pp. 647-677.

Richard, J.L. and J.R. Thurston. 1986. *Diagnosis of Mycotoxicoses*. Martinus Nijhoff Publishers, Dordrecht.

Riley, R.T. and W.P. Norred. 1996. Mechanistic toxicology of mycotoxins. In: *The Mycota VI. Human and Animal Relationships* (D.H. Howard and J.D. Miller, eds). Springer-Verlag, Berlin, pp. 193-211.

Riley, R.T., D.E. Goeger and W.P. Norred. 1995. Disruption of calcium homeostasis: The cellular mechanism of cyclopiazonic acid toxicity in laboratory animals. In: *Molecular Approaches to Food Safety* (M. Eklund, J.L. Richard and K. Mise, eds). Alaken, Inc., Fort Collins, pp. 461-480.

Riley, R.T. 1998. Mechanistic interaction of mycotoxins: Theoretical considerations. In: *Mycotoxins in Agriculture and Food Safety* (K.K. Sinha and D. Bhatnagar, eds). Marcel Dekker, New York, NY, pp. 227-223.

Robbins, J.E., J.K. Porter and C.W. Bacon. 1985. Occurrence and clinical manifestations of ergot and fescue toxicoses. In: *Diagnosis of Mycotoxicoses* (J.L. Richard and J.R. Thurston, eds). Martinus Nijhoff Publishers, Dordrecht, pp. 61-74.

Schwerdt, G., R. Freudinger, S. Silbernagl and M. Gekle. 1999. Ochratoxin A – binding proteins in rat organs and plasma and in different cell lines of the kidney. Toxicology 135:1-10.

Sewram, V., J.J. Nair, T.W. Nieuwoudt, W.C. Gelderblom, W.F. Marasas and G.S. Shephard. 2001. Assessing chronic exposure to fumonisin mycotoxins: the use of hair as a suitable noninvasive matrix. J. Anal. Toxicol. 25:450-455

Stoev, S.D., D. Goundasheva, T. Mirtcheva and P.G. Mantle. 2000a. Susceptibility to secondary bacterial infections in growing pigs as an early response in ochratoxicosis. Exper. Toxicol. Path. 52:287-296.

Stoev, S.D., G. Anguelov, I. Ivanov and D. Pavlov. 2000b. Influence of ochratoxin A and an extract of artichoke on the vaccinal immunity and health in broiler chicks. Exper. Toxicol. Path. 52:43-55.

Trenholm, H.L., D.B. Prelusky, J.C. Young and J.D. Miller. 1988. Reducing mycotoxins in animal feeds. Agriculture Canada Publication 1827E. Communications Branch, Agriculture Canada, Ottawa.

WHO. 2000. Environmental Health Criteria 219: Fumonisin B_1 (W.H.O. Marasas, J.D. Miller, R.T. Riley and A. Visconti, eds). International Programme on Chemical

Safety, United Nations Environmental Programme, The International Labour Organisation, and the World Health Organisation, Geneva, Switzerland.

Yang, G.H., B.B. Jarvis, Y.J. Chung and J.J. Pestka. 2000. Apoptosis induction by the satratoxins and other trichothecene mycotoxins: relationship to ERK, p38 MAPK and SAPK/JNK activation. Toxicol. Appl. Pharmacol. 164(2):149-160.

Zanic-Grubisic, T., R. Zrinski, I. Cepelak, J. Petrik, B. Radic and S. Pepeljnjak. 2000. Studies of ochratoxin A-induced inhibition of phenylalanine hydroxylase and its reversal by phenylalanine. Toxicol. Appl. Pharmacol. 167:132-139.

Zhou, H.R., Z. Islam and J.J. Pestka. 2003. Rapid, sequential activation of mitogen-activated protein kinases and transcription factors precedes proinflammatory cytokine mRNA expression in spleens of mice exposed to the trichothecene vomitoxin. Toxicol. Sci. 72:130-142.

Zollner, P., J. Jodlbauer, M. Klienova, H. Kahlbacher, T. Kuhn, W. Hochsteiner and W. Linder. 2002. Concentration levels of zearalenone and it metabolites in urine, muscle tissue and liver samples of pigs fed with mycotoxin-contaminated oats. J. Ag. Fd. Chem. 50:2494-2501.

14

EFFECTS OF MYCOTOXINS IN RUMINANTS

Jean-Pierre Jouany[1] and Duarte E. Diaz[2]
[1]*INRA, Herbivore Research Unit, Clermont-Theix, France*
[2]*University of Puerto Rico, Mayagüez, Puerto Rico*

Introduction

Mycotoxins are toxic, biologically diverse, secondary fungal metabolites produced by several fungi, particularly by species of *Aspergillus, Fusarium* and *Penicillium*. Apart from the threat to public health, mycotoxins are associated with significant economic losses for both crops and animals, including ruminants. Although several hundred mycotoxins have been described in the scientific literature, less than 10 have been extensively studied since the discovery of aflatoxin in the early 1960s. Mycotoxin-producing moulds are ubiquitous in nature and are commonly in contact with forages and cereals in the field, during harvest, drying and transport as well as during storage. Depending on the specific crop, mycotoxin, and environment, contamination with mycotoxins may be more likely at certain phases of production and handling, but contamination is rarely associated with just one mycotoxin. Because the same fungus is able to produce several mycotoxins and several fungi can produce the same mycotoxin, it is difficult to correlate mycotoxins to the presence of moulds in feeds. The presence of moulds does not directly indicate presence of mycotoxins, but if toxigenic moulds are present and the proper conditions exist (stress), there is a potential for mycotoxin production. Conversely, the absence of moulds does not guarantee the absence of mycotoxin, because the toxin may be present long after the death of the toxin-producing moulds.

Ruminant diets generally include both forages and concentrate, which may increase the risk of mycotoxins in comparison with animals that do not consume forages. The multiplicity of ingredients in complex diets may increase the probability of multiple mycotoxin contamination but decrease the risk of high mycotoxin concentrations because any one feed ingredient is diluted in the final diet. New techniques for preservation of wet forages, such as silages and wrapped bales, have been developed over the past 20-30 years. Preserved forages are more likely to harbour moulds and associated toxins than dry forages when anaerobic conditions are not strictly controlled. Many agricultural and food industry by-products such as fruit pulp, beet pulp and brewery wastes commonly used in ruminant diets are often handled in wet form, which means that moulds can grow and mycotoxins be produced during the storage and transport phases. Furthermore, grazing systems cannot be considered completely safe from mycotoxin contamination. Fresh grasses can be contaminated with mycotoxins (Erasmuson *et al.*, 1994; Sporsen *et al.*, 1995) including fungal endophytes that produce toxins such as ergovaline, lolitrem B and peramine (Lewis and Clements, 1986). Prohibition of antifungal agents in the expanding 'organic' agriculture sector leads to products that may have an increased likelihood of contamination with mycotoxins (Malmauret *et al.*, 2002).

Ruminants have a large compartmentalised digestive tract. The first and largest compartment, the rumen or reticulorumen, harbours a dense and diversified microbial population including bacteria (10^{11}/mL), protozoa (10^6/mL) and fungi (10^4 zoospores/mL). This active microbial ecosystem plays a central role in feed digestion since the microbes are essential for fibre digestion, B-vitamin production and production of amino acids. Rumen microbes help protect the animal against the development of pathogens in the ruminant digestive tract (McIntosh *et al.*, 2002) and are involved in the metabolism of toxins from plants (Cheeke and Palo, 1995) or filamentous fungi (Yiannikouris and Jouany, 2002). It has been widely accepted that ruminants are less susceptible to mycotoxin poisoning than monogastrics (Kurmanov, 1977). However, rumen metabolites of the parent mycotoxin may be equally or more toxic. In some cases, the production of toxic metabolites from innocuous compounds can occur. Because of this, differences between parent mycotoxins and their metabolites are much greater in ruminants than non-ruminants.

The intent of this chapter is to review the available literature on the effects of mycotoxins on ruminant animals used in agricultural production. The major mycotoxins will be discussed in detail and special attention will be given to rumen metabolism, absorption, excretion and animal toxicity.

Aflatoxins

Aflatoxins are produced primarily by the fungi *Aspergillus flavus* and *Aspergillus parasiticus*. They are a group of difuranocoumarin derivatives of which aflatoxin B_1, B_2, G_1, and G_2 are the most important. Aflatoxin B_1 (AFB_1), the most important of the four main derivatives, has been described as the most potent naturally occurring carcinogen (Squire, 1989). Because of its carcinogenic potential, aflatoxin has been studied throughout the world.

Interpretation of results from aflatoxin studies requires special attention, because some studies utilised mixtures of aflatoxins either from naturally contaminated feedstuffs or from culture material, while others used purified AFB_1. Naturally contaminated feedstuff and culture material tend to be more toxic as was demonstrated by Applebaum *et al.* (1982) who noted that aflatoxin produced from culture material was more toxic to dairy cattle than pure aflatoxin. The difference in toxicity is most likely due to a combination of aflatoxin metabolites or the presence of other mycotoxins in culture material and naturally contaminated feeds.

RUMEN METABOLISM OF AFLATOXINS

Conflicting results have been reported regarding the rumen biodegradation of AFB_1. Some authors found that up to 42% of AFB_1 is degraded when incubated *in vitro* with rumen fluid (Engel and Hagemeister, 1978), while others saw no major degradation (Kiessling *et al.*, 1984; Mathur *et al.*, 1976; Westlake *et al.*, 1989). Certain aflatoxin metabolites such as aflatoxicol and aflatoxin M_1 (AFM_1) have been detected in rumen contents (Trucksess *et al.*, 1983). However, a well controlled *in vitro* study which was designed to simulate rumen conditions and using ^{14}C-AFB_1 showed that rumen microbes produced only aflatoxicol (Auerbach *et al.*, 1998). This metabolite was shown to be of microbial origin, since it was not produced when the incubation was carried out with autoclaved rumen fluid. However, the final radioactivity of aflatoxicol represented only 0.6% of the initial radioactivity. Regarding the presence of AFM_1 in rumen contents when experiments were carried out *in vivo*, Auerbach *et al.* (1998) suggested that AFM_1

produced in the liver could "recirculate to the rumen via the rumino-hepatic pathway".

The addition of AFB_1 at 9.5 ng/mL of rumen contents did not alter the *in vitro* digestion of alfalfa, and did not impair the subsequent production of volatile fatty acids (VFAs) (Auerbach *et al.*, 1998). In contrast, Westlake *et al.* (1989) measured a large decrease in the digestive capacity of rumen microbes when AFB_1 was dosed at 1 μg/mL (-50% in rumen digestibility of alfalfa) and 10 μg/mL (-67% in rumen digestibility of alfalfa). Such highly divergent results can be explained by differences in toxin concentrations.

Because ruminal degradability of AFB_1 is minor and the toxicity of the main rumen-produced metabolite (aflatoxicol) is similar to that of the parent molecule, it can be concluded that the ruminant is not efficiently protected against aflatoxin toxicity.

ABSORPTION AND EXCRETION OF AFLATOXINS

Lipophilic mycotoxins with a small molecular weight such as AFB_1 are absorbed in the digestive tract by a passive diffusion mechanism. Studies with laboratory animals and primates show that absorption of AFB_1 is complete and rapid (Coulombe and Sharma, 1985; Dalezios *et al.*, 1972; Wogan, 1966). Aflatoxin B_1 absorption occurs chiefly at the duodenal level in rats, and appears to proceed in two phases indicating that two intestinal sites are involved, as indicated by kinetic studies (Kumagai, 1989). There is a rapid absorption of [^{3}H]-AFB_1 in the digestive tract of cows, and this explains the quick transfer of AFM_1 to milk (Polan *et al.*, 1974).

Around 90% of AFB_1 present in blood is found in plasma (Wong and Hsieh, 1980). The fraction associated with proteins is chiefly linked to albumin. Covalent bonds such as Schiff bases between lysine from albumin and certain derivatives from AFB_1 are involved in adduct formation with blood proteins. The chemical bond is partly reversible and compounds having a high affinity for albumin, such as phenylbutazone, can displace the bound fraction.

Blood-circulating AFM_1 is excreted in milk or in urine. After oral administration AFB_1 and metabolites are quickly recovered in urine and milk, while the fraction found in faeces is generally low, which confirms the intense and rapid absorption of AFB_1 in the digestive tract and its active metabolism in the liver. Pettersson (1998) established a relationship between the amount of AFB_1 ingested and the corresponding amount of metabolites in milk, based on 10 determinations collected from five trials carried out in controlled conditions:

$$AFM_1 \text{ (ng/kg milk)} = 10.95 + 0.787X$$
$$(X = \mu g\ AFB_1 \text{ ingested per day});\ r^2 = 0.915$$

Diaz *et al.* (2004) and Kiermeier (1973) have indicated that aflatoxins appear in milk within 12 hrs following oral administration of AFB_1. The peak concentration in milk occurs after 24 hrs. Clearance is also very rapid, since aflatoxin disappears from the milk four days after cessation of oral administration (Diaz *et al.*, 2004; Frobish *et al.*, 1986). These results confirm the rapid absorption and metabolism of aflatoxins in ruminants.

AFM_1, and to a lesser extent AFM_2, are the main metabolites found in milk from AFB_1 and AFB_2. These metabolites are the hydroxylated forms of the parent toxins. Other aflatoxin metabolites such as AFM_4, AFQ_1 and aflatoxicol molecules have also been detected in milk (Wood, 1991).

Most authors consider that the rate of transfer of aflatoxins to milk is between 1 and 3%, with a mean transfer of around 1.7% (Diaz *et al.*, 2004; Frobish *et al.*, 1986; van Egmond, 1989). The extent of transfer from feed to milk can vary widely between animals, from one day to another, and between milkings, because of nutritional or physiological factors such as rate of ingestion, rate of digestion, health of the animal, milk production, and individual sensitivity to mycotoxins or other stresses. A transfer of 6.2% has been measured in high-producing cows at peak lactation (Veldman *et al.*, 1992).

In a 2003 report, van Egmond (2003) determined that at least 99 countries in the world had specific mycotoxin control legislation. Sixty of these countries had AFM_1 legislation with many others having aflatoxin legislation for foodstuffs and/or animal feedstuffs. For example, in the US the Food and Drug Administration (FDA) limits both intake of aflatoxin-contaminated feed (20 ppb for lactating dairy cows) and AFM_1 levels in milk (0.05 ppb). Such worldwide legislation underscores the widespread occurrence of aflatoxin and global recognition of its importance in public health. Two major standards for AFM_1 in milk exist. In the US and in most Latin American countries the regulatory level for milk is 0.5 ppb while in the European Union the regulated level of AFM_1 is 0.05 ppb. The Joint Expert Committee on Food Additives (JECFA), a scientific advisory body of the Food and Agriculture Organization (FAO) and the World Health Organization (WHO) of the United Nations, has recently evaluated the discrepancy between these two standards (100-fold difference). Based on examination of human exposure to AFM_1 and a quantitative risk assessment study, they determined that there was no significant health benefit of lowering the regulatory from 0.5 ppb to 0.05 ppb (FAO, 2004).

Based on an average transfer of 1.7% of dietary AFB_1 to AFM_1 in milk, the regulatory limit of 0.5 ppb AFM_1 in milk will be reached when dietary dry matter contains in excess of 30 ppb AFB_1. Similarly, a concentration of 0.05 ppb AFM_1 will occur when dietary dry matter contains in excess of 3 ppb AFB_1. The low European regulatory limit of 0.05 ppb AFM_1 in milk makes compliance difficult for those regions of Europe where feedstuffs naturally contain aflatoxin. Samples of corn silage containing AFB_1 at 25 to 40 ppb were found in Italy (Vallone and Dragoni, 1997); and consequently there were situations where milk did not comply with the EU maximum allowable level of 0.05 ppb. In recent years, drought, high temperature and insect damage were conducive to *A. flavus* growth and the subsequent production of aflatoxins on maize in the southern countries of Europe. Based on the fact that aflatoxin is produced at significant levels in certain geographical areas of Europe, the European Food Safety Authority (EFSA) Journal (2004) noted that "the overall assumptions that aflatoxins occur only in imported feedingstuffs used in the concentrates for dairy cattle, and that subsequently a strict monitoring of imported feed materials would provide sufficient protection with respect to animal health and towards undesirable AFM_1 concentrations in milk, is challenged by these recent findings".

ANIMAL TOXICITY

The amount of aflatoxin required to affect performance and impair health is much greater than the dietary amounts associated with milk residues. Although no level of aflatoxin can be considered safe, the degree of toxicity is related to toxin level, duration of feeding, and the intensity of other stresses affecting the animal. Acute aflatoxicosis is characterised by liver lesions, leading to congestion and bleeding (Pier, 1992). Aflatoxin exposure causes fatty acid accumulation in the liver, kidneys and heart and may be responsible for encephalopathies and oedemas (Pfohl-Leszkowicz, 2000). Aflatoxins act as a DNA intercalating agent by binding to guanine bases and leading to cell death or its transformation into a tumour (Riley, 1998).

Garrett *et al.* (1968) showed that gain and intake of beef cattle were affected at 700 ppb AFB_1, but not at 100 or at 300 ppb; however, at 100 ppb AFB_1 liver weights were significantly increased. Similarly, Helferich *et al.* (1986) found no effect on growth rates and feed intakes in beef cattle consuming 60 and 300 ppb aflatoxin but did see a decrease in feed intake and rate of gain for animals consuming 600 ppb aflatoxin. The beef cattle consuming 600 ppb aflatoxin also developed minor liver lesions both during the consumption of the feed and after 7 days of non-contaminated feed consumption. In a field study, Jones and Ewart (1979) observed 140 cows receiving a diet containing aflatoxins. Dietary concentration of AFB_1 was about 20 ppb. Animals had reduced feed consumption and milk production. Feed consumption improved 3-4 days after the aflatoxin-contaminated ingredient was removed. Milk production increased within the next 5-8 days. In another field study, Guthrie (1979) noted a decline in reproductive efficiency in lactating dairy cattle consuming diets containing 120 ppb aflatoxin. Milk production increased over 25% when cows were changed to an aflatoxin-free diet. Patterson and Anderson (1982) and Marsi *et al.* (1969) also suggested that 100 ppb may reduce milk production. Mertens *et al.* (1977) saw a drastic decrease in milk production for cows consuming 100 or 300 μg AFB_1/kg BWt but the reduced milk production lagged approximately two days behind the decrease in feed intake, suggesting a cascade effect of aflatoxin on milk production through reduction of feed intake. These findings were latter supported by Applebaum and Marth (1982). Applebaum *et al.* (1982) showed that impure aflatoxin produced by a culture reduced milk production, but equal amounts of pure aflatoxin did not. Several studies suggest that naturally contaminated feeds are more toxic than would be expected from the concentrations of assayed mycotoxins, suggesting the presence of unidentified toxins. As with other mycotoxins, additive or synergistic interactions with one or many other mycotoxins is cause for concern and further study.

Data on toxicity in sheep are variable, but overall it appears that sheep are sensitive to aflatoxin. In probably the only long-term mycotoxin exposure study, Lewis *et al.* (1967) reported that feeding an aflatoxin-contaminated diet ranging from 1.75 to 2.4 ppm for five years resulted in few if any pathological changes. The authors did report nasal neoplasms (abnormal growth tissue) in the animals receiving the contaminated diets. Strikingly different results were seen in a field report by Suliman *et al.* (1987) where serum biochemical changes, hepatic lesions, and death were associated with a ration contaminated with only 0.75 ppm aflatoxin. Harvey *et al.* (1991) saw similar effects in growing lambs consuming 2.6 ppm aflatoxin (0.08 mg/kg BW). In this study the author noted reduced feed intake and body weight gain and an increase in serum aspartate aminotransferase (AST) and γ-glutamyl-transferase (GGT), which are indicative of liver damage. Edrington *et al.* (1994) fed 2.5 mg aflatoxin/kg of feed to growing lambs for 35 days. These animals had reduced feed intake, weight gain and feed conversion efficiency. Interestingly, feed intake and feed conversion, but not average daily gain (ADG) were still affected at 32 days post-aflatoxin feeding. Activities of AST and GGT were also significantly affected after the 35 days of treatment. Fernandez *et al.* (1996) saw that an 8-day detoxification period in lambs was sufficient for a return to normal haematological parameters while it was insufficient for biochemical effects to be alleviated. More recently Fernandez *et al.* (1997; 2000) conducted a series of experiments to investigate the immunological effects of aflatoxin. In both experiments with lambs consuming 2 ppm dietary aflatoxin, there was failure of acquired immunity. The principal immunological effects associated with aflatoxin are related to suppression of cell-mediated immunity and to a lesser extent humoral immunity (CAST, 2003). Aflatoxins are therefore capable of lowering resistance to bacterial and parasitic infections (Pier, 1992).

Zearalenone

Zearalenone (ZEA) is a phenolic resorcylic acid lactone produced primarily by *F. graminearium* and several other species of *Fusarium*. Zearalenone is a natural contaminant of corn and wheat but it may also occur in oats, rice, rye, sorghum, barley, wheat and other feedstuffs (Weidenborner *et al.*, 2000). Tanaka *et al.* (1990) found that 44% of 500 samples collected from 19 different countries were positive for ZEA with an average concentration of 45 ppm. Zearalenone produces potent hyper-oestrogenic responses in susceptible animals.

RUMEN METABOLISM OF ZEARALENONE

More than 90% of ZEA is degraded *in vitro* by rumen microorganisms to α-zearalenol (Kiessling *et al.*, 1984), which Hagler *et al.* (1979) considered about four times more oestrogenic than the parent toxin and, to lesser extent, to ß-zearalenol, which has a low affinity for oestrogenic receptors but is toxic to endometrial cells (Tiemann *et al.*, 2003). The rate of ZEA disappearance has been reported to be greatest in the first hour of the *in vitro* incubation (Moschini *et al.*, 1999). Protozoa have been reported about nine times more active than bacteria in ZEA degradation (Kiessling *et al.*, 1984). About twice as much α-zearalenol as ß-zearalenol was generated by protozoa. Zeranol (or α-zearalanol) is also formed by hydrogenation of α-zearalenol in the rumen and is detected in bovine bile following oral administration of ZEA (Kennedy *et al.*, 1998). The authors showed that α-zearalenol concentration always exceeds zeranol concentration in urine at least 5-fold. This result may be used to differentiate a zeranol treatment as an oestrogenic growth promoter and a natural contamination through zearalenone. ZEA, zearalenol and oestradiol compete together for the same cytosolic receptor site (Kiang *et al.*, 1978). A hydroxysteroid dehydrogenase, which is usually involved in the metabolism of steroids, is also used for the conversion of ZEA to zearalenol in the liver (Kiessling and Pettersson, 1978; Olsen *et al.*, 1981). This enzyme remains available for liver steroid metabolism in ruminants when the conversion to zearalenol has already been performed in the rumen. Therefore, in the ruminant ZEA may be converted to zearalenol in the rumen and liver.

Zeralenol had no toxic effect on the rumen bacterium *Butyrivibrio fibrisolvens*, which is one of the predominant rumen bacteria involved in the degradation of mycotoxins (Westlake *et al.*, 1987a).

A mixture of deoxynivalenol and ZEA fed to rumen-canulated wethers at concentrations of approximately 4.6 and 0.34 ppb in a total mixed ration (TMR) decreased the *in sacco* rate of straw degradation, whereas no changes occurred when alfalfa was incubated in the rumen (Danicke, 2002). No effect on milk production was demonstrated after the administration of ZEA at concentrations from 385 to 1925 ppb in feed over seven weeks (Shreeve *et al.*, 1979). At autopsy, no residues of ZEA were detected in the organs or muscles, although concentrations of nearly 2 ppm were present in experimental feed.

ABSORPTION AND EXCRETION OF ZEARALENONE

Ruminal reduction of ZEA to zearalenol will increase polarity, affecting absorption and excretion of the toxin (Kiessling *et al.*, 1984). The polar metabolites are more water-soluble than the parent toxin, reducing absorption in the digestive tract and increasing the rate of excretion via the enterohepatic cycle, and then in urine. *In fine*, fewer toxins will be in contact with animal tissues and ruminants may thus be partly protected. The absorption of ZEA in the rat small intestine follows first-order kinetics, with an absorption rate constant k_a of 9.27 ± 0.69/h (Kuiper-Goodman *et al.*, 1987). No quantitative results are available on the rate of absorption of ZEA or ZEA metabolites in ruminants.

Oral daily doses of 50 to 165 mg to dairy cows gave no toxin or metabolites in milk or blood (Prelusky *et al.*, 1990). A daily ingestion of 544.5 mg of ZEA for 21 days by dairy cows leads to the presence of the parent toxin and α-zearalenol in milk, with a cumulative transfer of 0.06% of the oral dose. The rate of transfer is highly dependent on the dose administered, falling to 0.016% and 0.008% following the administration of single doses of 1.8 g and 6 g of ZEA, respectively. A trace amount of ß-zearalenol and two other compounds were detected in milk. The rate of carryover of ZEA and its metabolites into milk is therefore low, and presents no real risk to consumers of dairy products.

ANIMAL TOXICITY

The primary effects of ZEA are reproductive problems, which include decreases in embryo survival, oedema and hypertrophy of the genitalia in pre-pubertal females, decreases in the amounts of luteinising hormone (LH) and progesterone produced affecting the morphology of uterine tissues, feminisation of young males due to decreased testosterone production, and infertility. In ruminants, symptoms have included vaginitis, vaginal secretions, abortions, infertility and mammary gland enlargement of virgin heifers.

Reported field observations of ZEA-related reproductive problems in ruminants are not uncommon (Kallela and Ettala, 1984; Khamis *et al.*, 1986; Mirocha *et al.*, 1968; Mirocha *et al.*, 1974; Roine *et al.*, 1971). One field study (Coppock *et al.*, 1990) reported total reproductive failure with diets containing approximately 750 ppb ZEA and 500 ppb deoxynivalenol.

Weaver *et al.* (1986b) fed non-lactating dairy cows 250 mg/day ZEA and saw no negative effect except that corpora lutea were smaller in treated cows. In a similar study with virgin heifers receiving 31.25 to 500 mg/day ZEA, conception rates were depressed by about 25% although no other negative effects were observed (Weaver *et al.*, 1986a). New Zealand workers (Smith *et al.*, 1995; Sprosen and Towers, 1995; Towers *et al.*, 1995a; 1995b) have related urinary ZEA and ZEA metabolites (zearalenone, zearalanone, α- and ß-zearalenol and α- and ß-zearalanol), to reproductive disorders in sheep and dairy cattle. In sheep, ZEA metabolites were related to lower conception, reduced ovulation, and increased twinning rates. Dairy herds with low fertility were found to have higher levels of blood and urinary ZEA metabolites when consuming pastures containing higher levels of these metabolites. In addition, within herds, individual cows were examined by palpation and those that were determined to be cycling had lower concentrations of ZEA metabolites than did cows that were not cycling. Reproductive problems in dairy cattle were noted with ZEA metabolite concentrations of approximately 400 ppb in the pasture samples.

Zearalenone consumption by ruminants can result in false positives when testing for growth promotants containing the ZEA metabolite, zeranol. Zeranol is included in the 1988 EU ban of hormonal growth promotants. Zeranol can occur naturally in both bile (Kennedy *et al.*, 1995) and urine (Erasmuson *et al.*, 1994) of animals consuming ZEA without administration of zeranol. Analytical methods have been developed based on α-zearalenol:zeranol:taleranol ratios to differentiate between zeranol derived from ZEA consumption and zeranol administration (Kennedy *et al.*, 1998).

Trichothecenes

Trichothecenes are produced primarily by several common moulds belonging to the genus *Fusarium*. They are composed of a tetracyclic sesquiterpene skeleton containing a six-member oxane ring, a stable epoxide group in positions 12 and 13, and a 9,10 olefinic bond. The main trichothecenes encountered in ruminant feeds belong to group A (diacetoxyscirpenol or DAS, T-2 toxin) or group B (deoxynivalenol or DON, nivalenol or NIV).

RUMEN METABOLISM OF TRICHOTHECENES

The pathways of ruminal trichothecene bioconversion involve both de-esterases and de-epoxidases giving rise to several metabolites (Swanson *et al.*, 1987), as indicated in Figure 1. Rumen microbes can degrade the DON molecule by opening the epoxy to form a diene called 12,13-de-epoxydeoxynivalenol or DOM-1 (Coté *et al.*, 1986; King *et al.*, 1984). It is likely that a specific epoxide reductase may be involved in this transformation. Since the epoxide group on the trichothecene nucleus explains the cytotoxicity and vomiting activity (Sato and Ueno, 1977), its bioconversion to a diene group reduces the toxicity of DON in ruminants. Generally speaking, epoxides are reactive towards thiol groups, which is why epoxide reductases involved in the transformation of DON by rumen microorganisms are probably of thiol-type (Ueno and Matsumoto, 1975). Anaerobic incubation of NIV with cow rumen fluid produced de-epoxide NIV in the same way (Hedman and Pettersson, 1997). Such bioconversion can also occur in the large intestine of monogastrics (Galtier and Alvinerie, 1976). However, because most of the microbial activity occurs downstream from the major absorption sites in the intestine, any protective effect of microbial metabolism for monogastric animals is low (Jouany, 2001). Surprisingly, Nakamura *et al.* (1998) demonstrated that DOM-1 could be re-oxidised to the parent DON molecule after incubation with mouse faeces, whereas such oxidation is not mediated by the hepatic cytochrome P450 system.

When T-2 toxin is added to rumen fluid, the carbon-4 of T-2 toxin is deacetylated to HT-2 toxin, which is then further deacetylated at carbon-15 to T-2 triol (Yoshizawa *et al.*, 1981; Yoshizawa *et al.*, 1985). Isovaleryl de-esterification at carbon-8 will lead to the production of neosolaniol (Westlake *et al.*, 1987b). Munger *et al.* (1987) showed that under conditions of inhibited esterase activity, acetylation at carbon-3 can occur with bovine rumen fluid. Acetylation of T-2 or other trichothecenes in the rumen *in situ* may result in more lipophilic metabolites, for which the absorption rate and toxicity remain unknown.

Westlake *et al.* (1989) reported that toxin-degrading activity was higher in protozoa than in bacteria fractions, and was absent from cell-free preparations. Ninety percent of T-2 toxin and DAS degradation is achieved by protozoa (Hussein and Brasel, 2001). T-2 toxin induced an 83% decrease in protozoal protein during a 6 hr incubation period, whereas bacterial protein increased by 14% during the same period. This negative effect on eukaryotic cells is due to T-2 binding to ribosomes and the subsequent inhibition of protein synthesis (Mackie and White, 1990). Rumen bacteria known to have esterase activity, including *B. fibrisolvens*, *Selenomonas ruminantium* and *Anaerovibrio lipolytica*, are probably involved in deacetylation reactions in the rumen. The efficiency of detoxification can be questioned if bacteria are sensitive to T-2 toxin. Even at a concentration of 1 mg/mL, no effect on *B. fibrisolvens* growth or lag time was observed (Westlake *et al.*, 1987a), thus indicating that this predominant species of rumen bacteria is not disturbed by T-2 toxin or DON and may be involved in the protection of ruminants against trichothecenes. Two rumen bacteria, *Ruminococcus albus* and *Methanobrevibacter ruminantium* are important for rumen cellulolysis and metabolism of hydrogen during anaerobic fermentation, respectively. May *et al.* (2000) reported that the growth of both organisms was inhibited by fusaric acid at concentrations as low as 15 μg/mL (84 μM) but not by DON, even at 100 μg/mL (338 μM). Neither bacterium was able to adapt to fusaric acid. Inhibition of *R. albus* began prior to any growth, while *M. ruminantium* doubled twice before inhibition occurred.

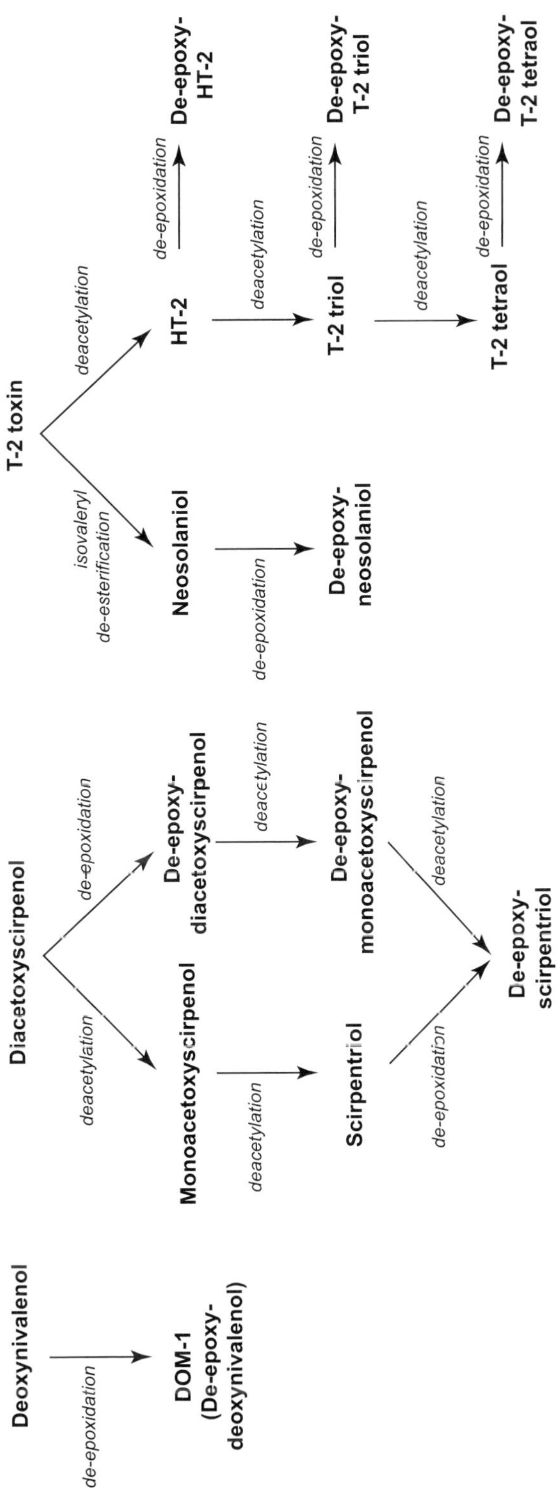

Figure 1. Possible pathways for ruminal bioconversion of trichothecenes (adapted from Mackie and White, 1990).

ABSORPTION AND EXCRETION OF TRICHOTHECENES IN RUMINANTS

Deoxynivalenol is one of the most polar trichothecenes, with a low molecular weight of 296.3. It contains one primary and two secondary hydroxyl groups, and is soluble in water. Such properties will influence its absorption and excretion rate. Preluski *et al.* (1984) showed that following a single oral dose of 920 mg DON to two cows, maximum blood levels were 200 and 90 ng/mL serum, and occurred 4.7 and 3.5 hrs post-administration, respectively. At 24 hrs post-administration, less than 2 ng/mL was detected. This indicates that approximately 30% of DON and its metabolites were absorbed during the 24 hr period. Free and conjugated DON accounted for 24 and 46% of total levels in serum, respectively, thus indicating that metabolites are probably absorbed in larger amounts than the parent molecule. Only extremely low amounts of DOM-1 (less than 4 ng/mL) were detected in the milk, and the authors concluded that DON given orally to dairy cows was not efficiently transferred to milk. After feeding three dairy cows contaminated corn to supply a concentration of 66 mg DON/kg total diet, Côté *et al.* (1986) observed that DON was not detected in milk. Approximately 20% of ingested DON was excreted in urine and faeces as unconjugated DOM-1 (96%) and DON (4%). Unconjugated DOM_1 and DON were not present in urine and faeces 72 hrs after the last oral exposure. Charmley *et al.* (1993) conducted an experiment on 18 primiparious cows, which were blocked into three groups provided with 0, 3 or 6 mg DON/kg total diet over a 10 week period. The toxin was supplied through contaminated wheat and contaminated ear corn. No detectable amounts of DON or DOM-1 were found in milk, indicating that absorption of the toxin and its metabolite in the digestive tract was low, and that transfer into milk was insignificant.

ANIMAL TOXICITY

Deoxynivalenol

Deoxynivalenol is commonly referred to as **vomitoxin**. In pigs DON has been associated with feed refusal, diarrhoea, emesis, reproductive failure, and death (Coté *et al.*, 1984; Vesonder *et al.*, 1978). In cattle, DON has been associated with reduced feed intake (Trenholm *et al.*, 1984) and lower milk production (Whitlow and Hagler, 1987). Clinical data from 300 herds representing about 40,000 cow records showed that DON was associated with a loss in milk production but did not establish a cause and effect relationship (Whitlow and Hagler, 1987). Field observations have led researchers to conclude that DON is a useful marker for problem feeds (Gotlieb, 1997; Seglar, 1997).

Charmley *et al.* (1993) demonstrated a 13% (2.85 kg) numerical decrease in 4% fat corrected milk production (statistics not available) utilising 18 mid-lactation dairy cows (average 19.5 kg milk) consuming diets shown to contain no common mycotoxins other than DON, which was fed at dietary concentrations of 0, 2.7 or 6.4 ppm. While the decrease in actual milk production (1.35 kg) was not statistically significant ($P = 0.24$), the decrease in fat test (3.92% vs. 3.04%) was significant ($P = 0.05$). Noller *et al.* (1979) utilised 54 lactating dairy cows in a 21 day feeding experiment using corn grain contaminated with *Gibberella zeae* (*Fusarium graminearum*) and containing approximately 500 ppb of ZEA. DON was present in corn harvested from the same field at 12 to 13 ppm, but the corn used in this study was not directly analysed. Neither dry matter intake nor milk production (average 22.9 kg) was affected by addition of this grain to the diet. However, compared with controls, cows that received this grain at either 10% (about 1.25 ppm DON and 50 ppb ZEA) or 20% (about 2.50 ppm DON and 100 ppb ZEA) of the diet gained significantly less weight during the study (5.8 kg or 8.1 kg less for cows consuming the 10% or 20% diets over 21 days, respectively). Ingalls (1994) fed lactating dairy cows 0, 2.1, 6.3 or 8.5 ppm DON for 21 days, apparently without effect on feed intake (3.7 to 3.8% of BW) or milk production, which averaged about 34 kg daily.

Beef cattle and sheep appear to tolerate relatively large amounts of DON without obvious effects. Reports from Nebraska indicated similar feed intakes, average daily gains and feed efficiencies when sheep (8.5 ppm dietary DON) or cattle (1 ppm dietary DON), consuming contaminated diets were compared with those consuming diets containing no detectable DON (DeHaan *et al.*, 1984). DiCostanzo *et al.* (1996) indicated that feeding up to 18 ppm dietary DON did not affect intake, daily gain, feed efficiency or carcass characteristics of 415 kg steers fed for 166 days. Other recent feeding experiments suggest that beef cattle can tolerate large concentrations of DON in a feedlot situation without effects on dry matter intake, average daily gain, or feed efficiency (Boland *et al.*, 1994; Windels *et al.*, 1995).

Data suggest that beef cattle and sheep are more tolerant to DON than are dairy cattle. This difference may be related to the higher level of stress to which dairy cattle are exposed. For dairy cattle, early lactation is very stressful during which time they are immune suppressed (Kehrli *et al.*, 1989a, b). The trichothecenes are immunosuppressive and may be more toxic in animals with low levels of immunity. Additionally, the high levels of dry matter intake in dairy cattle in comparison to beef and sheep contribute to a faster rumen turnover rate, less exposure time in the rumen, and perhaps less rumen microbial degradation of mycotoxins.

T-2 toxin

T-2 toxin, a *Fusarium* mycotoxin, has been associated with gastroenteritis, intestinal haemorrhages (Mirocha *et al.*, 1976; Petrie *et al.*, 1977) and death (Hsu *et al.*, 1972; Kosuri *et al.*, 1970). Weaver *et al.* (1980) demonstrated that T-2 toxin was associated with feed refusal and gastrointestinal lesions in a cow, but failed to show a haemorrhagic syndrome. Serum immunoglobulins and certain complement proteins were lowered in calves receiving T-2 toxin (Mann *et al.*, 1984). Other data demonstrated a reduction in white blood cell and neutrophil counts in calves (Gentry *et al.*, 1984). A calf intubated with T-2 developed hindquarter ataxia, knuckling of the rear feet, listlessness and anorexia (Weaver *et al.*, 1980).

While data with cattle are limited, the toxicity of T-2 toxin in laboratory animals is well-documented (Wannemacher *et al.*, 1991). Field observations suggest that T-2 is a severe gastrointestinal irritant, which can cause haemorrhage and necrosis of the intestinal tract (Gotlieb, 1997). Diarrhoea is usually present but may not be haemorrhagic. With high levels of T-2, there can be congestion and irritation to the liver, lungs and heart (Kosuri *et al.*, 1970).

Fumonisins

Fumonisins are a group of mycotoxins produced by the moulds *Fusarium verticillioides* (formerly *F. moniliforme*) and *F. proliferatum*. Fumonisin B_1 (FB_1) is the most prevalent in nature. After their initial discovery in 1988 (Gelderblom *et al.*, 1988), fumonisins have been described as occurring worldwide in feedstuffs and foodstuffs (Shephard *et al.*, 1996). Marasas (1996) conducted a literature review of studies reporting natural occurrence of fumonisins showing that over 25 countries around the world published reports of natural occurrence of fumonisin in both feedstuffs and foodstuffs.

RUMEN METABOLISM OF FUMONISIN B_1

Little is known of the metabolic fate of FB_1

following oral ingestion in ruminants. FB_1 is apparently poorly metabolised in the rumen, since only 10-18% of the initial amount added to an *in vitro* rumen system disappeared after a 72 hr period of incubation (Caloni *et al.*, 2000; Gurung *et al.*, 1999). No hydrolysed metabolites such as aminopolyol or aminopentol were detected at the end of the *in vitro* incubation (Caloni *et al.*, 2000). Thus, rumen metabolites produced from the degraded fraction of FB_1 remain unknown at the present time. Although the rumen microbes have only limited action against FB_1, ruminants are considered to be more tolerant to this toxin than monogastrics.

Rumen digestion seems to be unaffected by FB_1. *In vitro* dry matter digestibility of feeds was not significantly changed by FB_1 contamination, indicating that microbial efficiency was unaffected by the toxin at dietary concentrations up to 100 mg/kg (Gurung *et al.*, 1999) Caloni *et al.* (2000) observed no significant effect of FB_1 at 1 μg/mL in rumen fluid on the pattern of fermentation estimated *in vitro*. Qualitative and quantitative VFA productions were not modified by FB_1.

ABSORPTION AND EXCRETION OF FUMONISIN B_1

No FB_1 or known metabolites were detected in plasma after dosing cows orally with 1 or 5 mg/kg BW (Prelusky *et al.*, 1995). Following single intravenous injection of 0.05 or 0.20 mg FB_1/kg BW, the same authors observed that the toxin was rapidly cleared from the blood, falling below detectable levels within 120 min of dosing. No known metabolites were detected in the plasma. The same pharmacokinetic parameters were observed in animals given the low and the high doses of FB_1, which suggests that FB_1 distribution and blood elimination are not dose dependent at the toxin concentrations tested. Systemic absorption based on plasma levels and accumulation of ^{14}C-labelled compounds in tissues has been estimated to be less than 1% of the initial dose (Prelusky *et al.*, 1996).

A study conducted with beef cattle fed corn grain contaminated with a mixture of FB_1 and FB_2 obtained from *F. verticillioides* culture material indicated that most of the fumonisins were found unmetabolised in the faeces, while only trace quantities were detected in blood and urine (Smith and Thakur, 1996). Dairy cows dosed with pure FB_1 either orally (1.0 to 5.0 mg FB_1/kg BW) or intravenously (0.05 and 0.20 mg FB_1/kg BW) had no detectable residues of the parent toxin, the aminopentol hydrolysis product of FB_1 or their conjugates in milk (Prelusky *et al.*, 1996; Scott *et al.*, 1994).

Several researchers have concluded that the carryover of fumonisins in milk is extremely low, approximately 0.05% (Richard *et al.*, 1996; Scott *et al.*, 1994). In consequence, there seems to be no risk of fumonisin contamination in milk and milk products.

ANIMAL TOXICITY

Fumonisins have been shown to cause leukoencephalomalacia in horses (Marasas *et al.*, 1988; Ross *et al.*, 1993), pulmonary oedema in swine (Colvin and Harrison, 1992; Osweiler *et al.*, 1992) and liver cancer in rats (Gelderblom *et al.*, 1988). Fumonisins may also pose a potential risk to humans as they were correlated with the prevalence of human oesophageal cancer in southern Africa (Rheeder *et al.*, 1992; Sydenham *et al.*, 1990). Little information is currently available on the effects of fumonisins on ruminants. Kriek *et al.* (1981) concluded that the target organs for fumonisin toxicity in sheep were the liver and the kidneys. A study by Edrington *et al.* (1995) suggested that daily doses of 11.1 mg FB/kg BW were toxic to sheep weighing 32 kg and consuming approximately 750 g/d of feed (470 ppm). The lambs in this study had increased AST and GGT activities. Osweiler *et* al. (1993) also reported increases in AST and GGT activities in feeder calves fed 148 ppm fumonisin for a total of 31 days (short term study). Although Osweiler et al. (1993) did not observe differences in dry matter intake or body weight gain, a negative trend was apparent after 31 days of exposure to the fumonisin-contaminated feed. More recently Diaz et al. (2000) fed dairy cattle (Holsteins and Jerseys) a diet naturally contaminated with 100 ppm fumonisin for approximately 7 days prior to parturition through 70 days postpartum. These animals had lower milk production (6 kg/cow/day), which was closely correlated with reduced feed consumption (5.2 kg/cow/day). Additionally, cows consuming the contaminated diets had elevated serum AST and GGT activities. Dairy cattle may be more susceptible to fumonisin toxicity than beef cattle and sheep because of the intense demands of milk production, especially in early lactation.

Ochratoxin A

Ochratoxin A (OTA) is a phenylalanine-dihydroisocoumarin produced primarily by *Penicillium verrucosum* and *Aspergillus ochraceus*. Ochratoxins occur in a wide variety of feedstuffs/foodstuffs including cereals, beans, groundnuts, dried fruits, wine, beer and coffee. Ochratoxin A has been reported to affect cattle (Vough and Glick, 1993), but it is rapidly degraded in the rumen and thus thought to be of lesser consequence (Sreemannarayana et al., 1988).

RUMEN METABOLISM OF OCHRATOXIN A

The ochratoxin A molecule associates a methyl-isocoumarin unit (ochratoxin-α) with L-ß-phenylalanine via a peptide bond, which can be easily hydrolysed by proteases produced by rumen microbes. Thus, OTA in the rumen is converted into phenylalanine and OTA-α, the latter being less toxic but retaining some genotoxicity (Föllmann et al., 1995). It may also be esterified to produce ochratoxin C of similar toxicity (Galtier and Alvinerie, 1976). The average amount of OTA-α formed in the rumen relative to the disappearance of OTA can be 100%, regardless of diet (Müller et al., 2001). According to Hult et al. (1976), as much as 60% of dietary OTA was degraded as OTA-α during *in vitro* incubation with rumen contents. Protozoa are considered the main agents involved in rumen OTA degradation (Kiessling et al., 1984) to OTA-α. This could explain why ruminants are much less sensitive to the acute toxicity of OTA (Ribelin et al., 1978).

Müller (1995) estimated from *in vitro* data that an adult cow is able to degrade 33 to 72 mg of OTA per day, which corresponds to 3 to 7 mg/d for sheep. However, sheep have been shown to be sensitive to contaminated concentrate feeds when dosed at 20 mg OTA/kg BW corresponding to an intake in the range of 6-14 mg/d. According to Höhler et al. (1999), differences in OTA detoxification in the rumen could be related to the accessibility of microbes to the toxin. The use of a pure crystallised form of OTA may have enhanced microbial access and thus breakdown, whereas OTA fed as part of the naturally contaminated feed may be partially protected in the matrix of feed. Analysis of OTA in a naturally contaminated feed matrix or in rumen fluid may be biased due to extraction, recovery and quantification. The type of diet, which influences rumen microbial composition and activity, may affect the extent of OTA degradation. According to Xiao et al. (1991a; 1991b), metabolism of OTA in the rumen is much lower in grain-fed than in hay-fed ruminants. The respective ruminal half-lives were 2.7 or 0.63 hrs for OTA and 1.9 or 0.9 hrs for OTA-α when animals were fed a grain or hay diet, respectively. The authors calculated that OTA bioavailability was 4.3 times greater in sheep fed grain than in sheep fed hay. Feeding the complete grain diet resulted in a severe pH drop (pH = 5.6), which led to a significant decrease in the number of protozoa and to a lower OTA degradation rate. Conversely, Özpinar *et al.* (2002) reported that adding a reasonable amount of starch decreased the OTA half-life to 1.9 hrs compared with 3.7 hrs with the control (lower starch) diet, and that adding cellulose increased OTA half-life to 4.5 hrs. They confirmed that protozoa are more active in metabolizing OTA than bacteria. Müller *et al.* (1998; 2001) also showed that increasing the amount of concentrate in the diet accelerated OTA degradation. They observed that hydrolysis of OTA occurred in the rumen via first order kinetics, and that OTA-α was not subsequently degraded.

The capacity of rumen fluid to degrade OTA follows the kinetics of protozoa concentration in the rumen. Ruminal degradation of OTA decreased for the first 1.5 hrs after feeding, thereafter increased to reach a peak 5 hrs after exposure (Kiessling *et al.*, 1984). *In vivo*, when pure OTA was added to the diet at 2 ppm and fed to a fistulated sheep, no OTA was found in the rumen fluid after feeding. However, with 5 ppm dietary OTA, rumen fluid concentrations were 29 and 14 ng/ml at 0.5 and 1 hr post-feeding, respectively. OTA-α was detected in all rumen fluid samples. Neither OTA nor OTA-α was found in peripheral blood samples. Höhler *et al.* (1999) adapted three groups of four sheep to mixed diets containing 70% concentrates and 30% hay and one of three concentrations of OTA (0, 2, 5 ppm). OTA was produced by a culture of *Aspergillus ochraceus* grown on wheat, which was included in the diet. The authors demonstrated that OTA hydrolysis in the digestive tract of sheep is lowered when animals are fed concentrate-rich diets.

The rumen and its microbial population can therefore be considered efficient in protecting ruminants against the toxic effects of OTA. The role of rumen microbes becomes quantitatively apparent when comparing the Lowest Observed Adverse Effect Level (LOAEL) of OTA given orally (13 ppm) to cattle with that of 1 ppm given intravenously (Cheeke, 1998). However, attention must be paid to a possible accumulation in tissues of the unexcreted fraction of the parent toxin, leading to harmful effects on ruminants in case of chronic feeding of OTA. Willis *et al.* (1980) pointed out that in general, long-term feeding of a subacute dose will be more critical than a single dose for mycotoxin poisoning.

The toxicity of OTA to rumen microbes has been investigated only for *Butyrivibrio fibrisolvens* CE51 (Westlake *et al.*, 1987a), as this bacterial species is considered to be involved in the rumen mycotoxin detoxification process (Mackie and White, 1990). The results indicated that the specific growth of *B. fibrisolvens* was not affected by OTA. Since no clear effect of OTA was observed on digestibility in sheep when present in the diet at 5 ppm (Höhler *et al.*, 1999), the toxin appears to be harmless to rumen microbes and digestion.

ABSORPTION AND EXCRETION OF OCHRATOXIN A AND OCHRATOXIN-α IN RUMINANTS

According to Marquardt and Frohlisch (1992), OTA is absorbed passively in its non-ionised form, which occurs at acidic pH values but not at neutral values, the pK value for dissociation of the phenolic hydroxy group in the dihydroisocoumarin ring being 7.04. Starch-containing diets therefore promote OTA absorption more than do high fibre diets. These observations also suggest that the forestomachs and the abomasum are the main sites of OTA absorption in the digestive tract.

Höhler *et al.* (1999) indicated that feeding sheep a mixed diet contaminated with 3 concentrations of OTA (0, 2, 5 ppm), gave significant concentrations of OTA and OTA-α in blood serum (10 and 3 ng/mL with 2 ppm, and 80 and 15 ng/mL with 5 ppm). According to the authors, ruminal and intestinal microbes degrade substantially less OTA when a concentrate-rich diet is utilised. Total excretion of OTA-α accounted for 70% of

OTA intake (9% in faeces, 61% in urine), whereas only 1% and 3.8% of OTA intake were found undegraded in faeces and urine, respectively. Blank *et al.* (2003) fed sheep various doses of OTA (0, 9.5, 19, 28.5 μg OTA/kg BW/day) added in the form of contaminated wheat to diets consisting of 70% concentrate and 30% grass silage. They detected significant concentrations of undegraded OTA in the serum of sheep at all levels of dietary OTA (from 1.5 through to 18 μg/kg BW/day). Serum concentrations of OTA accumulated with time of exposure and were linearly related to the administered dose of OTA. The proportion excreted in urine remained constant (6-8% of the ingested dose). Besides OTA, small amounts of OTA-α were detected in the serum (from 0.5 through to 1.6 μg/kg BW/day), increasing linearly with increasing doses. Chronic feeding of OTA would therefore raise the serum concentration of OTA and OTA-α. The authors concluded that even at moderate levels of OTA intake, quite significant amounts of mycotoxin are absorbed by ruminants, and may accumulate in tissues.

Ochratoxin A can persist in the blood of animals and humans for a long time due to its ability to bind plasma proteins and to be reabsorbed in the nephron (Dahlmann *et al.*, 1998). However, accumulation of OTA in the blood of ruminants is generally lower than in other animals. This could be explained by rumen degradation of OTA, and also by a lower reabsorption in nephron segments due to a higher urinary pH in ruminants, since reabsorption depends on luminal pH (Blank *et al.*, 2003). Such mechanisms can be highlighted to account for the short elimination half-life of 17.3 hrs in ruminants compared to 510 hrs in the rhesus monkey, and around 100 hrs in pigs and rats (Kuiper-Goodman and Scott, 1986; Xiao *et al.*, 1991a).

A survey conducted in Great Britain on 50 farm milk samples and 50 commercial milk samples coming from organic or traditional farms revealed no detectable OTA (FSA, 2001). The same result was reported by Shreeve *et al.* (1979) on a limited number of animals treated with controlled doses of OTA. On the contrary, Skaug (1999) detected OTA (>10 ng/L) in 6/40 samples from traditional farms and 5/47 from organic farms. Maximal concentrations were 28 and 58 ng/L in organic and traditional farms, respectively. The author concluded that daily intake in high-milk consumers, such as children, can exceed the allowed daily dose (5 ng/kg of live weight).

Such inconsistency between reports could be explained by differing diet composition. Microbial detoxification in the rumen is likely lowered when the protozoa population is reduced, for example in high starch diets. Genetics in high producing dairy cows could also be a factor in the high rate of mycotoxin transfer in milk.

ANIMAL TOXICITY

The main clinical patterns associated with acute ochratoxicosis in rats, pigs and poultry are anorexia, weight loss, vomiting, increased body temperature, conjunctivitis, polyuria, bloody mucus from the rectum, dehydration, prostration and death (Marquardt and Frohlich, 1992). In chronic toxicosis, OTA elicits polydipsia and hepatotoxicity (Yiannikouris and Jouany, 2002). It has teratogenic effects on embryos in the mouse and rat, but not in pigs and chicks (Juskiewicz *et al.*, 1982). Its immunotoxicity has been described in poultry (Dwivedi and Burns, 1985) and pigs (Szczech *et al.*, 1973) as well as incidence of renal adenomas and carcinomas in rats (Marquardt and Frohlich, 1992). Carlton and Krogh (1979) summarised two experiments with acute OTA toxicosis. In the first experiment, lambs dosed with 1 mg OTA/kg died less than 24 hrs after administration. Upon evaluation, animals had pulmonary oedema and friable livers. The foetuses of these lambs also showed marked lesions. In the second experiment, goats were dosed daily with 1, 2, or 3 mg OTA/kg. Goats on the high dose died after 6 days of exposure while those receiving either 1 or 2 mg OTA/

kg showed no obvious sign of toxicity for the duration of the experiment (14 days). Due to effective detoxification of OTA by rumen microbes, ruminants are more resistant than non-ruminant species. However, modern dairy operations include large quantities of concentrates in diets, which in turn affects the rumen environment, modifies the rumen microbial population, and perhaps alters OTA degradation in the rumen. The literature suggests that within a normal range, additional dietary grain can enhance OTA degradation in the rumen, if the protozoa population is not adversely affected. If grain is overfed and causes a low rumen pH, the protozoal population may be decreased and result in decreased OTA degradation.

Patulin

Patulin is a cyclic γ-lactone produced by *Penicillium*, *Aspergillus*, and *Byssochlamys*. Patulin can be commonly found in silages (Escoula, 1974) and has been incriminated as a possible ruminant toxin in Europe and New Zealand (Lacey, 1991).

RUMEN METABOLISM OF PATULIN

Patulin has been utilised as an antibiotic for medical or veterinarian therapy (Ciegler, 1977; Madhyastha *et al.*, 1994; Scott, 1974), but its toxicity toward living organisms limited application. It inhibits the growth of both Gram(+) and Gram(-) bacteria, and the growth of some protozoa. *Bacillus brevis* is very sensitive to the antibiotic action of patulin (Wooters and Speijers, 2003). Patulin is also cytotoxic against free protozoa such as *Tetrahymena pyriformis* (LD_{50} = 0.32 μg/mL) (Madhyastha *et al.*, 1994).

Several *in vitro* experiments have been conducted to test the antimicrobial effect of patulin on digestion and fermentation in the rumen. Escoula (1992) showed that patulin had a negative dose-dependent effect on VFA production, more specifically on acetate production and microbial protein synthesis, when concentrations increased from 20 to 300 μg/mL. Tapia *et al.* (2002) measured the multiple effects of patulin on digestion and fermentation in a continuous artificial rumen. They showed that digestion of a mixed diet made of lucerne hay (38%), corn silage (28%), corn grain (27%) and soyabean meal (5%), was lowered from 27 to 43% when the dietary concentration of patulin increased from 30 to 90 ppm. The decrease in total VFA production was significant ($P<0.05$) only at 90 ppm and, among the VFAs, only acetate production was altered at this dose. Such a result indicates that patulin had a negative effect on digestion of the cellulosic fraction. The significant decrease in branched VFAs such as isobutyrate and isovalerate would show a reduction of protein degradation at the highest doses (60 and 90 ppm). The outflow of microbial protein was lowered by patulin independently of the dose, whereas its negative effect on the efficiency of microbial protein synthesis (g microbial N/kg fermented OM) was reached only at the highest dose of 90 ppm.

Using batch *in vitro* microsystems to simulate rumen function, Morgavi *et al.* (2003) found that patulin significantly altered lucerne hay degradation at a dosage of 25 μg/mL. The negative effect increased until the dosage of 50 μg/mL was reached, then stabilised until 100 μg/mL. The authors indicated that VFA and gas production decreased as the dose of patulin increased. They confirmed that, among VFAs, only acetate was reduced by patulin addition, and concluded that cellulolysis was partly inhibited by the toxin.

Patulin is an electrophilic molecule, which exerts its deleterious action by attaching through covalent bonds to the thiol groups of proteins and glutathione (Ciegler, 1977; Riley and Showker, 1991). The affinity to thiol compounds can be used to detoxify patulin (Lindroth and von Wright, 1990). Other reducing agents such as ascorbic acid are able to detoxify patulin in contaminated waste or in apple juice (Brackett and Marth, 1979; Fremy *et al.*, 1995). Morgavi *et al.* (2003) compared the detoxifying activity of several reducing agents in an *in vitro* rumen simulating system. They showed that only compounds with a thiol group such as cysteine or glutathione, can detoxify patulin in the rumen. Glutathione dosed at 4 mM totally eliminated the toxic effect of patulin at the concentration of 50 μg/mL. The same dose of cysteine was less efficient since rumen fermentation remained partly inhibited by patulin. Cystine, which is a dimeric form of cysteine where the SH-groups are not free, had no protective effect against patulin. The same study showed that other reducing agents such as ascorbic or ferulic acids, which are present in ruminant feeds, did not protect the rumen microbial ecosystem against patulin. In conclusion, Morgavi *et al.* (2003) recommended the addition of cysteine, which is a non-essential amino acid, to the ruminant diet to protect animals against the toxic effect of patulin.

ABSORPTION AND EXCRETION OF PATULIN

Patulin is a small water-soluble polar molecule, which is quickly absorbed and metabolised. It will therefore be mainly excreted in urine. These physico-chemical characteristics indicate that it would not be stored in hydrophobic cellular structures and, therefore, its excretion in milk would be low. Its strong tendency to bind with thiol compounds can however change its behaviour in animal (or human) cells and increase its half-life. To our knowledge, patulin or its metabolites have never been detected in milk. Moreover, effects of patulin on performance of dairy or beef cattle have never been clearly defined. Only acute toxicoses cause severe reduction in feed intake, milk production, and then death of the animal.

ANIMAL TOXICITY

Serious injuries with nervous symptoms have been observed in ruminants fed patulin-contaminated diets (Moreau and Moreau, 1960). The death of more than 100 cows receiving malted barley or silage has been ascribed to patulin (Ciegler, 1977; Ukai *et al.*, 1954). Haemorrhagic syndromes were noted in cattle fed poor quality silages contaminated with patulin (Syrett, 1979). The presence of patulin-producing moulds like *Byssochlamys nivea* in silages is quite common (Escoula, 1977). Numerous field reports of outbreaks of disease in both cattle and sheep related to feedstuffs contaminated with *Aspergillus clavatus* (Gilmour *et al.*, 1989; Kellerman *et al.*, 1984; Kellerman *et al.*, 1976). Although *A. clavatus* can produce patulin, the etiology of the disease was not determined in these reports. More recently, Sabater-Vilar *et al.* (2004) implicated patulin from *A. clavatus* as the causative agent of several field outbreaks in Belgium. Animals consuming the contaminated feedstuffs had severe neurotoxicosis consisting of tremors, ataxia, paresis, recumbency and in some animals death resulted.

Conclusion

Mycotoxin producing moulds are ubiquitous in nature and are commonly in contact with forages and cereals in the field before harvest, during harvesting, drying and transport and during storage. Ruminants, because they are fed complex diets that generally include both forages and concentrate, may have an increased probability of multiple mycotoxin contamination. However, they may also have a decreased risk of high mycotoxin concentrations due to the dilution of these ingredients in the final diet. Although a large volume of information is available on the effects of mycotoxins on ruminants, more research is required to better understand mycotoxin occurrence in ruminant feeds and to better comprehend the subclinical effects caused by these toxins.

References

Applebaum, R.S. and E.H. Marth. 1982. Use of sulphite or bentonite to eliminate aflatoxin M1 from naturally contaminated whole milk. Z. Lebensm. Unters. Frosch. 174:303-305.

Applebaum, R.S., R.E. Brackett, D.W. Wiseman and E.H. Marth. 1982. Responses of dairy cows to dietary aflatoxin: Feed intake and yield, toxin content and quality of milk of cows treated with pure and impure aflatoxin. J. Dairy Sci. 65:1503-1508.

Auerbach, H., R.F.M. Maas, H.J.M. Op Den Camp, A. Pol and J. Fink-Gremmels. 1998. Biodegradation of aflatoxin B_1 by bovine rumen microorganisms *in vitro* and its effects on rumen fermentation. Rev. Méd. Vét. 149:573.

Blank, R., J.P. Rolfs, K.H. Sudekum, A.A. Frohlich, R.R. Marquardt and S. Wolffram. 2003. Effects of chronic ingestion of ochratoxin A on blood levels and excretion of the mycotoxin in sheep. J. Agric. Food Chem. 51:6899-6905.

Boland, E.W., V.L. Anderson, H.H. Casper, P.T. Berg and D.V. Dhuyvetter. 1994. The effects of increasing vomitoxin (DON) levels from scab infested barley fed to feedlot steers. In: *Beef Production Field Day*, North Dakota State University, Ag. Expt. Sta., pp. 34.

Brackett, R.E. and E.H. Marth. 1979. Patulin in apple juice from roadside stands in Wisconsin. J. Food Prot. 42:862-863.

Caloni, F., M. Spotti, H. Auerbach, H.O. Camp, J.F. Gremmels and G. Pompa. 2000. *In vitro* metabolism of fumonisin B_1 by ruminal microflora. Vet. Res. Com. 24:379-387.

Carlton, W.W. and P. Krogh. 1979. Ochratoxins. In: *Mycotoxins in Animal Feeds and Grains Related to Animal Health*, Rockville, MD, USA, pp. 165-287.

CAST. 2003. Mycotoxins: Risk in plant, animal and human systems. Council for Agricultural Science and Technology, Ames, Iowa.

Charmley, E., H.L. Trenholm, B.K. Thompson, D. Vudathala, J.W.G. Nicholson, D.B. Prelusky and L.L. Charmley. 1993. Influence of level of deoxynivalenol in the diet of dairy cows on feed intake, milk production and its composition. J. Dairy Sci. 76:3580-3587.

Cheeke, P.R. 1998. Mycotoxins in cereal grains and supplements. In: *Natural Toxicants in Feeds, Forages and Poisonous Plants*. Interstate Publishing Inc., Danville, IL, USA, pp. 87-136.

Cheeke, P.R. and R.T. Palo. 1995. Plant toxins and mammalian herbivores: Co-evolutionary relationships and antinutritional effects. In: *Recent Developments in the Nutrition of Herbivores* (E. Journet, E. Grenet, M-H. Farce, M. Theriez and C. Demarquilly, eds). INRA Editions, Versailles, France, pp. 87-136.

Ciegler, A. 1977. Patulin. In: *Mycotoxins in Human and Animal Health* (J.V. Rodricks, C.W. Hesseltine and M.A. Mehlman, eds). Pathotox. Publ., Park Forest South, IL, USA, pp. 609-624.

Colvin, B.M. and L.R. Harrison. 1992. Fumonisin-induced pulmonary edema and hydrothorax in swine. Mycopathologia 117:79-82.

Coppock, R.W., M.S. Mostrom, C.G. Sparling, B. Jacobsen and S.C. Ross. 1990. Apparent zearalenone intoxication in a dairy herd from feeding spoiled acid-treated corn. Vet. Hum. Toxicol. 32:246-248.

Coté, L-M., A.M. Dahlem, T. Yoshizawa, S.P. Swanson and W.B. Buck. 1986. Excretion of deoxynivalenol and its metabolite in milk, urine and feces of lactating dairy cows. J. Dairy Sci. 69:2416-2423.

Coté, L-M., J. Nicoletti, S.P. Swanson and W.B. Buck. 1986. Production of deepoxy-deoxynivalenol (DOM-l), a metabolite of deoxynivalenol, by *in vitro* rumen incubation. J. Agric. Food Chem. 34:458-460.

Coté, L.M., J.D. Reynolds, R.F. Vesonder, W.B.

Buck, S.P. Swanson, R.T. Coffey and D.C. Brown. 1984. Survey of vomitoxin-contaminated feed grains in midwestern United States and associated health problems in swine. J.A.V.M.A. 184:189-192.

Coulombe, R.A. and R.P. Sharma. 1985. Clearance and excretion of intratracheally and orally administered aflatoxin B_1 in the rat. Food Chem. Toxicol 23:827-830.

Dahlmann, A., W.H. Dantzler, S. Silbernagl and M. Gelke. 1998. Detailed mapping of ochratoxin A reabsorption along the rat nephron *in vivo*: The nephrotoxin can be reabsorbed in all nephron segments by different mechanisms. JPET 286:157-162.

Dalezios, J.I., R.I. Keriger and D.P.H. Hseih. 1972. The *in vivo* metabolism of aflatoxin B_1 in rhesus monkey liver. Fed. Proc. 31:733.

Danicke, S. 2002. Effects of fusarium toxin contaminated wheat grain and of a detoxifying agent on rumen physiological parameters and *in sacco* dry matter degradation of wheat straw and lucerne hay in wethers. J. Anim. Feed Sci. 11:437-451.

DeHaan, K., R. Stock, D. Brink, T. Klopkenstein and N. Schneider. 1984. Scabby wheat influence on performance. MP-47, Nebraska Beef Cattle Report, Nebraska.

Diaz, D.E., W.M. Hagler Jr., J.T. Blackwelder, J.E. Eve, B.A. Hopkins, K.L. Anderson, F.T. Jones and L.W. Whitlow. 2004. Aflatoxin binders II: Reduction of aflatoxin M_1 in milk by sequestering agents of cows consuming aflatoxin in feed. Mycopathologia 157:233-241.

Diaz, D.E., B.A. Hopkins, L.M. Leonard, W.M. Hagler, Jr. and L.W. Whitlow. 2000. Effect of fumonisin on lactating dairy cattle. J. Dairy Sci. 83(abstr.):1171.

DiCostanzo, A., L. Johnston, H. Windels and M. Murphy. 1996. A review of the effects of molds and mycotoxins in ruminants. Prof. Anim. Scientist 12:138-150.

Dwivedi, P. and R.B. Burns. 1985. Immunosuppressive effects of ochratoxin A in young turkeys. Avian Pathol. 14:213-218.

Edrington, T.S., R.B. Harvey and L.F. Kubena. 1994. Effect of aflatoxin in growing lambs fed ruminally degradable or escape protein sources. J. Anim. Sci. 72:1274-1281.

Edrington, T.S., C.A. Kamps Holtzapple, R.B. Harvey, L.F. Kubena, M.H. Elissalde and G.E. Rottinghaus. 1995. Acute hepatic and renal toxicity in lambs dosed with fumonisin-containing culture material. J. Anim. Sci. 73:508-515.

EFSA. 2004. Opinion of the scientific panel on contaminants in the food chain on a request (request # EFSA-q-2003-035) from commission related to aflatoxin B_1 as undesirable substance in animal feed. The EFSA Journal 39:1-27.

Engel, V.G. and H. Hagemeister. 1978. Untersuchungen über den verbleib von aflatoxin B_1 im verdauungstrakt von kühen. Milchwissenenschaft 33:21-23.

Erasmuson, A.F., B.G. Scahill and D.M. West. 1994. Natural zeranol (α-zearalanol) in the urine of pasture-fed animals. J. Agric. Food Chem. 42:2721-2725.

Escoula, L. 1974. Moisissures toxinogenes des fourrages ensiles. I. Presence de patuline dans le fronts de coupe d´ensilages. Ann. Rech. Vet. 5:423-432.

Escoula, L. 1977. Moisissures des ensilages et conséquences toxicologiques. Fourrages 69:97-114.

Escoula, L. 1992. Patulin production by *Penicillium granulatum* and inhibition of rumen flora. J. Environ. Pathol. Toxicol. Oncol. 11:45-48.

FAO. 2004. Worldwide regulations for mycotoxins in food and feed in 2003. 81, FAO, Rome, Italy.

Fernandez, A., M. Fernandez, M.C. Sanz, M.T. Verde and J.J. Ramos. 1997. Serological serum protein fraction and responses to *Brucella melitensis* in lambs fed aflatoxins. Vet. Hum. Toxicol. 39:137-140.

Fernandez, A., M. Fernandez, M.T. Verde and M.C. Sanz. 2000. Effect of aflatoxin on performance, hematology and clinical immunology in lambs. Can. J. Vet. Res. 64:53-58.

Fernandez, A., J.J. Ramos, M.C. Sanz and D.

Fernandez de Luco. 1996. Alterations in the performance, haematology and clinical biochemistry of growing lambs fed with aflatoxin in the diet. J. Appl. Toxicol. 16:85-91.

Föllmann, W., I.E. Hillebrand, E.E. Creppy and H.M. Bolt. 1995. Sister chromatid exchange frequency in culture isolated porcine urinary bladder epithelial cells (PUBEC) treated with ochratoxin A and α. Arch. Toxicol. 69:280-286.

Fremy, J.M., M.J.J. Castegnaro, E. Gleizes, M. De Meo and M. Laget. 1995. Procedures for destruction of patulin in laboratory wastes. Food Addit. Contamin. 12:331-336.

Frobish, R.A., B.D. Bradley, D.D. Wagner, P.E. Long-Bradley and H. Hairston. 1986. Aflatoxin residues in milk of dairy cows after ingestion of naturally contaminated grain. J. Food Prot. 49:781-785.

FSA. 2001. Survey of milk for mycotoxins. 17/01, Food Standards Agency.

Galtier, P. and M. Alvinerie. 1976. *In vitro* transformation of ochratoxin A by animal microbial floras. Ann. Rech. Vet. 7:91-98.

Garrett, W.N., H. Heitman and A.N. Booth. 1968. Aflatoxin toxicity in beef cattle. Proc. Soc. Exp. Biol. Med. 127:188.

Gelderblom, W.C., K. Jaskiewicz, W.F. Marasas, P.G. Thiel, R.M. Horak, R. Vleggaar and N.P. Kriek. 1988. Fumonisins - novel mycotoxins with cancer-promoting activity produced by *Fusarium moniliforme*. Appl. Environ. Microbiol. 54:1806-1811.

Gentry, P.A., M.L. Ross and P.K-C. Chan. 1984. Effect of T-2 toxin on bovine hematological and serum enzyme parameters. Vet. Hum. Toxicol. 26:24-28.

Gilmour, J.S., D.M. Inglis, J. Robb and M. Maclean. 1989. A fodder mycotoxicosis of ruminants caused by contamination of a distillery by-product with *Aspergillus clavatus*. Vet. Rec. 124:133-135.

Gotlieb, A. 1997. Causes of mycotoxins in silages. In: *Silage: Field to feedbunk*. Northeast Regional Agricultural Engineering Service, New York, pp. 213.

Gurung, N.K., D.L. Rankins, Jr. and R.A. Shelby. 1999. *In vitro* ruminal disappearance of fumonisin B_1 and its effects on *in vitro* dry matter disappearance. Vet. Hum. Toxicol. 41:196-199.

Guthrie, L.D. 1979. Effects of aflatoxins in corn on production and reproduction of dairy cattle. J. Dairy Sci. 62 (suppl. 1):134-142.

Hagler Jr., W.M., C.J. Mirocha, S.V. Pathre and J.C. Behrens. 1979. Identification of the naturally occurring isomer of zearalenol produced by *Fusarium roseum* "gibbosum" in rice culture. Appl. Environ. Microbiol. 37:849-853.

Harvey, R.B., L.F. Kubena, T.D. Phillips, D.E. Corrier, M.H. Elissalde and W.E. Huff. 1991. Diminution of aflatoxin toxicity to growing lambs by dietary supplementation with hydrated sodium calcium aluminosilicate. Am. J. Vet. Res. 52:152-156.

Hedman, R. and H. Pettersson. 1997. Transformation of nivalenol by gastro-intestinal microbes. Arch. Tierernahr. 50:321-329.

Helferich, W.G., R.L. Baldwin and D.P.H. Hsieh. 1986. [^{14}C]-aflatoxin B_1 metabolism in lactating goats and rats. J. Anim. Sci. 62:697-705.

Höhler, D., K.H. Sudekum, S. Wolffram, A.A. Frohlich and R.R. Marquardt. 1999. Metabolism and excretion of ochratoxin A fed to sheep. J. Anim. Sci. 77:1217-1223.

Hsu, I.C., C.B. Smalley, F.M. Strong and W.E. Ribelin. 1972. Identification of T-2 toxin in moldy corn associated with a lethal toxicosis in dairy cattle. Appl. Microbiol. 24:648.

Hult, K., A. Teiling and S. Gatenberg. 1976. Degradation of ochratoxin A by a ruminant. Appl. Environ. Microbiol. 32:443-444.

Hussein, H.S. and J.M. Brasel. 2001. Toxicity, metabolism and impact of mycotoxins on humans and animals. Toxicology 167:101-134.

Ingalls, J.R. 1994. Influence of DON on feed consumption by dairy cows. In: *Western Nutr. Conf.*, Winnipeg, MB, Canada, pp. 129.

Jones, M.G.S. and J.M. Ewart. 1979. Effects on milk production associated with consumption of decorticated extracted groundnut meal contaminated with aflatoxin. Vet. Rec. 21:492-493.

Jouany, J-P. 2001. The impact of mycotoxins on performance and health of dairy cattle. In: *Science and Technology in the Feed Industry, Proceedings of Alltech's 17th Annual Symposium* (K.A. Jacques and T.P. Lyons, eds). Nottingham University Press, pp. 191-223.

Juskiewicz, T., J. Piskorska-Pliszezynska and H. Wisniewska. 1982. Ochratoxin in laying hens: Tissue deposition and passage into eggs. In: *Proc. V. International IUPAC Symp. Mycotoxins and Phycotoxins*, Vienna, Austria. pp. 122.

Kallela, K. and E. Ettala. 1984. The oestrogenic *fusarium* toxin (zearalenone) in hay as a cause of early abortions in the cow. Nord. Vet. Med. 36:305-309.

Kehrli, M.E., Jr., B.J. Nonnecke and J.A. Roth. 1989a. Alterations in bovine lymphocyte function during the periparturient period. Am. J. Vet. Res. 50:215-221.

Kehrli, M.E., Jr., B.J. Nonnecke and J.A. Roth. 1989b. Alterations in bovine neutrophil function during the periparturient period. Am. J. Vet. Res. 50:207-214.

Kellerman, T.S., S.J. Newsholme, J.A. Coetzer and G.C.A. Van der Westhuizen. 1984. A tremorgenic mycotoxicosis of cattle caused by maize sprouts infested with *Aspergillus clavatus*. Onderst. J. Vet. Res. 51:271-274.

Kellerman, T.S., J.G. Pienaar, G.C.A. Van der Westhuizen, L.A.P. Anderson and T.W. Naud. 1976. A highly fatal tremorgenic mycotoxicosis of cattle caused by *Aspergillus clavatus*. Onderst. J. Vet. Res. 43:147-157.

Kennedy, D.G., S.A. Hewitt, J.D.G. McEvoy, J.W. Currie, A. Cannavan, W.J. Blanchflower and C.T. Elliot. 1998. Zeranol is formed from *Fusarium* spp. toxins in cattle *in vivo*. Food Addit. Contam. 15:393-400.

Kennedy, D.G., J.D.G. McEvoy, W.J. Blanchflower, S.A. Hewitt, A. Cannavan, W.J. McCaughey and C.T. Elliott. 1995. Possible naturally occurring zeranol in bovine bile in Northern Ireland. J. Vet. Med. Series B 42:509-512.

Khamis, Y., H.A. Hammad and N.A. Hemeida. 1986. Mycotoxicosis with oestrogenic effect in cattle. Zuchthygiene 21:233-236.

Kiang, D.T., B.J. Kennedy, S.V. Pathre and C.J. Mirocha. 1978. Binding characteristics of zearalenone analogs to oestrogen receptors. Cancer Res. 38:3611-3615.

Kiermeier, F. 1973. Ueber die aflatoxin-m-ausscheidung in kuhmilch in abhängigkeit von der aufgenommenen aflatoxin-B_1-menge. Milchwissenschaft 128:683-685.

Kiessling, K-H. and H. Pettersson. 1978. Metabolism of zearalenone in rat liver. Acta Pharmacol. Toxicol. 43:285-290.

Kiessling, K-H., H. Pettersson, K. Sandholm and M. Olsen. 1984. Metabolism of aflatoxin, ochratoxin, zearalenone and three trichothecenes by intact rumen fluid, rumen protozoa and rumen bacteria. Appl. Environ. Microbiol. 47:1070-1073.

King, R.R., R.E. McQueen, D. Levesque and R. Greehalgh. 1984. Transformation of deoxynivalenol (vomitoxin) by rumen microorganisms. J. Agric. Food Chem. 32:1181-1183.

Kosuri, N.R., M.D. Grove, S.G. Yates, W.H. Tallent, J.J. Ellis, I.A. Wolff and R.E. Nichols. 1970. Response of cattle to mycotoxins of *Fusarium tricinctum* isolated from corn and fescue. J. Am. Vet. Med. Assoc. 157:938-940.

Kriek, N.P., T.S. Kellerman and W.F. Marasas. 1981. A comparative study of the toxicity of *Fusarium verticillioides* (= *F. moniliforme*) to horses, primates, pigs, sheep and rats. Onderstepoort J. Vet. Res. 48:129-131.

Kuiper-Goodman, T. and P.M. Scott. 1986. Risk assessment of the mycotoxin ochratoxin A. Biomed. Environ. Sci. 2:179-248.

Kuiper-Goodman, T., P.M. Scott and H. Watanabe. 1987. Risk assessment of the mycotoxin zearalenone. Reg. Toxicol. Pharmacol. 7:353-306.

Kumagai, S. 1989. Intestinal absorption and excretion of aflatoxin in rats. Toxicol. Appl. Pharmacol. 97:88-97.

Kurmanov, L.A. 1977. *Fusariotoxicosis in*

Cattle and Sheep. Marcel Dekker Inc., New York, USA.

Lacey, J. 1991. Natural occurrence of mycotoxins in growing and conserved forage crops. In: *Mycotoxins and Animal Foods* (J.E. Smith and R.E. Henderson, eds). CRC Press, Boca Raton.

Lewis, G., L.M. Markson and R. Allcroft. 1967. The effect of feeding toxic groundnut meal to sheep over a period of five years. Vet. Rec. 312-314.

Lewis, G.G. and R.O. Clements. 1986. A survey of ryegrass endophyte (*Acremonium loliae*) in the UK and its apparent ineffectuality on seedling pest. J. Agric. Food Chem. 107:633-637.

Lindroth, S. and A. von Wright. 1990. Detoxification of patulin by adduct formation with cysteine. J. Environ. Path. Toxicol. Onc. 10:254-259.

Mackie, R.I. and B.A. White. 1990. Recent advances in rumen microbial ecology and metabolism: Potential impact on nutrient output. J. Dairy Sci. 73:2971-2995.

Madhyastha, S., R.R. Marquadt, A. Mast, J. Borsa and A.A. Frohlich. 1994. Comparison of toxicity of different mycotoxins to several species of bacteria and yeast: Use of *Bacillus brevis* in a disc diffusion assay. J. Food Prot. 57:48-53.

Malmauret, L., D. Parent-Massin, J-L. Hardy and P. Verger. 2002. Contaminants in organic and conventional foodstuffs in France. Food Addit. Contam. 19:524-532.

Mann, D. D., G.M. Buening, G.D. Osweiler and B.S. Hook. 1984. Effect of subclinical levels of T-2 toxin on the bovine cellular immune system. Can. J. Comp. Med. 48:308-312.

Marasas, W.F. 1996. Fumonisins: history, world-wide occurrence and impact. Adv. Exp. Med. Biol. 392:1-17.

Marasas, W.F., T.S. Kellerman, W.C. Gelderblom, J.A. Coetzer, P.G. Thiel and J.J. van der Lugt. 1988. Leukoencephalomalacia in a horse induced by fumonisin B_1 isolated from *Fusarium moniliforme*. Onderstepoort J. Vet. Res. 55:197-203.

Marquardt, R.R. and A.A. Frohlich. 1992. A review of recent advances in understanding ochratoxicosis. J. Anim. Sci. 70:3968-3988.

Marsi, M.S., V.C. Garcia and J.R. Page. 1969. The aflatoxin M_1 content of milk from cows fed known amounts of aflatoxin. Vet. Rec. 84:146.

Mathur, C.F., R.C. Smith and G.E. Hawkins. 1976. Growth and morphology of *Streptococcus bovis* and of mixed rumen bacteria in the presence of aflatoxin B_1, *in vitro*. J. Dairy Sci. 59:455-458.

May, H.D., Q. Wu and C.K. Blake. 2000. Effects of the *Fusarium* spp. mycotoxins fusaric acid and deoxynivalenol on the growth of *Ruminococcus albus* and *Methanobrevibacter ruminantium*. Can. J. Micro. 46:692-699.

McIntosh, F.M., K.N. Stanley, C.S. Stewart and C.J. Newbold. 2002. The role of protozoa in the survival of *E. coli*, verocytotoxin encoding bacteriophage and their vtec lysogens in the rumen. Reprod. Nutr. Develop. 42(suppl.1):S34.

Mertens, D.R. and R.D. Wyatt. 1977. Acute aflatoxicosis in lactating dairy cattle. J. Dairy Sci. 60 (Suppl. 1):153-154.

Mirocha, C.J., J. Harrison, A.A. Nichols and M. McClintock. 1968. Detection of fungal oestrogen (zen) in hay associated with infertility in dairy cattle. Appl. Microbiol. 16:797.

Mirocha, C.J., S.V. Pathre, B. Schauerhamer and C.M. Christensen. 1976. Natural occurrence of *Fusarium* toxins in feedstuff. Appl. Environ. Microbiol. 32:553-556.

Mirocha, C.J., B. Schauerhomer and S.V. Pathre. 1974. Isolation, detection and quantitation of zearalenone in maize and barley. J. Assoc. Off. Anal. Chem. 57:1004.

Moreau, C. and M. Moreau. 1960. Un danger pour le bétail nourri de plantules fourragères cultivés en germoir: La pullulation d'une moisissure toxique; *Aspergillus clavatus*; cause des accidents mortels. C. R. Acad. Agric. 46:441-445.

Morgavi, D.P., H. Boudra, J-P. Jouany and D. Graviou. 2003. Prevention of patulin toxicity on rumen microbial fermentation by SH-containing reducing agents. J. Agric. Food Chem. 51:6906-6910.

Moschini, M., F. Masoero and T. Bertuzzi. 1999. *In vitro* rumen metabolism of zearalenone. In: *Recent Progress in Animal Production Science*. Piacenza, Italy, pp. 363-365.

Muller, H.M., C. Lerch, K. Muller and W. Eggert. 1998. Kinetic profiles of ochratoxin A and ochratoxin-α during *in vitro* incubation in buffered forestomach and abomasal contents from cows. Nat. Toxins 6:251-258.

Müller, H.M., K. Müller and H. Steingass. 2001. Effect of feeding regime on the metabolism of ochratoxin A during the *in vitro* incubation in buffered rumen fluid from cows. Arch. Tierernahr. 54:265-279.

Muller, K. 1995. Einfluss der fütterung und anderer faktoren auf den umsatz von ochratoxin A in pansenflüssigkeit *in vitro* und *in vivo*. Ph.D. Thesis, Hohenheim University, Stuttgart, Germany.

Munger, C.E., G.W. Ivie, R.J. Christopher, B.D. Hammock and T.D. Phillips. 1987. Acetylation/deacetylation reactions of T-2, acetyl T-2, HT-2 and acetyl HT-2 toxins in bovine rumen fluid *in vitro*. J. Agric. Food Chem. 35:354-358.

Nakamura, K., H. Okumura, Y. Sugiura and Y. Ueno. 1998. Epoxidation of deepoxy-deoxynivalenol in mice. Rev. Méd. Vét. 149:579.

Noller, C.H., M. Stob and J. Tutie. 1979. Effect of feeding *Gibberella zeae*-infected corn on feed intakes, body weight gain and milk production of dairy cows. J. Dairy Sci. 62:1003.

Olsen, M., H. Pettersson and K-H. Kiessling. 1981. Reduction of zearalenone to zearalenol in female rat liver by a 3 α-hydroxysteroid dehydrogenase. Acta Pharmacol. Toxicol. 48:157-161.

Osweiler, G.D., M.E. Kehrli, J.R. Stabel, J.R. Thurston, P.F. Ross and TM. Wilson. 1993. Effects of fumonisin-contaminated corn screenings on growth and health of feeder calves. J. Anim. Sci. 71:459-466.

Osweiler, G.D., P.F. Ross, T.M. Wilson, P.E. Nelson, S.T. Witte, T.L. Carson, L.G. Rice and H.A. Nelson. 1992. Characterization of an epizootic of pulmonary edema in swine associated with fumonisin in corn screenings. J. Vet. Diagn. Invest. 4:53-59.

Özpinar, H., T. Bilal, I. Abas and C. Kutay. 2002. Umsatz von ochratoxin A in pansenflüssigkeit *in vitro*. Facta Universitatis. Series Medicine and Biology 9:66-69.

Patterson, D.S.P. and P.H. Anderson. 1982. Recent aflatoxin feeding experiments in cattle. Vet. Rec. 110:60.

Petrie, L., J. Robb and A.F. Stewart. 1977. The identification of T-2 toxin and its association with a hemorrhagic syndrome in cattle. Vet. Rec. 101:326.

Pettersson, H. 1998. Carryover of aflatoxin from feedingstuffs to milk. Department of Animal Nutrition and Management, Swedish University of Agricultural Sciences, Upsala, Sweden.

Pfohl-Leszkowicz, A. 2000. Risques mycotoxicologiques pour la santé des animaux et de l'homme. Cha. Nutr. Diet. 35:389-397.

Pier, A.C. 1992. Major biological consequences of aflatoxicosis in animal production. J. Anim. Sci. 70:3964-3967.

Polan, C.E., J.R. Hayes and T.C. Campbell. 1974. Consumption and fate of aflatoxin B_1 by lactating cows. J. Agric. Food Chem. 22:635-638.

Prelusky, D.B., M.E. Savard and H.L. Trenholm. 1995. Pilot study on the plasma pharmacokinetics of fumonisin B_1 in cows following a single dose by oral gavage or intravenous administration. Nat. Toxins 3:389-394.

Prelusky, D.B., P.M. Scott, H.L. Trenholm and G.A. Lawrence. 1990. Minimal transmission of zearalenone to milk of dairy cows. J. Environ. Sci. Health Part B Pestic. Food Contam. Agric. Wastes 25:87-103.

Prelusky, D.B., H.L. Trenholm, G.A. Lawrence and P.M. Scott. 1984. Nontransmission of deoxynivalenol (vomitoxin) to milk following oral administration to dairy cows. Environ. Sci. Health B19:593-609.

Prelusky, D.B., H.L. Trenholm, B.A. Rotter, J.D. Miller, M.E. Savard, J.M. Yeung and P.M. Scott. 1996. Biological fate of fumonisin B_1 in food-producing animals. Adv. Exp. Med. Biol. 392:265-278.

Rheeder, J.P., W.F.O. Marasas, P.G. Thiel, E.W. Sydenham, G.S. Shephard and D.J. van Schalkwyk. 1992. *Fusarium moniliforme* and fumonisins in corn in relation to human esophageal cancer in Transkei. Phytopathology 82:353-357.

Ribelin, W.E., K. Fukushima and P.E. Still. 1978. The toxicity of ochratoxin A to ruminants. Can. J. Comput. Med. 42:172-176.

Richard, J.L., G. Meerdink, C.M. Maragos, M. Tumbleson, G. Bordson, L.G. Rice and P.F. Ross. 1996. Absence of detectable fumonisins in the milk of cows fed *Fusarium proliferatum* (matsushima) nirenberg culture material. Mycopathologia 133:123-126.

Riley, R.T. 1998. Mechanistic interactions of mycotoxins: Theoretical considerations. In: *Mycotoxins in Agriculture and Food Safety* (K.K. Sinha and D. Bhatnagar, eds). Marcel Dekker, Inc., New York. pp. 227-253.

Riley, R.T. and J.L. Showker. 1991. The mechanism of patulin's cytotoxicity and the antioxidant activity of indole tetramic acids. Toxicol. Appl. Pharmacol. 109:108-126.

Roine, K., E.L. Korpinen and K. Kallela. 1971. Mycotoxins as a probable cause of infertility in dairy cows. Nord. Vet. Med. 23:628.

Ross, P.F., A.E. Ledet, D.L. Owens, L.G. Rice, H.A. Nelson, G.D. Osweiler and T.M. Wilson. 1993. Experimental equine leukoencephalomalacia, toxic hepatosis and encephalopathy caused by corn naturally contaminated with fumonisins. J. Vet. Diagn. Invest. 5:69-74.

Sabater-Vilar, M., R.F.M. Mass, H. De Bosschere, R. Ducatelle and J. Fink-Gremmels. 2004. Patulin produced by an *Aspergillus clavatus* isolated from feed containing malting residues associated with a lethal neurotoxicosis in cattle. Mycopathologia 00:1-8.

Sato, N. and Y. Ueno. 1977. *Mycotoxins in Human and Animal Health*. Pathotox Publishers, Park Forest South, IL, USA.

Scott, P.M. 1974. Patulin. In: *Mycotoxins* (I.F.H. Purchase, ed). Elsevier Scientific Publishing Company, Amsterdam, The Netherlands, pp. 383-403.

Scott, P.M., T. Delgado, D.B. Prelusky, H.L. Trenholm and J.D. Miller. 1994. Determination of fumonisins in milk. J. Environ. Sci. Health Part B Pestic. Food Contam. Agric. Wastes 29:989-998.

Seglar, B. 1997. Case studies that implicate silage mycotoxins as the cause of dairy herd problems. In: *Silage: Field to feedbunk*. Northeast Regional Agricultural Engineering Service, Ithaca, pp. 242.

Shephard, G.S., P.G. Thiel, S. Stockenstrom and E.W. Sydenham. 1996. Worldwide survey of fumonisin contamination of corn and corn-based products. J. AOAC Int. 79:671-887.

Shreeve, B.J., D.S. Patterson and B.A. Roberts. 1979. The 'carry-over' of aflatoxin, ochratoxin and zearalenone from naturally contaminated feed to tissues, urine and milk of dairy cows. Food Cosmet. Toxicol. 17:151-152.

Skaug, M.A. 1999. Analysis of Norwegian milk and infant formulas for ochratoxin A. Food Addit. Contam. 16:75-78.

Smith, J., C. Wesselink, J. Parr, J.M. Sporsen, E.A. Fowke and N.R. Towers. 1995. Effects of zearalenone on ewe pregnancy rates. In: *Toxinology and Food Safety Research Report* (L.L. Garthwaite, ed). Ruakura Research Center, Hamilton, New Zealand, pp. 41-43.

Smith, J.S. and R.A. Thakur. 1996. Occurrence and fate of fumonisins in beef. Adv. Exp. Med. Biol. 392:39-55.

Sporsen, J.M., J.A. Armstrong, I. Garthwaite and M.E. di Menna. 1995. 'Zearalenone' production by pasture fungi other than *fusarium*. In: *Toxinology and Food Safety Research Report* (L.L. Garthwaite, ed). Ruakura Research Center, Hamilton, New Zealand. pp. 44-45.

Sprosen, J.M. and N.R. Towers. 1995. Urinary

zearalenone metabolite concentrations in herds with fertility problems. In: *Toxinology and Food Safety Research Report* (L.L. Garthwaite, ed). Ruakura Research Centre, Hamilton, New Zealand, pp. 45-46.

Squire, R.A. 1989. Ranking animal carcinogens: A proposed regulatory approach. Science 214:887-891.

Sreemannarayana, O., A.A. Frohlich, T.G. Vitti, R.R. Marquart and D. Abramson. 1988. Studies of the tolerance and disposition of ochratoxin A in young calves. J. Animal Sci. 66:1703.

Suliman, H.B., A.F. Mohamed, N.A. Awadelsied and A.M. Shommein. 1987. Acute mycotoxicosis in sheep: Field cases. Vet. Hum. Toxicol. 29:241-243.

Swanson, S.P., H.D. Rood, J.C. Behrens and P.E. Sanders. 1987. Preparation and characterization of the deepoxy trichothecenes: deepoxy HT-2, depoxy T-2 triol, deepoxy T-2 tetraol, deepoxy 15-monoacetoxyscirpenol, deepoxy scirpenol. Appl. Environ. Microbiol. 53:2821-2826.

Sydenham, E.W., W.C. Gelderblom, P.G. Thiel and F.O. Marasas. 1990. Evidence for the natural occurrence of fumonisin B_1, a mycotoxin produced by *Fusarium moniliforme*, in corn. J. Agric. Food Chem. 38:285-290.

Syrett, R. 1979. Moulds and mycotoxins in animal foodstuffs - reports of ADAS microbiologists. In: *3rd Meeting on Mycotoxins in Animal Disease*, pp. 4-7.

Szczech, G.M., W.W. Carlton and J. Tuite. 1973. Ochratoxin A toxicosis in swine. Vet. Pathol. 10:347-351.

Tanaka, T., S. Yamamoto, A. Hasegawa, N. Aoki, J.R. Besling, Y. Sugiura and Y. Ueno. 1990. A survey of the natural occurence of *Fusarium* mycotoxins, deoxynivalenol, nivalenol and zearalenone, in cereals harvested in The Netherlands. Mycopathologia 110:19-22.

Tapia, M.O., M.D. Stern, R.L. Koski, A. Bach and M.J. Murphy. 2002. Effects of patulin on rumen microbial fermentation in continuous culture fermenters. Anim. Feed Sci. Technol. 97:239-246.

Tiemann, U., T. Viergutz, L. Jonas and F. Schneider. 2003. Influence of the mycotoxins α- and ß-zearalenol and deoxynivalenol on the cell cycle of cultured porcine endometrial cells. Reprod. Toxicol. 5506:1-10.

Towers, N.R., J.M. Sporsen and W. Webber. 1995a. Zearalenone metabolites in cycling and non-cycling cows. In: *Toxinology and Food Safety Research Report*, (L.L. Garthwaite, ed). Ruakura Research Center, Hamilton, New Zealand, pp. 46-47.

Towers, N.R., C. Wesselink, E.A. Fowke and J.M. Sporsen. 1995b. Plasma vs urinary 'zearalenone' concentrations as indicators of 'zearalenone' intake. In: *Toxinology and Food Safety Research Report*, (L.L. Garthwaite, ed). Ruakura Research Center, Hamilton, New Zealand, p. 41.

Trenholm, H.L., R.M.G. Hamilton, D.W. Friend, B.K. Thompson and K.E. Hartin. 1984. Feeding trials with vomitoxin (deoxynivalenol)-contaminated wheat: Effects on swine, poultry and dairy cattle. J. Am. Vet. Med. Assoc. 185:527-531.

Trucksess, M.W., J.L. Richard, L. Stoloff, J.S. McDonald and W.C. Brumley. 1983. Absorption and distribution patterns of aflatoxicol and aflatoxin B_1 and M_1 in blood and milk of cows given aflatoxin B_1. Am. J. Vet. Res. 44:1753-1756.

Ueno, Y. and H. Matsumoto. 1975. Inactivation of some thiol-enzymes by trichothecene mycotoxins from *Fusarium* species. Chem. Pharm. Bull. 23:2439-2442.

Ukai, T., Y. Yamamoto and T. Yamamoto. 1954. Studies on the poisonous substance from a strain of *Penicillium* (hori-yamamoto strain). II. Culture method of hori-yamamoto strain and chemical structure of its poisonous substance. J. Pharmacol. Jpn. 74:450-454.

Vallone, L. and I. Dragoni. 1997. Investigation of mycotoxin (aflatoxin B_1) occurrence in corn silage trench. Atti della Societa Italiana delle Scienze Veterinarie 51:237-238.

van Egmond, H.P. 1989. *Mycotoxins in Dairy Products*. Elsevier Science Pub. Co., New

York.

van Egmond, H.P. 2003. Natural toxins: Risks, regulations and the analytical situation in Europe. Anal. Bioanal. Chem. 378:1152-1160.

Veldman, A., J.A.C. Meijst, G.J. Borggreve and J.J. Heeres-vander Tol. 1992. Carry-over of aflatoxin from cow's food to milk. Anim. Prod. 55:163-168.

Vesonder, R.F., A. Ciegler, R.F. Rogers, K.A. Burbridge, R.J. Bothast and A.H. Jensen. 1978. Survey of 1977 crop year preharvest corn for vomitoxin. Appl. Environ. Microbiol. 36:885-888.

Vough, L.R. and I. Glick. 1993. Round bale silage. Silage production from seed to animal. Northeast Regional Agricultural Engineering Service, Ithaca, pp. 117.

Wannemacher, R.W., Jr., D.L. Brunner and H.A. Neufeld. 1991. Toxicity of trichothecenes and other related mycotoxins in laboratory animals. In: *Mycotoxins and Animal Foods* (J.E. Smith and R.S. Henderson, eds). CRC Press, Boca Raton.

Weaver, G.A., H.J. Kurtz, J.C. Behrens, T.S. Robison, B.E. Seguin, F.Y. Bates and C.J. Mirocha. 1986a. Effect of zearalenone on the fertility of virgin dairy heifers. Am. J. Vet. Res. 47:1395-1397.

Weaver, G.A., H.J. Kurtz, C.J. Mirocha, F.Y. Bates, J.C. Behrens, T.S. Robison and S.P. Swanson. 1980. The failure of T-2 mycotoxin to produce hemorrhaging in dairy cattle. Can. Vet. J. 21:210.

Weaver, G.A., H.J. Kutz, J.C. Behrens, T.S. Robison, B.E. Sequin, F.Y. Bates and C.J. Mirocha. 1986b. Effect of zearalenone on dairy cows. Am. J. Vet. Res. 47:1826-1828.

Weidenborner, M., C. Wieczorek, S. Appel and B. Kunz. 2000. Whole wheat and white wheat flour - the mycobiota and potential mycotoxins. Food Microbiol. 17:103-107.

Westlake, K., R.I. Mackie and M.F. Dutton. 1987a. Effects of several mycotoxins on specific growth rate of *Butyrivibrio fibrisolvens* and toxin degradation *in vitro*. Appl. Envirom. Microbiol. 53:613-614.

Westlake, K., R.I. Mackie and M.F. Dutton. 1987b. T-2 toxin metabolism by ruminal bacteria and its effect on their growth. Appl. Environ. Microbiol. 53:587-592.

Westlake, K., R.I. Mackie and M. Dutton. 1989. *In vitro* metabolism of mycotoxins by bacterial, protozoal and ovine ruminal fluid preparations. Anim. Feed Sci. Technol. 25:169-178.

Whitlow, L.W. and W.M. Hagler, Jr. 1987. The association of productivity in dairy cows with deoxynivalenol. In: *Recent Developments in the Study of Mycotoxins*, pp. 3-13.

Willis, R.M., J.J. Mulvihill and J.H. Hoofnagle. 1980. Attempted suicide with purified aflatoxin. Lancet 8179:1198-1199.

Windels, H.F., A. DiCostanzo and R.D. Goodrich. 1995. Effect of deoxynivalenol from barley on performance and health of large frame crossbred steers. B-417, Minnesota Cattle Feeder Rep., St. Paul, MN, USA.

Wogan, G.N. 1966. Chemical nature and biological effects of the aflatoxins. Bacteriol. Rev. 2:460-470.

Wong, S.A. and D.P.H. Hsieh. 1980. The comparative metabolism and toxicokinetics of aflatoxin B_1 in the monkey rat and mouse. Toxicol. Appl. Pharmacol. 55:115-125.

Wood, G.E. 1991. Aflatoxin M_1. In: *Mycotoxins and Phytoalexins* (R.P. Sharma and D.K. Salunke, ed). CRC Press Inc., Boca Raton, FL, USA, pp. 145-164.

Wooters, M.F.A. and G.J.A. Speijers. 2003. Patulin. WHO Food Additives series IPCS INCHEM 35:1-22.

Xiao, H., R.R. Marquardt, A.A. Frohlich, G.D. Phillips and T.G. Vitti. 1991a. Effect of a hay and a grain diet on the bioavailability of ochratoxin A in the rumen of sheep. J. Anim. Sci. 69:3715-3723.

Xiao, H., R.R. Marquardt, A.A. Frohlich, G.D. Phillips and T.G. Vitti. 1991b. Effect of a hay and grain diet on the rate of hydrolysis of ochratoxin A in the rumen of sheep. J. Anim. Sci. 69:3706-3714.

Yiannikouris, A. and J-P. Jouany. 2002. Mycotoxin in feeds and their fate in animals: A review. Anim. Res. 51:81-99.

Yoshizawa, T., C.J. Mirocha, J.C. Behrens and S.P. Swanson. 1981. Metabolic fate of T-2 toxin in a lactating cow. Food Cosmet. Toxicol. 19:31-39.

Yoshizawa, T.C., K. Okamoto, T. Sakamoto and K. Kuwamura. 1985. *In vivo* metabolism of T-2 toxin; a trichothecene mycotoxin. On the formation of deepoxydation products. Proc. Jpn. Assoc. Mycotoxicol. 21:9-12.

15

MYCOTOXIN SEQUESTERING AGENTS: PRACTICAL TOOLS FOR THE NEUTRALISATION OF MYCOTOXINS

Duarte E. Diaz[1] and Trevor K. Smith[2]
[1]*University of Puerto Rico, Mayagüez, Puerto Rico;* [2]*University of Guelph, Guelph, Ontario, Canada*

Introduction

Ever since the first report of a mycotoxicosis in the early 1960s researchers all over the world have been meticulously researching ways to eliminate or minimise the effects of these inevitable contaminants. Due to the wide range of mycotoxins that can contaminate animal feeds and their variable chemical compositions, protection against mycotoxicosis is a relatively difficult task.

Among all the available approaches to control mycotoxin contamination the simplest strategy is based on the prevention of the formation of mycotoxins in feeds (Lopez-Garcia and Park, 1998). Even with current technologies it is very difficult to predict or to prevent their occurrence either preharvest or during storage and feed processing (Wood, 1992). Once ingredients become contaminated with mycotoxins, elimination of the contaminated product is the most effective method of avoiding the problems related to their ingestion (CAST, 2003). Unfortunately, due to the difficulties in obtaining a representative sample it is quite difficult to accurately determine the level of contamination of a specific feedstuff. The impracticality and cost associated with the complete substitution of these ingredients means that this practice is not performed as frequently as recommended. Therefore, mycotoxins are often present in animal feedstuffs and pose a substantial economic loss to agricultural animal industries. Moreover, mycotoxins represent a public health concern as animals consuming mycotoxin-contaminated diets may leave residues of these toxins in animal-derived food products.

Mycotoxin **decontamination** refers to methods by which the mycotoxins are removed or neutralised from the contaminated feed while mycotoxin **detoxification** refers to methods by which the toxic properties of the mycotoxins are removed. These strategies include physical, chemical and biological methods. Chemical procedures like treatment with acid/base solutions or other chemicals, ammoniation, ozonation, and reaction with food grade additives such as sodium bisulfite have proven effective in degrading and detoxifying aflatoxin contaminated feedstuffs (CAST, 2003). Biological methods primarily involving the degradation of the toxin by microorganisms are receiving increasing interest among researchers and have shown positive results (Volkl *et al.*, 2004). Physical procedures like sorting, thermal inactivation, irradiating or extracting contaminated products have been attempted with variable success (CAST, 2003).

Any detoxification or decontamination method for mycotoxin-contaminated feedstuffs should fulfil the following prerequisites (Sinha, 1998).

1. Be effective in removing, destroying and inactivating the mycotoxin,
2. Do not produce toxic or carcinogenic/ mutagenic residues in the treated products

nor in the food products derived from animals consuming these treated feedstuffs,

3. Do not alter the nutritive properties of the feedstuffs or affect feed acceptability,
4. Be economically and technologically feasible so as not to alter significantly the cost of the final product.

Many of the physical, chemical and biological methods found in the literature are effective in reducing, destroying or inactivating different mycotoxins but rarely do they meet the other equally important requisites. Nutritional approaches like supplementation of nutrients or additives with protective properties against toxicity, and the addition of non-nutritive sequestrants capable of reducing mycotoxin bioavailability can in fact decontaminate or detoxify contaminated feedstuffs while meeting the other four criteria (Galvano *et al.,* 2001). The utilisation of nutrients with antioxidant properties to reduce the oxidative effects of mycotoxins has been reviewed in another chapter of this book. Dietary supplementation with non-nutritive mycotoxin-sequestering agents is by far the most practical and most widely studied method for reducing the effects of mycotoxin exposure. An effective sequestering agent is one that prevents or limits toxin absorption from the gastrointestinal tract of the animal. Ideally, a sequestering agent should be effective against several mycotoxins as feedstuffs are commonly contaminated with more than one mycotoxin. To remain practical, mycotoxin-sequestering agents should also be reasonably priced and should not occupy a large portion of the complete diet. Additionally, mycotoxin-sequestering agents should be free of impurities, off-flavours and odours.

The objective of this chapter is to summarise the information available on the utilisation of non-nutritive sequestering agents to minimise the toxicological effects of mycotoxins and to reduce the potential carryover of the toxins into the human food chain. The principal classes of sequestering agents, silicate minerals, activated charcoals, polymers, chlorophyll products and yeast-derived products, will be discussed.

One of the most complex aspects of evaluating sequestering agents is in properly classifying these products. Care must be taken not to generalise any conclusions for a given adsorbent group, as all products under a specific classification will not always have the same composition or efficacy. For example 'clays' vary in structure and mineral composition while yeast used in production of 'yeast cell wall products' differ in cell wall composition. In either case the physico-chemical properties that determine sequestrant efficacy can vary markedly among products in the same category.

In vitro evaluation of sequestering agents

A major obstacle in the evaluation of mycotoxin sequestering agents is related to the inadequacy of *in vitro* models in predicting response in the animal. Several researchers have shown that ability of a compound to sequester mycotoxin *in vitro* does not necessarily correlate with *in vivo* response. Rotter *et al.* (1989) selected an activated charcoal that showed a high capacity to sequester ochratoxin A at different pH values *in vitro*. This activated charcoal was then fed to chicks consuming diets contaminated with ochratoxin A. The authors concluded that the addition of the charcoal did not reduce ochratoxin A toxicosis. Similarly, Dwyer *et al.* (1997) tested three mineral clays (an acidic phyllosilicate, a neutral phyllosilicate and a zeolite), which had previously shown *in vitro* sequestration of cyclopiazonic acid (CPA). The researchers found that none of the three mineral clays were effective in reducing the toxic effects of CPA in broiler chicks when added to the diet.

Experiments with swine and cattle have

shown similar results. Piva *et al.* (in press) found that an activated charcoal that sequestered fumonisin B_1 effectively *in vitro* did not have any protective properties in growing piglets consuming fumonisin B_1 from a culture material. Diaz *et al.* (2004) fed several sequestering agents (three bentonites, an activated charcoal and a yeast glucomannan polymer), all of which had previously shown high aflatoxin B_1 sequestration ability *in vitro*, to lactating dairy cows and measured the change in milk aflatoxin M_1 before and after dietary inclusion of the different sequestering agents. The authors found that two of the tested sequestering agents, an activated carbon and a calcium bentonite, had little or no effect in reducing the aflatoxin M_1 concentration.

In vitro analysis of mycotoxin sequestration is however a powerful tool for screening potential mycotoxin sequestering agents. If a sequestering agent does not absorb a mycotoxin *in vitro*, it has little or no chance to do so *in vivo*. These laboratory techniques can be very useful in identifying potential dietary sequestering agents and in helping to determine the mechanisms and conditions favourable for sequestration to occur.

The simplest and most widely utilised *in vitro* method measures absorption of purified toxin preparations in an aqueous medium. In these systems, a known amount of mycotoxin is reacted with a known amount of test compound in an aqueous solution. The difference between the amount of toxin still present in the liquid after separation and the amount of adsorbed toxin is determined. Due to the relative insolubility of mycotoxins, these tests are generally conducted at very low mycotoxin concentrations. Unfortunately, because of their simplicity these methods have been widely used to evaluate sequestering agents and are the basis of most claims for the effectiveness of certain commercial products.

More elaborate *in vitro* systems utilised to evaluate mycotoxin sequestering agents are based on a two-step process. These systems determine the strength of the mycotoxin-sequestrant complex by first measuring the quantity of toxin sequestered, and then measuring the separation of the mycotoxin from the sequestrant after exposure to a second solvent system. Efficacy is evaluated by comparing the initial sequestration ('weak binding') with that after desorption ('strong binding') (Dawson *et al.*, 2001). While such tests are useful in demonstrating the potential of a specific sequestering agent, they are not very efficient for comparisons among different types of compounds. Additionally, these tests provide little information on how a specific compound should be utilised in a practical field situation.

Absorption isotherms have been effectively utilised to evaluate mycotoxin sequestering agents (Grant and Phillips, 1998; Ramos and Hernandez, 1996). The amount of mycotoxin absorbed per unit of weight is plotted against the concentration of the mycotoxin in solution at a constant temperature and under stable conditions. This system takes into account that sequestering of mycotoxins is a reversible process that can be characterised as a chemical equilibrium. As a result, absorption is a concentration-dependent process influenced by mycotoxin concentration, the amount of sequestrant, and the relative affinity of the compound for the mycotoxin (Dawson *et al.*, 2001).

It is extremely important that any *in vitro* results be supported by *in vivo* experiments utilising the species for which the product is intended and with feeds contaminated with levels of mycotoxins that are commonly found in the field.

Activated charcoal

Activated charcoal (AC) is an amorphous form of carbon heated in the absence of air and then treated with oxygen to open millions of pores between the carbon atoms. The carbon source is selected from a variety of materials such as the shells of nuts, wood (coal), moss, etc. This highly absorbent powder has been commonly utilised as a medical treatment for severe intoxications since the 19th century (Huwig *et al.*, 2001).

Activated charcoal may be an efficacious product to decontaminate foodstuffs containing aflatoxins or other mycotoxins, although it is probably more suited to use as an antidote for severe toxicity. The sequestrant properties of AC are dependent on many factors including pore size, surface area, structure of the mycotoxin, and dose (Galvano *et al.*, 2001). Since *in vivo* data on the effectiveness of AC are variable, careful consideration is needed when evaluating AC utilisation. The specific surface area of activated charcoals varies from a low of 500 m^2/g for lignin-based AC to more than 3,500 m^2/g for superactivated charcoals (Ramos *et al.*, 1996a).

Hatch *et al.* (1982) effectively utilised AC as an antidote for lethal doses of aflatoxin B_1 (AFB_1) in goats. Animals consuming the AC 8 hrs after ingestion of 3 mg AFB_1/kg BW had no visible liver lesions and had a lower percentage of hepatic damage (3%) compared to animals receiving no antidote (25%). Following this discovery, Dalvi and Ademoyero (1984) evaluated the ability of AC to prevent functional alterations of the liver in chickens. When chickens were fed a mixture of 6 mg AFB_1/kg BW and 200 mg of AC/kg BW, functional alterations to the liver (quantity of hepatic cytochrome P450, the rate of microsomal metabolism of benzphetamine and the activity of serum glutamic oxalacetic transaminase (SGOT)) were reduced or prevented.

Dalvi and McGowan (1984) fed broiler chicks 10 ppm aflatoxin for 8 weeks. One group of birds received 0.1% AC in the drinking water, while no AC was added to the drinking water of the control group. Improvement in feed consumption and weight gain was observed in the charcoal-treated birds. Activated charcoal was also able to reduce the effect of AFB_1 on the microsomal cytochrome P450 and the activities of benzphetamine demethylase and SGOT. The authors proposed that AFB_1 sequestered by the AC becomes unavailable for gastrointestinal absorption.

Kubena and colleagues (1990b) had very different results. Using broiler chicks they found that activated charcoal included at 0.5% of the diet was not effective in reducing toxicity of purified AFB_1 or of aflatoxins from naturally contaminated material and that AC may have even enhanced the toxic effect. The authors were not able to determine a mechanism through which the activated charcoal would enhance aflatoxin toxicity, but hypothesised that it may be due to chemical alteration of the sites and the rates of sorption and desorption of aflatoxin. Urinary excretion of aflatoxin M_1, an indicator of absorbed aflatoxin, was reduced by adding 0.5% AC to feedstuffs consumed by turkey poults (Edrington *et al.*, 1996). In this experiment however, the performance-associated effects produced by aflatoxin were not ameliorated by dietary AC inclusion.

Bonna *et al.* (1991) fed a diet containing two levels of aflatoxin (low 34 ppb, and high 102 ppb) to mink for a 77 day period. The addition of 1.0% AC to the diet containing the aflatoxins reduced mortality by 50% in the 102 ppb aflatoxin group and prevented mortality in the 34 ppb group. The increased survival time of the mink was noticeable in this experiment. The AC reduced or eliminated the occurrence of histopathological lesions in the liver but could not relieve the mink of the effects of aflatoxicosis as effectively as a hydrated sodium calcium aluminosilicate (HSCAS) at a 0.5% concentration used in the same experiment. The authors suggested using higher concentrations of AC or superactivated charcoals to improve efficacy.

Galvano *et al.* (1996) conducted an experiment on late lactation dairy cattle to study the potential of activated charcoal and HSCAS to reduce the carryover of aflatoxin from feed to milk. Using 12 Friesian cows, the authors compared two activated charcoals and an HSCAS at 2.0% of the diet DM. One of the two activated charcoals significantly reduced transfer of aflatoxin B_1 to aflatoxin M_1 in milk while the second AC was ineffective in reducing this carryover. Diaz *et al.* (2004) saw no significant reduction in milk aflatoxin M_1 concentrations when AC was added at 0.25% of the diet to late lactation Holstein cows.

Buck and Bratich (1986) observed a 70% survival rate (7/10) in rats orally dosed with 6 times the LD_{50} for T-2 toxin (25 ppm) when a commercially available superactivated charcoal was administered immediately after toxin exposure. However, two other AC products tested did not protect the rats exposed to the lethal dose of the toxin. Similarly, Poppenga *et al.* (1987) dosed swine intravenously 3.6 ppm (3 x LD_{50}), and unexpectedly found that AC was an important part of an oral therapy for acute T-2 toxin exposure. They hypothesised that AC could be sequestering T-2 toxin eliminated in the bile thereby reducing reabsorption of the toxin. Additionally, the authors also theorised that AC may aid animals by binding endotoxins produced by intestinal microbes. Such endotoxins may be more easily absorbed during toxicoses due to the damage to the intestinal epithelium commonly seen in acute T-2 toxin toxicity.

Activated charcoal was ineffective when directly mixed in a complete chicken diet contaminated with 4 ppm ochratoxin A (Rotter *et al.*, 1989). The authors concluded that the addition of AC was an impractical method of reducing ochratoxin A toxicity in poultry chronically exposed to this toxin. Similarly, Solfrizzo *et al.* (2001) found that sphinganine:sphingosine ratios (Sa:So), a powerful biomarker for fumonisin B_1 exposure, were not modified in rats consuming 2% AC plus fumonisin B_1 as compared to the observed change in Sa:So in the fumonisin B_1-fed group. More recently Piva *et al.* (in press) found no positive effect of AC on feed efficiency, on serum concentrations of cholesterol, free sphinganine, sphingosine-1-phosphate and sphinganine-1-phosphate or in activities of γ-glutamyltransferase (GGT) and glutamic oxaloacetic transaminase (GOT) in pigs consuming diets containing 30 ppm fumonisin B_1 from culture material. Interestingly, the authors saw that histopathological effects in liver, lung, heart, pancreas, spleen, intestine and lymph nodes were more severe in animals given fumonisin B_1 plus AC, suggesting that the AC treatment worsened the toxic effects of fumonisin B_1. The authors hypothesised that the activated carbon could also be sequestering important nutrients such as vitamins, minerals, and essential amino acids, all of which play important roles in the overall health and performance of animals.

Activated charcoal is a relatively nonspecific sequestrant with great potential as an antidote for acute exposure to several mycotoxins. However, the great variability in the results of long-term exposure experiments and its potential for also sequestering important nutrients diminish its overall practical effectiveness for routine dietary inclusion.

Silicate minerals

The largest and most complex class of mycotoxin sequestering agents are the silicate minerals. Within this group there are two important subclasses, the phyllosilicate subclass and the tectosilicate subclass. The mineral clays belonging to the phyllosilicate subclass include important sequestrants like the montmorillonite/smectite group, the kaolinite group and the illite (or clay-mica) group. The tectosilicates include the important and highly studied zeolites.

MONTMORILLONITE/BENTONITE

Montmorillonite is the main constituent of bentonite. Based on interchangeable cation composition, bentonites can be classified as calcium, magnesium, potassium or sodium bentonites. Due to their montmorillonite content, bentonites swell and form thixotropic gels. Montmorillonites and hence bentonites are used as suspension-stabilising additives in coatings, as bonding agents for foundry sands and foundry washes. As a result of their ion exchange capabilities, these products have been widely utilised as mycotoxin sequestering agents.

Masimango *et al.* (1978) demonstrated early on that several bentonites had the ability to bind to aflatoxin B_1 in a buffer solution. Moreover, Dvorak (1989) showed that a bentonite was effective in sequestering aflatoxin B_1 in different liquid media such as water, saline solution, serum of pigs, stomach fluid of pigs, and bovine rumen fluid. Winfree and Allred (1992) and Ramos and Hernandez (1996) also demonstrated the high capacity of bentonites to sequester aflatoxins in several different solutions. These authors hypothesised that bentonites would sequester aflatoxins *in vivo,* thus reducing aflatoxicosis.

In a series of experiments on growing swine Lindemann *et al.* (1993) demonstrated that the addition of a sodium bentonite (0.5%) to diets contaminated with 800 ppb aflatoxin B_1 improved average daily feed intake and increased average daily gain. Bentonite supplementation significantly improved concentrations of blood urea, total protein, albumin and activities of aspartate aminotransferase (AST), alkaline phosphatase (ALP) and γ-glutamyl aminotransferase (GGT), all of which had been significantly altered by the mycotoxin. Schell *et al.* (1993a; 1993b) also saw a positive effect of bentonite on growing pigs consuming aflatoxin-contaminated diets without negatively affecting mineral metabolism. Similarly, addition of a montmorillonite (5 g/kg) to aflatoxin-exposed rats prevented the degenerative changes in hepatic and renal tissue thus offering protection against serum biochemical parameters affected by the toxin (Abdel-Wahhab *et al.,* 2002). These authors postulated that the bentonite formed a complex with the toxin that prevented the absorption of the aflatoxin across the intestinal epithelium.

ZEOLITES

Zeolites are a group of silicates consisting of interlocking tetrahedrons of SiO_4 and AlO_4. The aluminosilicate framework is negatively charged and attracts and holds positive cations within the structure. Zeolites have large pores that provide space for large cations such as sodium, potassium, barium and calcium and even relatively larger molecules such as water, ammonia, carbonate ions and nitrate ions. In some zeolites, the spaces are interconnected to form long and wide channels of varying sizes depending on the mineral composition. These channels facilitate the movement of the resident ions and molecules in and out of the structure. Zeolites are characterised by their ability to lose and absorb water without damage to the crystalline structure. There are about 45 naturally ocurring minerals that are recognised as members of the zeolite group, which can be subdivided into silicate zeolites, synthetic zeolites and phosphate minerals that have a zeolite structure. The complexity of this combined group is extensive with over 120 structural variations.

Since zeolites were identified as effective *in vitro* sequestrants of aflatoxin B_1 (Dvorak, 1989) several researchers have studied their *in vivo* response in different animal models. Scheideler (1993) fed a natural zeolite to broilers consuming 2.5 ppm aflatoxin B_1. The zeolite added at 1% of the diet alleviated the growth depression and reduced the increase in liver lipid concentration caused by the aflatoxin. The zeolite did however decrease serum phosphorus (P) and chloride (Cl) concentrations independently of aflatoxin exposure. Sova *et al.* (1991) showed that zeolite inclusion at 5% of the diet significantly reduced heterophilia and lymphopenia in broilers consuming an aflatoxin-contaminated diet. The authors however noted that the zeolite did not protect the animals against the acute alteration of liver parenchyma caused by aflatoxin B_1.

Harvey *et al.* (1993) fed growing broiler chickens a diet contaminated with 3.5 ppm total aflatoxins with or without the inclusion of five different commercially available zeolites. Three of five tested zeolites were not effective in alleviating the decrease in body weight gain associated with the aflatoxins. The other two products were clearly effective in alleviating the aflatoxin-associated problems. More recently Zaghini *et al.* (1998) fed layers 2.5 ppm aflatoxin B_1 with or without the inclusion of a zeolite. Zeolite (clinoptilolite) at 2% of the diet was effective in reducing liver concentrations of aflatoxin B_1. Results of a study with pregnant rats were not, however, as positive. Clinoptilolite addition resulted in maternal liver lesions more severe than with aflatoxin B_1 alone. The authors suggested that the clinoptilolite may interact with dietary components that modulate aflatoxicosis (Mayura *et al.,* 1998). Abdel-Wahhab and Nada (1998) found a significant reduction in aflatoxin M_1 output in urine of rats exposed to aflatoxin. The authors concluded that by tightly binding the aflatoxin, the zeolite reduces intestinal absorption of the toxin and hepatic metabolism into aflatoxin M_1 and hence excretion in urine. Similarly, Parlat *et al.* (1999) showed that the addition of clinoptilolite significantly reduced the negative effects of aflatoxin on feed consumption and body weight gain in Japanese quail. Clinoptilolite at 1% of the diet however did not reduce the negative effects in broilers exposed to cyclopiazonic acid (Dwyer *et al.,* 1997).

HSCAS

A hydrated sodium calcium aluminosilicate (HSCAS) from natural zeolite (Novasil™, Engelhard Corp., Cleveland, OH) is the most widely studied mycotoxin-sequestering agent among the mineral clays. Interest in its potential as a mycotoxin sequestrant was based on ability of the silicate mineral to sequester positively charged or cationic compounds (Ramos and Hernandez, 1997). Two early studies (Davidson *et al.,* 1987; Phillips *et al.,* 1987) identified this product, which is sold as a feed additive for reducing moisture in feedstuffs and improving flow properties during storage and transport, as also being effective in reducing aflatoxicosis in poultry. Based on a series of *in vitro* screening tests of several silicate minerals Phillips *et al.* (1988; 1987) determined that this HSCAS was the most effective mineral clay for aflatoxin sequestration due to its high affinity for and stable association with aflatoxin B_1. Later on, Phillips *et al.* (1990a; 1990b) determined the that the ß-carbonyl portion of the aflatoxin molecule bound to the uncoordinated edge site containing aluminium ions in this HSCAS. Moreover, the authors determined that the maximum quantity of aflatoxin that could be bound was 200-232 nmol per milligram of HSCAS and that the reaction reached equilibrium after 30 minutes. What followed were a series of experiments that provided evidence of the potential of this HSCAS as an aflatoxin sequestering agent. Ramos and Hernandez (1997) reviewed over 20 publications, which demonstrated the *in vivo* capacity of this aflatoxin sequestering agent in poultry (chickens and turkeys), swine, ruminants (dairy cattle, lambs, and goats) and mink.

As this HSCAS was able to sequester aflatoxin in several animal models, its ability to bind other mycotoxins has also been studied. Bursian *et al.* (1992) showed that although addition of 0.5% NovaSil® to a diet contaminated with zearalenone did not alleviate the oestrogenic effect of the toxin, it did mitigate some of the adverse effects of zearalenone on reproductive performance of female mink. Similarly, Huff *et al.* (1992) concluded that addition of 0.5% HSCAS to a diet contaminated with ochratoxin A and aflatoxin, alone or combination, had little effect on toxicity caused by either ochratoxin

A alone or the combination of aflatoxin and ochratoxin. Kubena *et al.* (1990a) conducted a series of experiments in which the HSCAS, at 0.5% of the diet, showed no reduction in T-2 toxin and diacetoxyscirpenol (DAS) toxicosis in broilers. More recently, Danicke *et al.* (2004) saw no reduction in deoxynivalenol (DON) concentrations in serum and no positive effect on performance in growing pigs consuming diets containing approximately 2.7 mg DON/kg during the starter period and 2.8 mg DON/kg during the grower period. The authors concluded that HSCAS was ineffective in sequestering DON and reducing its intestinal absorption. Based on the identified mechanism for the formation of the aflatoxin-HSCAS complex, is not unexpected that this product would be effective in adsorbing other mycotoxins.

Cholestyramine

Cholestyramine is an anion exchange acid-binding resin utilised in human medicine for absorbing bile acids in the gastrointestinal tract and as a therapy to reduce cholesterol. Cholestyramine has been shown *in vitro* to sequester ochratoxin A (Bauer, 1994) and zearalenone (Ramos *et al.*, 1996b). Using rats as an experimental model, Madhyastha *et al.* (1992) found that cholestyramine lowered ochratoxin A (OTA) concentrations in plasma. Rats consuming 1 ppm OTA or 3 ppm OTA plus cholestyramine had lower enzyme activities (GGT and NAG) as well as higher concentrations of the unmetabolised OTA. The authors hypothesised that cholestyramine reduced OTA absorption by either reducing concentrations of bile acids, which may be important in OTA absorption in the upper gastrointestinal tract, or by direct absorption of the OTA molecule. Since the chemical structure of OTA is quite similar to that of bile acids, it is plausible that both mechanisms are involved. Due to the high cost of this product, its practicality is questionable.

Chlorophyllin

Chlorophyllin, a water-soluble derivative of the green plant pigment chlorophyll, has been shown effective in reducing the toxicity associated with aflatoxin. Several researchers have demonstrated the chemoprotective properties of chlorophyllin (Atroshi *et al.*, 2002; Breinholt *et al.*, 1995; Dashwood *et al.*, 1998). Researchers theorised that the antioxidant properties of this chlorophyll derivative were responsible for its chemoprotective abilities. More recently researchers identified that a complex between chlorophyllin and aflatoxin was actually the main mechanism responsible for chemoprotection. Arimoto-Kobayashi *et al.* (1997) reported that chlorophyllin, when mixed with chitosan (a polyglucosamine), establishes an insoluble salt-like material that can form a non-covalent bond with polycyclic compounds like aflatoxin.

Breinholt (1999) utilised rainbow trout to further elucidate the mechanism of activity of chlorophyllin. The authors determined that the formation of the complex between chlorophyllin and AFB_1 was highly effective in reducing the absorption of AFB_1 by rainbow trout and that *in situ* target organ protection was an insignificant mechanism of chemoprotection. Based on these reports chlorophyllin can be considered effective in reducing aflatoxicosis. Because of the specific mechanism of adsorption chlorophyllin would only be expected to be effective for aflatoxin and other mycotoxins having a polycyclic structure.

Yeast cell wall-derived sequestering agents

Initial research with a yeast cell culture preparation based on a *Saccharomyces cerevisiae* strain 1026 in which performance was improved in poultry (Stanley *et al.*, 1993), led researchers to hypothesise that the yeast culture had the ability to sequester mycotoxins. Studies with whole yeast cells led to identification of specific cell wall components in yeast that interacted with mycotoxins. Modification in manufacturing techniques have allowed the production of a specific modified yeast cell wall preparation with the ability to adsorb a range of mycotoxins (Dawson *et al.*, 2001). This preparation, commercially available as Mycosorb® (Alltech Inc., Nicholasville KY, USA), has been studied in detail with over 20 published reports in the scientific literature on both mode of action and animal response.

The adsorptive capacity of the carbohydrate complexes in the yeast cell wall offers an interesting alternative to inorganic adsorbents. Mineral clay-based compounds and activated carbon are generally used at high concentrations (>1.0% of the diet) in animal feeds. In animals such as broilers and young pigs, where feed intake may limit economic performance, it is important that non-nutritive feed additives take as little space in the diet as possible. Additionally, high levels of inclusion could provide excessive sequestrant capacity that may decrease the bioavailability of important micronutrients. In contrast, the high toxin affinity and adsorptive capacity of the yeast cell wall-derived adsorbents allow use at much lower concentrations (<0.1% of the diet). Since these preparations can be used at low concentrations (1-2 kg/T), their inclusion would have minimal impact on diet nutrient density. In addition, the types of toxin-cell wall complexes formed (described below) indicate activity more specific to toxins and would not be expected to greatly influence the availability of dietary micronutrients.

MODE OF ACTION

Dawson *et al.* (2001) utilised multiple isotherm equations to evaluate the adsorption of aflatoxin B_1 to the yeast cell wall sequestering agent, Mycosorb® (MS). In these studies, MS was effective in adsorbing aflatoxin at very low concentrations. This reflects the high affinity of MS for aflatoxin and suggested that this material would be effective at the low mycotoxin concentrations found in the gastrointestinal tract. A combination of a high affinity for aflatoxins and high adsorption capacity makes the use of low concentrations of MS attractive for controlling aflatoxicosis.

Further *in vitro* studies have helped to characterise the interactions between the yeast cell wall-derived adsorbent and specific mycotoxins. Serial elution of the aflatoxins adsorbed to yeast glucomannan clearly demonstrated that the adsorption of the mycotoxin is a concentration-dependent reversible process and that aflatoxins are not modified during adsorption (Dawson *et al.*, 2001). This is important, since it clearly established the basic kinetic mechanisms associated with adsorption. Aflatoxin adsorption is also influenced by the relative acidity (pH) in the aqueous environment. Maximum adsorption occurs at a pH of approximately 4.0. In addition, adsorption is also influenced by the relative phosphate concentration in an aqueous environment. Maximum adsorption was observed in a buffer that contained 0.5 M phosphate. Both pH and phosphate concentration optima are consistent with those found in the gastrointestinal tract and suggest that the conditions in the gastrointestinal tract would enhance adsorption and not decrease the mycotoxin-adsorbent interactions.

Recently a series of experiments have utilised zearalenone (ZEA) to help explain the mode of action in the formation of the mycotoxin-MS complex. The kinetics of adsorption can be described using a concentration-dependent model (Figure 1). Yiannikouris *et al.* (2003; 2004a) determined that two types of interaction were responsible for the formation of the complex. These consist firstly of stable interactions between glucan and toxin molecules, and secondly of less stable glucan-toxin interactions. Adsorption is associated with ß-D-glucans in yeast cell wall material; and weak non-covalent bonds are involved in the formation of the ZEA-cell wall complexes. This indicates that the chemical interaction between ß-D-glucans and ZEA is not the result of covalent linkage but is better described as ***toxin adsorption***. Nuclear Magnetic Resonance (NMR) and molecular modeling studies showed that the hydroxyl groups of ZEA are involved in these interactions and that there is geometric complementarity between ZEA and the single helix of ß-(1,3)-D-glucans (Figure 2). Modeling studies were further utilised to confirm the NMR data and made it possible to visualise the formation of hydrogen bonds involving not only the hydroxyl groups but also the lactone and ketone groups of ZEA, which interact with the ß-D-glucopyranose units of ß-(1,3)-D-glucans. The authors also utilised molecular mechanic analyses to help confirm the existence of less stable interactions, called 'stacking' interactions, involving van der Waals forces between the aromatic ring of ZEA and the ring structures of ß-D-glucopyranose units. Analysis of the side groups of some ß-(1,6)-D-glucan units allowed the authors to correlate macroscopic observations *in vitro* with adsorption capacity, which confers on these molecules the ability to stabilise the complex formed. These chains increase the stability of the complex formed between ZEA and ß-(1,3)-D-glucan by increasing the contribution of van der Waals forces to the potential energy of the complex. The physical and chemical properties of the yeast cell wall determine its ability to adsorb mycotoxins. Additionally, the highly complex organisation of the yeast cell wall increases adsorptive ability of the glucans.

Animal trials

Supplementation with the yeast glucomannan polymer Mycosorb® (MS) has been found to be beneficial in reducing the individual and combined adverse effects of several mycotoxins in poultry. Raju and Devegowda (2000) demonstrated that the inclusion of MS improved the body weight gains and antibody titres suppressed by aflatoxin, ochratoxin and

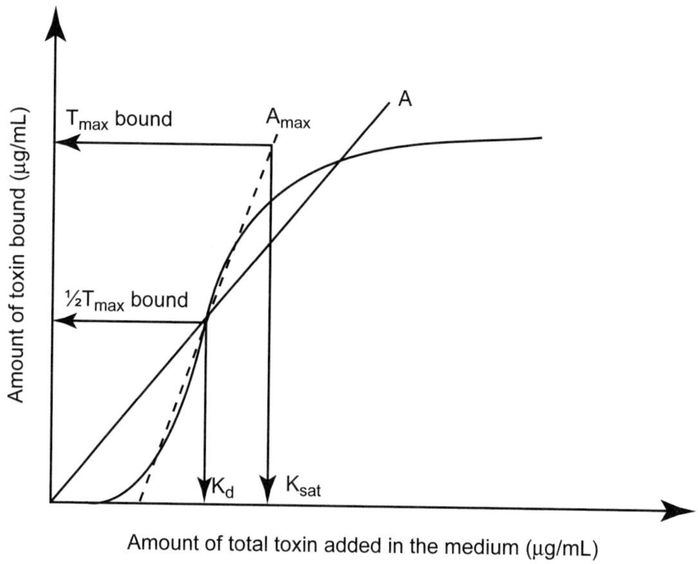

Figure 1. Adsorption curve demonstrating concentration-dependent interaction of the active ß-glucan fraction of yeast with a mycotoxin. K_{sat} = saturation point (µg/mL), A = affinity rate (%), and A_{max} = maximal affinity (%).

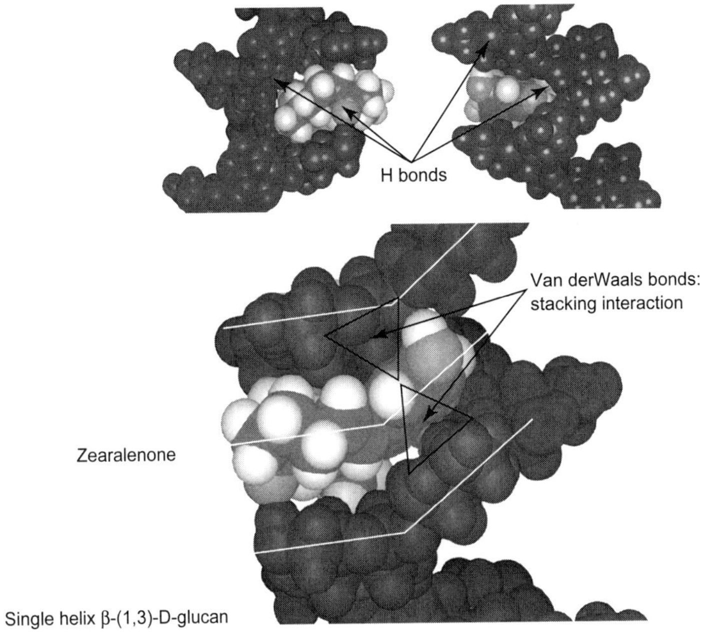

Figure 2. Computer-generated view of the energy-minimized structure of the docking of the most favourable conformation of zearalenone into the single helix of ß-(1,3)-D-glucan chain. Arrows indicate hydrogen bonds. Lines highlight the steric complementarity between zearalenone and ß-(1,3)-D-glucan. (Adpated from Yiannikouris *et al.*, 2004b)

T-2 toxin. The polymer also improved serum biochemical and haematological parameters. Similarly, broilers consuming a mixture of mycotoxins (168 ppb aflatoxin, 8.4 ppb ochratoxin A, 54 ppb zearalenone and 32 ppb T-2 toxin) had significantly lower body weight and feed consumption, which were alleviated by the inclusion of MS at 0.05% of the diet (Aravind *et al.*, 2003). Dvorska and Surai (2001) evaluated the antioxidant status of quail consuming T-2 toxin plus either a zeolite (3% of diet) or MS (0.1% of the diet). The authors concluded that, unlike the tested zeolite, the glucomannan polymer was able to significantly inhibit liver antioxidant and vitamin A depletion caused by the T-2 toxin. They stated that toxin adsorption by the polymer could be inhibiting toxin absorption from the intestine as well as preventing the participation of T-2 toxin in development of oxidative stress in the intestine. Due to antioxidant depletion in the liver, susceptibility to lipid peroxidation increased more than 2-fold in the test animals. Inclusion of MS in the T-2 toxin-contaminated diet significantly decreased tissue susceptibility to lipid peroxidation in comparison to diets containing toxin only, although the inclusion of the MS was unable to completely mitigate the powerful stimulating effect of T-2 toxin on lipid peroxidation.

Aurofusarin, a secondary metabolite of *Fusarium graminearum,* is responsible for significant changes in fatty acid and antioxidant profiles of quail eggs. The inclusion of MS in the diet of quail consuming aurofusarin prevented changes in fatty acid and antioxidant composition in the egg yolk as well as helped to maintain vitamin E concentration in the egg yolk and liver of the newly hatched quail (Dvorska *et al.*, 2003). Furthermore, retinol, retinyl esters and carotenoids in the liver of newly hatched chicks from eggs contaminated with aurofusarin were also increased due to MS in the diet. Consequently, lipid peroxidation in tissues of newly hatched quail was significantly reduced (Dvorska and Surai, 2004).

Swamy *et al.* (2002a) noted that a blend of *Fusarium* toxins in contaminated corn and wheat (up to 8.3 mg DON/kg) in diets fed to slow-growing broiler chickens resulted in growth depression only in the finisher period. Addition of MS at 2.0 kg/T of feed partially prevented the growth depression. A subsequent study of more rapidly growing broilers indicated that feeding contaminated grains depressed growth and feed intake in the grower period and that this could also be prevented by supplementation with the MS yeast cell wall polymer (Swamy *et al.*, 2004). Raju and Devegowda (2002) fed broilers diets containing aflatoxin (2 mg/kg) and T-2 toxin (1 mg/kg). Broilers consuming the mycotoxin-contaminated diets had significantly reduced feed intake, weight gain and feed efficiency, weights of thymus and bursa of Fabricius, antibody titres against Newcastle disease and infectious bursal disease. The MS glucomannan (1 kg/T of feed) significantly improved body weight, feed intake, antibody titres, weights of the thymus and the bursa of Fabricius.

Chowdhury and Smith (2004) evaluated the effects of feeding laying hens diets naturally contaminated with several *Fusarium* toxins with or without the inclusion of MS. The authors found an initial reduction in feed intake (month 1), which was followed by increased feed intake (months 2 and 3) and large reductions in the utilisation of feed for egg production. Supplementation with 2.0 kg MS/T prevented these effects as well as adverse effects on egg production, egg shell weight, blood uric acid concentrations and biliary immunoglobulin A concentrations.

Starter pigs (10 kg initial BW) fed graded levels of *Fusarium* toxin-contaminated grains for three weeks had depressed weight gain and feed intake in the first week of the experimental period when approximately 4.6 mg DON/kg was fed (Swamy *et al.*, 2002b). Supplementation of 2.0 kg MS/T could not prevent the severe depression in growth rate and feed consumption. However, MS did prevent some of the mycotoxin-induced neurochemical changes, including depression of norepinephrine concentration in the pons. The depression in norepinephrine concentration was thought to be caused by the mycotoxin fusaric acid. Hence it is possible that the glucomannan polymer is effective in reducing its absorption from the intestine. Mycosorb® addition also prevented elevation in blood concentrations of immunoglobulins A and M in pigs fed contaminated grains. In a similar study with starter pigs, the authors concluded that the greatest factor contributing to reduced growth rate when *Fusarium* toxin contaminated grains were fed was reduced feed intake (Swamy *et al.*, 2003). Although supplementation with yeast cell wall polymer prevented some of the toxin-induced changes in metabolism, the feeding of 2.0 kg MS/T could not completely prevent appetite suppression.

Conclusions

Mycotoxin contamination of crops and the ensuing consumption of these contaminated ingredients by animals is an inevitable part of animal production systems. The most widely utilised strategy available to reduce or impede this exposure is the addition of materials that sequester the toxin and inhibit intestinal toxin absorption. The available data suggest that care must be taken when examining potential sequestering agents. *In vitro* tests, although informative, should not be used as the sole indicator of sequestrant efficacy. Additionally, generalisation should be avoided, as sequestrant compounds differ in efficacy even within the same category. Moreover, the chemical complexity of mycotoxins means that the effectiveness of a compound in sequestering one mycotoxin does not mean equal ability to sequester other mycotoxins. Therefore, selection of a mycotoxin sequestering agent must be based on published scientific data, gathered under controlled scientific conditions, with the target animal species exposed to the levels and varieties of mycotoxin contamination seen under practical conditions.

References

Abdel-Wahhab, M.A. and S.A. Nada. 1998. Effect of phyllosilicate clay and zeolite on the metabolic profile of aflatoxin B_1 in Sprague-Dawley rats. Rev. Med. Vet. 149:663.

Abdel-Wahhab, M.A., S.A. Nada and F.A. Khalil. 2002. Physiological and toxicological responses in rats fed aflatoxin-contaminated diet with or without sorbent materials. Anim. Feed Sci. Technol. 97:209-219.

Aravind, K.L., V.S. Patil, G. Devegowda, B. Umakantha and S.P. Ganpule. 2003. Efficacy of esterified glucomannan to counteract mycotoxicosis in naturally contaminated feed on performance and serum biochemical and hematological parameters in broilers. Poult. Sci. 82:571-576.

Arimoto Kobayashi, S., N. Harada, R. Tokunaga, I. Odo Jun and H. Hayatsu. 1997. Adsorption of mutagens to chlorophyllin-chitosan, an insoluble form of chlorophyllin. Mutation Res. 381:243-249.

Atroshi, F., A. Rizzo, T. Westermarck and T. Ali-Vehmas. 2002. Antioxidant nutrients and mycotoxins. Toxicology 180:151-167.

Bauer, J. 1994. Moglichkeiten zur etgiftung mykotoxin-haltiger futtermittel. Monatsh. Veterinarmed 49:175-181.

Bonna, R.J., R.J. Aulerich, S.J. Bursian, R.H. Poppenga, W.E. Braselton and G.L. Watson. 1991. Efficacy of hydrated sodium aluminosilicate and activated charcoal in reducing the toxicity of dietary aflatoxin to mink. Arch. Environ. Contam. Toxicol. 20:441-447.

Breinholt, V., D. Arbogast, P. Loveland, C. Pereira, R. Dashwood, J. Hendricks and G. Bailey. 1999. Chlorophyllin chemoprevention in trout initiated by aflatoxin B1 bath treatment: an evaluation of reduced bioavailability vs. target organ protective mechanisms. Toxicol. Appl. Pharmacol. 158:141-151.

Breinholt, V., J. Hendricks, C. Pereira, D. Arbogast and G. Bailey. 1995. Dietary chlorophyllin is a potent inhibitor of aflatoxin B-1 hepatocarcinogenesis in rainbow trout. Cancer Res. 55:57-62.

Buck, W.B. and P.M. Bratich. 1986. Activated charcoal: preventing unnecessary death by poisoning. Vet. Med. 81:73-77.

Bursian, S.J., R.J. Aulerich, J.K. Cameron, N.K. Ames and B.A. Steficek. 1992. Efficacy of hydrated sodium calcium aluminosilicate in reducing the toxicity of zearalenone to mink. J. Appl. Toxicol. 12:85-90.

CAST. 2003. Mycotoxins: risk in plant, animal and human systems. 139, Council for Agricultural Science and Technology, Ames, Iowa.

Chowdhury, S.R. and T.K. Smith. 2004. Effects of feeding blends of grains naturally contaminated with *Fusarium* mycotoxins on performance and metabolism of laying hens. Poult. Sci. 83:1849-1856.

Dalvi, R.R. and A.A. Ademoyero. 1984. Toxic effect of aflatoxin B_1 in chickens given feed contaminated with *Aspergillus flavus* and reduction of the toxicity by activated charcoal and some chemical agents. Avian Dis. 28:61-69.

Dalvi, R.R. and C. McGowan. 1984. Experimental induction of chronic aflatoxicosis in chickens by purified aflatoxin B_1 and its reversal by activated charcoal, phenobarbital and reduced glutathione. Poult. Sci. 63:485-491.

Danicke, S., H. Valenta, S. Doll, M. Ganter and G. Flachowsky. 2004. On the effectiveness of a detoxifying agent in preventing fusario-toxicosis in fattening pigs. Anim. Feed Sci. Technol. 114:141-157.

Dashwood, R., T. Negishi, H. Hayatsu, V. Breinholt, J. Hendricks and G. Bailey. 1998. Chemopreventive properties of chlorophylls towards aflatoxin B_1: a review of the antimutagenicity and anticarcinogenicity data in rainbow trout. Mutation Res.

399:245-253.

Davidson, J.N., J.G. Babish, K.A. Delaney, D.R. Taylor and T.D. Phillips. 1987. Hydrated sodium aluminosilicate decreases the bioavailability of aflatoxin in the chicken. Poult. Sci. 66(abstr.):89.

Dawson, K.A., J. Evans and M. Kudupoje. 2001. Understanding the adsorption characteristics of yeast cell wall preparations associated with mycotoxin binding. In: *Science and Technology in the Feed Industry: Proceedings of Alltech's 17th Annual Symposium* (K.A. Jacques and T.P. Lyons, eds). Nottingham University Press, pp. 169-182.

Diaz, D.E., W.M. Hagler, Jr., J.T. Blackwelder, J.E. Eve, B.A. Hopkins, K.L. Anderson, F.T. Jones and L.W. Whitlow. 2004. Aflatoxin binders II: reduction of aflatoxin M1 in milk by sequestering agents of cows consuming aflatoxin in feed. Mycopathologia 157:233-241.

Dvorak, M. 1989. Ability of bentonite and natural zeolite to adsorb aflatoxin from liquid media. Vet. Med. (Praha) 34:733-741.

Dvorska, J.E. and P.F. Surai. 2001. Effects of T-2 toxin, zeolite and Mycosorb on antioxidant systems of growing quail. Asian-Aust. J. Anim. Sci. 14:1752-1757.

Dvorska, J.E. and P.F. Surai. 2004. Protective effect of modified glucomannans against changes in antioxidant systems of quail egg and embryo due to aurofusarin consumption. Asian-Aust. J. Anim. Sci. 17:434-440.

Dvorska, J.E., P.F. Surai, B.K. Speake and N.H.C. Sparks. 2003. Protective effect of modified glucomannans against aurofusarin-induced changes in quail egg and embryo. Comp. Biochem. Physiol. C 135:337-343.

Dwyer, M.R., L.F. Kubena, R.B. Harvey, K. Mayura, A.B. Sarr, S. Buckley, R.H. Bailey and T.D. Phillips. 1997. Effects of inorganic adsorbents and cyclopiazonic acid in broiler chicks. Poult. Sci. 76:1141-1149.

Edrington, T.S., A.R. Sarr, L.F. Kubena, R.B. Harvey and T.D. Phillips. 1996. Hydrated sodium calcium aluminosilicate (HSCAS), acidic HSCAS and activated charcoal reduce urinary excretion of aflatoxin M_1 in turkey poults. Lack of effect by activated charcoal on aflatoxicosis. Toxicol. Lett. 89:115-122.

Galvano, F., A. Pietri, T. Bertuzzi, G. Fusconi, M. Galvano, A. Piva and G. Piva. 1996. Reduction of carryover of aflatoxin from cow feed to milk by addition of activated carbons. J. Food Prot. 59:551-554.

Galvano, F., A. Piva, A. Ritieni and G. Galvano. 2001. Dietary strategies to counteract the effects of mycotoxins: a review. J. Food Prot. 64:120-131.

Grant, P. G. and T. D. Phillips. 1998. Isothermal adsorption of aflatoxin B_1 on HSCAS clay. J. Agric. Food Chem. 46:599-605.

Harvey, R.B., L.F. Kubena, M.H. Elissalde and T.D. Phillips. 1993. Efficacy of zeolite core compounds on the toxicity of aflatoxin growing broiler chickens. Avian Dis. 37:67-73.

Hatch, R.C., J.D. Clark, A.V. Jain and R. Weiss. 1982. Induced acute aflatoxicosis in goats: treatment with activated charcoal or dual combinations of oxytetracycline, stanozol and activated charcoal. Am. J. Vet. Res. 43:644-648.

Huff, W.E., L.F. Kubena, R.B. Harvey and T.D. Phillips. 1992. Efficacy of hydrated sodium calcium aluminosilicate to reduce the individual and combined toxicity of aflatoxin and ochratoxin A. Poult. Sci. 71:64-69.

Huwig, A., S. Freimund, O. Kappeli and H. Dutler. 2001. Mycotoxin detoxification of animal feed by different adsorbents. Toxicol. Lett. 122:179-188.

Kubena, L.F., R.B. Harvey, W.E. Huff, D.E. Corrier, T.D. Phillips and G.E. Rottinghaus. 1990a. Efficacy of a hydrated sodium calcium aluminosilicate to reduce the toxicity of aflatoxin and T-2 toxin. Poult. Sci. 69:1078-1096.

Kubena, L.F., R.B. Harvey, T.D. Phillips, D.E. Corrier and W.E. Huff. 1990b. Diminution of aflatoxicosis in growing chickens by the dietary addition of a hydrated sodium calcium aluminosilicate. Poult. Sci. 69:727-735.

Lindemann, M.D., D.J. Blodgett, E.T. Kornegay

and G.G. Schurig. 1993. Potential ameliorators of aflatoxicosis in weanling/ growing swine. J. Anim. Sci. 71:171-178.

Lopez-Garcia, R. and D.L. Park. 1998. Effectiveness of postharvest procedures in management of mycotoxin hazards. In: *Mycotoxins in Agriculture and Food Safety* (D. Bhatnagar, ed). Marcel Dekker, Inc., New York, NY, pp. 45-65.

Madhyastha, S.M., A.A. Frohlich and R.R. Marquardt. 1992. Effect of dietary cholestyramine on the elimination pattern of ochratoxin A in rats. Food Chem. Toxicol. 30:709-714.

Masimango, N., J. Remacle and J.L. Ramaut. 1978. The role of adsorption in the elimination of aflatoxin B_1 from contaminated media. Eur. J. Appl. Microbiol. Biotechnol. 6:101-105.

Mayura, K., M.A. Abdel-Wahhab, K.S. McKenzie, A.B. Sarr, J.F. Edwards, K. Naguib and T.D. Phillips. 1998. Prevention of maternal and developmental toxicity in rats via dietary inclusion of common aflatoxin sorbents: potential for hidden risks. Toxicol. Sci. 41:175-182.

Parlat, S.S., A.O. Yildiz and H. Oguz. 1999. Effect of clinoptilolite on performance of Japanese quail (*Coturnix coturnix japonica*) during experimental aflatoxicosis. Brit. Poult. Sci. 40:495-500.

Phillips, T.D., B.A. Clement, L.F. Kubena and R.B. Harvey. 1990a. Detection and detoxification of aflatoxins: prevention of aflatoxicosis and aflatoxin residues with hydrated sodium calcium aluminosilicate. Vet. Hum. Toxicol. 32:15-19.

Phillips, T.D., L.F. Kubena, R.B. Harvey, D.R. Taylor and N.D. Heidelbaugh. 1987. Mycotoxin hazards in agriculture: new approach to control. J. Am. Vet. Med. Assoc. 12:1617.

Phillips, T.D., L.F. Kubena, R.B. Harvey, D.R. Taylor and N.D. Heidelbaugh. 1988. Hydrated sodium calcium aluminosilicate: a high affinity sorbent for aflatoxin. Poult. Sci. 67:243-247.

Phillips, T.D., A.B. Sarr, B.A. Clement, L.F. Kubena and R.B. Harvey. 1990b. Prevention of aflatoxicosis in farm animals via selective chemisorption of aflatoxin. In: *Mycotoxins, Cancer and Health* (G.A. Bray and D.H. Ryan, eds). Pennington Center Nutrition Series No. 1., Louisiana State University Press, Baton Rouge, LA, pp. 223-237.

Piva, A., G. Cassadei, G. Pagliuca, E. Cabassi, F. Galvano, M. Solfrizzo, R.T. Riley and D.E. Diaz. In press. Inability of activated carbon to prevent the toxicity of culture material containing fumonisin B1 when fed to weaned piglets. J. Anim. Sci.

Poppenga, R.H., G.R. Lundeen, V.R. Beasley and W.B. Buck. 1987. The assessment of a general therapeutic protocol for the treatment of acute toxicosis in swine. Vet. Hum. Toxicol. 29:237-239.

Raju, M.V.L.N. and G. Devegowda. 2000. Influence of esterified-glucomannan on performance and organ morphology, serum biochemistry and haematology in broilers exposed to individual and combined mycotoxicosis (aflatoxin, ochratoxin and T-2 toxin). Brit. Poult. Sci. 41:640-650.

Raju, M.V.L.N. and G. Devegowda. 2002. Esterified glucomannan in broiler chicken diets contaminated with aflatoxin, ochratoxin and T-2 toxin: evaluation of its binding ability (*in vitro*) and efficacy as immunomodulatory. Asian-Aust. J. Anim. Sci. 15:1051-1056.

Ramos, A.J. and E. Hernandez. 1996. *In vitro* aflatoxin adsorption by means of a montmorillonite silicate. A study of absorption isotherms. Anim. Feed Sci. Technol. 62:263-269.

Ramos, A.J. and E. Hernandez. 1997. Prevention of aflatoxicosis in farm animals by means of hydrated sodium calcium aluminosilicate addition to feedstuffs: a review. Anim. Feed Sci. Technol. 65:197-206.

Ramos, A.J., J. Fink-Gremmels and E. Hernandez. 1996a. Prevention of toxic effects of mycotoxins by means of non-nutritive adsorbent compounds. J. Food Prot. 59:631-641.

Ramos, A.J., E. Hernandez, J.M. PlaDelfina and

M. Merino. 1996b. Intestinal absorption of zearalenone and *in vitro* study of non-nutritive sorbent materials. Int. J. Pharm. 128:129-137.

Rotter, R.G., A.A. Frohlich and R.R. Marquardt. 1989. Influence of dietary charcoal on ochratoxin A toxicity in Leghorn chicks. Can. J. Vet. Res. 53:449-453.

Scheideler, S.E. 1993. Effect of various types of aluminosilicates and aflatoxin B1 on the aflatoxin toxicity, chick performance and mineral status. Poult. Sci. 72:282-288.

Schell, T.C., M.D. Lindemann, E.T. Kornegay and D.J. Blodgett. 1993a. Effect of feeding aflatoxin-contaminated diets with or without clays to weanling and growing pigs on performance, liver function and mineral metabolism. J. Anim. Sci. 71:1209-1218.

Schell, T.C., M.D. Lindemann, E.T. Kornegay, D.J. Blodgett and J.A. Doerr. 1993b. Effectiveness of different types of clay for reducing the detrimental effects of aflatoxin-contaminated diets on performance and serum profiles of weanling pigs. J. Anim. Sci. 71:1226-1231.

Sinha, K.K. 1998. Detoxification of mycotoxins and food safety. In: *Mycotoxins in Agriculture and Food Safety* (K.K. Sinha and D. Bhatnagar, eds). Marcel Dekker, Inc., New York, NY, pp. 45-65.

Solfrizzo, M., M.R. Carratu, G. Avantaggiato, F. Galvano, A. Pietri and A. Visconti. 2001. Ineffectiveness of activated carbon in reducing the alteration of sphingolipid metabolism in rats exposed to fumonisin-contaminated diets. Food Chem. Toxicol. 39:507-511.

Sova, Z., H. Pohunkova, H. Reisnerova, A. Slamova and K. Haisl. 1991. Hematological and histological response to the diets containing aflatoxin B1 and zeolite in broilers of domestic fowl. Acta. Vet. Brno 60:31-40.

Stanley, V.G., R. Ojo, S. Woldesenbet and D.H. Hutchinson. 1993. The use of *Saccharomyces cerevisiae* to suppress the effects of aflatoxicosis in broiler chicks. Poult. Sci. 72:1867-1872.

Swamy, H.VL.N., T.K. Smith, P.F. Cotter, H.J. Boermans and A.E. Sefton. 2002a. Effects of feeding blends of grains naturally contaminated with *Fusarium* mycotoxins on production and metabolism in broilers. Poult. Sci. 81:966-975.

Swamy, H.V.L.N., T.K. Smith, N.A. Karrow and H.J. Boermans. 2004. Effects of feeding blends of grains naturally contaminated with *Fusarium* mycotoxins on growth and immunological parameters of broilers chickens. Poult. Sci. 83:533-543.

Swamy, H.V.L.N., T.K. Smith, E.J. Macdonald, H.J. Boermans and E.J. Squires. 2002b. Effects of feeding a blend of grains naturally contaminated with *Fusarium* mycotoxins on swine performance, brain regional neurochemistry and serum chemistry and the efficacy of a polymeric glucomannan mycotoxin adsorbent. J. Anim. Sci. 80:3257-3267.

Swamy, H.V.L.N., T.K. Smith, E.J. Macdonald, N.A. Karrow, B. Woodward and H.J. Boermans. 2003. Effects of feeding a blend of grains naturally contaminated with *Fusarium* mycotoxins on growth and immunological measurements of starter pigs and the efficacy of a polymeric glucomannan mycotoxin adsorbent. J. Anim. Sci. 81:2792-2803.

Volkl, A., B. Vogler, M. Schollenberger and P. Karlovsky. 2004. Microbial detoxification of mycotoxin deoxynivalenol. J. Basic Microbiol. 44:147-156.

Winfree, R.A. and A. Allred. 1992. Bentonite reduces measurable aflatoxin B1 in fish feed. Progr. Fish-Cult. 54:157-162.

Wood, G.E. 1992. Mycotoxins in foods and feeds in the United States. J. Anim. Sci. 70:3941-3949.

Yiannikouris, A., J. Francois, L. Poughon, C.G. Dussap, G. Bertin, G. Jeminet and J.P. Jouany. 2004a. Alkali extraction of ß-D-glucans from *Saccharomyces cerevisiae* cell wall and study of their adsorptive properties toward zearalenone. J. Agric. Food Chem. 52:3666-3673.

Yiannikouris, A., G. Andre, Alain Buleon, G.

Jeminet, I. Canet, J. Francois, G. Bertin and J-P Jouany. 2004b. Comprehensive conformational study of key interactions involved in zearalenone complexation with ß-D-glucans. Biomacromolecules 5:2176 2185.

Yiannikouris, A., L. Poughon, X. Cameleyre, C-G. Dussap, J. Francois, G. Berting and J-P. Jouany. 2003. A novel technique to evaluate interactions between *Saccharomyces cerevisiae* cell wall and mycotoxins: application to zearalenone. Biotech. Letters 25:783-788.

Zaghini, A., P. Roncada, P. Anfossi and L. Rizzi. 1998. Aflatoxin B_1 oral administration to laying hens: effects on hepatic MFO activities and efficacy of a zeolite to prevent aflatoxicosis B_1. Rev. Med. Vet. 149:668.

INDEX